AF602085

Advances in Water Security

Series editor
Ali Fares, Prairie View, USA

More information about this series at http://www.springer.com/series/13753

Ali Fares
Editor

Emerging Issues in Groundwater Resources

Editor
Ali Fares
Prairie View A&M University
College of Agriculture and Human Sciences
Prairie View, TX, USA

Advances in Water Security
ISBN 978-3-319-32006-9 ISBN 978-3-319-32008-3 (eBook)
DOI 10.1007/978-3-319-32008-3

Library of Congress Control Number: 2016947216

Printed on acid-free paper

This Springer imprint is published by Springer Nature
The registered company is Springer International Publishing AG Switzerland

Preface

Groundwater resources are decreasing at alarming rates and face increasing pressure due to climate variability, population pressure, and increases in energy demands. Furthermore, emerging groundwater contamination and pollution risks are seriously reducing available groundwater resources. The topics discussed throughout this book are grouped into five sections: (1) sea level rise, climate change, and food security, (2) emerging contaminants, (3) technologies and decision support systems, (4) surface water-groundwater interactions, and (5) economics, and energy production and development. Sea level rise in coastal areas and food security risks as a result of climate change are discussed in Chaps. 1 through 3. Chapters 4 through 6 cover emerging groundwater contamination issues as a result of hydraulic fracturing for energy production and wastes from pharmaceutical and wastewater treatment plants. Chapters 7 and 8 include different approaches and decision support systems for quantifying groundwater resources. Surface water–groundwater interactions are discussed in Chaps. 9 through 12. Finally, Chaps. 13 and 14 discuss groundwater management approaches for agricultural production and assess potential environmental impacts associated with the use of groundwater resources for energy production.

This book is unique and different from other groundwater hydrology books in that it uses a holistic approach in investigating the risks related to groundwater resources. This book is the first book in the newly initiated book series: *Advances in Water Security*. This book is of interest to a wider audience in academia, governmental and nongovernmental organizations, and environmental entities.

Overall, the book will greatly contribute to better understanding emerging risks to groundwater resources and should help responsible stakeholders make informed decisions in this regard.

Prairie View, TX Ali Fares, Ph.D.

Contents

Contributors

Ahmet Dogan Civil Engineering Department, Hydraulics Lab, Yıldız Technical University, Istanbul, Turkey

Alana Kleven Department of Geology & Geophysics, University of Hawaii-Manoa, Honolulu, HI, USA

Ali Fares College of Agriculture and Human Sciences, Prairie View A&M University, Prairie View, TX, USA

Chittaranjan Ray Nebraska Water Center, Nebraska Innovation Campus, Lincoln, USA

Dawit T. Ghebreyesus Institute Center for Water and Environment (iWATER), Masdar Institute of Science and Technology, Masdar City, Abu Dhabi, United Arab Emirates

Donato Sollitto Department of Bari, IRSA-CNR—Water Research Institute of the National Research Council, Bari, Italy

Dorina Murgulet Department of Physical and Environmental Sciences, Texas A&M University-Corpus Christi, Corpus Christi, TX, USA

Emanuele Barca Department of Bari, IRSA-CNR—Water Research Institute of the National Research Council, Bari, Italy

E. Smith University of South Carolina, North Inlet Winyah Bay National Estuarine Research Reserve, Georgetown, SC, USA

Florence Thomas Hawaii Institute of Marne Biology, University of Hawaii-Manoa, Honolulu, HI, USA

Giuseppe Passarella Department of Bari, IRSA-CNR—Water Research Institute of the National Research Council, Bari, Italy

Haimanote K. Bayabil College of Agriculture and Human Sciences, Prairie View A&M University, Prairie View, TX, USA

Henrietta Dulai Department of Geology & Geophysics, University of Hawaii-Manoa, Honolulu, HI, USA

I. Kisekka Water and Irrigation Management, Kansas State University, Southwest Research-Extension Center, Manhattan, KS, USA

J. Aguilar Water and Irrigation Management, Kansas State University, Southwest Research-Extension Center, Manhattan, KS, USA

Kathleen Ruttenberg Department of Geology & Geophysics, University of Hawaii-Manoa, Honolulu, HI, USA

Department of Oceanography, University of Hawaii-Manoa, Honolulu, HI, USA

Lorenzo De Carlo Department of Bari, IRSA-CNR—Water Research Institute of the National Research Council, Bari, Italy

L. Peterson Coastal Carolina University, School of Coastal and Marine Systems Science, Conway, SC, USA

Maria Clementina Caputo Department of Bari, IRSA-CNR—Water Research Institute of the National Research Council, Bari, Italy

Marouane Temimi Institute Center for Water and Environment (iWATER), Masdar Institute of Science and Technology, Masdar City, Abu Dhabi, United Arab Emirates

Matteo D'Alessio Nebraska Water Center, Nebraska Innovation Campus, Lincoln, USA

Mohammad Safeeq Sierra Nevada Research Institute, University of California at Merced, Merced, CA, USA

USDA Forest Service, PSW Research Station, Fresno, CA, USA

Rachna Tewari Department of Agriculture, Geosciences, and Natural Resources, The University of Tennessee at Martin, Martin, TN, USA

Ram L. Ray Cooperative Agricultural Research Center, College of Agriculture and Human Sciences, Prairie View, A&M University, Prairie View, TX, USA

Rebecca Briggs Department of Oceanography, University of Hawaii-Manoa, Honolulu, HI, USA

Ripendra Awal College of Agriculture and Human Sciences, Prairie View A&M University, Prairie View, TX, USA

Rita Masciale Department of Bari, IRSA-CNR—Water Research Institute of the National Research Council, Bari, Italy

Robert W. Taylor Department of Earth and Environmental Studies, Montclair State University, Montclair, NJ, USA

Ruopu Li Department of Geography and Environmental Resources, Southern Illinois University-Carbondale, Carbondale, IL, USA

R. Peterson Coastal Carolina University, School of Coastal and Marine Systems Science, Conway, SC, USA

Stefanie A. Brachfeld Department of Earth and Environmental Studies, Montclair State University, Montclair, NJ, USA

Sushant K. Singh Department of Earth and Environmental Studies, Montclair State University, Montclair, NJ, USA

S. Libes Coastal Carolina University, School of Coastal and Marine Systems Science, Conway, SC, USA

Chapter 1
Effects of Climate Change and Sea Level Rise on Coastal Water Resources

Dorina Murgulet

Abstract A better understanding of the relevant climate-change drivers for coastal areas has been realized in the past few years. A significant increase in atmospheric CO_2 concentration is predicted to virtually occur and as a result, more CO_2 is absorbed by surface waters, decreasing seawater pH and carbonate saturation. Sea surface temperatures are also essentially certain to rise significantly, although less than the global mean temperature rise. Globally, SLR derived from thermal expansion due to warming of oceans and the melting of ice caps, glaciers, and ice sheets (i.e., Greenland and Antarctica) act together as major factors contributing to RSLR. The rise will not be spatially uniform, with possible intensification of ENSO and time variability which suggests greater change in extremes with important implications for coral reefs. In most cases there will be significant regional variations in the changes, and any impacts will be the result of the interaction between climate change drivers (i.e., CO_2 concentrations, SST, SLR, storm intensity, frequency, and track, wave conditions and runoff) and other drivers of change, leading to diverse effects and vulnerabilities. The direct influences of sea level rise (SLR) on coastal water resources are associated with seawater encroachment into surface waters and coastal aquifers. Inundation as a result of increases in mean sea level will have devastating impacts on unprotected low-lying areas, especially as a result of storm events which are expected to intensify. Seawater intrusion caused by natural and human-derived factors will be exacerbated by SLR. However, some coastal areas, especially on some arid coasts, receiving more precipitation may be less impacted by seawater intrusion as a result of increased aquifer recharge. Climate change adverse impacts on freshwater supplies at a global scale are most likely to be more visible in developing countries with large extents of coastal lowland, small island states, semi-arid and arid coasts, and large coastal cities particularly in the Asia-Pacific region, reflecting both natural and socio-economic aspects that increase the risk levels. Thus, it is difficult to identify future coastal areas with stressed freshwater resources, particularly where there is a seasonal water demand stress and poor management. Assessments of RSLR-related coastal impacts, adaptation, and

D. Murgulet (✉)
Department of Physical and Environmental Sciences, Texas A&M University-Corpus Christi, 6300 Ocean Drive, Corpus Christi, TX 78412, USA
e-mail: dorina.murgulet@tamucc.edu

A. Fares (ed.), *Emerging Issues in Groundwater Resources*, Advances in Water Security, DOI 10.1007/978-3-319-32008-3_1

mitigation, require information related to climate-induced GMSLR, regional variations, and non-climate-related sea level changes and freshwater stressors.

1 Introduction

The global debate on whether climate is changing and what are the main contributing factors have recently shifted towards how to deal with changes that are now foreseeable. Although climate has changed throughout the Earth's history as a result of natural cycles, a general consensus exists now among scientists that the trend in change since the beginning of the industrial revolution (circa ~ 1880) is mainly the result of anthropogenic factors such as the increase of greenhouse gasses released into the atmosphere from the combustion of fossil fuels (Neelin 2011). There are several indicators measured globally over many decades that show that the Earth's climate is warming (Fig. 1.1). Current indicators show that if the change in climate occurs gradually, the impact is expected to be minor by 2025, with some areas being more affected than others. Impacts are projected to strongly intensify during the decades following 2025 (UNEP 2003).

The effects and coping strategies in coastal areas, are of particular interest because these locales provide important habitats and ecosystem services, have growing human populations, and include large economic centers such as New York, London, Tokyo, and Mumbai (Nicholls 2011). Among the major effects of climate change, SLR is caused by thermal expansion of water and the melting of

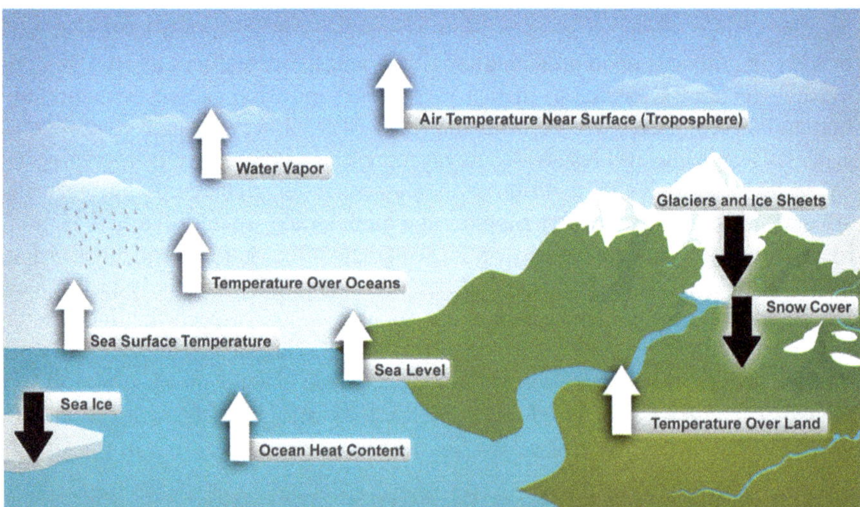

Fig. 1.1 Long-term measured global indicators that show that the Earth's climate is warming. *White arrows* indicate increasing trends-indicators expected to increase in a warming world; *black arrows* indicate decreasing trends-those indicators expected to decrease in a warming world (Figure source: NOAA-NCDC 2014)

land-based ice (i.e., ice sheets). The glacio-eustatic mechanism has been shown to be the most important control on changes in sea level during the Neogene and Quaternary geologic times (Chappell and Shackleton 1986; Matthews and Poore 1980). Sea level rise has been increasing at around 1.8 mm/year (global average) for the past century (Leuliette and Miller 2009; Blunden and Arndt 2014) and the rates will accelerate in response to global warming resulting in a rise of about 600–900 mm by the year 2100 (Leatherman et al. 1995). Also, sea levels are expected to rise for the next few centuries because of the considerable thermal inertia characteristic to the Earth's climate system (Fig. 1.2) (Schmidt and Wolfe 2009).

A rise in sea level of this magnitude will indisputably have a dramatic effect on low-lying coastal areas and islands. The effects and coping strategies in coastal areas, are of particular interest because these locales provide important habitats and ecosystem services, have growing human populations, and include large economic centers such as New York, London, Tokyo, and Mumbai (Nicholls 2011). Coastal areas are facing increasing vulnerability to higher frequency of storm surges, salt water intrusion, coastal erosion, and loss of coastal habitats (Nicholls 2011).While the extent and severity of impacts may vary considerably, the general treats to

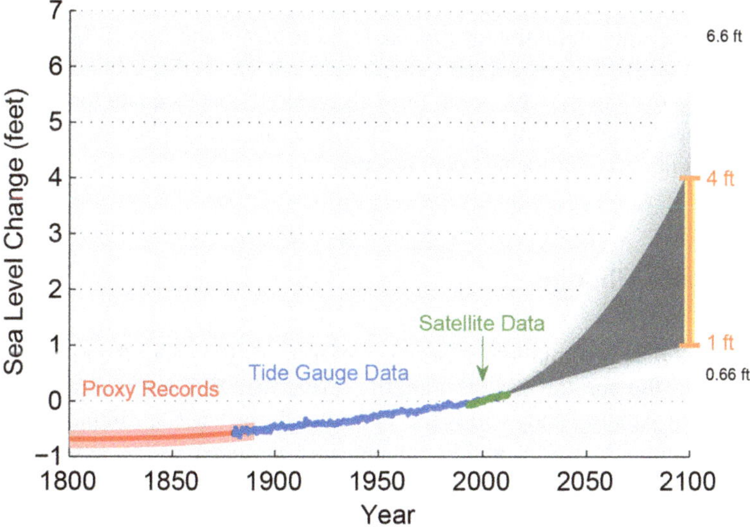

Fig. 1.2 Estimated, observed, and possible future amounts of global sea level rise (GSLR) from 1800 to 2100, relative to the year 2000. Shown in *red* are estimates from proxy data (for example, based on sediment records) (1800–1890, *pink* band shows uncertainty), in *blue*—tide gauge data for 1880–2009, and in *green* are shown satellite observations from 1993 to 2012. The future scenarios range from 0.66 to 6.6 ft in 2100. The graph shows the range of possible scenarios based on scientific studies not climatic models. The large currently projected range of SLR of 1–4 ft by 2100, presented by the *orange line*, reflects uncertainty related to how glaciers and ice sheets will react to the warming ocean and atmosphere and changing winds and currents. The trend shows year-to-year variations (Figure source: NOAA- NCDC 2014)

coastal areas are flooding, erosion, wetland inundation and saltwater intrusion and/or salinization.

Though generalizations of climate change and SLR effects can be made on a global scale, the major factors of concern are unique to different areas of the world. For instance, estimates of future relative sea level rise (RSLR) and vulnerability of coastal ecosystems and human populations to SLR vary for different regions due to differences in land motion and different rates of growing populations and development. The impacts are expected to render freshwater availability around the world. The impact of warmer temperatures will intensify the hydrologic cycle through changes in rates of precipitation and ET and an increase likelihood of severe weather patterns, higher flooding events (Fig. 1.3), and more droughts. This has indirect adverse effects on the flux and storage of water in surface and subsurface reservoirs (Taylor et al. 2012). Increases in evaporation rates will enhance salinization of soils (i.e., irrigation fields or flooded coastal areas) and recharging waters percolating to aquifers. Furthermore, the indirect effects of climate on water availability through changes in land practices (i.e. changes in agricultural practices and irrigation sources and demand) may have greater impacts than the direct impacts of climate change (Taylor et al. 2012). Groundwater overdraft (i.e., groundwater production rates exceed supply to the aquifer) induced by growing concentrations in urban areas along the coasts result in water quality degradation due to saltwater intrusion into aquifers. Nevertheless, rising sea levels may be enhancing the saltwater intrusion process (Abd-Elhamid and Javadi 2008; Langevin and Zygnerski 2012). To add, most recent research shows that groundwater exploration and recycling into the ocean through precipitation accounts for a rise in sea level of approximately 5 mm*year−1 (Taylor et al. 2012). Current sea level predictions do not account for this phenomenon.

2 Climate Change

Climate has changed multiple times in the Earth's history, but for reasons totally unrelated to the current pattern. Patterns reveal that SST has a major effect on precipitation patterns and worldwide atmospheric and ocean circulation. As the planet continues to warm, additional moisture is added to the atmosphere and changes in precipitation patterns follow. This trend has been observed in the Earth's past especially during the Eocene (50 Mya-present) time when rising CO_2 levels, in combination with the influence of changing ocean currents lead to increasing temperatures and changes in precipitation patterns (Sloan and Rea 1995) (Fig. 1.4). Changes in these patterns dictate where human societies formed and settled (Ma and Xie 2013). For instance, Hester et al. (1997) show how changes in climatic patterns affected lake levels in North America, which forced relocation of early Holocene people.

The American Southwest is particularly sensitive to changes in climatic patterns as seen during the late Holocene (11.7 kya—present) when it experienced long

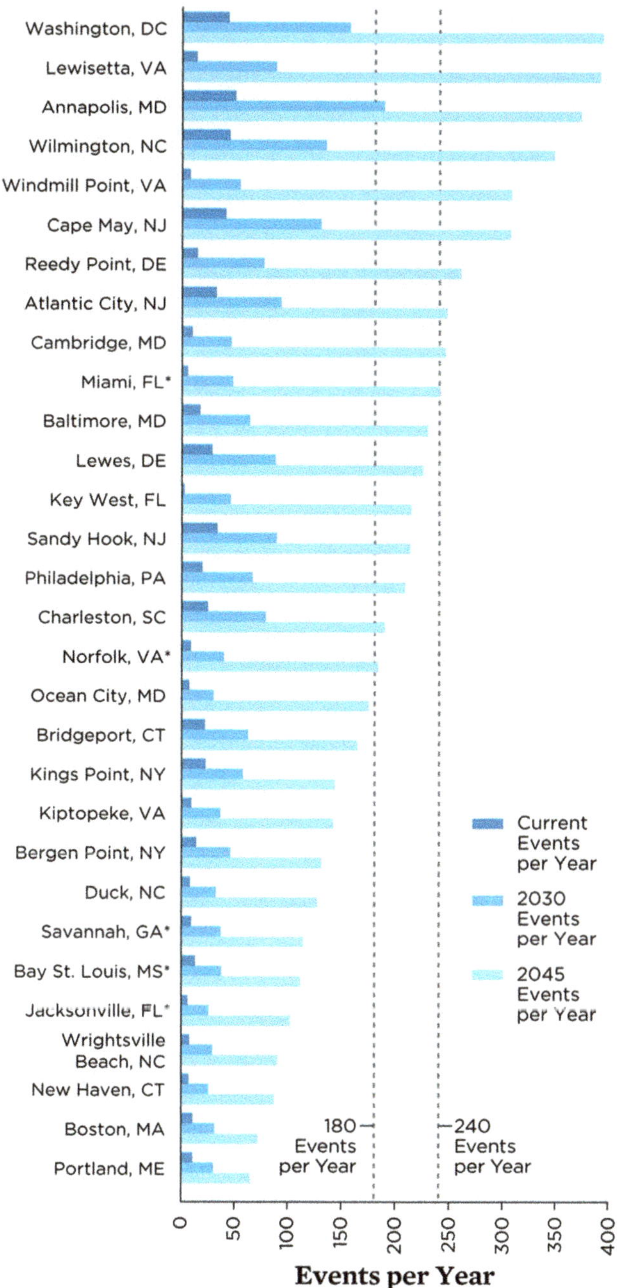

Fig. 1.3 According to the Union of Concerned Scientists, there are at least 30 locations (shown in the figure) expected to experience at least 20 tidal flood events per year by the 2030, by the 2045, one third of 52 locations analyzed can expect 180 or more flooding events per year (UCS 2014)

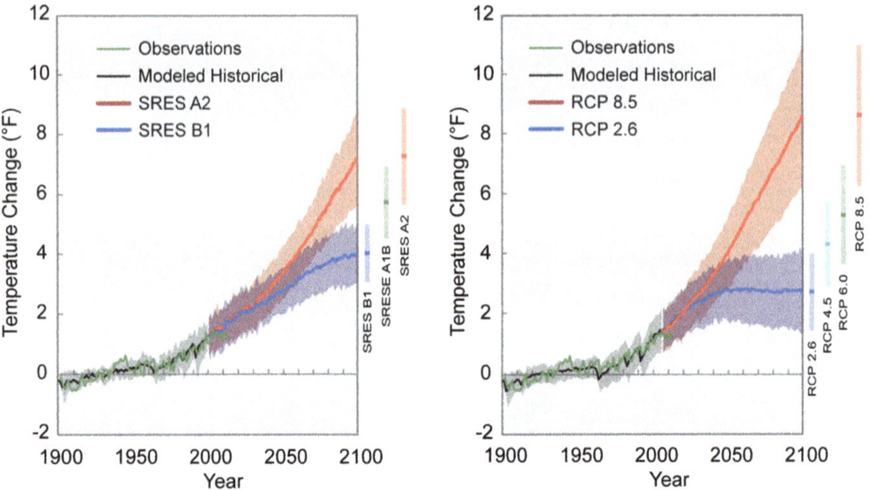

Fig. 1.4 Increases in temperature as dictated by emissions. In the figure, each line represents a central estimate of global average temperature rise (relative to the 1901–1960 average) for a specific emissions pathway. Shading indicates the range (5th–95th percentile) of results from a suite of climate models. Projections in 2099 for additional emissions pathways are indicated by the bars to the *right* of each panel. In all cases, temperatures are expected to rise, although the difference between lower and higher emissions pathways is substantial. The panel to the (*left*) shows two main scenarios as presented in the SRES—Special Report on Emissions Scenarios: A2 assumes continued increases in emissions throughout this century, and B1 assumes much slower increases in emissions beginning now and significant emissions reductions beginning around 2050. The (*right*) panel shows more modern analyses derived from the most recent generation of climate models (CMIP5) using the most recent emissions pathways (RCPs—Representative Concentration Pathways). Compared to the old projections (SRES), these new projections explicitly consider climate policies (includes both lower- RCP 2.6 and higher- RCP 8.5 pathways) that would result in emissions reductions. The lowest emissions pathway assumes immediate and rapid reductions in emissions that would lead to a warming of about 2.5 °F of warming in this century while the highest pathway which is roughly the result of a continuation of the current path of global emissions increase, is projected to cause an increase of more than 8 °F warming by 2100 (a higher end of the prediction is more than 11 °F) (Source: NOAA- NCDC 2014)

periods of drought due to reduced precipitations (Seager et al. 2007). Ratios of oxygen stable isotopes signals ($\delta^{18}O$) recorded on speleothems were used as a proxy to reconstruct precipitation patterns which show that moisture from the Gulf of Mexico (GOM) has had a profound influence on the Southwest (Feng et al. 2014). Most notably, there was an increase in GOM-derived precipitation in the region during a period of warming known as the interstadial Bølling-Allerød period, which occurred between ca. 14.7–12.9 ka and a decrease during the Younger Dryas cooling period (ca. 12.9–11.5 ka) (Feng et al. 2014; Obbink et al. 2009). Isotope records reveal phenomena that have influenced past precipitations patterns and provide a predictive model for how they will affect future climatic behavior. For instance, in the American Southwest, predictions indicate that if climate continues to warm water availability will become scarcer due to lower precipitation rates (Feng et al. 2014). This region provides a good model for other regions influenced by GOM-derived moisture and by synoptic or greater sized atmospheric patterns such as the effect El Niño and La Niña events play on many parts of North America (Neelin 2011).

Currently there is a good understanding of the causes of most of the Earth's climate cycles, as described by the Milankovitch cycles (Hays et al. 1976; Raymo and Nisancioglu 2003) and as revealed by the geological/stratigraphical records (i.e., describe geological and oceanographic changes as a result of plate tectonics) (Weedon 2005). Planetary orbit cycles have had a profound effect on radiative forcing, hence, greatly affecting the climate. With a good understanding of changes that occurred pre-industrial age, better understanding of anthropogenic effects upon climate is possible. The theory behind this assertion is related to the Milankovitch cycles indicating that periodic changes in the Earth's obliquity or tilt, over a 41,000 year cycle has effected glacial-interglacial cycles during the last three million years (Hays et al. 1976). Furthermore, Earth's precession or the direction of Earth's spin axis (over ca. 19 and 23 ka cycles) and eccentricity are considered: the degree to which the Earth's orbit is elliptical, in 100 and 400 ka periods (Raymo and Nisancioglu 2003) have affected climate, both during the Quaternary, and in earlier epochs. A modelling approach as presented by Erb et al. (2013) evaluates the mechanism by which the evolution of Earth's climate system can be modelled, especially in terms of our most recent past, the Quaternary Period (2.5 mya to present) primarily because this period reflects relatively current chances and forcings and has most abundant and consistent data. Many of the modeling observations seem to be in consensus with stable oxygen isotope $\delta^{18}O$ signals and along with evidence from stratigraphic records (Weedon 2005) demonstrate the role of orbitally forced climate change for past and future climate change predictions.

2.1 *Climatic Models*

The use of models in climate science is greatly misunderstood, and in general, many in the non-scientific public arena have expressed doubts and disbelief in the models, as no one model has proven itself 100 % accurate. The purpose of developing and using climate models is to explain, understand and predict past, current, and future climate phenomena (Neelin 2011). Zhao (2014) studied modeling of convection as related to a General Climate Model (GCM) of cloud formation and feedbacks. While the model demonstrates increases in precipitation with warming, relating this to cloud feedbacks has proven difficult.

Current models cannot predict with accuracy the feedback sensitivity clouds impart to the climate system as a whole. While many factors are at play concerning clouds and climate, two are most important. For instance, water vapor is a highly effective greenhouse gas that enters the atmosphere as a result of feedback to increase heat which increases water vapor that in turn stimulates a further increase in heat; projected increases in Earth's temperature are dictated by the different anthropogenic-derived amounts of heat-trapping gases released into the atmosphere (NOAA-NCDC 2014) (Fig. 1.4). Secondly, low, thick clouds primarily reflect incoming solar radiation, thereby having a cooling effect on the Earth's surface. On the other hand, the high, thin clouds primarily transmit incoming solar radiation and also trap some of the outgoing infrared radiation emitted by the Earth and radiate it back contributing to warming of the Earth surface. The portion of the solar energy that is reflected back to space, called the albedo, is different for different parts of the world. Even though the balance between the cooling and warming actions of clouds is considered to be very close, if averaging the effects of all the clouds around the globe (i.e., cloud albedo forcing), cooling predominates (i.e., negative forcing) (Graham 1999). Furthermore, a portion of the Earth's longwave radiation absorbed by clouds is reemitted to the outer space and another portion back toward the surface. The intensity of the emission is dependent upon cloud characteristics such as temperature, thickness, and makeup of the particles that form the cloud. A cold cloud top will reduce the longwave emission to space, and contrary to the momentary albedo forcing of the cloud, by trapping it beneath the cloud top and will increase the temperature of the Earth's surface. This so called "cloud greenhouse forcing" if considered separately has a warming effect or "positive forcing" on the Earth's climate (Graham 1999).

Due to complications and uncertainties associated with modeling atmospheric physics and the extreme variability of cloud formation, determining the effects of clouds on climate feedbacks in respect to climate change has proven to be difficult (Zhao 2014). It has also caused much debate in the field of climate science. For example, Spencer and Braswell (2011) suggested that the process by which clouds form could be causing climate change, as opposed to the more widely accepted view that climate change is affecting the way clouds form. This concept would mean that the climate system as a whole is less sensitive to the effects that greenhouse gasses have on climate change. While the climate research community does not generally accept this view, it demonstrates how the results of models are inconclusive and can be easily criticized.

3 Climate Change and Sea Level Rise

The subject of climate change and SLR has risen to prominence in the last couple decades; focus has shifted from identifying causes for environmental changes to observing more closely their impacts and potential strategies to mitigate and/or adapt to these changes. Analyses of past records related to the planet's land and ocean temperatures are utterly important to understanding GSLR associated with climate change. Research shows that climate change is expected to cause major changes in SLR, sea surface temperatures, coastal storms, and wave and runoff characteristics. Climate-induced SLR is produced by thermal expansion of seawater and melting of land-based ice. Coastal areas are vulnerable to many coastal hazards, including climate-induced SLR. Sea level rise changes are exacerbated by increases

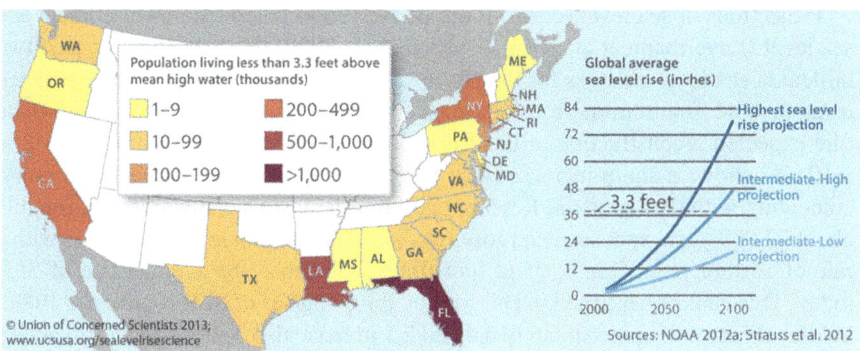

Fig. 1.5 Coastal US states at risk from GSLR. Depending on future emission, global warming, and land ice melting rates, global average sea level could rise to or above the 3.3 foot mark within this century (UCS 2013; NOAA 2012; Strauss et al. 2012)

in storm events and subsidence caused by anthropogenic activities such as groundwater withdrawals. It is estimated that over 200 million people are vulnerable to flooding during extreme events produced by storm surges while 20 million people live below normal high tide levels (Nicholls 2011). In the United States, although all coastal states are at risk of flooding due to increased sea level and global warming, Florida, Louisiana, New York, and California have the highest number of people residing in areas less than 3.3 f. above high tide (Fig. 1.5). For effective management to mitigate and reduce the impacts of SLR on coastal ecosystems and human populations, a better scientific understanding of the climate change and SLR and the impacts they have on systems is necessary.

3.1 Past and Future Trends: Evidences from Instrumental and Proxy Records and Model Simulations

Improvements have been made to oceanographic observations and satellite measurements over the last century, robust data exists beginning with the late nineteenth century to the early twentyfirst century. Church and White (2011) summarize the global average sea level from satellite altimeter data and coastal sea level measurements, to provide a synthesis of SLR during this time frame. The study showed that GMSL (Global Mean Sea Level) has risen by 100–200 mm at the rate of 1–2 mm/year over the past century. Furthermore, the rate of GMSL rise during the past two decades is approximately 3.2 mm/year, roughly two times higher than the average speed of the preceding 80 years. These trends correlate well with the global mean temperature and sea surface temperature (SST) measurements over the same time frame (Arndt et al. 2010), showing the relationship between climate change as the SLR.

Projections of sea level scenarios are derived from paleo records of climate and sea level (Levermann et al. 2013; Grinsted et al. 2009). For instance, the past two millennia sea level changes for the North Carolina coastal zone was reconstructed using benthic foraminifera from salt marshes (Kemp et al. 2011). The overall trend (the expected reconstruction error of the global mean sea level record was within ±10 cm) shows a sharp increase of SLR in the middle of the nineteenth century, coinciding with the Industrial Revolution. After the last glacial maximum, roughly about 22,000 years ago, sea level rose by about 120 m in about 10,000 years with a rate of 12 mm/year. This is about four times faster than the average rate of SLR today. This change happened in episodes of abrupt pulses of SLR (Fairbanks 1989). Sea-level jumps of approximately 0.4 and 2.1 m have also been related to an abrupt cooling in the Northern Hemisphere known as the 8.2 kyr event during the last interglacial period, the Holocene epoch. It has been suggested that this event was due the freshwater release from Lake Agassiz into the North Atlantic that perturbed the North Atlantic Meridional Overturning Circulation (Törnqvist and Hijma 2012).

Stratigraphic information from the Exxon Production Research Company (EPR) provide an unprecedented record of sea level change on a longer time scale, from Triassic (~250 Ma) to present (Haq et al. 1987a, b; Vail et al. 1977). Even though the amplitude of sea level change is poorly constrained using this records, the data show that causal mechanisms for the large, rapid sea level changes for the past 250 million years interval can only be explained by glacioeustasy linked with climate change (Miller et al. 2003). Furthermore, past changes in the sea level could explain the transgression and regression of the sea, which is associated with eustatic (global) sea level changes, tectonics and sediment supply (Lyell 1830).

Various predictive modeling studies developed semi-empirical models to project future sea level scenarios incorporating paleo records. Levermann et al. (2013) has combined paleo-evidence with simulations from physical models to estimate the future regional sea level projections at a multi-millennial time scale. They proposed a SLR of approximately 2.3 m per degree centigrade (1 °C) within the next 2000 years. Using a semi-empirical approach Rahmstorf (2007) projected a rise in sea level of 0.5–1.4 m above the 1990 level by the 2100. These are in close agreement with efforts conducted by Grinsted et al. (2009) which estimated a SLR between 0.9 and 1.3 m over the next century. The inertia gained by the Earth's climate system and feedbacks will cause the current SLR to continue even if greenhouse emissions are reduced. The fifth IPCC report on climate change predicts a global rise in sea level of 52–98 cm by the 2100, with a best estimate around 74 cm. Even with a minimum emission scenario, an estimated mean of 44 cm SLR is expected (IPCC 2013).

3.2 Impacts of Sea Level Rise

The relative SLR (RSLR) takes into account the sum of global, regional, and local components of sea level change. The drivers of these components are climate change and changing ocean dynamics, non-climatic processes such as tectonic land movements (i.e., uplift or subsidence), glacial isostatic adjustment (GIA), and natural/anthropogenic-induced subsidence. RSLR varies from one place to another and is only partly the result of climate change. Its impacts are dependent on the physical and socioeconomic characteristics of each region (Nicholls 2011). There are five main impacts associated with the RSLR. These include flooding and submergence, ecosystem alteration, erosion, salinization, and rising water tables. Immediate effects include: submergence (also known as the Bathtub effect), coastal flooding and saltwater intrusion. Longer-term impacts are: wetland loss and change, salinization of surface waters, soils and shallow aquifers, beach erosion, and saltwater intrusion in deeper aquifers. Flooding and related physical impacts have the most significant effect on the socioeconomics of an area. Flooding often results in damage of coastal infrastructure, built environments, agricultural areas, and human deaths caused by drowning.

Other examples of socioeconomic impacts of SLR include erosion leading to loss of beachfront buildings and related infrastructure, adverse effects on tourism and recreation, and indirect impacts on human health and quality of life (Nicholls 2011). The future health status of the coastal population, its capacity to cope with climate hazards such as control infectious diseases, and other public health measures dictate the potential impacts of climate change and SLR on coastal populations. For instance, many coastal communities rely on marine resources for food; both human health and the economy of these communities are at risk in terms of both food supply and quality. Temperature changes have also impacts on marine ecological processes which can also affect human health associated with cholera, and other enteric pathogens (Vibrio parahaemolyticus), shellfish and reef fish poisoning , and HABs (Pascual et al. 2000; Lipp et al. 2002; Hunter 2003). Although, there is no convincing evidence of the impacts of observed climate change on coastal disease patterns (Kovats and Haines 2005), connections may exist between rainfall changes associated with ENSO and cholera risk in Bangladesh (Pascual et al. 2000) and malaria epidemics in coastal regions of Venezuela and Colombia (Kovats et al. 2003).

4 Global Climate Change and Coastal Water Resources

Mounting evidence indicates that we are in a climate change period caused by increasing atmospheric concentrations of greenhouse gases (Intergovernmental Panel on Climate Change (IPCC) 2014). This phenomenon can have profound effects on the hydrologic cycle through precipitation, soil moisture, increasing temperatures and ET. General Circulation Models (GCMs) fail to accurately predict changes on mean precipitation but an agreement exists on extreme changes of temperature and precipitation as a result of an intensification of the hydrologic system (Bates et al. 2008). Although more rain and ET is expected, the extra precipitation will not be equally distributed around the globe. Some parts of the word may experience significant flooding while others may see significant decreases in precipitation and season changes (i.e. timing alteration for the wet and dry seasons). Consequently, it is imperative to understand the impact of climate change on hydrological processes and water resources. The most updated 100-year warming trend (1906–2005) as reported by IPCC is 0.74 °C ± 0.18 °C (IPCC 2014). Research has shown that the mean sea level has been increasing at a rate of 3.2 millimeters (mm) per year over the past 20 years (NOAA 2014). As seawater migrates farther inland it can cause destructive erosion, flooding of wetlands, contamination of aquifers and agricultural soils, and can have disturbing effects on coastal habitats (NOAA 2014).

4.1 A Global Perspective

With both, anthropogenic factors and an overall increase in temperatures worldwide, the security of water resources is becoming a topic of increase interest. In particular, water resources in the coastal regions are of major concern given the necessary quantities to accommodate the growing needs for human and industrial consumption. The effects of climate change on the world's oceans are felt directly by coastal communities. Climate change effects tend to shift the availability of freshwater around the globe—decreasing the availability in some areas, while increasing the availability in others (Vörösmarty et al. 2013). However, as the population is increasing in different areas, water availability tends to decrease worldwide. Thus, best evaluations of water resources should take into account both, the increased human use and climate-driven shifting availability. Additionally, climate change is altering ecosystems across the globe affecting the water resource services they provide to humans (e.g., recreation, fishing) (Hoegh-Guldberg and Bruno 2010).

Globally, climate change-derived freshwater supply problems are expected to affect the developing countries with large areas of law-land, semiarid or arid coasts, and coastal megacities. At great risk will be the Asia-Pacific region and small island states as a result of both natural and socio-economic factors (Alcamo and Henrichs 2002; Ragab and Prudhomme 2002). Consideration of water use efficiency is particularly important in areas where agriculture is a large consumer, like in the Nile delta and Asian megadeltas and areas with large freshwater demands associated with increasing population (i.e., Asia-Pacific region). The IPCC-SRES emissions scenarios estimate a significant impact on water resources by the 2050s based on different scenarios of water stresses (Arnell et al. 2004). Critical regions with higher sensitivity to water stresses caused by either increases in water withdrawal or decreases in water available have been identified in parts of coastal areas from western coasts of Latin America and the Algerian coast (Alcamo and Henrichs 2002).

The effects of climate change vary across the globe, and different concerns are associated with different regions. For instance, the Artic is facing many of the effects of climate change, including the melting of glaciers and permafrost, changes in seasonal precipitation, and increased evapotranspiration (Evengard et al. 2011). It is not only the total amount of water available (i.e., reduced freshwater flows) that influences coastal Arctic ecosystems and human populations, but also the quality of the resource that is affected in many ways. For example, the water released as permafrost thaws is turbid, and includes pollutants otherwise less mobile. China is faced with a wide variety of stressors based solely on limited water resources **due to** ever increasing population. Similarly to the observed trend worldwide, China has seen a steady increase in temperatures over the past 50 years. Increases in temperature and heatwave frequency caused the retreat of glaciers and a slight change in seasonal trends. The consequences are flooding events such as that from 1998 which left 21 million hectares of land inundated and destroyed five million homes in the Yangtze basin (Piao et al. 2010). While the glacier retreat has provided some aid in maintaining a healthy water budget, poor conduits, low recharge rates, and underdeveloped infrastructures lead to losses of that the majority of freshwater discharge from the depleted glaciers to the neighboring countries. To cope with climate changes China has made adaptations in agricultural realms such as modifications in crop cycles, irrigation, and crop types. Furthermore, the Mediterranean experiences various stressors on coastal water resources, some of which include increasing population, tourism, and living standards and development of irrigated agriculture. Climate change predictions for this area include changes in land use covers, riverine flows, more extreme hot and severe weather episodes, decreased precipitation and increases in temperature (García-Ruiz et al. 2011). These potential changes will be a catalyst for change in the region due to increase in water resource demands and a shortage of water availability.

4.2 Climate Change and Impacts on Coastal Natural Systems

Coastal areas are dynamic systems that undergo changes at different time and spatial scales as a result of oceanographic and geomorphologic processes (Cowell et al. 2003) but human pressures may dominate over these natural pressures. Costal landforms affected by short-term perturbations like storm surges tend to return to their previous pre-disturbed morphology adjusting towards a dynamic equilibrium which sometimes results in different conditions in response to varying wave energy and sedimentation rates (Woodroffe 2003). This natural variability of coasts makes it challenging to determine the climate change-derived impacts. For instance, recent erosion of beaches is observed around the world as a result of changes in wind patterns, development which reduces sediment accretion, and offshore bathymetric changes (Pirazzoli et al. 2004; Regnauld et al. 2004). It is difficult in this case to determine if the observed changes are caused by external factors like climate change, short-term disturbance of natural climate variability, or by how much they exceed an internal threshold (i.e., changes in sedimentation patterns) (IPCC 2014). Furthermore, climate-related ocean-atmosphere oscillations such as the El Niño Southern Oscillation (ENSO) also influence Earth's climate. During an El Niño event, the persistent tropical Trade Winds, that cause the upwelling of cold subsurface water of the coast of Peru during the La Niña, are weakened reducing the upwelling of cold water. It is at the peak of an El Niño event, that the Pacific Ocean water is warmer than normal and the heat is transferred from the ocean to the atmosphere increasing the near-surface temperature (World Meteorological Organization 2014). Recent research indicates that dominant wind patterns and the ENSO-derived storms may have a great impact on coastal dynamics, perturbing the geomorphologic changes in eastern Australia, mid-Pacific, and Oregon (Ranasinghe et al. 2004) and Oregon (Allan et al. 2003; Confalonieri et al. 2007) and groundwater levels in mangrove ecosystems in Micronesia and Australia (Drexler 2001; Rogers et al. 2005). It is likely that coasts also respond to longer term variations such as the decadal oscillations. For instance, long-term monitoring of the south-east Australian beach reveals a relationship with the Pacific Decadal Oscillation (PDO) (McLean and Shen 2006) and storm frequencies on Atlantic coasts were correlated the North Atlantic Oscillation (NAO) (Tsimplis et al. 2005, 2006). Similar periodic fluctuations on coasts around the Indian Ocean are correlated with the Indian Ocean Dipole (IOD) (Saji et al. 1999).

Wetlands, among the most productive ecosystems on Earth, have been abundant in the past across the US. However, in the past few decades they have been extensively drained and filled to be used for agricultural, industrial, commercial, and residential uses. Development can damage and ultimately remove wetlands as it can modify the amount of sediment delivered to coastal areas and exacerbate erosion. For instance, the US Environmental Protection Agency estimates that more than 200 million acres of wetland existed in the lower 48 states in the 1600s and only about 100 million acres still existed in the 2000s. Historic wetland loss varied across the states with the greatest losses in California with more than 90 %, Mississippi with 80 %, and Louisiana with loses of 30,000–40,000 acres of coastal area per year. Growing recognition of their importance as habitats and as buffers of impacts associated with climate change and SLR lead to restoration efforts across the US. Consequently, the net wetland acreage has been observed to increase between 1998 and 2004 (Dahl 2005).

4.2.1 Impacts on Surface Coastal Systems

Most of the world's shorelines have been retreating for the past century and climate change and SLR are among the underlying causes (Leatherman 2001; Eurosion 2004; Confalonieri et al. 2007). For instance, one half or more of the shorelines in Mississippi and Texas have eroded at rates between 3.1 and 2.6 m/year since the

1970s and about 90 % of the Louisiana shoreline eroded at a rate of 12 m/year (Morton et al. 2004) while even higher rates of 30 % are reported for Nigeria. Coastal steepening and squeeze are also noticed around the world (Taylor et al. 2004). Human development and acceleration in SLR will greatly exacerbate beach erosion around the globe, although the response will vary locally dependent on the sediment budget (Confalonieri et al. 2007). Deltaic landforms will also be affected by both human development and rising sea levels. Sediment starvation due to artificial infrastructures such as dams and navigation and flood control and alteration of tide direction are the result of human activities. Changes on surface water inflows and sediment loads can have tremendous impacts on the ability of a delta to cope with climate change impacts. For instance, in south-east Louisiana, the subsiding of the Mississippi delta due to human development (i.e., increased groundwater use and salinity increase and sediment starvation and increase in water levels of coastal marshes) occurred at a very fast rate; consequently, 1565 km^2 of coastal marshes and adjacent lands were covered by water between 1978 and 2000 (Barras et al. 2003; Confalonieri et al. 2007). Although much of these land losses are episodic as demonstrated during the landfall of Hurricane Katrina, if the current global and local stressors will continue along with the projected increases in sea level and tropical storms, losses will be exacerbated and may become permanent (Barras et al. 2003).

Global mean SLR will generally cause an increase in surface water levels and salinities in estuarine systems. Estuarine plant and animal populations may be displaced as a result of rising sea levels or they may persist if migration is not blocked and if the rate of change is not faster than the rate at which these natural communities can adapt or migrate. The greatest impact of climate change on estuaries may result from changes in physical mixing between freshwater and seawater as a result of altered runoff rates. For example, freshwater inflows into estuaries influence nutrient loads, water residence times, vertical stratification, salinity, and govern phytoplankton growth. Higher freshwater inflows reduce residence times and increase vertical stratification and vice versa (Moore et al. 1997). The effect of altered residence times can have a high impact on phytoplankton production rates, which can increase fourfold per day. In waters with very low residence times, phytoplankton are generally removed from the system as fast as they grow, reducing the estuary's susceptibility to eutrophication and harmful algal bloom (HABs). Changes in temperatures could also affect algal production and light availability, dissolved oxygen and carbon for estuarine health. While temperature plays an important factor on estuarine physiographic processes, feedbacks and interactions between temperature changes and independent physical and biogeochemical processes complicate ecological predictions (Lomas et al. 2002). Acidification and carbonate saturation of coastal waters will lead to enhanced dissolution of nutrients and carbonate minerals in sediments and reduced calcification rates. Coral reefs have experienced increasing bleaching rates resulting in coral mortality on a massive scale during the last two decades suggesting that coral reef degradation will intensify within the twentyfirst century (Silverman et al. 2009).

The increasing trend in atmospheric CO_2 concentrations and the induced global warming has significant impacts on the hydrologic cycle. Melting of ice sheets can have an impact on the Meridional Overturning Circulation (MOC), which fuels productivity in the oceans. The observed rapid melting of the Greenland Ice Sheet, for instance, may cause a significant weakening to the MOC as it creates a freshwater influx that can interfere with the sinking of cool saline water. This freshwater flux combined with an increase in global temperatures causes a weak MOC and leads to ocean stratification and shallower mixing (Hu et al. 2010). In the short term, the higher temperatures and higher concentrations of atmospheric carbon could fuel a boom in phytoplankton productivity. However, the shallower mixing, weak MOC, and stratification will ensure that this is short-lived. Once the nutrients are depleted, productivity will cease and a massive carbon sequestration event will take place as sea life dies and accumulates on the ocean floors (Keeling 2007).

4.3 Groundwater and Climate Change

The rate of natural aquifer recharge is highly dependent on the current climatic conditions, land-use/land-cover and underlying geology (Taylor et al. 2012). Climate change is among the factors threatening groundwater availability and sustainability around the world; it directly affects groundwater resources as it impacts aquifer replenishment through changes in precipitation rates, land practices, and water demands. Impacts on precipitation, evapotranspiration (ET), soil humidity, and surface water levels ultimately directly affect groundwater systems since aquifers recharge mainly from precipitation percolating downward through soils and from interaction with surface water bodies. The relationship between climate and groundwater in any given area is complicated by land practices, demographics, and by shifts in rain-fed and irrigated agriculture. This is a very sensitive issue in many coastal areas where groundwater is the main source of freshwater available for human consumption and irrigation. Of particular concern are coastal areas where groundwater overdraft and the aquifer seawater intrusion have been observed owing to large increases on population and industry demands. In this areas it is expected that the SLR of 1.8 mm year^{-1} may have contributed to migration of the freshwater–seawater interface inland (Bindoff et al. 2007).

4.3.1 Water Issues that Affect Groundwater in Coastal Areas

In terms of groundwater sustainability in coastal areas, the most far-reaching factors are: salinization (that affects surface water as well) and trends in withdrawals. Climate change can indirectly exacerbate the intensity of these factors.

- *Trends in withdrawals*: Freshwater withdrawals (i.e., groundwater extraction) have been increasing over the past few decades despite of environmental requirements (i.e., saltwater intrusion, freshwater inflows to surface waters to maintain ecological diversity, etc.). Groundwater resources are at high risk of being depleted as population and industrial and agricultural developments are increasing. Worldwide, an order of magnitude increase in groundwater use has been observed during the second half of the twentieth century with projections of further increase (Burke and Villholth 2007). For instance, in the US groundwater withdrawals have been steadily increasing between 1950 and 1980 (at about 0.7 billion gallons per day) and remained almost constant until 2005 (Hutson et al. 2004). Increased groundwater use leads to development of "fossil" groundwater (i.e., deep aquifers containing water that recharged thousands of years ago) resources bringing water to the surface otherwise not available for evaporation. This new pool of water has been shown to contribute to SLR through evaporation to the atmosphere and return to the ocean as precipitation (Taylor et al. 2012). Nevertheless, paleohydrological patterns reveal the existence of long-term responses of groundwater to climate forcing in major aquifers in arid and semi-arid zones around the world which are independent of human activities (Taylor et al. 2012). Furthermore, as previously mentioned, aquifers depletion in coastal aquifers leads to the encroachment of saltwater rendering the freshwater resource available for domestic uses.
- *Salinization*: A major water sustainability treat in coastal areas, aquifer salinization can be the result of a combination of factors such as: groundwater overabstraction (Fig. 1.6), climate change (i.e., increased storm events) and SLR which accelerate the saltwater intrusion (Fig. 1.7). In general, there are three scenarios to consider: rise in sea levels due to climate change, decline in groundwater levels due to over-pumping and a combination of SLR and overpumping (Abd-Elhamid and Javadi 2008). Seawater intrusion is a function of multiple factors such as reduction of groundwater recharge and lowering of seaward fluxes, coastal topography, and groundwater withdrawals from coastal aquifers (Oude Essink et al. 2010; Ferguson and Gleeson 2012). Analytical models suggest that the seawater intrusion effect on coastal aquifers due to SLR and climatic forcing is negligible compared to the effects of decreased recharge and groundwater overdraft (Murgulet and Tick 2008; Abd-Elhamid and Javadi 2011; Ferguson and Gleeson 2012; Langevin and Zygnerski 2012). Most seawater intrusion problems have been observed predominately in areas with unsustainable groundwater withdrawals in response to fast increases in population rates (i.e. areas such as Bangkok, Jakarta, Bangladesh, Gaza, U.S. the Coastal Plain aquifer system and the Central & West Coast Basins, etc.) (Yakirevich et al. 1998; Taniguchi 2011; Johnson and Whitaker 2004; Murgulet and Tick 2008).

It is expected that the effects of SLR will be more severe in the low-lying coastal areas compared to those in more elevated zones where saltwater contamination may occur from surface infiltration due to increasing intensity of storm surges and

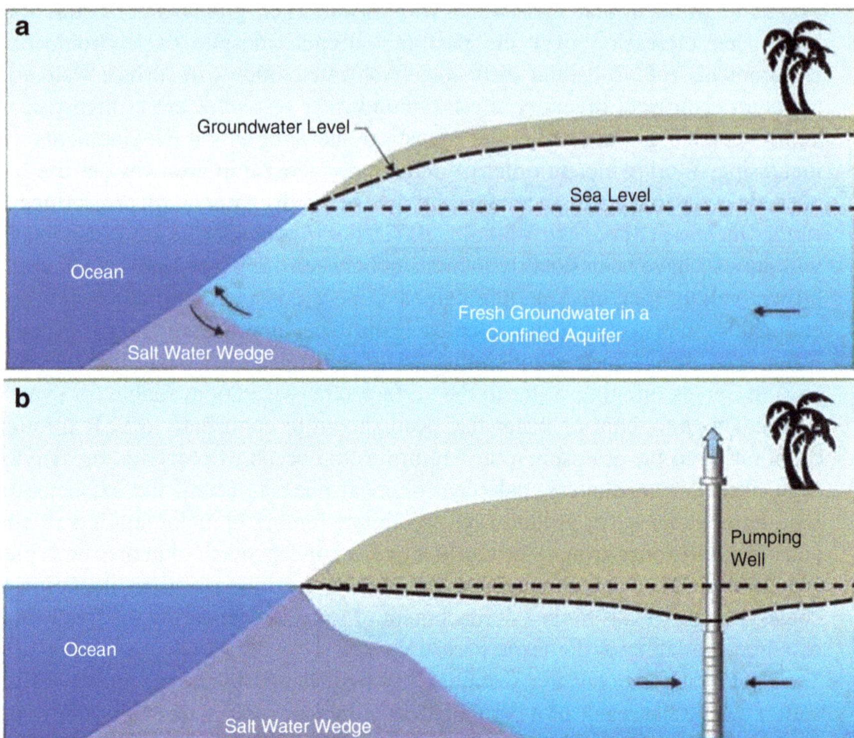

Fig. 1.6 Model showing the hypothetical location of the saltwater-freshwater interface in response to pumping: (**a**) an unconfined coastal aquifer with groundwater levels above sea level equilibrium level (no pumping wells and no seawater intrusion); and (**b**) the same aquifer under a coastal aquifer pumping scenario; the saltwater edge is being pushed inland, closer to the pumping well to reach equilibrium. If sea-level rise will occur, the saltwater intrusion process would be faster (Source: Kan-Rice 2014)

flooding (Figs. 1.3 and 1.8) (Ferguson and Gleeson 2012; Murgulet and Tick 2008). Storm surge and high tides intensify the risks of to SLR at local to regional scales in coastal areas (Fig. 1.8). Furthermore, increases in evaporation rates will enhance salinization of soils (i.e., irrigation fields or flooded coastal areas) and recharging waters percolating to aquifers. Nevertheless, recent simulations show that under a wide range of hydrogeologic conditions and population density scenarios, coastal aquifers are more vulnerable to groundwater extraction than the predicted sea-level rise. Aquifers with very low hydraulic gradients however, could have higher risks of contamination due to SLR and will be impacted by saltwater inundation (e.g., surface contamination) before saltwater intrusion (Ferguson and Gleeson 2012).

Salinized groundwater can only become available for uses through energy-intensive desalination or by dilution. For instance, mixing of fresh groundwater

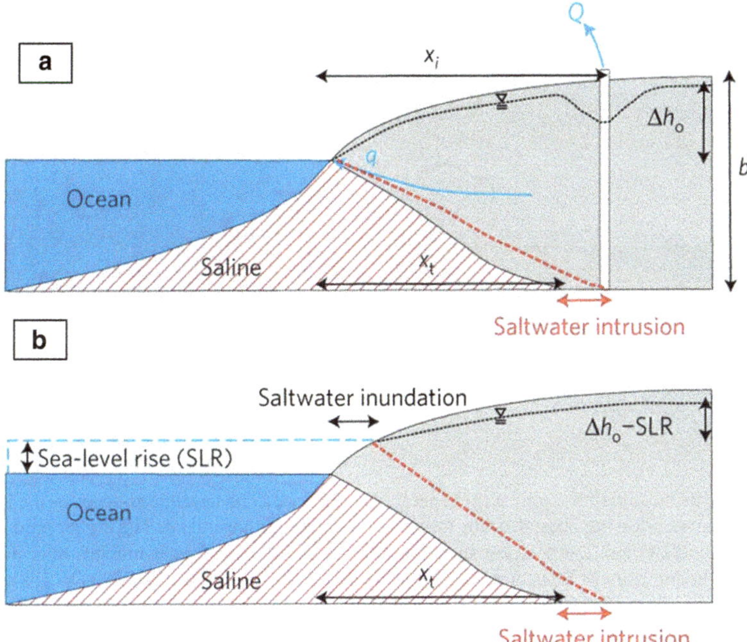

Fig. 1.7 Conceptual model showing the relationship between the saltwater-freshwater interface, groundwater pumping, and SLR in: (**a**) an unconfined coastal aquifer under current conditions with pumping; and (**b**) the same aquifer under saltwater intrusion, saltwater inundation due to sea level rise, scenario with no pumping. Included in the diagrams are simulation variables used to simulate the impact of pumping, sea level rise and inundation: discharge per unit coastline (q), groundwater extraction rate (Q), aquifer thickness (b), the difference in hydraulic head between the inland boundary of the flow system and the coast before sea-level rise (Δho), the distance from the coastline to the well (x_i) and the toe of the saltwater wedge (x_t). The fresh aquifer water before extraction or sea-level rise is shown in the *gray color area* with the water table shape shown by the *dashed line*. In the sea-level rise scenario, the upper portion of the saltwater-freshwater interface moved further inland as a result of saltwater inundation (Source: Ferguson and Gleeson 2012)

(~0 salinity) with just 3 to 4 percent (%) sea water (or groundwater of equivalent salinity) will reduce fresh groundwater availability for many uses, and it becomes unfit for any purpose other than cooling and flushing once the saltwater content rises to 6 % or more (UNEP 2003). Aquifer desalinization is a slow process because of long time residence times given that no natural attenuation is involved other than dilution and dispersion. Aquifer recovery requires displacement of water in storage which in fractured-flow systems with porous matrix or systems characterized by large secondary porosities is difficult to attain because of the trapped, relatively immobile water that has entered by diffusion from the fracture/mobile network (Morris et al. 2003).

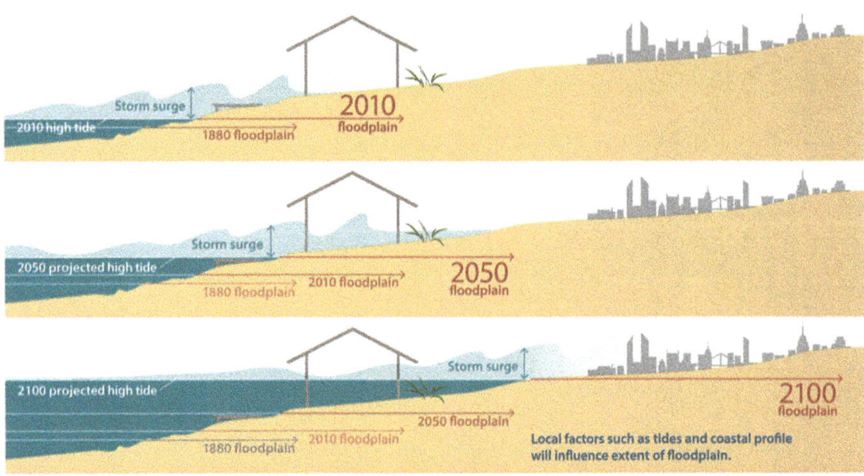

Fig. 1.8 The baseline for storm surge is set by the sea level. The baseline changes as the local sea level increases allowing coastal storm surges to penetrate farther inland. The higher predicted sea levels (e.g., 2050 and 2100), storm surges and flooding will affect areas much further inland. In addition, the intensity of tides, natural and artificial infrastructures, and the relief of coastal lands also influence the extent of local flooding (UCS 2013)

5 Coping with Climate Change/Sea Level Rise and Changing Water Resources in Coastal Areas

5.1 Management Practices

A combination of mitigation and adaptation measures is necessary to address the impacts of SLR. Mitigation, for instance, aims to reduce greenhouse gas emissions and increase sinks, minimizing climate change through climate policy. Adaptation aims to reduce the impacts of SLR through behavioral changes, ranging from individual action to coastal management policy. For example, the upgrading of defense and warning systems and the use of land management can be used to adapt to SLR. Coastal communities, in particular, and national and international governments will have to adopt or develop appropriate responses to cope with the inevitable climate change induced-SLR effects since large populations and regions will be affected if no action is taken. Generally accepted are the following measures that humanity can take: (1) stay in place and do nothing, (2) stay in place and mitigate the problems or (3) be proactive and relocate from the affected areas to unaffected areas or manage the source of the issue (Reuveny 2007). Socioeconomics dictate to a large extent the ability to mitigate or relocate; on a large scale, for developed countries will be more feasible to stay and mitigate or relocate whole groups of people as oppose to less developed countries. The effects of climate change such as SLR are apparent along the California coastline (Caldwell et al. 2013). The San Diego region is one of the hearts of economic activity and culture in southern California and the worst impacts caused by SLR are projected to occur in this region (ICLEI 2014). To start development and implementation of a solution, a Public Agency Steering Committee, consisting of staff from the five bayfront cities, the San Diego Unified Port District, and the San Diego County Regional Airport Authority, gathered to create the "Sea Level Rise Adaptation Strategy for San Diego Bay." The steps adopted by this committee for adaptation are: understanding the problem, setting the goals, developing plans, implementing the plans, and monitoring the program. Additionally, the vulnerability to SLR is assessed based upon: exposure, sensitivity, and adaptive capacity. The vulnerability assessment helps define four aggressive management practices that are applied in the area: hard defense (seawalls and levees), soft defense (natural habitat), accommodation, and withdrawal (ICLEI 2014).

- *Hard Defense*

One of the best examples of hard defense strategies is the San Francisco Bay which has been protected for more than 30 years by a 4-mile-long concrete seawall that kept water away and turned the 700 acres tidal marshes around the bay into developable land. However, the water level around San Francisco is predicted to rise by 38 cm inches by 2050 (BCDC 2009). To protect this area which is prone to inundation giving the rise in sea level, the city's Capital Planning Committee (2014) has drafted a proposal to construct a levee on the shoreline and a tidal gate at the entrance to Mission Creek. Yet another example is the East Coast with New York City facing a real threat from SLR (Gornitz and Hartig 2002). Hurricane Sandy brought waters from Flushing Bay into the LaGuardia Airport, on October 29, 2012 (Freedman 2013). Strategies such as a partial seawall or levees and elevated runways might be applied to mitigate such impacts. The importance of seawalls as a hard defense is observed by Irish et al. (2013) which analyzed the damages caused by Hurricane Sandy by using beach and dune profiles and community conditions for two regions before and after the hurricane: one region had seawall protection which lessened the impacts of the hurricane; another region without seawall protection experienced a significantly higher number of damages (i.e., destroyed residences). Other examples of hard defenses are dikes, a natural or artificial slope or wall used to regulate water levels, also called levee. In Europe for instance, the Dutch have fought back against storm surges by constructing dikes for centuries. However, because those hard defenses are now threatened by SLR and decreased funding for maintenance, the efforts are reoriented towards developing soft defenses such as restoration of natural habitats (Inman 2010).

- *Soft Defense*

Lately an increased interest from scientists and policy-makers is directed towards the potential use of natural or sustainable infrastructures as a passive defense strategy against climate change because natural infrastructures are relatively tractable, flexible, and cost-effective (Tam 2009). Soft defenses, particularly coastal habitats such as salt marshes, sea grass beds, and mangroves, are of interest to scientists in recent years because of the ecosystem services they provide.

These habitats, called blue carbon due to their ability to absorb carbon dioxide and to reduce global warming (McLeod et al. 2011), are also recognized as ideal infrastructures to buffer storm surges and wave erosions (Bakker et al. 2002). In the Netherlands for instance, the historical restored salt marsh sites efficiently accrued sediments that counteracted the sedimentation deficit due to erosion. Furthermore, coral reefs provide protection against erosion and wave damage, and these protections are valued at US $16.9 million per year. Ecosystem services provided by the Mississippi Delta wetlands are worth US $12–47 billion per year. In the Nam Dinh Province of Vietnam, the total cost of restoring mangroves is about US $166 per hectare, but the mangrove would provide benefits worth US $630 per hectare (IUCN 2003). In California, the California State Coastal Conservancy conducted a project (i.e., the Nearshore Linkages Project) to install oyster reefs and restore sea grass beds in order to protect shoreline from SLR (IPCC 2012). Restored or engineered oyster reefs could reduce wave energy while sea grass beds could enhance sedimentation. The Richardson Bay Audubon Center and Sanctuary and the County of Marin in California used gravel and cobbles to build an erosion-resistant beach on the Aramburu Island. This was successful restoration operation as the restored beach is more resilient to erosion and provides high quality habitats (IPCC 2012).

- *Planned Retreat and Relocation*

The impacts of climate change and SLR are already so severe in some parts of the world that a planned retreat or relocation is already necessary. The United Nations High Commissioner for Refugees in consultation with Brookings LSE and Georgetown University with funding from the European Union, Norway and Switzerland met in Sanremo, Italy, to draft a guidebook containing good practices and suggestions for planned relocations due to climate change and anticipated disasters (Ferris 2014). This consultation was developed to address the problems associated with climate change-induced displacement and to ensure that different entities and national systems develop and implement public policies, legislation, coordination arrangements, allocation of specific responsibilities and funding that will not resemble any forced relocations from history (Ferris 2014). Moreover, a guide for climate-induced relocation should also include strategies for mitigating potential conflicts between the migrants and the residents in the receiving areas (Reuveny 2007). In the absence of these arrangements, communities proactively seeking relocation in response to their declining environment are facing great difficulties.

Whereas large, more affluent cities like San Francisco, New York and San Diego have resources to invest in construction of hard defenses, smaller towns and villages must seek an alternative adaptation method. For instance, coastal tribal communities such as the Kivalina and Newtok, Alaska, are faced with an immediate threat from climate change and SLR: both communities have seen an increase in extreme storm events and amplified erosion, salt water intrusion, and alteration of infrastructure (e.g., waste treatment facilities) (Maldonado et al. 2013). The Army Corps of Engineers (USACE) estimated that these villages will be lost to erosion by 2020

(USACE 2006) and estimated a cost of relocation for each village to be US $80–US $200 million, respectively. Accumulated funds allocated for disaster reliefs after five extreme storm events for the Kivalina village, and the construction of a sea barrier that failed just before completion and the construction of a large rock revetment, have already surpassed the estimated amount for relocation (Maldonado et al. 2013). In 1992 the people of Kivalina voted to relocate but have not yet been able to because they do not have the necessary funds and there is no state and/or federal designated entity or process in place to provide the necessary support and assistance. Although at an international level a climate change fund exists (i.e., the Green Climate Fund), resources provided by developed countries are to be applied exclusively for climate change mitigation in developing countries (UN 2011). Furthermore, the federal government has a treaty-based trust relationship that requires the federal government to protect the interests and wellbeing of the tribes; Kivalina's efforts to get assistance in relocating were not successful (Randall 2013). However, in 2013 the Alaska governor submitted a $2.5 million capital budget amendment to legislature in order to authorize development of an eight-mile evacuation and access road for the Native Village of Kivalina (NGA 2013). The village of Newtok has faced the same barriers at the state and federal levels even though village leaders have already negotiated a land swap with the US Fish and Wildlife Service so the village has a tract of land nine miles inland. It is obvious from these examples that changes in infrastructure using hard defense strategies, planned retreat, and relocation are yet not sustainable solutions to SLR; they only provide a temporary solution to an ongoing and increasingly deteriorating problem.

5.2 Mitigation

- *Policy Changes*

Presently, there are many policies in the developed nations that address carbon emissions. *Cap and trade*, one of the most environmentally and economically practical market-based approaches to controlling the greenhouse gas emissions, is used to control pollution by providing incentives for reducing emissions. Permits are issued by governments to set limits to the amount of carbon emissions a company can release to the atmosphere. Companies are allowed to purchase or trade permits to get a larger allowance. Another way is to price carbon emissions through a carbon tax which helps offset some of the negative impacts on the environment by driving up the cost and reducing the consumption. The "Cash for Clunkers" program or the Car Allowance Rebate System (CARS) is a governmental program through which consumers traded qualifying vehicles—passenger cars or light trucks getting less than 18 miles per gallon and less than 25 years old—and receive a rebate towards the purchase of a newer, more fuel efficient vehicle (Shoshannah et al. 2010). Unfortunately, these policies have already been insufficient in meeting the goals of the American Clean Energy and Security Act, which planned to reduce greenhouse gas emissions to 14 % below 2005 levels by the year 2020, largely due to CO_2 levels continuing to increase steadily from 1.5 parts per million per year (ppm/year) between 1980 and 1999, 2.0 ppm/year for the 2000–2007 period, and 2.2 ppm in 2007 (Morrow et al. 2010; Dietz et al. 2009). Clearly, limiting carbon emissions has been ineffective thus far so better policies are necessary.

A different approach is to analyze the primary source of the stress on the environment, the exponential growth of the human population. In many third world countries the population has been growing by 3–10 % since 2005 and the main causes are believed to be the limited education or no access to family planning resources and unchecked instances of rape (Cohen 1995). On the contrary, China, the nation with the largest population, implemented the one-child policy since 1979. This reduced the number of births by about 250–300 million and reduced the total fertility rate from 2.9 in 1979 to 1.7 in 2004 (Hesketh et al. 2005). These reductions have eased some of the pressure on communities, state and the environment in a country that still carries a fifth of the population (Penny and Ching 1999). For effective management, mitigation, and adaptation techniques to the impacts of global climate change, a better understanding of the importance of human interactions with the environment is critical. Since human water use is a key driver in the hydrology of coastal aquifers, efforts to adapt to sea-level rise should include better water management practices.

References

Abd-Elhamid HF, Javadi AA (2008) Mathematical models to control saltwater intrusion in coastal aquifers. GeoCongress 2008:790–797. doi:10.1061/40972(311)98

Abd-Elhamid HF, Javadi AA (2011) Impact of sea level rise and over-pumping on seawater intrusion in coastal aquifers. J Water Clim Change 2(1):19–28. doi:10.2166/wcc.2011.053

Alcamo J, Henrichs T (2002) Critical regions: a model-based estimation of world water resources sensitive to global changes. Aquat Sci 64:352–362

Allan J, Komar P, Priest G (2003) Shoreline variability on the high-energy Oregon coast and its usefulness in erosion-hazard assessments. J Coastal Res S38:83–105

Arndt DS, Baringer MO, Johnson MR (2010) State of the climate in 2009. Bull Am Meteorol Soc 91(7):S1–S224

Arnell NW, Livermore MT, Kovats RS, Levy P, Nicholls RJ, Parry ML, Gaffin SR (2004) Climate and socio-economic scenarios for global-scale climate change impacts assessments: characterising the SRES storylines. Global Environ Change 14:3–20

Bakker JP, Esselink P, Dijikema KS, van Duin WE, de Jong DJ (2002) Restoration of salt marshes in the Netherlands: ecological restoration of aquatic and semi-aquatic ecosystems in the Netherlands (NW Europe). Dev Hydrobiol 166:29–51

Barras J, Beville S, Britsch D, Hartley S, Hawes S, Johnston J, Kemp P, Kin-Ler Q, Co-authors (2003) Historical and projected coastal Louisiana land changes: 1978–2050. Open File Report 03-334. U.S. Geological Survey, p 39

Bates BC, Kundzewicz ZW, Wu S, Palutikof JP (2008) Climate change and water technical paper of the intergovernmental panel on climate change VI, IPCC, 2008

BCDC, Bay Conservation and Development Commission (2009) Living with a rising bay: vulnerability and adaptation in San Francisco Bay and on the Shoreline. April 7. http://www.bcdc.ca.gov/BPA/LivingWithRisingBay.pdf. Accessed 15 Oct 2014

Bindoff N et al (2007) In: Solomon S et al (eds) Climate change (2007) the physical science basis. Cambridge University Press, Cambridge, pp 385–432

Blunden J, Arndt DS (2014) State of the climate in 2013. Bull Am Meteorol Soc 95(7):S1–S238

Burke J, Villholth K (2007) Groundwater, a global assessment of scale and significance. In: Molden D (ed) Agriculture. International Water Management Institute, London: Eartscan

Caldwell MR, Hartge E, Ewing L, Griggs G, Kelly R, Moser S, Newkirk S, Smyth R (2013) Assessment of climate change in the Southwest United States: a report prepared for the national climate assessment. Southwest climate alliance, Island Press, Washington DC, p 168–196

Chappell J, Shackleton NJ (1986) Oxygen isotopes and sea level. Nature 324:137–140

Church JA, White NJ (2011) Sea-level rise from the late 19th to the early 21st century. Surv Geophys 32:585–602. doi:10.1007/s10712-011-9119-1

Cohen JE (1995) Population growth and earth's human carrying capacity. Science 269:341–346

Confalonieri U, Menne B, Akhtar R, Ebi KL, Hauengue M, Kovats RS, Revich B, Woodward A (2007) Human health. Climate change 2007: impacts, adaptation and vulnerability. Contribution of Working Group II to the Fourth Assessment Report of the Intergovernmental Panel on Climate Change, Parry ML, Canziani OF, Palutikof JP, van der Linden PJ, Hanson CE (eds), Cambridge University Press. Cambridge, UK, 391–431

Cowell PJ, Stive MJF, Niedoroda AW, Vriend HJ, Swift DJP, Kaminsky GM, Capobianco M (2003) The coastal - tract (part 1): a conceptual approach to aggregated coastal modeling of low - order coastal change. J Coast Res 19:812–827

Dahl TE (2005) Florida's wetlands-an update on status and trends 1985 to 1996. U.S. Department of Interior, Fish and Wildlife Service, Washington, D.C., 80p

Dietz T, Gardner GT, Gilligan J, Stern PC, Vandenbergh MP (2009) Household actions can provide a behavioral wedge to rapidly reduce US carbon emissions. Proc Natl Acad Sci U S A 106(44)

Drexler JE (2001) Effect of the 1997-1998 ENSO-related drought on hydrology and salinity in a Micronesian wetland complex. Estuaries 24:343–358

Erb M, Broccoli A, Clement A (2013) The contribution of radiative feedbacks to orbitally driven climate change. J Climate 26:5897–5914

Eurosion (2004) Living with Coastal Erosion in Europe: Sediment and Space for Sustainability. Part-1Major Findings and Policy Recommendations of the EUROSION Project. Guidelines for implementing local information systems dedicated to coastal erosion management. Service contract B4-3301/2001/329175/MAR/B3 "Coastal erosion–Evaluation of the need for action". Directorate General Environment, European Commission, p 54

Evengard B, Berner J, Brubaker M, Mulvad G, Revich B (2011) Climate change and water security with a focus on the arctic. Glob Health Action 2011(4):1–4

Fairbanks RG (1989) 17,000-year glacio-eustatic sea level record: influence of glacial melt ingrates on the Younger Dryas event and deep-ocean circulation. Nature 342:637–642

Feng W, Hardt B, Banner J, Meyer K, James E, Musgrove M, Edwards R, Cheng H, Min A (2014) Changing amounts and sources of moisture in the U.S. southwest since the Last Glacial Maximum in response to global climate change. Earth Planet Sci Lett 401:47–56

Ferguson G, Gleeson T (2012) Vulnerability of coastal aquifers to groundwater use and climate change. Nature Clim Change 2:342–345

Ferris E (2014) Planned relocations, disasters and climate change: consolidating good practices and preparing for the future. United Nations High Commissioner for Refugees. March 12–14. http://www.unhcr.org/53c4d6f99.pdf. Accessed Sep 2014

Freedman A (2013) U.S. airports face increasing threat from rising seas. http://www.climatecentral.org/news/coastal-us-airports-face-increasing-threat-from-sea-level-rise-16126. Accessed Oct 2014

García-Ruiz J, López-Moreno JI, Vincente-Serrano SM, Lasanta-Martínez T, Beguería S (2011) Mediterranean water resources in a global change scenario. Earth Sci Rev 105:121–139
Gornitz V, Couch S, Hartig EK (2002) Impacts of sea level rise in New York City metropolitan area. Global Planet Change 32:61–88
Graham S (1999) The Earth's climate system constantly adjusts, NASA, Earth Observatory. http://earthobservatory.nasa.gov/Features/Clouds/. Accessed 26 Nov 2014
Grinsted A, Moore JC, Jevrejeva S (2009) Reconstructing sea level from paleo and projected temperatures 200 to 2100 AD. Clim Dyn 34:461–472
Haq BU, Hardenbol J, Vail PR (1987a) Chronology of fluctuating sea levels since the Triassic (250 million years ago to present). Science 235:1156–1167
Haq BU, Hardenbol J, Vail PR (1987b) Chronology of fluctuating sea levels since the Triassic (250 million years ago to present). Petrol Geol Mem 26:49–212
Hays JD, Imbrie J, Shackleton NJ (1976) Variations in the earth's orbit: pacemaker of the ice ages. Science 194:1121–1132
Hesketh T, Lu L, Xing ZW (2005) The effects of china's one-child family policy after 25 years. N Engl J Med 353:1171–1176
Hester T, Shafer H, Feder K (1997) Field methods in archaeology, 7th edn. McGraw-Hill, New York
Hoegh-Guldberg O, Bruno JF (2010) The impact of climate change on the world's marine ecosystems. Science 328:1523–1528
Hutson SS, Barber NL, Kenny JF, Linsey KS, Lumia DS, Maupin MA (2004). Estimated Use of Water in the United States in 2000. USGS Circular 1268. Denver, C0:U.S. Geological Survey
Hu A, Meehl GA, Weiqing H, Yin J (2010) Effect of the potential melting of the Greenland Ice Sheet on the Meridional Overturning Circulation and global climate in the future. Deep-Sea Research II
Hunter PR (2003) Climate change and waterborne and vectorborne disease. J Appl Microbiol 94:37–46
ICLEI, International Council for Local Environmental Initiatives (2014) Sea level rise adaptation strategy for San Diego Bay. http://www.icleiusa.org/static/San_Diego_Bay_SLR_Adaptation_Strategy_Complete.pdf. Accessed 14 Oct 2014
Inman MD (2010) Working with water. http://www.nature.com/climate/2010/1004/full/climate.2010.28.html. Accessed 14 Oct 2014
IPCC (2013) Climate change 2013: the physical science basis. Contribution of Working Group I to the Fifth Assessment Report of the Intergovernmental Panel on Climate Change. Stocker TF, Qin D, Plattner G-K, Tignor M, Allen SK, Boschung J, Nauels A, Xia Y, Bex V, Midgley PM (eds). Cambridge University Press, Cambridge, United Kingdom and New York, NY, USA, pp 1137–1215
IPCC (2014) Climate Change 2014, Mitigation of Climate Change. Contribution of Working Group III to the Fifth Assessment Report of the Intergovernmental Panel on Climate Change. Edenhofer O, Pichs-Madruga R, Sokona Y, Farahani E, Kadner S, Seyboth K, Adler A, Baum I, Brunner S, Eickemeier P, Kriemann B, Savolainen J, Schlömer S, Stechow C von, Zwickel T, Minx JC (eds). Cambridge University Press, Cambridge, United Kingdom and New York, NY, USA
IPCC, Intergovernmental Panel on Climate Change (2012) Managing the risks of extreme events and disasters to advance climate change adaptation. http://www.nature.org/ourinitiatives/regions/northamerica/unitedstates/california/ca-green-vs-gray-report-2.pdf. Accessed 25 Oct 2014
Irish J, Lynett P, Weiss R, Smallegan S, Chen W (2013) Curied relic seawall mitigates Hurricane Sandy's impacts. Coast Eng 80:79–82
IUCN, International Union for Conservation of Nature. 2003. Economic valuation of demonstration wetland sites in Vietnam. http://cmsdata.iucn.org/downloads/06_vietnam_economic_valuation_of_wetland_sites.pdf. Accessed Oct 2014

Johnson TA, Whitaker R (2004) Salt water intrusion in the coastal aquifers of Los Angeles County, California: Coastal Aquifer Management. In: Cheng AH, Ouazar D, Lewis Publishers, Chapter 2, p 24–48
Kan-Rice P (2014) Optimizing irrigation may ease groundwater overdraft in Pajaro Valley. Green Blog. http://ucanr.edu/blogs/blogcore/postdetail.cfm?postnum=11846. Accessed 30 Nov 2014
Keeling RF (2007) Deglaciation mysteries. Science 316(5830):1440–1441
Kemp AC, Horton BP, Donnelly JP, Mann ME, Vermeer M, Rahmstorf S (2011) Climate related sea-level variations over the past two millennia. Proc Natl Acad Sci U S A 108:11017–11022
Kovats RS, Haines A (2005) Global climate change and health: recent findings and future steps [invited commentary]. Can Med Assoc J 172(4):501–502
Kovats RS, Bouma MJ, Hajat S, Worrell E, Haines A (2003) El Nino and health. Lancet 362:1481–1489
Langevin CD, Zygnerski M (2012) Effect of sea-level rise on salt water intrusion near a coastal well field in Southeastern Florida. Ground Water 51(5):781–803. doi:10.1111/j.1745-6584. 2012.01008.x
Leatherman SP (2001) Social and economic costs of sea-level rise. In: Douglas BC, Kearney MS, Leatherman SP (eds) Sea-level rise, history and consequences. Academic Press, San Diego, pp 181–223
Leatherman SP, Chalfont R, Pendleton EC, McCandless TL, Funderburk S (1995) Vanishing lands: sea level, society and Chesapeake Bay. University of Maryland and US Fish and Wildlife Service, Annapolis, MD
Leuliette EW, Miller L (2009) Closing the sea level rise budget with altimetry, Argo, and GRACE. Geophys Res Lett 36:L04608. doi:10.1029/2008GL036010, issn: 0094-8276
Levermann A, Clark PU, Marzeion B, Milne GA, Pollard D, Radic V, Robinson A (2013) The multimillennial sea-level commitment of global warming. Proc Natl Acad Sci 110:13745–13750
Lipp EK, Huq A, Colwell RR (2002) Effects of global climate on infectious disease: the cholera model. Clin Microbiol Rev 15:757
Lomas MW, Glibert PM, Shiah F, Smith EM (2002) Microbial process and temperature in Chesapeake Bay: current relationships and potential impacts of regional warming. Glob Change Biol 8:51–70
Lyell C (1830) Principles of geology: being an attempt to explain the former changes of the Earth's surface, by reference to causes now in operation. John Murray, London
Ma J, Xie S (2013) Regional patterns of sea surface temperature change: a source of uncertainty in future projections of precipitation and atmospheric circulation. J Climate 26:2482–2501
Maldonado JK, Shearer C, Bronen R, Peterson K, Lazrus H (2013) The impact of climate change on tribal communities in the US: displacement, relocation, and human rights. Clim Change 120:601–614
Matthews RK, Poore RZ (1980) Tertiary $\delta^{18}O$ record and glacio-eustatic sea-level fluctuations. Geol J 8:501–504. doi:10.1130/0091-7613
McLean RF, Shen J-S (2006) From foreshore to foredune: foredune development over the last 30 years at Moruya Beach, New South Wales. Australia J Coastal Res 22:28–36
McLeod E, Chmura GL, Bouillon S, Salm R, Bjork M (2011) A blueprint for blue carbon: toward an improved understanding of the role of vegetated coastal habitats in sequestering CO2. Front Ecol Environ 9:552–560
Miller KG, Sugarman PJ, Browning JV, Kominz MA, Hernández JC, Olsson RK, Wright JD, Feigenson MD, Van Sickel W (2003) A Late Cretaceous chronology of large, rapid sea level changes: glacioeustasy during the greenhouse world. Geology 31:585–588
Moore MV, Pace ML, Mather JR, Murdoch PS, Howarth RW, Folt CL, Chen CY, Hemond HF, Flebbe PA, Driscoll CT (1997) Potential effects of climate change on freshwater ecosystems of the New England/mid-Atlantic region. Hydrol Process 11:925–947
Morris BL, Lawrence ARL, Chilton PJC, Adams B, Calow RC, Klinck BA (2003) Groundwater and its Susceptibility to Degradation: A Global Assessment of the Problem and Options for

Management. Early Warning and Assessment Report Series, RS. 03-3. United Nations Environment Programme, Nairobi, Kenya. http://www.unep.org/dewa/water/groundwater/pdfs/Groundwater_INC_cover.pdf. Accessed 25 Nov 2014

Morrow WR, Gallagher KS, Collantes G, Lee H (2010) Analysis of policies to reduce oil consumption and greenhouse gas emissions from the U.S. transportation sector. Energ Policy 38(3):1305–1320, http://belfercenter.ksg.harvard.edu/files/Policies%20to%20Reduce%20Oil%20Consumption%20and%20Greenhouse%20Gas%20Emissions%20from%20Transportation.pdf

Morton RA, Miller TL, Moore LJ (2004) National assessment of shoreline hange: Part1 Historical shoreline changes and associated coastal land loss along the U.S. Gulf Of Mexico. Open File Report 2004-1043. U.S. Geological Survey, 44pp

Murgulet D, Tick GT (2008) The extent of saltwater intrusion in southern Baldwin County. Alabama Env Geol 55(6):1235–1245. doi:10.1007/s00254-007-1068-0

Neelin J (2011) Climate change and climate modeling. Cambridge University Press, Cambridge

NGA (2013) Center for best practices, Front and Center March 29, 2013. http://www.nga.org/cms/home/nga-center-for-best-practices/front--center-newsletters/2013--front--center-newsletters/col2-content/front--center---march-29-2013.html. Accessed 25 Nov 2014

Nicholls RJ (2011) Planning for the impacts of sea level rise. Oceanography 24(2):144–157

NOAA (2012) Global sea level rise scenarios for the United States National Climate Assessment. National Oceanic and Atmospheric Administration. NOAA technical report OAR CPO-1. Silver Spring, MD. http://scenarios.globalchange.gov/sites/default/files/NOAA_SLR_r3_0.pdf.

NOAA-NCDC (2014) Highlights from the National Climate Assessment. NOAA National Climatic Data Center. http://nca2014.globalchange.gov/highlights. Accessed 28 Nov 2014

Obbink EA, Carlson AE, Klinkhammer GP (2009) Eastern North American freshwater discharge during the Bølling-Allerød warm periods. Geol J 38(2):171–174. doi:10.1130/G30389.1

Oude Essink GHP, van Baaren ES, de Louw PGB (2010) Effects of climate change on coastal groundwater systems: a modeling study in the Netherlands. Water Resour Res 46:W00F04

Pascual M, Rodo X, Ellner SP, Colwell R, Bouma MJ (2000) Cholera dynamics and El Niño Southern oscillation. Science 289:1766–1767

Penny K, Ching YC (1999) China's one child family policy; Education and debate. BMJ 319:992–994

Piao S, Ciais P, Huang Y, Shen Z, Peng S, Li J, Zhou L, Liu H, Ma Y, Ding Y, Friedlingstein P, Liu C, Tan K, Yu Y, Zhang T, Fang J (2010) The impacts of climate change on water resources and agriculture in China. Nature 467:43–51

Pirazzoli PA, Regnauldand H, Lemasson L (2004) Changes in storminess and surges in western France during the last century. Mar Geol 210:307–323

Ragab R, Prudhomme C (2002) Climate change and water resources management in arid and semi-arid regions: prospective and challenges for the 21st century. Biosyst Eng 81(1):3–34

Rahmstorf S (2007) A semi-empirical approach to projecting future sea-level rise. Science 315(5810):368–370. doi:10.1126/science.1135456

Ranasinghe R, McLoughlin R, Shortand AD, Symonds G (2004) The Southern oscillation index, wave climate, and beach rotation. Mar Geol 204:273–287

Randall SA (2013) Corporate responsibility and climate justice: a proposal for a U.S. financed relocation fund for federally recognized tribes imperiled by climate change. Fordham Environ Law Rev 25:10–45. http://law.famu.edu/download/file/Abate%20-%20Corporate%20Responsibility%20and%20Climate%20Justice.pdf. Accessed 25 Nov 2014

Raymo M, Nisancioglu K (2003) The 41 kyr world: Milankovitch's other unsolved mystery. Paleoceanography 18(1):1011, doi: 10.1029/2002PA000791. http://www.agu.org/pubs/sample_articles/cr/2002PA000791/0.shtml. Accessed 19 Oct 2014

Regnauld H, Pirazzoli PA, Morvan G, Ruz M (2004) Impact of storms and evolution of the coastline in western France. Mar Geol 210:325–337

Reuveny R (2007) Climate change-induced migration and violent conflict. Polit Geogr 26 (6):656–673
Rogers K, Saintilan N, Heinjis H (2005) Mangrove encroachment of salt marsh in Western Port Bay, Victoria: the role of sedimentation, subsidence, and sea-level rise. Estuaries 28:551–559
Saji NH, Goswami BN, Vinayachandran PN, Yamagata T (1999) A dipole mode in the tropical Indian Ocean. Nature 401:360–363
Schmidt G, Wolfe J (2009) Climate change: picturing the science. W. W. Norton & Company; Original edition. p 320. ISBN-10: 0393331253
Seager R, Ting M, Held I, Kushnir Y, Lu J, Vecchi G, Huang H, Harnik N, Leetmaa A, Lau N, Li C, Velez J, Naik N (2007) Model projections of an imminent transition to a more arid climate in southwestern North America. Science 316:1181–1184
Shoshannah ML, Keoleian GA, and Bolon KM (2010) The impact of 'Cash for Clunkers' on greenhouse gas emissions: a life cycle perspective. Environ Res Lett 5:044003, doi: 10.1088/1748-9326/5/4/044003. http://iopscience.iop.org/1748-9326/5/4/044003/. Accessed 25 Nov 2014
Silverman J, Lazar B, Cao L et al (2009) Coral reefs may start dissolving when atmospheric CO2 doubles. Geophys Res Lett 36:L05606. doi:10.1029/2008GL036282
Sloan L, Rea D (1995) Atmospheric carbon dioxide and early Eocene climate: a general circulation modeling sensitivity study. Paleogeogr Paleoclimatol Paleoecol 119:275–292
Spencer R, Braswell W (2011) On the misdiagnosis of climate feedbacks from variations in Earths radiant energy balance. Remote Sens 3(8):1603–1613
Strauss BH, Ziemlinski R, Weiss JL, Overpeck JT (2012) Tidally adjusted estimates of topographic vulnerability to sea level rise and flooding of the contiguous United States. Environ Res Lett 7:014033
Tam L (2009) Strategies for managing sea level rise. The Urbanist 487
Taniguchi M (2011) Groundwater and subsurface environments — human impacts in Asian Coastal Cities (Springer, 2011)
Taylor JA, Murdock AP, Pontee NI (2004) A macroscale analysis of coastal steepening around the coast of England and Wales. Geogr J 170:179–188
Taylor RG, Scanlon B, Döll P, Rodell M, van Beek R, Wada Y, Longuevergne L, Leblanc M, Famiglietti JS, Edmunds M, Konikow L, Green TR, Chen J, Makoto T, Bierkens MFP, MacDonald A, Fan Y, Maxwell RM, Yechieli Y, Gurdak JJ, Allen DM, Shamsudduha M, Hiscock K, Yeh PJ-F, Holman I, Treidel H (2012) Ground water and climate change. Nature Clim Change 3:322–329. doi:10.1038/nclimate1744
Törnqvist TE, Hijma MP (2012) Links between early Holocene ice-sheet decay, sea-level rise and abrupt climate change. Nat Geosci 5:601–606
Tsimplis MN, Woolf DK, Osborn TJ, Wakelin S, Wolf J, Flather RA, Shaw AGP, Woodworth PH, Challenor PG, Blackman D, Pert YZ, Jevre-jeva S (2005) Towards a vulnerability assessment of the UK and northern European coasts: the role of regional climate variability. Philos T Roy Soc A 363:1329–1358
Tsimplis MN, Shaw AGP, Flatherand RA, Woolf DK (2006) The influence of the North Atlantic Oscillation on the sea level around the northern European coasts reconsidered: the thermosteric effects. Philos T Roy Soc A 364:845–856
U.S. Army Corps of Engrs. (2006) Alaska village erosion technical assistance program. Alaska dist. p. 21–25. http://www4.nau.edu/tribalclimatechange/resources/docs/res_USArmyCorpEngAKVillErosionTechAssistProg.pdf. Accessed 23 Nov 2014
UCS (2013) Causes of sea level rise: what the science tells us (2013). Union of Concerned Scientists. http://www.ucsusa.org/global_warming/science_and_impacts/impacts/causes-of-sea-level-rise.html#.VHaLWcnt6Cj. Accessed 26 Nov 2014
UN (2011) Framework Convention on Climate Change. Report of the Conference of the Parties on its seventeenth session, held in Durban from 28 November to 11 December 2011. DIstr. General 15 March 2012. FCCC/CP/2011/Add. 1. http://unfccc.int/resource/docs/2011/cop17/eng/09a01.pdf. Accessed 25 Nov 2014

UNEP (2003) Water supply and sanitation coverage in UNEP regional seas, need for regional wastewater emission targets? Section II: targets and indicators for domestic sanitation & wastewater treatment: Discussion Paper. UNEP/GPA, The Hague, The Netherlands. http://esa.un.org/iys/docs/san_lib_docs/wet_section_ii_english.pdf. Accessed 22 Nov 2014

Vail PR, Mitchum RM Jr, Todd RG, Widmier JM, Thompson SIII, Sangree JB, Bubb JN, Hatlelid WG (1977) Seismic stratigraphy and global changes of sea level. In: Payton CE (ed) Seismic stratigraphy–applications to hydrocarbon exploration. American Association of Petroleum Geologists, Tulsa, Oklahoma

Vörösmarty CJ, Green P, Salisbury J, Lammers RB (2013) Global water resources: vulnerability from climate change and population growth. Science 289:284–288

Weedon G (2005) Time-series analysis and cyclostratigraphy: examining stratigraphic records of environmental cycles. Cambridge University Press, Cambridge

Woodroffe CD (2003) Coasts: form process and evolution. Cambridge University Press, Cambridge 623pp

World Meteorological Organization (2014) Significant natural climate fluctuations. http://www.wmo.int/pages/themes/climate/significant_natural_climate_fluctuations.php. Accessed 27 Nov 2014

Yakirevich A et al (1998) Simulation of seawater intrusion into the Khan Yunis area of the Gaza Strip coastal aquifer. Hydrogeol J 6:549–559

Zhao M (2014) An investigation of the connections among convection, clouds, and climate sensitivity in a global climate model. J Climate 27:1845–1862

Chapter 2
Assessing Groundwater Pollution Risk in Response to Climate Change and Variability

Ruopu Li

Abstract Groundwater systems worldwide are facing increasing pressure from climate variability and human activities. The climate can affect the aquifer directly and/or indirectly through various pathways, such as groundwater recharge, the change in soil biogeochemical processes, and land use activities. To date, there has been relatively little research that specifically addressed the impacts of climate change on groundwater quality. To address the research gap, this paper highlights the mechanisms occurring at various hydrogeologic phases (e.g., soil, unsaturated zones and aquifers) of groundwater contamination that are susceptible to the effects of climate change. Relevant studies on groundwater pollution risk that have addressed or can potentially incorporate climate change impacts are summarized and synthesized in this article. It has been found that change in groundwater quality as a direct result of climate change is subject to various uncertainties affecting credible future projections. The change in land use activities, indirectly affected by climate change, may play more important roles in determining future patterns of groundwater pollution risks. Research and policy recommendations are provided to cope with the challenges to sustainable water quality management in a changing climate.

1 Introduction

As the second largest fresh water resource on Earth, groundwater is the principal source to meet 36 % of worldwide demand for drinking water (Döll et al. 2012). In the United States, approximately 40 % of the public water supply, serving over 74 million people, is withdrawn from groundwater. Over 97 % of those persons residing in rural areas of the U.S. use groundwater for drinking (National Research Council 2000). It is likely that population growth and increased climate variability

---

R. Li (✉)
Department of Geography and Environmental Resources, Southern Illinois University-Carbondale, Carbondale, IL, USA
e-mail: Ruopu.Li@siu.edu

A. Fares (ed.), *Emerging Issues in Groundwater Resources*, Advances in Water Security, DOI 10.1007/978-3-319-32008-3_2

will, in the near future, lead to greater dependence on groundwater for public water supply (Hall et al. 2008).

Increasing evidence of groundwater contamination in recent years, coupled with concerns about human health and ecological effects of contaminants such as nitrates and pesticides, has heightened pressure on public agencies to better manage groundwater (National Research Council 2000; Sampat 2000). The application of fertilizers and pesticides on croplands, for example, has resulted in deterioration in drinking water quality worldwide (Spalding and Exner 1993). Nitrate contamination of groundwater has been linked with various fatal diseases, such as fatal blue baby syndrome, increasing incidence of gastric cancer and elevated non-Hodgkin's lymphoma (Karkouti et al. 2005; Knobeloch et al. 2000). Apart from human health, changes in groundwater quality also have negative impacts on groundwater-dependent species such as sightless and non-pigmented crawfish and cavefish (Butscher and Huggenberger 2009).

Climate change has been regarded as one of the most significant global change that our society is facing. It has been widely agreed that global warming will continue affect our societal vulnerability (Jiménez Cisneros et al. 2014). Observed and predicted alterations of climate such as earlier onset of spring, longer growing seasons, spatial and temporal changes in precipitation patterns (Feng and Hu 2007), and higher mean soil temperatures may lead to a profound modification to our surface and groundwater systems around the world (Arnell 1998). For example, groundwater recharge in the U.S. High Plain Aquifer may decline by over 20 % if the mean temperature increases by 2.5 °C (Hall et al. 2008). Groundwater levels are expected to shift around 0.5 m under the future climate change scenarios in south-central British Columbia, Canada (Scibek et al. 2007). Increasing frequency of droughts and floods contributed by climate variability is expected to cause growing dependence on groundwater for public water supplies. In general, groundwater resource provides a buffer for our water security against the climate extremes. For water resource managers, it is critical to understand that the availability of groundwater is not an issue of quantity alone, but also implies sufficient quality to meet demands (Butscher and Huggenberger 2009). It is important to take measures and develop strategies that could mitigate the impacts of climate change on our fresh groundwater reserves based on scientifically defensible information.

In recent years, there have been growing research interests in understanding the potential impacts of climate change on fresh water resources. However, most of the research efforts have been focused on climate-stressed surface water systems (Rehana and Mujumdar 2012). Compared with the surface water, research on the responses of groundwater systems to climate change, especially in a coupled human-environment system, has still been relatively scarce (Goderniaux et al. 2009; Green et al. 2011). Unsurprisingly, surface water is more accessible to be observed, monitored, measured and analyzed, and its dynamics with climate change is also more responsive. Among climate change-related research on groundwater, the focus was dominantly on climate change and its impact on groundwater quantity instead of groundwater quality (Kumar 2012). For example, in the most recent Intergovernmental Panel on Climate Change (IPCC) Fifth Assessment

Report (AR5), very limited information was provided concerning the climate change's impacts on groundwater quality (Jiménez Cisneros et al. 2014). In the report, the major groundwater quality concern was directed to the elevated concentrations of fecal coliforms associated with extreme rainfall events (Jean et al. 2006). In a report by Clifton et al. (2010), the effects of global warming on increased groundwater salinization and seawater intrusion were concerned.

The current research gap in climate and groundwater quality may be due to the following reasons: (1) the effects of climate change on groundwater quality deterioration is confounded by anthropogenic factors such as agricultural land use and land management (Wilby et al. 2006); (2) the physical dynamics of groundwater system in response to climate change are often indirect and complicated (Jyrkama and Sykes 2007); (3) considerable uncertainties exist in association with climate change projections (Bloomfield et al. 2006; Toews and Allen 2009); (4) long-term observation data of groundwater quality are still scarce or unavailable (Fogg and LaBolle 2006).

This paper focuses on the climate change and its impacts on the groundwater pollution risk affected by diffusive sources from human activities such as farm chemical applications. Although many coastal regions are facing increased groundwater salinity from saltwater intrusion, vulnerability of aquifers to seawater rise due to climate change is beyond the scope of this paper. A comprehensive review of seawater-related issues is available (Ferguson and Gleeson 2012). The objective of this paper is to summarize and synthesize current research progress and new frontier regarding modeling practices and long-term projection on the climate-affected groundwater degradation, and to provide suggestions to policy makers for integrating the consideration of climate change into groundwater quality management. It is well recognized that there has been a few excellent reviews on climate and groundwater published in recent years (such as Green et al. 2011; Kløve et al. 2014), but focused perspective dedicated to groundwater pollution risk in response to climate change is still scant.

2 How Climate Change Affects Groundwater Quality

As climate continues to affect our surface water systems at unprecedented rates, the reliability of groundwater as a fresh water resource becomes critically important for water security of our society. Buried underground, the groundwater is often regarded a resource that is protected from direct impacts of climate change. However, groundwater resource is still expected to be prone to long-term climate change through various pathways (Green et al. 2011; Stuart et al. 2011; Taylor et al. 2013). Contamination of groundwater involves various physical and geochemical processes, potentially subject to climate change, along with the transport of leachate from the pollution sources. Figure 2.1 shows a schematic representation

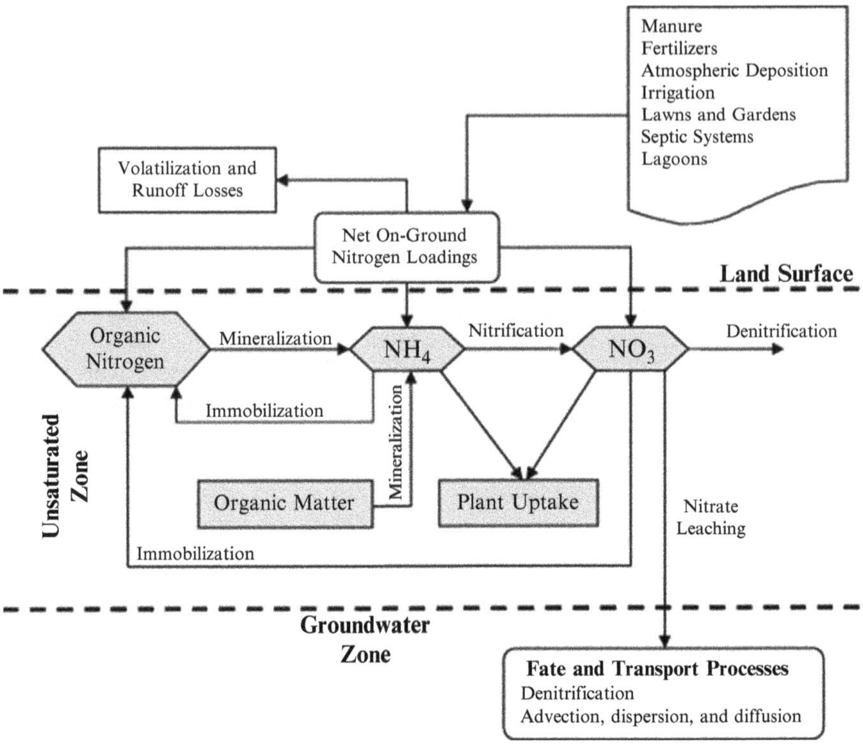

Fig. 2.1 A conceptual representation of the processes that contribute to nitrate occurrences in groundwater (the figure cited from Almasri and Kaluarachchi 2007)

of the processes that contribute to nitrate occurrences in groundwater. There has been increasing evidence that nitrate and pesticide concentrations has been a serious challenge in many parts of the world (Spalding and Exner 1993; Stuart et al. 2011; Babiker 2004; Bloomfield et al. 2006; Obeidat et al. 2013).

Climate change can affect the physical and geochemical processes that affect groundwater quality in many ways, either directly or indirectly (Green et al. 2011). Figure 2.2 shows a conceptual framework that illustrates an overview of the paths where climate variability and change could make this precious resource precarious. Climate change and variability can influence the quality of the groundwater system directly through aquifer replenishment, indirectly through the modification of land use that results in varied recharge patterns and pollutant loadings, shifts in biogeochemical processes and concentrated channels in the soil and unsaturated zone. The following sections provide details about these pathways. Table 2.1 provides a summary of the impacts of climate change on factors that may affect groundwater quality along the pathways.

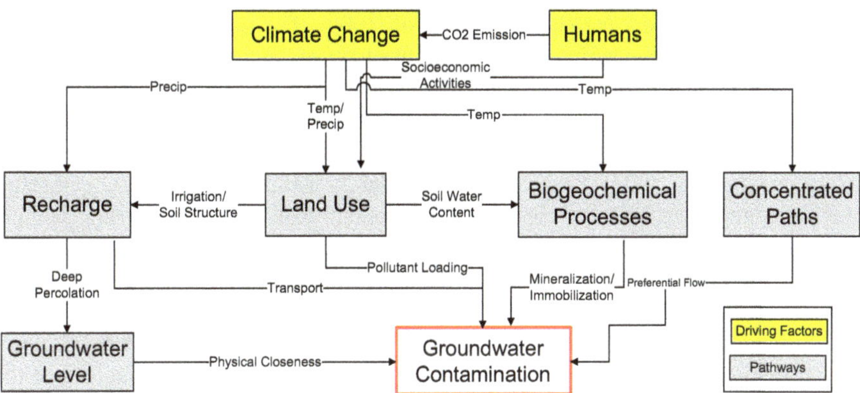

Fig. 2.2 Pathways that climate change can affect groundwater quality

Table 2.1 Pathways that climate change affect components of groundwater systems

Climate scenarios	Pathways	Potential impacts	References
Changes in temperature and precipitation	Recharge	Changes in transport of leachate; shifts in seasonal and annual groundwater levels	Eckhardt and Ulbrich (2003), Scibek and Allen (2006), Ali et al. (2012)
Warmer temperature	Land use	Favorable conditions for arable crops that require higher fertilizer and pesticide application (e.g., corn); earlier and extended growing seasons	Bloomfield et al. (2006), Li and Merchant (2013)
Warmer temperature	Biochemical processes	Lower soil organic matter; higher fertilizer application rate for nutrient compensation; fast decomposition for preferential paths; enhanced denitrification	Dalias et al. (2001), Zhu and Fox (2003), Rivett et al. (2008)
Extreme rainfalls	Concentrated paths	Increased incidence of microbial contamination through flooding wells	Jean et al. (2006)
Warmer temperature and less precipitation	Concentrated paths	Increases in the size and frequency of crack formation in soils, which acts as preferential flow channels for contaminants	Bloomfield et al. (2006), Stuart et al. (2011)

2.1 Impacts of Climate Change on Groundwater Recharge

Assessing the vulnerability of aquifers to pollution is inextricably related to the understanding of the recharge mechanism, as the transport of contaminant to aquifers occurs as part of the recharge process (Foster 1998). Changes in recharge rates and mechanisms can contribute to the mobilization of pollutants stored in the soil and unsaturated zone (Bloomfield et al. 2006). The magnitude and timing of groundwater recharge shift as a first response to climate change (Scibek and Allen 2006; Ali et al. 2012), in accordance with seasonal mean and annual groundwater levels (Kløve et al. 2014). Higher precipitation, for example, would generally be expected to produce greater aquifer recharge rates (Rosenberg et al. 1999; Rosenzweig et al. 2007). Reduced soil frost due to warming temperature can result in more recharge and increase risk of contaminant leaching during winters (Okkonen et al. 2010).

Many modeling techniques have been used to determine the potential impacts of climate change on groundwater recharge (Chen et al 2002; Croley and Luukkonen 2003; Eckhardt and Ulbrich 2003; Holman 2006; Scibek and Allen 2006). Groundwater quality in response to climate change requires not only the reliability of the downscaled projections from the global circulation model (GCMs), but also accurate estimation of groundwater recharge (Jyrkama and Sykes 2007).

Rosenberg et al. (1999) used HUMUS (the Hydrologic Unit Model of the United States) and SWAT (Soil Water Assessment Tool) to assess plausible impacts of climate change on water yields and recharge of the Ogallala Aquifer. They applied three GCMs projections and projected different degrees of reduction in groundwater recharge. The modeling results showed that increasing precipitation projected by GCMs may not result in rising groundwater recharge since the elevated evapotranspiration caused by warmer temperature can consume the surplus water.

Eckhardt and Ulbrich (2003) found 3.5–7 % decreases in annual groundwater recharge in a central European low mountain range under B1 and A2 scenarios based on a modified SWAT model, SWAT-G, and climate change scenarios developed from the Europe ACACIA project. The model predicted a decline in recharge due to the reduction in snow-rainfall ratio and increases in temperature and potential evapotranspiration.

Scibek and Allen (2006) used a one-dimensional model, Hydrologic Evaluation of Landfill Performance (HELP), coupled with statistically downscaled GCMs, to estimate future recharge conditions under climate change scenarios in a semi-arid region of south-central British Columbia, Canada. A transient groundwater flow model, developed using MODFLOW software, was then used to quantify the impacts of climate change on groundwater levels in an unconfined aquifer. The changes in groundwater recharge were estimated to range between −0.1 and 0.5 m, and changes in water levels relative to historic period were less than 0.5 m. It has been found that the computed recharge was sensitive to the selection of GCM scenarios and hydrogeologic parameters such as soil and aquifer permeability. But some technical challenges still existed in this study. For example, the potential changes in snow melting, growing periods and annual groundwater extraction, as a result of global warming, were not considered.

Jyrkama and Sykes (2007) projected heightened groundwater recharge in the Grand River watershed in Ontario, Canada as a result of future changes in climate conditions. The focus of the study was the projection of spatially explicit recharge patterns based on HELP3 model and ArcView GIS. The projected alteration in recharge exhibited high spatial variability in the study area.

It is noted that the change in recharge rates is also inextricably linked with land use and cover change in response to climate change. Changes in vegetation cover and structure, for example, can alter components of the hydrologic cycle such as interception, infiltration and transpiration, and thus have significant impacts on groundwater recharge (Le Maitre et al. 1999). Increased frequency of drought may provoke widespread of conversion from dryland farming practices to irrigation. This can create potential enhanced artificial recharge. More details on land use are available in the following section.

2.2 *Impacts of Land Use Change (LUC) on Groundwater Quality as a Result of Climate Change*

Climate is probably the most important controlling factor in determining general types of crops that can be grown within certain geographic locations (Bloomfield et al. 2006). Alteration of climate patterns can create favorable or unfavorable conditions for certain arable crops, and thus may cause geographical shifts in the dominant crops. For example, warmer climate may make some lands in high latitudes more suitable for arable production. In recent decades, the northern Great Plains has witnessed change in cropping patterns driven by warmer climate and biofuel demands (Li et al. 2012). In Britain, southern England is expected to become more suitable for crops such as maize and sunflowers (Bloomfield et al. 2006). In addition, the global warming can also extend the growing season and allow earlier onset of farming. Eckhardt and Ulbrich (2003) projected that the growing period would start from 1 week to 1 month earlier than the baseline period depending on the scenarios of greenhouse gas emission.

Groundwater resources are indirectly prone to climate change through conversion of land use. The current groundwater quality concerns largely originate from the land use that is associated with either non-point sources such as fertilizer application or more concentrated sources such as dairy lagoons, leakage of urban sewage or septic tank systems (Arnade 1999; MacQuarrie et al. 2001). For example, drought, desertification and land degradation associated with increased climate variability may force more rural population at risk to move into urban areas (Tacoli 2009; Warn and Adamo 2014), thus potentially creating more chances of leakage from the sewage systems. But this concentrated source is more incidental and less widespread than the non-point source related to agriculture. Arable crops have frequently been found to relate to nitrate loadings in groundwater (Goulding 2000; Babiker 2004; Panagopoulos et al. 2006; Li et al. 2014). The linkage between the climate and groundwater resources in recent century has been complicated by land use, including the expansion of agriculture in the form of rain-fed and irrigated (Taylor et al. 2013). Lima et al. (2011) predicted an increase in groundwater vulnerability in a basin of Argentina using a modified DRASTIC model that incorporates future agricultural expansion scenarios from the Dyna-CLUE model.

Furthermore, it has been found that the indirect climate impacts through farming decisions often have more significant effects on groundwater quality than the direct impacts through recharge (Li and Merchant 2013; Taylor et al. 2013). For example, elevated nitrate concentration is typically a major groundwater quality concern in the U.S. High Plains, where a large fraction of the wells exceed the maximum contaminant level of 10 mg/L NO_3-N (Scanlon et al. 2006). Increased recharge associated with conversion of natural vegetation to irrigated cropland has resulted in large-scale groundwater contamination caused by enhanced flushing of nitrate and chloride, which had accumulated in the unsaturated zone for centuries (Scanlon et al. 2006). Fast expansion of corn, partially driven by warmer temperature and higher corn prices, was expected to contribute to elevated groundwater pollution risks in this region (Li and Merchant 2013). A shift of cropping pattern from rotational corn to continuous corn may contribute to elevated NO_3-N loading to groundwater (Owens et al. 1995; Plourde et al. 2013). Both conversion of crop types and extended growing period may bring in more loadings of farm chemicals.

2.3 *Shifts in Biogeochemical Processes*

Global warming is likely to influence temperature-dependent reaction rates and reduction–oxidation (redox) reactions that directly affect many biogeochemical processes, including soil organic carbon decomposition and nitrate attenuation (Dalias et al. 2001; Green et al. 2011; Kløve et al. 2014). On one hand, global warming can amplify microbial activities and accelerate soil organic decomposition, contributing to the release of nutrients such as nitrate and phosphorus. Increase in ambient temperature is widely believed to contribute to the reduction in organic matter contents (Kirschbaum 1995), lowering the effective nitrate in soil and potentially triggering higher rate of fertilizer application. In addition, the fast decomposition of some plant root systems (e.g., alfalfa) can create preferential flow paths and channels for rapid transport of leachate in the deep soil and vadose zone (Le Maitre et al. 1999; Zhu and Fox 2003). On the other hand, the climate change may directly or indirectly affect denitrification processes. Studies have found that higher temperature can lead to increase in the rate of denitrification, because warming groundwater temperature tends to reduce the dissolved oxygen concentration and create more preferential anaerobic conditions for denitrification (Rivett et al. 2008). Therefore, the net effect of climate change on the biogeochemical processes is largely uncertain and needs more research. Up to now, the availability of relevant monitoring data for the long-term groundwater temperature is still very limited. Scarce research has been found regarding the magnitude of the impacts of climate change on groundwater quality through this pathway.

2.4 Concentrated Paths

Climate variability and change may affect the contamination process in a more concentrated manner in the future. The significance of these paths may vary depending on the geographic locations. For example, microbial contamination can occur through well locations during the extreme rainfall events (Jean et al. 2006). It is expected that more frequent and extended drought may lead to growing reliance on groundwater resource and thus increase the drilling of wells for agricultural irrigation, municipal and industrial purposes. Thus, more opportunities for the point-source contamination could take place through the wells during flooding events. This is a valid concern especially in many developing countries. Moreover, drier and warmer climate conditions in some areas could result in increases in the size and frequency of crack formation in soils, which will create preferential flow channels that enable rapid movement of ground pollutants through the soil profiles and directly into the aquifers (Bloomfield et al. 2006; Greve et al. 2010; Stuart et al. 2011).

3 Modeling Practices

To quantify the effects of climate change on groundwater contamination process, we need models relying on physically based descriptions of relevant processes. The risk of groundwater pollution is typically assessed through simple index models (Aller et al. 1985), statistical models (Nolan et al. 2002), or process-based models (Tiktak et al. 2006). Traditionally, groundwater pollution risk assessments have been focused on the existing conditions of groundwater vulnerability to contamination. Groundwater pollution risk was modeled based on a 'static' concept (Butscher and Huggenberger 2009) and analyses of historic or current hydrogeology conditions. The dynamics caused by climate were seldom considered. Since the risk of groundwater pollution is strongly dependent on factors such as depth-to-water, recharge and land use conditions that are susceptible to climate impacts, the patterns of the risk may shift accordingly (Li and Merchant 2013). In this study, we reviewed a few examples of applications or potential applications for modeling climate-affected groundwater quality risk.

Index models are usually formulated as equations using a weighted linear combination of factors in a Geographic Information Systems (GIS) environment to compute a pollution potential index, which qualitatively indicates the vulnerability of groundwater to contamination (Li et al. 2014). Among numerous index models (Aller et al. 1985; Foster 1998; Van Stempvoort et al. 1993; Civita 1993; Goldscheider 2003), The DRASTIC model and its modified versions have been used with exceptional frequency (Rundquist et al. 1991; Guo et al. 2007; Li et al. 2014). DRASTIC, is derived from the seven variables used in the model: **D**epth-to-water table; **R**echarge (net); **A**quifer media; **S**oil media (texture);

Topography (slope); **I**mpact of the vadose zone; and, **C**onductivity (hydraulic) of the aquifer (Eq. 2.1). Ratings and weightings are commonly based on expert knowledge (Merchant 1994) or statistical relationships between variables and actual pollutant concentration (Panagopoulos et al. 2006). Index methods are often attractive because they are conceptually simple, usually require fewer datasets, and are easily implemented in GIS. Their performance, however, is often reported as mixed (Neukum et al. 2008; Tesoriero and Voss 1997). Major drawbacks of index methods include (1) the subjectivity inherent in determination of the rating scales and weighting coefficients, (2) linear model formulation used to represent non-linear hydrogeologic processes, and (3) the interpretation of the results which are expressed as dimensionless pollution potential index values (Merchant 1994). Therefore, groundwater pollution risk under scenarios of climate change using index methods ought to be interpreted cautiously.

$$\textbf{\textit{Pollution Potential}} = D_rD_w + R_rR_w + A_rA_w + S_rS_w + T_rT_w + I_rI_w + C_rC_w \qquad (2.1)$$

where D is Depth to Water, R is Net Recharge, A is Aquifer Media, S is Soil Media, T is Topography (Slope), I is Impact of the Vadose Zone, C is Conductivity of the Aquifer, r is rating, and w is weight.

In practice, a few studies have employed the index models to quantify the groundwater pollution risks under different climate scenarios. Ducci (2005) proposed that patterns of regional groundwater pollution vulnerability will vary between drought, average, and wet periods. And, an index model was employed to demonstrate that the groundwater vulnerability can be sensitive to climate conditions. Li and Merchant (2013) implemented a modeling framework that employs four sub-models, including climate change scenarios, a LUC model, recharge and groundwater level models, and a modified DRASTIC model, linked within a GIS framework (Fig. 2.3). The approach was used to evaluate the groundwater pollution risks under scenarios of future climate and LUCs in North Dakota. The northward shifts in a cropping pattern of rotational corn-soybeans may lead to more lands susceptible to elevated fertilizer application. Such changes may have significant implications for groundwater quality. The results showed that areas with high vulnerability will expand northward and/or northwestward in Eastern North Dakota under different scenarios. This research features a LUC model, which has been used to spatially explicitly model the expansion of corn and soybeans affected by the climate change. Again, the result of this research can be deemed as empirical and qualitative in nature due to the limitation of the index models.

Statistical models can evaluate groundwater vulnerability based on statistical relationships between observed groundwater contamination and related predictor variables. Predictions of potential climate effects on groundwater quality can be derived from statistical relationships based on long-term data sets. Statistical models generate coefficients that best fit observed water quality data, and thus reduce the subjectivity involved in assigning factor ratings and weights needed by

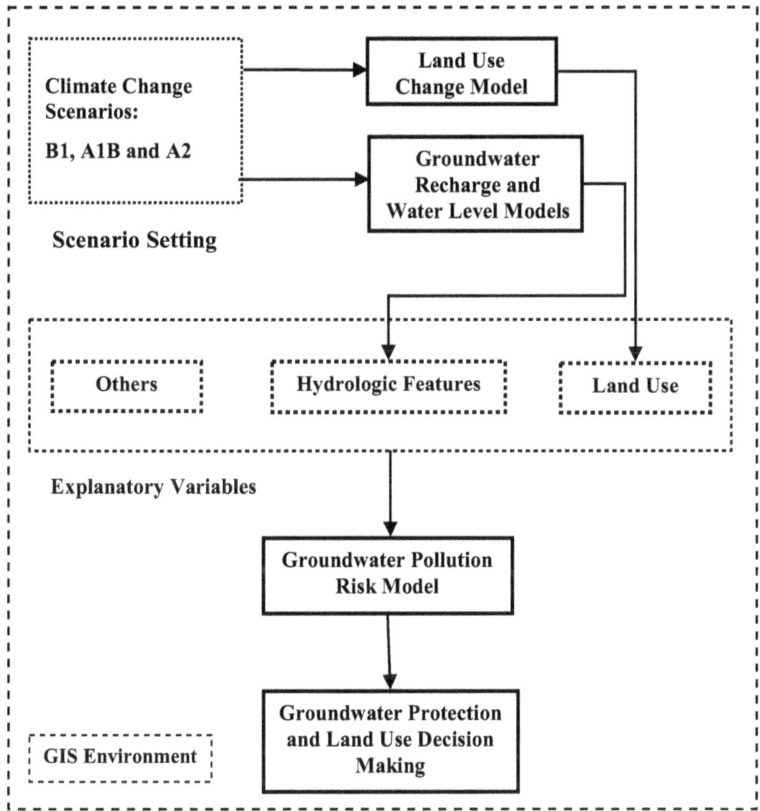

Fig. 2.3 Modeling framework employed to simulate the groundwater pollution risk affected by climate change (Li and Merchant 2013)

index models (Focazio 2002). Such models may employ geo-statistical delineation of contaminated areas (Almasri and Kaluarachchi 2007), logistic regression analysis of relationships between predictor variables and observed contamination occurrence (Nolan et al. 2002; Gurdak and Qi 2006), or weights of evidence (WofE) modeling, a Bayesian-probabilistic approach (Raines et al. 2000; Masetti et al. 2007). Logistic regression has been used with especially great frequency among statistical models, since groundwater pollution risk can readily be computed as a statistical probability of pollutant concentration above certain threshold. For instance, Nolan et al. (2002) used logistic regression to predict the probability of nitrate contamination of recently recharged groundwater in the conterminous United States. The drawbacks of this approach include the requirement for large datasets of groundwater quality and hydrogeologic factors to generate robust statistical relationships. When required data are not available at sufficient spatial, temporal and/or categorical accuracy, the resulting statistical relationships between predictor variables and groundwater contamination may be unreliable. It assumes spatio-temporal stationarity, which may not hold true for an evolving environment with climate alteration. Thus, statistical models need to be cautiously evaluated if it is used for a climate change impact study.

Process-based modeling methods account for physical processes of water movement and the associated fate and transport of contaminants in a quantitative manner, and hence can produce relatively accurate estimates of pollutant concentration. For example, Tiktak et al. (2006) used EuroPEARL, a one-dimensional, mechanistic pesticide leaching model, in a GIS to map pesticide leaching in Europe. Sinkevich et al. (2005) implemented the Generalized Preferential Flow Transport Model (GPFM) to locate areas with high risk of contamination by agrochemicals. Almasi and Kaluarachchi (2007) employed a soil nitrogen dynamic model to estimate nitrate leaching to the aquifer and then used the MODFLOW and MT3D to simulate the fate and transport processes of nitrate in the groundwater system (Fig. 2.4). Although this study did not consider climate change, its modeling framework could potentially be adapted to simulate the impacts of climate change on the groundwater quality by simulating the climate change effects on the input factors such as land use and recharge.

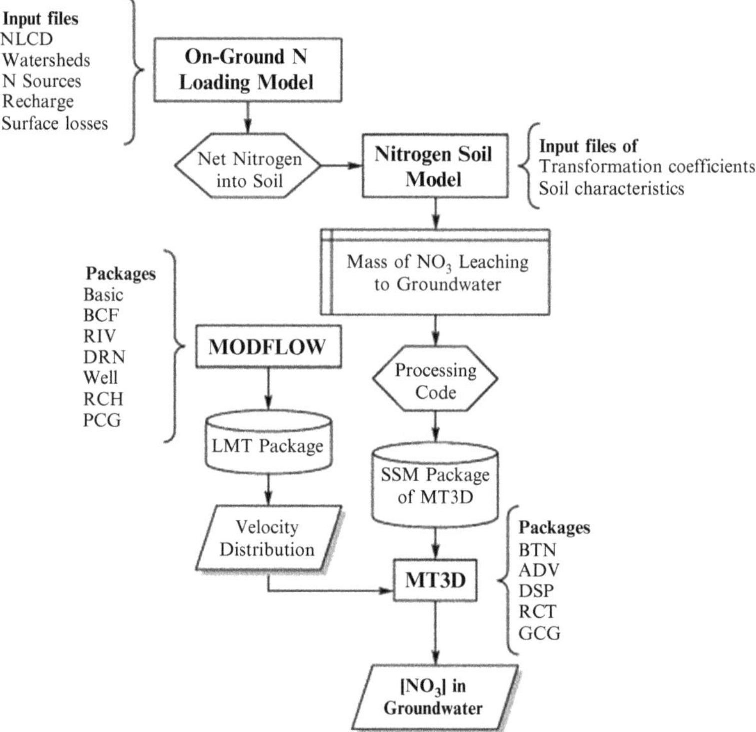

Fig. 2.4 Flowchart for modeling nitrate contamination in groundwater using physical models (the figure was cited from Almasri and Kaluarachchi 2007)

Butscher and Huggenberger (2009) analyzed a karst aquifer system in Switzerland based on a lumped parameter model, and found that groundwater vulnerability depends on climate-affected recharge conditions. The Karst system represents the groundwater systems that are rapidly responding and most susceptible to the climate effects. In this research, the temporal variation in Karst groundwater vulnerability is modeled as a dynamic response to variable precipitation and evaporation conditions. It is based on a lumped parameter model in which flow can be attributed to three distinct processes representing the recharge system, the conduit flow system, and the diffuse flow system. Two indices were developed to represent the vulnerability relative to the fecal bacteria and persistent contaminants that correspond to the faster conduit flow system and the slower diffuse flow system respectively. The simulation results showed that hot and dry climate, which are very likely to occur in the study area, help alleviate the pollution risk of short-lived contaminants but exacerbate the risk of persistent contaminant.

Process-based models take into account important physical, chemical, geological and biological processes, which are described by deterministic equations and can produce accurate estimates of contaminant concentration. Thus, it is suitable for quantification of the climate change impacts on groundwater quality. However, the process-based methods are generally technically complex and constrained for large regional analyses since the data on key physical parameters are usually not available. The sensitivities of modeled groundwater quality to the parameterization and climate change scenarios need to be carefully evaluated.

4 Research and Policy Suggestions

4.1 Research Limitations and Opportunities

Groundwater quality is affected by many complex parameters and processes, which are directly or indirectly influenced by the climate change. Through a review of existing related studies, we have found that climate variability and change can affect the groundwater quality through multiple pathways such as recharge processes and LUC. Although groundwater quality is important to meet future demands for clean drinking water, very few studies have been directed to specifically address the groundwater pollution risk affected by the climate change and variability so far. This identified research gap may not be occasional, partially owing to a variety of uncertainties.

Climate change impact study typically involves projection disparity in GCMs, uncertainties associated with downscaling techniques of the GCMs, and parameter and structural uncertainties in the hydro-physical models (Wilby et al. 2006). First of all, large uncertainties are inherent in the choice of GCMs. Uncertainties concerning future greenhouse gas emissions, climate sensitivity and regional climate change responses were typically incorporated into the construction of climate change scenarios (Eckhardt and Ulbrich 2003).Various studies have confirmed that climate change projections based on current global and regional climate models contribute to the major uncertainties in responses of groundwater system (Bloomfield et al. 2006; Scibek and Allen 2006; Wilby et al. 2006; Toews and Allen 2009). For example, projections of GCMs can vary significantly, especially along the U.S. Great Plains (Li and Merchant 2013). The outputs from different GCM models show that the boundary where no change in precipitation is oriented more or less west-to-east in this region. To deal with this uncertainty, most studies adopt more than one projection for each emission scenario (Butscher and Huggenberger 2009; Kløve et al. 2014) or a multi-model mean of climate projections (Li and Merchant 2013). Secondly, the application and usability of GCM products for hydrologic studies were often constrained by their relatively coarse spatial and temporal resolutions (Wood et al. 2004). Dynamical and statistical downscaling are the most commonly used approaches to resolve GCM outputs at finer spatial and temporal resolutions (Fowler et al. 2007). But the reliability of GCM outputs theoretically degrades with increases in temporal and spatial resolutions, affecting accurate representation of climate extremes (Fowler et al. 2007). Thirdly, the uncertainties associated with the choice of parameterization and model structures could be formidably high. The model development is technically challenging and normally requires considerable expertise in hydrogeology. The modeling results could be sensitive to the models in use. For example, Eckhardt and Ulbrich (2003) has found that different versions of SWAT model contributed to a large disparity in projected groundwater recharge in response to climate change in a central European low mountain range. There has been little information in many studies regarding which type of uncertainties dominated the modeling outputs. Ideally, a comprehensive test of key sources of uncertainty to determine their relative significance to the impacts can be conducted in an integrated modeling approach (Wilby et al. 2006).

In addition, the research on climate-related groundwater pollution risk may be affected by the observation of groundwater quality. In many areas, groundwater quality data were unavailable or collected sporadically. Apart from those fast-responsive groundwater systems such as karst aquifers, the groundwater quality may well lag behind climate induced effects in some places due to longer residence time of the newly recharged groundwater. For instance, the observed nitrate concentration in groundwater does not necessarily reflect the current risk, since the nitrate from the most recent recharge leaching may still be deposited in the soil profiles or vadose zones. Increase in precipitation may help mobilize and facilitate the release of these contaminants to aquifers. The observed incidence of groundwater contamination often depends on many site-specific factors such as the type, characteristics and quantities of applied farm chemicals, and detailed hydrogeologic parameters (such as subsurface redox conditions and preferential flow), which may not be mappable at the scales where the models were developed. For future research, it is indispensable to initiate long-term data collection efforts in those areas likely to be affected by climate change. Multiple studies have indicated a possible way to reduce model uncertainties is to improve the necessary database by establishing more monitoring networks (Butscher and Huggenberger 2009; Green et al. 2011; Kløve et al. 2014).

Future research may focus on the following areas: (1) indirect impacts of climate change on groundwater quality, because indirect impacts such as LUC may be even greater than the direct effects of climate change; (2) the spatio-temporal interactions between the agricultural landscape, climate change and policy options; (3) the development of physically based models that can account for the key processes influencing nitrate fate and transport in groundwater and their interactions involving climate change.

4.2 Policy Recommendations

To cope with the potential impacts of climate change on groundwater quality, climate perspectives should be incorporated for the preparation of future water resource management plans. To make scientifically defensible decisions, water resource managers and policy makers should start with better understanding of the relationship between climate change and groundwater, and overcome 'fears' of climate change. First, climate has always been evolving and should never be regarded as 'static'. Second, climate may not necessarily result in negative consequences for specific regions. For example, wetter climate does not necessarily lead to increase recharge (Taylor et al. 2013) and elevated groundwater contamination. Third, impacts of climate change on groundwater quality may be largely indirect through its effects on human systems. To develop effective management strategies for ensuing protection of groundwater quality, we should not only focus on the direct change in hydrologic-related factors but also those indirect changes such as phonological change, LUC and management decisions. For example, increased winter rainfall may stimulate plant disease, and thus require greater use of pesticides (Bloomfield et al. 2006). Therefore, effective communication between the scientific community and policy makers can help foster scientifically defensible policies and actions.

At current stage, projecting impacts of climate change on water resources still represents one of the most difficult problems that water resource managers are facing (Goderniaux et al. 2009). Although subject to a multitude of uncertainties, groundwater modeling practices could still benefit the groundwater quality management in the context of potential climate change, aiding in selecting and prioritizing sites for future groundwater monitoring and protection. For the areas projected to have elevated groundwater pollution risks, appropriate agricultural policies/practices may be imperative to prevent groundwater contamination, such as strict irrigation scheduling and fertilizer recommendations for irrigated crops.

It is noted that, in some cases, managing groundwater quantity and quality with challenges from climate change is in an apparent paradox. For example, the groundwater recharge in the High Plain Aquifer was projected to decline based on three GCM scenarios (Rosenberg et al. 1999). A reduction of recharge helps diminish the percolation of potentially contaminated recharge water to the aquifer, but it meanwhile reduces the availability of aquifer replenishment and affects the sustainable use of the aquifer. The increased intensity of rainfall and alteration in magnitude, timing and forms of precipitation (e.g., early snow melting and lower snow-rain ratio) are expected to increase the runoff and decrease the availability of deep percolation, and hence reduce the chance of pollutant leachate to arrive at the aquifer. However, this risk reduction could be at the cost of decreased groundwater availability. Thus, it is very useful to integrate both groundwater quantity and quality and prioritize the manage objectives when the water resource managers draft plans to cope with climate change.

5 Summary

The groundwater systems are increasingly susceptible to the impacts of climate change and variability. Various pathways can potentially direct the climate change effects to the aquifer vulnerability, including altered recharge patterns, the change in soil biogeochemical processes, land use activities and some concentrated paths. It has been found that indirect effects of climate on groundwater can be greater than direct impacts on recharge. For example, the change in land use activities, indirectly affected by climate change, may play more important roles in future patterns of groundwater pollution risk. The complexity in mechanisms along these pathways causes technical challenges in understanding the response of groundwater quality to climate change.

Compared with surface and groundwater quantity, there has been comparatively little research that specifically addressed the impacts of climate change on groundwater quality. Previous studies have been conducted based on the index models and/or physical models. The results of related modeling practices are subject to various uncertainties hurdling credible future projections. The magnitude and nature of uncertainties vary depending on the choice of the GCM projections, downscaling techniques and hydrological models in use. To cope with the challenges to sustainable water quality management in a changing climate, climate perspectives should be taken into consideration for the preparation of future water management plans.

References

Ali R, McFarlane D, Varma S, Dawes W, Emelyanova I, Hodgson G, Charles S (2012) Potential climate change impacts on groundwater resources of South-Western Australia. J Hydrol 475:456–472

Aller L, Robert S, Kerr Environmental Research (1985) DRASTIC: a standardized system for evaluating ground water pollution potential using hydrogeologic settings. Robert S. Kerr Environmental Research Laboratory, Office of Research and Development,

U.S. Environmental Protection Agency. http://books.google.com/books?id=J7OQQgAACAAJ
Almasri MN, Kaluarachchi JJ (2007) Modeling nitrate contamination of groundwater in agricultural watersheds. J Hydrol 343:211–229
Arnade LJ (1999) Seasonal correlation of well contamination and septic tank distance. Ground Water 37:920–923
Arnell NW (1998) Climate change and water resources in Britain. Clim Change 39:83–110
Babiker I (2004) Assessment of groundwater contamination by nitrate leaching from intensive vegetable cultivation using geographical information system. Environ Int 29:1009–1017
Bloomfield JP, Williams RJ, Gooddy DC, Cape JN, Guha P (2006) Impacts of climate change on the fate and behaviour of pesticides in surface and groundwater—a UK perspective. Sci Total Environ 369:163–177
Butscher C, Huggenberger P (2009) Modeling the temporal variability of karst groundwater vulnerability, with implications for climate change. Environ Sci Technol 43:1665–1669
Chen Z, Grasby S, Osadetz K (2002) Predicting average annual groundwater levels from climatic variables: an empirical model. J Hydrol 260:102–117
Civita M (1993) Groundwater vulnerability maps: a review. Proceedings IX symposium pesticide chemistry, mobility and degradation of xenobiotics. Biagini, Lucca, Italy, p 832
Clifton C, Evans R, Hayes S, Hirji R, Puz G, Pizarro C (2010) Water and climate change: impacts on groundwater resources and adaptation options. Water Working Notes 1:1–76
Croley TE, Luukkonen CL (2003) Potential effects of climate change on ground water in Lansing, Michigan. J Am Water Resour Assoc 39:149–163
Dalias P, Anderson J, Bottner P, Coûteaux M-M (2001) Long-term effects of temperature on carbon mineralisation processes. Soil Biol Biochem 33:1049–1057
Döll P, Hoffmann-Dobrev H, Portmann FT, Siebert S, Eicker A, Rodell M, Strassberg G, Scanlon BR (2012) Impact of water withdrawals from groundwater and surface water on continental water storage variations. J Geodyn 59–60:143–156
Ducci D (2005) Influence of climate changes on vulnerability maps. Presented at the first Conference and Workshop on "Groundwater and Climate Change" held in Norwich, UK, University of East Anglia, 4–6 April 2005
Eckhardt K, Ulbrich U (2003) Potential impacts of climate change on groundwater recharge and streamflow in a central european low mountain range. J Hydrol 284:244–252
Feng S, Hu Q (2007) Changes in winter snowfall/precipitation ratio in the contiguous United States. J Geophys Res 112:D15109. doi:10.1029/2007JD008397
Ferguson G, Gleeson T (2012) Vulnerability of coastal aquifers to groundwater use and climate change. Nature Clim Change 2:342–345
Focazio MJ (2002) Assessing ground-water vulnerability to contamination: providing scientifically defensible information for decision makers. US Dept. of the Interior, US Geological Survey
Fogg GE, LaBolle EM (2006) Motivation of synthesis, with an example on groundwater quality sustainability: motivation of synthesis. Water Resour Res 42:W03S05
Foster SSD (1998) Groundwater recharge and pollution vulnerability of British aquifers: a critical overview. Geol Soc London Spec Publ 130:7–22
Fowler HJ, Blenkinsop S, Tebaldi C (2007) Linking climate change modelling to impacts studies: recent advances in downscaling techniques for hydrological modelling. Int J Climatol 27:1547–1578
Goderniaux P, Brouyère S, Fowler HJ, Blenkinsop S, Therrien R, Orban P, Dassargues A (2009) Large scale surface–subsurface hydrological model to assess climate change impacts on groundwater reserves. J Hydrol 373:122–138
Goldscheider N (2003) Karst groundwater vulnerability mapping: application of a new method in the Swabian Alb, Germany. Hydrogeol J 13:555–564
Goulding K (2000) Nitrate leaching from arable and horticultural land. Soil Use Manage 16:145–151

Green TR, Taniguchi M, Kooi H, Gurdak JJ, Allen DM, Hiscock KM, Treidel H, Aureli A (2011) Beneath the surface of global change: impacts of climate change on groundwater. J Hydrol 405:532–560

Greve A, Andersen MS, Acworth RI (2010) Investigations of soil cracking and preferential flow in a weighing lysimeter filled with cracking clay soil. J Hydrol 393:105–113

Guo Q, Wang Y, Gao X, Ma T (2007) A new model (DRARCH) for assessing groundwater vulnerability to arsenic contamination at basin scale: a case study in Taiyuan Basin, Northern China. Environ Geol 52:923–932

Gurdak JJ, Qi SL (2006) Vulnerability of recently recharged ground water in the High Plains Aquifer to nitrate contamination. USGS Sci Investig Rep 2006–5050:39

Hall ND, Stuntz BB, Abrams RH (2008) Climate change and freshwater resources. Nat Resour Environ 30–35

Holman IP (2006) Climate change impacts on groundwater recharge- uncertainty, shortcomings, and the way forward? Hydrogeol J 14:637–647

Jean J-S, Guo H-R, Chen S-H, Liu C-C, Chang W-T, Yang Y-J, Huang M-C (2006) The association between rainfall rate and occurrence of an enterovirus epidemic due to a contaminated well. J Appl Microbiol 101:1224–1231

Jiménez Cisneros BE, Oki T, Arnell NW, Benito G, Cogley JG, Döll P, Jiang T, Mwakalila SS (2014) Freshwater resources. Field CB, Barros VR, Dokken DJ, Mach KJ, Mastrandrea MD, Bilir TE, Chatterjee M, Ebi KL, Estrada YO, Genova RC, Girma B, Kissel ES, Levy AN, MacCracken S, Mastrandrea PR, White LL (eds) Climate change 2014: impacts, adaptation, and vulnerability. Part A: global and sectoral aspects. Contribution of Working Group II to the Fifth Assessment Report of the Intergovernmental Panel of Climate Change. Cambridge University Press, Cambridge, United Kingdom and New York, NY, USA, p 229–269

Jyrkama MI, Sykes JF (2007) The impact of climate change on spatially varying groundwater recharge in the grand river watershed (Ontario). J Hydrol 338:237–250

Karkouti K, Djaiani G, Borger MA, Beattie WS, Fedorko L, Wijeysundera D, Ivanov J, Karski J (2005) Low hematocrit during cardiopulmonary bypass is associated with increased risk of perioperative stroke in cardiac surgery. Ann Thorac Surg 80:1381–1387

Kirschbaum MUF (1995) The temperature dependence of soil organic matter decomposition, and the effect of global warming on soil organic C storage. Soil Biol Biochem 27:753–760

Kløve B, Ala-Aho P, Bertrand G, Gurdak JJ, Kupfersberger H, Kværner J, Muotka T, Mykrä H, Preda E, Rossi P, Uvo CB, Velasco E, Pulido-Velazquez M (2014) Climate change impacts on groundwater and dependent ecosystems. J Hydrol 518:250–266

Knobeloch L, Salna B, Hogan A, Postle J, Anderson H (2000) Blue babies and nitrate-contaminated well water. Environ Health Perspect 108:675–678

Kumar CP (2012) Climate change and its impact on groundwater resources. Int J Eng Sci 1 (5):43–60

Le Maitre DC, Scott DF, Colvin C (1999) Review of information on interactions between vegetation and groundwater. Water SA 25(2):137–152

Li R, Merchant JW (2013) Modeling vulnerability of groundwater to pollution under future scenarios of climate change and biofuels-related land use change: a case study in North Dakota, USA. Sci Total Environ 447:32–45

Li R, Guan Q, Merchant J (2012) A geospatial modeling framework for assessing biofuels-related land-use and land-cover change. Agr Ecosyst Environ 161:17–26

Li R, Merchant JW, Chen X-H (2014) A geospatial approach for assessing groundwater vulnerability to nitrate contamination in agricultural settings. Water Air Soil Pollut 225:2214. doi:10.1007/s11270-014-2214-4

Lima ML, Zelaya K, Massone H (2011) Groundwater vulnerability assessment combining the drastic and dyna-clue model in the argentine pampas. Environ Manage 47:828–839

MacQuarrie KTB, Sudicky EA, Robertson WD (2001) Numerical simulation of a fine-grained denitrification layer for removing septic system nitrate from shallow groundwater. J Contam Hydrol 52:29–55

Masetti M, Poli S, Sterlacchini S (2007) The use of the weights-of-evidence modeling technique to estimate the vulnerability of groundwater to nitrate contamination. Nat Resour Res 16:109–119
Merchant JM (1994) GIS-based groundwater pollution hazard assessment: a critical review of the DRASTIC model. Photogramm Eng Remote Sens 60:1117–1127
National Research Council (2000) Investigating groundwater systems on regional and national scales. The National Academies Press, Washington, DC
Neukum C, Hötzl H, Himmelsbach T (2008) Validation of vulnerability mapping methods by field investigations and numerical modelling. Hydrogeol J 16:641–658
Nolan BT, Hitt KJ, Ruddy BC (2002) Probability of nitrate contamination of recently recharged groundwaters in the conterminous United States. Environ Sci Technol 36:2138–2145
Obeidat M, Awawdeh M, Al-Mughaid H (2013) Impact of a domestic wastewater plant on groundwater pollution, north Jordan. The Revista Mexicana de Ciencias Geologicas 30:371–384
Okkonen J, Jyrkama M, Kløve B (2010) A conceptual approach for assessing the impact of climate change on groundwater and related surface waters in cold regions (Finland). Hydrogeol J 18:429–439
Owens LB, Edwards WM, Shipitalo MJ (1995) Nitrate leaching through lysimeters in a corn-soybean rotation. Soil Sci Soc Am J 59:902
Panagopoulos GP, Antonakos AK, Lambrakis NJ (2006) Optimization of the DRASTIC method for groundwater vulnerability assessment via the use of simple statistical methods and GIS. Hydrogeol J 14:894–911
Plourde JD, Pijanowski BC, Pekin BK (2013) Evidence for increased monoculture cropping in the central United States. Agr Ecosyst Environ 165:50–59
Raines GL, Bonham-Carter GF, Kemp LD (2000) Predictive probabilistic modeling using ArcView GIS. ArcUser 3:45–48
Rehana S, Mujumdar PP (2012) Climate change induced risk in water quality control problems. J Hydrol 444–445:63–77
Rivett MO, Buss SR, Morgan P, Smith JWN, Bemment CD (2008) Nitrate attenuation in groundwater: a review of biogeochemical controlling processes. Water Res 42:4215–4232
Rosenberg N, Epstein D, Wang D, Vail L, Srinivasan R, Arnold J (1999) Possible impacts of global warming on the hydrology of the ogallala aquifer region. Clim Change 42:677–692
Rosenzweig C, Casassa G, Karoly DJ, Imeson A, Liu C, Menzel A, Rawlins S, Root TL, Seguin B, Tryjanowski P (2007) Assessment of observed changes and responses in natural and managed systems. Climate change 2007: impacts, adaptation and vulnerability. Contribution of Working Group II to the Fourth Assessment Report of the Intergovernmental Panel on Climate Change. Parry ML, Canziani OF, Palutikof JP, van der Linden PJ, Hanson CE (eds), Cambridge University Press, Cambridge, UK, 79–131
Rundquist DC, Peters AJ, Di L, Rodekohr DA, Ehrman RL, Murray G (1991) Statewide groundwater-vulnerability assessment in Nebraska using the drastic/GIS - model. Geocarto Int 6:51
Sampat P (2000) Groundwater shock: the polluting of the world's major freshwater stores. World Watch 13:10–22
Scanlon BR, Keese KE, Flint AL, Flint LE, Gaye CB, Edmunds WM, Simmers I (2006) Global synthesis of groundwater recharge in semiarid and arid regions. Hydrol Process 20:3335–3370
Scibek J, Allen DM (2006) Modeled impacts of predicted climate change on recharge and groundwater levels. Water Resour Res 42, doi: 10.1029/2005WR004742
Scibek J, Allen DM, Cannon AJ, Whitfield PH (2007) Groundwater–surface water interaction under scenarios of climate change using a high-resolution transient groundwater model. J Hydrol 333:165–181
Sinkevich MG, Walter MT, Lembo AJ, Richards BK, Peranginangin N, Aburime SA, Steenhuis TS (2005) A GIS-based ground water contamination risk assessment tool for pesticides. Ground Water Monitor Remediat 25:82–91

Spalding RF, Exner ME (1993) Occurrence of nitrate in groundwater—a review. J Environ Qual 22:392

Stuart ME, Gooddy DC, Bloomfield JP, Williams AT (2011) A review of the impact of climate change on future nitrate concentrations in groundwater of the UK. Sci Total Environ 409:2859–2873

Tacoli C (2009) Crisis or adaptation? Migration and climate change in a context of high mobility. Environ Urban 21:513–525

Taylor RG, Scanlon B, Döll P, Rodell M, van Beek R, Wada Y, Longuevergne L, Leblanc M, Famiglietti JS, Edmunds M, Konikow L, Green TR, Chen J, Taniguchi M, Bierkens MFP, MacDonald A, Fan Y, Maxwell RM, Yechieli Y, Gurdak JJ, Allen DM, Shamsudduha M, Hiscock K, Yeh PJ-F, Holman I, Treidel H (2013) Ground water and climate change. Nat Clim Change 3:322–329

Tesoriero AJ, Voss FD (1997) Predicting the probability of elevated nitrate concentrations in the puget sound basin: implications for aquifer susceptibility and vulnerability. Ground Water 35:1029–1039

Tiktak A, Boesten JJTI, van der Linden AMA, Vanclooster M (2006) Mapping ground water vulnerability to pesticide leaching with a process-based metamodel of EuroPEARL. J Environ Qual 35:1213

Toews MW, Allen DM (2009) Evaluating different GCMs for predicting spatial recharge in an irrigated arid region. J Hydrol 374:265–281

Van Stempvoort D, Ewert L, Wassenaar L (1993) Aquifer vulnerability index: a GIS compatible method for groundwater vulnerability mapping. Can Water Resour J 18:25–37

Warn E, Adamo SB (2014) The impact of climate change: migration and cities in South America. Int Meteor Organiz Bull 63(2). http://www.wmo.int/bulletin/en/content/impact-climate-change-migration-and-cities-south-america

Wilby RL, Whitehead PG, Wade AJ, Butterfield D, Davis RJ, Watts G (2006) Integrated modelling of climate change impacts on water resources and quality in a lowland catchment: River Kennet, UK. J Hydrol 330:204–220

Wood AW, Leung LR, Sridhar V, Lettenmaier DP (2004) Hydrologic implications of dynamical and statistical approaches to downscaling climate model outputs. Clim Change 62:189–216

Zhu Y, Fox RH (2003) Corn–soybean rotation effects on nitrate leaching. Agron J 95:1028

Chapter 3
Deficit Irrigation as a Strategy to Cope with Declining Groundwater Supplies: Experiences from Kansas

I. Kisekka and J. Aguilar

1 Declining Groundwater Supplies in Kansas

The High Plains aquifer shown in Fig. 3.1 underlies 453,248 km^2 in parts of eight states (Nebraska, Texas, Kansas, Colorado, New Mexico, Oklahoma, Wyoming, and South Dakota) making it the biggest aquifer in North America (Fig. 3.1). The predominant formation in the High Plains aquifer is the Ogallala Formation of the Miocene/Pliocene age (Sophocleous 2012), thus this aquifer is commonly referred to as the Ogallala aquifer. This predominantly unconfined aquifer comprises of silt, sand, gravel and clay, and varies in saturated thickness and permeability (Buchanan et al. 2009). For example, the predevelopment aquifer thickness and depth to water in Kansas varies from 15 m to over 90 m, and 8 m to over 105 respectively (High Plain Atlas). Most of the water from the aquifer is paleo or fossil water associated with the events of the last ice age when glaciers covered the Great Plains.

The Ogallala aquifer is the principal source of irrigation water for very productive agricultural areas of the US Great Plains including western Kansas, earning this region the nickname "breadbasket of the world" (Opie 2000). About 27 % of the irrigated cropland in the United States overlies the High Plains aquifer (Kansas Water Office 2015). In Kansas, the High Plains aquifer consists of lithologically different yet hydraulically interconnected Ogallala aquifer in the west, the shallower and geologically younger Great Bend Prairie and the Equus Beds aquifers in south central Kansas and the associated alluvial aquifers (Kansas Water Office 2015; High Plains Atlas). The Ogallala portion of the aquifer is shared by all users, domestic, industrial and agriculture, but is heavily developed for irrigation. Over the last four decades, groundwater levels in the aquifer have been declining due to

I. Kisekka (✉) • J. Aguilar
Water and Irrigation Management, Kansas State University,
Southwest Research-Extension Center, Manhattan, KS, USA
e-mail: ikisekka@ksu.edu

A. Fares (ed.), *Emerging Issues in Groundwater Resources*, Advances in Water Security, DOI 10.1007/978-3-319-32008-3_3

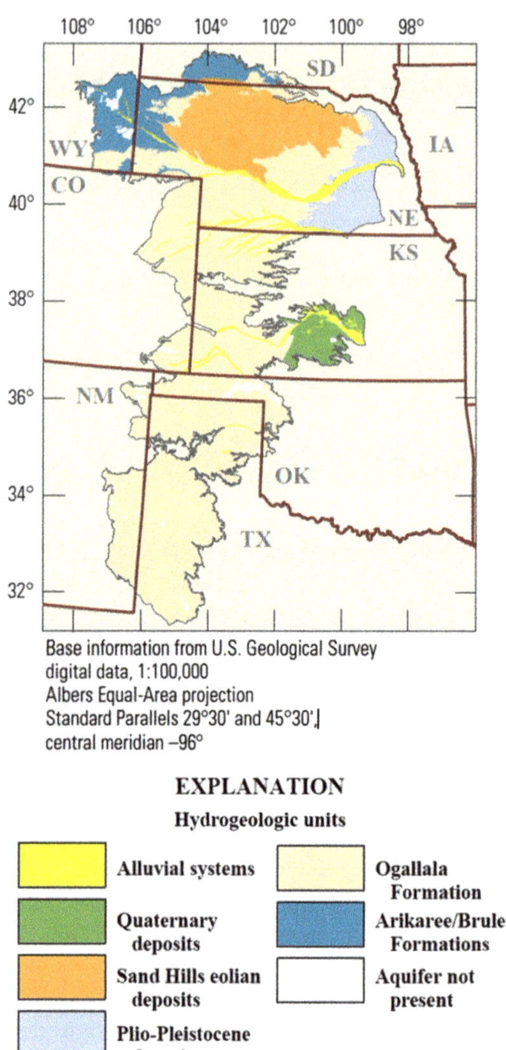

Fig. 3.1 Map of spatial extent of the High Plains aquifer (Credits Gurdak and Roe 2009)

water withdrawals exceeding annual recharge. Recharge rates in many parts of the Ogallala aquifer are relatively low (Gurdak and Roe 2009) with areas recording annual recharge rates as low as 1.27 mm (Miller and Appel 1997). Areas that have experienced significant decline (>13 m) in water level have very low recharge in addition to having low precipitation and high evapotranspiration rate. Currently the aquifer is only 10 % sustainable in many parts of western Kansas and Panhandles of Texas and Oklahoma. In some locations in western Kansas, the Ogallala aquifer is projected to be depleted in less than 50 years as shown in Fig. 3.2.

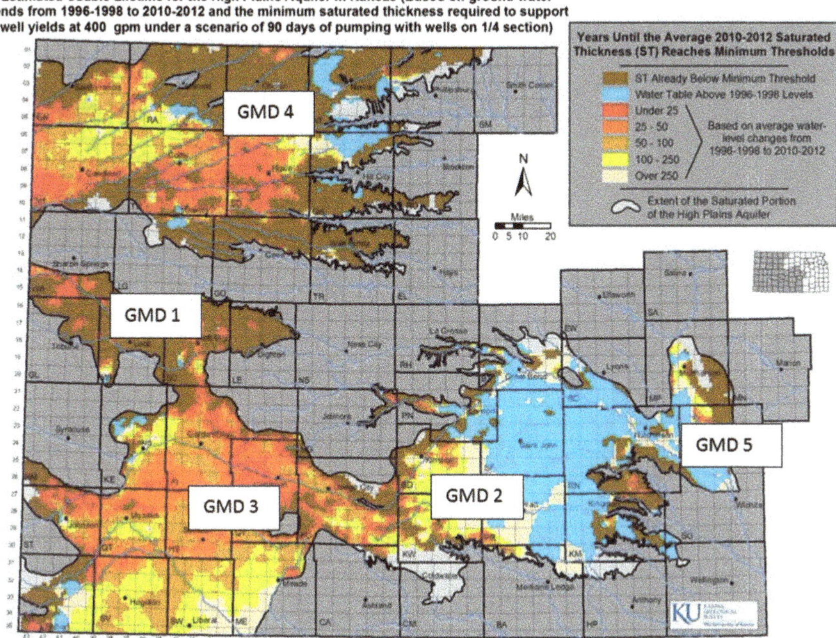

Fig. 3.2 Projected life of the Ogallala aquifer in Kansas, were GMD refers to Groundwater Management District (Credits Kansas Geological Survey, The University of Kansas)

2 Irrigation Trends in Kansas

Irrigation in Kansas dates back to 1880s when several irrigation ditch companies were established along the Arkansas River (Erhart 1969). More rapid irrigation development is attributed to the period after the Second World War from 1945 to 1980 mainly due to favorable laws encouraging groundwater development, advances in technology (engines, gear heads and deep well turbine pumps) and availability of energy (Rogers and Lamm 2012). Figure 3.3 shows increase in irrigated acreage from 1945 to the late 1970s and peaking at approximately 1.416 million hectares in early 1980s. Irrigation system types have also evolved substantially. For example in the early 1970s only 18 % of the 728,434 hectares were under sprinkler irrigation while the rest was under surface irrigation (Rogers and Lamm 2012). Today of the roughly 1.214 million irrigated hectares, over 90 % are irrigated by center pivot. Figure 3.4 illustrates increased adoption of center pivots as shown by the shapes of the fields changing from square to circular in southwest Kansas between 1972 and 2011. The rest of the irrigated area is under surface irrigation with a small portion (approximately 6070 hectares) under subsurface drip irrigation (Rogers et al. 2008). The total area under SDI is increasing steadily as well capacities continue to decline. These changes in irrigation application

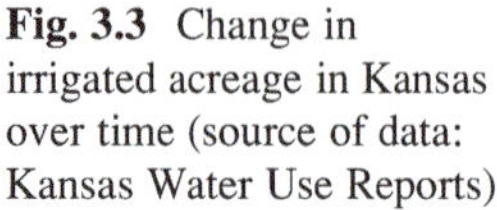

Fig. 3.3 Change in irrigated acreage in Kansas over time (source of data: Kansas Water Use Reports)

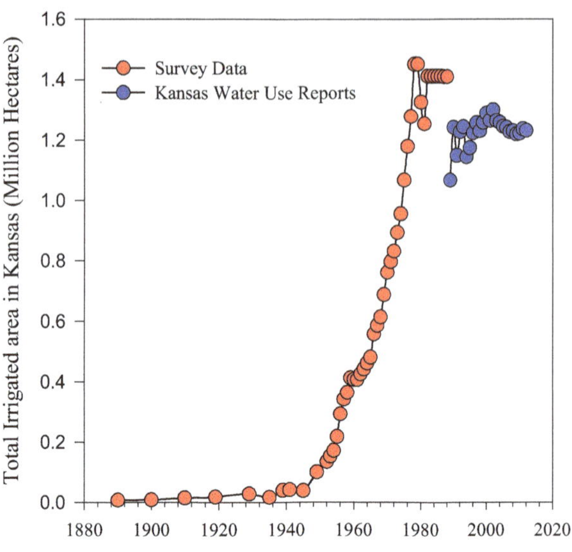

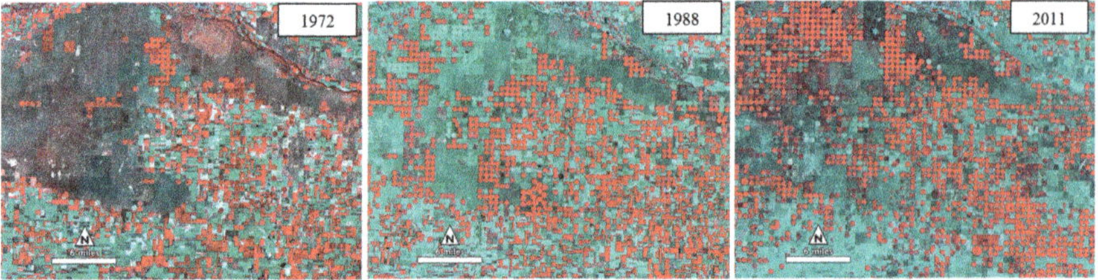

Fig. 3.4 Change in irrigation system type in southwest Kansas as shown by the change in field shapes from square to circular over the 3 years of 1972, 1988 and 2011 (image source: http://www.nasa.gov)

technologies have been credited for improving irrigation water use efficiency due to reductions in amounts of unproductive water lost to evaporation, percolation and runoff. However, improvements in irrigation application efficiency have not resulted in direct water conservation since groundwater level trends have continued to decline. For example Fig. 3.5 shows that average depth to water in southwest Kansas has increased by approximately 18 m. On an individual well basis groundwater level declines will be different due to variability in the aquifers saturated thickness. In order to extend the usable life of the aquifer, the latest focus is on water conservation and one of the key approaches identified by the Kansas Water Plan for conserving Ogallala water is deficit irrigation.

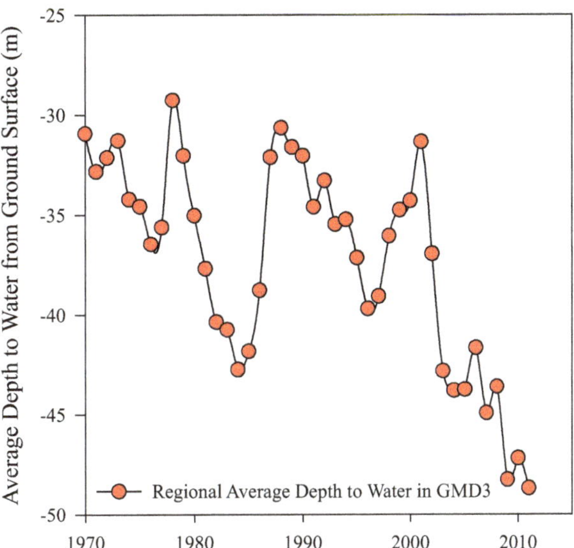

Fig. 3.5 Average depth to water in groundwater management district 3 covering southwest Kansas

3 Deficit Irrigation

Deficit irrigation is defined as the application of water below full crop water evapotranspiration. Under diminished well capacities where full crop water needs cannot be met, deficit irrigation management strategies that maximize economic returns and water productivity and not necessarily yield are attractive to producers. Usually the goal when working with limited water is to capture and store every possible source of water in the production system (Schneekloth et al. 2003). In Kansas, these sources of water include rainfall, snowfall and irrigation. In order to cope with declining groundwater supplies, many producers are combining dryland soil water conservation practices such as reduced tillage and residue management with deficit irrigation management strategies in order to reduce risk, increase crop water productivity, and increase nutrient use efficiency among other benefits. No-till and strip till coupled with residue management have been proven to increase available soil water by reducing soil water loss from tillage operations, reducing soil water evaporation, enhancing infiltration into the soil, increasing snow capture and reducing runoff (Smika and Wicks 1968; Norwood 1992; Nielsen et al. 2002; Klocke et al. 2009).

Other agronomic practices that can be implemented in conjunction with deficit irrigation include rotating high water use crops such as corn with winter crops such as wheat, and canola; timely termination of irrigation towards the end of the season to promote use of previously stored soil water; minimizing water stress during critical growth stages; choosing drought tolerant varieties and implementing irrigation scheduling to optimally apply irrigation water. Choosing the right set of strategies to maximize net returns of irrigated cropping systems is complex and

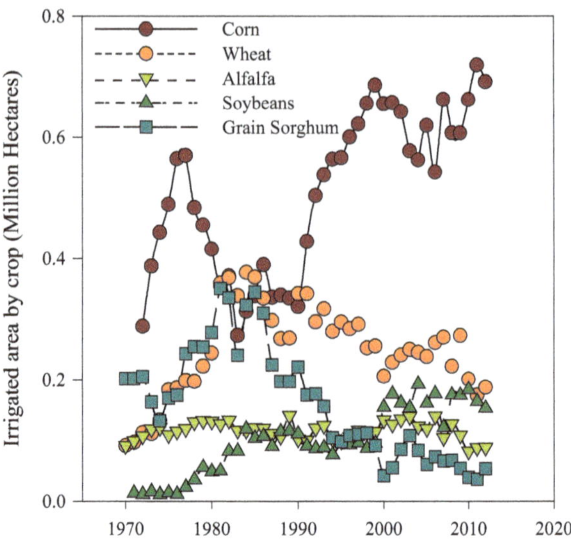

Fig. 3.6 Major irrigated crops in Kansas by area (source of data: Kansas Water Use Reports)

varies from one producer to the other and is mainly influenced by the level of risk the producer can accept under erratic climate conditions of the Central High Plains and changes in market prices. Lamm et al. (2014) provides a review of basic deficit irrigation concepts and when and how they could be implemented for grains, and oil crops. Corn is the dominant irrigated crop in Kansas (Fig. 3.6) and therefore we will focus the rest of the discussion on deficit irrigation management strategies for corn although deficit irrigation could be applied to other row crops.

4 Deficit Irrigation Research for Corn in Kansas

Corn response to limited water has been studied in Kansas for several years. Lamm et al. (2014) used experimental data from 1989 to 2004 near Colby (northwest Kansas) to demonstrate that yield and water productivity (yield per unit of crop water use) for irrigated corn peaked at approximately 80 to 90 % of full irrigation under subsurface drip irrigation (Figs. 3.7 and 3.8). These results provide evidence that opportunities exist to conserve water under moderate deficit irrigation without substantial reductions in yields and profitability. A groundwater conservation effort has been initiated in northwest Kansas called the Sheridan 6 Local Enhancement Management Area (LEMA) where producers have voluntarily agreed to reduce their average water use by 20 % over a five year period. Klocke et al. (2015) also used long term experimental data from Garden City (southwest Kansas) to show that corn response to irrigation varied by year depending on climatic conditions and other abiotic stress factors such as hail damage (Fig. 3.9). For example a 22 % reduction in irrigation water applied in 2011 which was a drought year resulted in

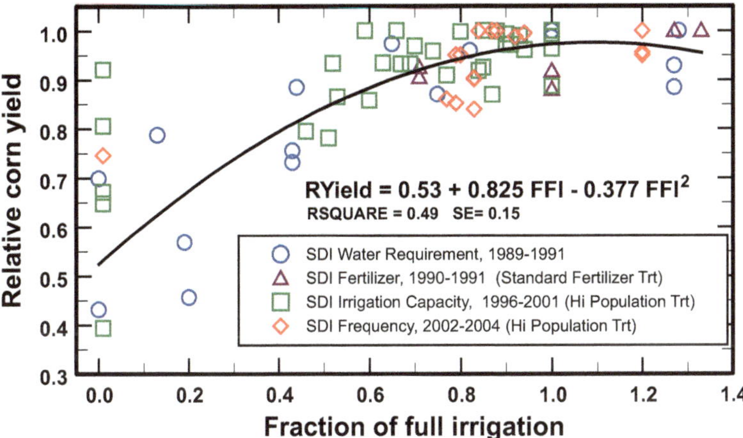

Fig. 3.7 Relative yield of drip irrigated corn in northwest Kansas (Lamm et al. 2014)

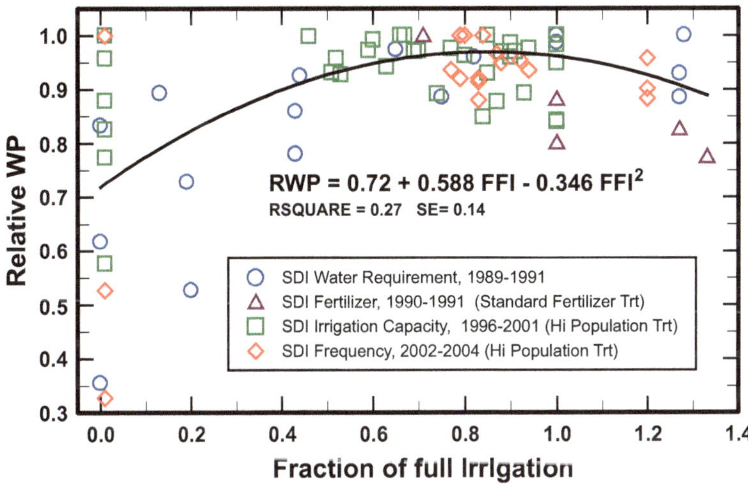

Fig. 3.8 Relative water productivity of subsurface drip irrigated corn in northwest Kansas (Lamm et al. 2014)

only a 16 % reduction in yield, while a 16 % reduction in irrigation in a wet year resulted in less 1 % reduction compared to a fully irrigated treatment (Fig. 3.7).

It has been reported by others (Howell et al. 1989; Lamm et al. 1993; Klocke et al. 2009) that deficit irrigation of corn is complicated to implement since reductions in irrigation water directly result in yield losses given that corn is a high water use response crop. Although this is true, it is worth noting that the deep silt loams in western Kansas have high water holding capacities compared to other arable soils which allow the corn crop to supplement its water needs by extracting soil water from deeper soil layers if there was sufficient soil water at planting.

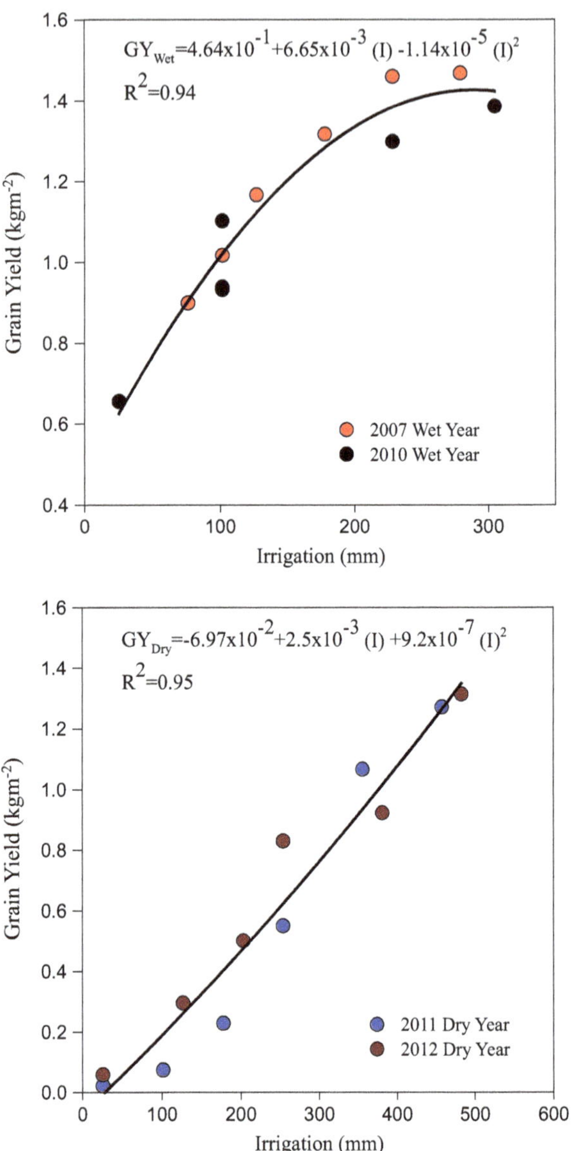

Fig. 3.9 Corn grain yield response to irrigation in wet and dry hail years at Garden City Kansas

Looking to the future, the choice will be whether to concentrate the water by irrigating fewer land area (mainly adopted by risk averse producers who prefer lower but consistent net returns) or spread the water by irrigating large land area (producers who are willing to take more risk in anticipation of high net returns during years with above normal rainfall).

Besides water allocation, other research on deficit irrigated corn in Kansas has explored strategic irrigation water management such as the effect of late spring

Table 3.1 (Adapted from Schlegel et al. 2012): Net returns to land, irrigation equipment, and management from preseason irrigation at three well capacities and seeding rates at Tribune, KS 2006–2009

Well capacity	Preseason irrigation	Seeding rate (seeds ha^{-1})		
		55,575	67,925	80,275
mm day^{-1}		Net returns \$ ha^{-1}		
3	No	571	588	529
	Yes	704	741	734
4	No	716	699	645
	Yes	793	869	882
5	No	1025	1109	1198
	Yes	1030	1131	1215

preseason irrigation on yield and net returns of corn under low well capacities. Schlegel et al. (2012) reported that preseason irrigation was profitable at low well capacities for corn (Table 3.1). These results demonstrate that as well capacities diminish, late spring preseason irrigation used to build up the soil profile could be essential to buffer the crop between in-season irrigation events or rainfall. Other research has explored the effect of soil water monitoring based irrigation scheduling on economic net returns. For example Kisekka et al. (2016) utilized a cropping systems simulation model (DSSAT-CSM) and 64 years of historical weather data for Garden City and reported that deficit irrigated corn scheduled by triggering irrigation at plant available water (PAW) between 40 and 50 % produced the highest long term average economic net returns over a range of corn prices, as shown in Fig. 3.10. These irrigation triggers resulted in median seasonal irrigations of 356–406 mm; assuming an average irrigation season of 90 days for corn in western Kansas that would translate into median capacities of 5 mm/day. The simulations were based on assuming a full 1.2 m soil water profile at planting.

5 Adaptation of Crop Production to Water Limited Environments

In limited water environments, a crop can either experience intermittent or terminal water stress. The latter is common in rainfed systems where the crop could be subjected to continuous stress between emergence and physiological maturity, due to long periods with minimal rainfall. Intermittent water stress on the other hand involves the crop going through cycles of stress and recovery, for example due to rainfall and deficit irrigation. For this discussion, we only attempt to summarize some of the important concepts and practices that can be used to adapt deficit irrigated grain production to limited water environments. Grain yield can be maximized in water limited environments by increasing total crop water use or evapotranspiration (ET) and harvest index (HI) while minimizing soil evaporation

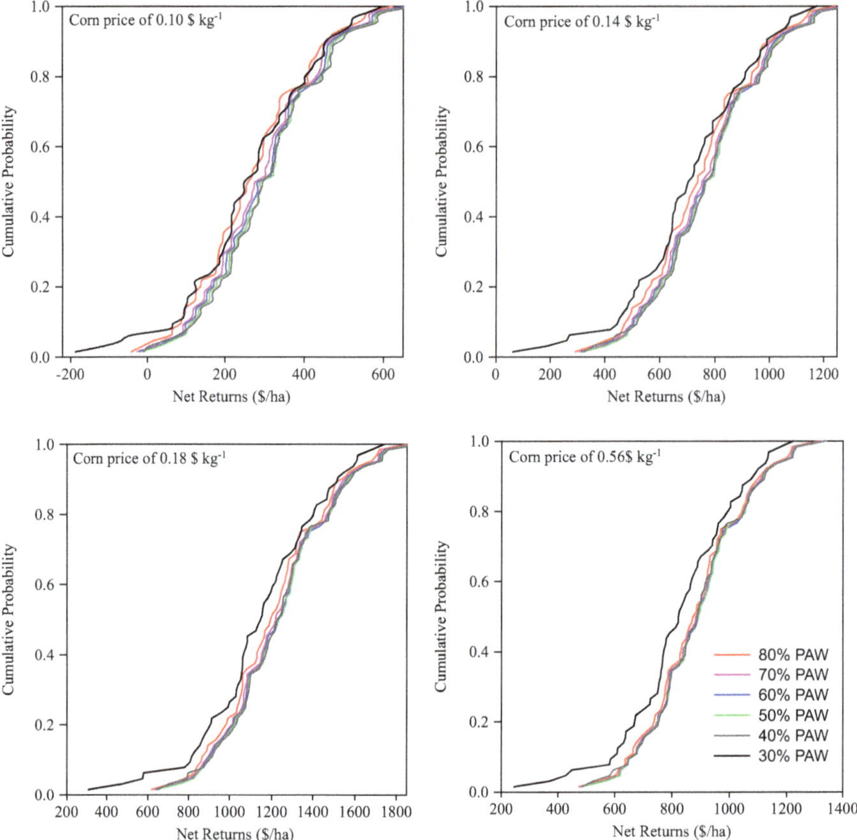

Fig. 3.10 Effect of plant available water irrigation scheduling trigger on corn net returns at Garden City Kansas (Kisekka et al. 2015)

contribution to total crop water use as shown in Eq. (3.1) proposed by Stewart and Peterson (2014):

$$GY = ET * \frac{T}{ET} * \frac{1}{TR} * HI \tag{3.1}$$

where GY is dry grain yield (kg ha^{-1}), ET is total crop water use (soil water evaporation plus transpiration) between seeding and harvest (kg ha^{-2}), T/ET is the fraction of the ET that is transpired by the crop, TR is the transpiration ratio or the kg of water transpired to produce 1 kg of above ground biomass and HI is harvest index (kg dry grain/kg aboveground dry biomass). Each of the components in Eq. (3.1) and how they can be managed to maximize grain yield are discussed in an example below:

In southwest Kansas average grain yield of fully irrigated corn averages 14,000 kg ha^{-1} in wet years and 13,000 kg ha^{-1} in dry years. Reduction in irrigation is usually accompanied by reductions in yield. The reductions in yield vary based on seasonal rainfall, for deficit irrigated corn at 80 % ET level yield reductions could range from 7 % in wet years to over 16 % in drought years (Fig. 3.7). To increase GY in Eq. (3.1), the ET, T/ET and HI terms need to be increased through changes in management or genotypic adaption. TR is usually a more stable variable for a given crop in a given environment.

Sufficient water at planting ensures good seedling establishment and buffers the crop from early drought stress during vegetative growth which could have substantial negative impacts on leaf area index, biomass accumulation and grain number for corn. For example in dryland systems soil water at plant can account for as high as 50 % of seasonal ET in years with low precipitation (Stewart and Peterson 2014). Encouraging deep soil water extraction might also increase total seasonal ET in water limited environments. Selecting hybrids with ability to develop deep rooting systems to enhance soil water extraction in deep soil layers should be preferred. In Kansas irrigation shortly after planting is not a recommended practice since it can result in shallow rooting systems (Lamm et al. 1996). Many producers prefer to have soil water in the top 1.2 m of the soil profile between 0 and 50 % at planting. Steward and Peterson (2014) noted that having sufficient water at planting reduced risk and improved yields and recommended taking soil profile soil water readings at planting. Optimum fertilizer management (matching nitrogen, phosphorus and potassium application rates to available water resources) could also enhance plant growth and increase soil water extraction and enhance maximization of ET. Therefore maximizing ET requires taking a dynamic cropping systems approach to management with recognition of physical and legal constraints on production systems such as diminished well capacity and water right restrictions.

The fraction of ET that is transpired by the crop (T/ET) could be increased by adopting management practices that reduce soil water evaporation (E). Klocke et al. (2009) measured soil water evaporation in corn in southwest Kansas and reported that E was reduced by nearly 50 % compared to bare soil when the soil surface was covered with corn Stover and wheat stubble. Ritchie and Burnett (1971) noted that T/ET is greatly affected by LAI; for dry soil surface T/ET ranged from 0 when LAI is 0, to 0.5 when LAI was 1 up to T/ET of 1 when LAI was 3 or greater. These findings imply that management practices that result in high LAI such as planting high population densities particularly for corn could increase T/ET, and improve grain yields at high to moderate irrigation capacities but for low irrigation capacities a balance has to be achieved between reducing HI and net returns due to increased input costs and the marginal yield increases as demonstrated by Schlegel et al. (2012).

Transpiration ratio (TR) is related to VPD. Tanner and Sinclair (1983) showed that biomass accumulation was directly proportional to transpiration. Sinclair (2009) noted that the rate of transpiration through the stomata was directly influenced by the difference in vapor pressure inside and outside the leaf or vapor pressure deficit (VPD). Stewart and Peterson (2014) noted that VPD was much higher in arid and semi-arid environments compared to humid environments which increases the amount of water required to produce one unit of biomass in these environments. The TR for C4 crops such as corn averages 220 kg H_2O kg above ground biomass^{-1} when growing in an average climate of VPD of 2 kPa (Sinclair and Weiss 2010). This estimate is close to a TR calculated for full and deficit irrigated corn for 2011 growing season at Garden City Kansas ranging from 204 to 243 kg H_2O kg above ground biomass^{-1} respectively. Stewart and Peterson (2014) explained that the transpiration environment of a fully irrigated crop at high population density was different from a similar dryland crop at low population density due to differences in microclimate. Given this background there appears to be an opportunity to control microclimate in deficit irrigated corn by increasing plant population, and making timely deficit irrigation applications which could lower VPD and reduce TR by increasing LAI and radiation use efficiency.

Harvest index is strongly affected by water stress, as the level of stress increases HI decreases. Prihar and Stewart (1990) reported HI for corn under stress free conditions to range from 0.58 to 0.6 representing the genetic HI for corn. HI for corn grown in Garden City in 2011 ranged from 0.45 to 0.37 for the full and deficit irrigated respectively. The slightly lower HI for the fully irrigated corn probably indicates that the crop might have experienced some level of stress along its life cycle since 2011 was a severe drought year. Generally Steiner et al. (1994) showed that HI increased with grain yield therefore any management strategy that maximizes ET would increase HI. Although reducing plant density in water limited environments could have a positive effect on HI, its negative effect on water use efficiency may negate its benefits on HI.

Given that Eq. (3.1) is linear it is evident that the independent variables are intertwined and increasing one would affect other terms as demonstrated in Stewart and Peterson (2014). However, ET is the major driving factor and all management strategies need to focus on maximizing ET. Debaeke and Aboudrare (2004) reviewed strategies that can be used to maximize ET some of which have already be discussed including: (1) increasing soil water at planting, (2) increasing plant soil water extraction, (3) reducing the contribution of soil water evaporation to total crop water use, (4) optimizing seasonal water use patterns e.g., planting short season varieties to match seasonal rains, (5) selection of crops that tolerate stress and (6) irrigating at critical growth stages.

6 Decision Support Tools for Implementing Deficit Irrigation

Other than agronomic and economic considerations, producers have to make daily decisions on how to manage irrigation with limited water. There are producers who prefer conserving water by dividing their production into various crops without substantially affecting their overall net return. This is particularly important and useful in irrigated areas where water supplies are limited, such as some areas in western Kansas. To help producers manage their limited water resources, K-State Research and Extension established the Mobile Irrigation Lab website (www.bae.ksu.edu/mobileirrigationlab), which hosts a suite of web-based and downloadable tools that can help producers make pre-season decisions (Crop Water Allocator), in-season decisions (Crop Yield Predictor), and daily irrigation (KanSched) decisions.

An online tool developed by K-State Research-Extension called Crop Water Allocator (CWA) calculates net economic return for selected combinations of crops in a given land split and water allocated to each crop (Klocke et al. 2006). The decision support tool examines each possible combination of crops selected for every possible combination of water allocation by 10 % increments of the water supply with the intention of helping farmers choose the best combination of crops with the highest net return within the cropping season (Fig. 3.11). The backbone of this tool are the different yield versus ET production functions generated by Kansas Water Budget model and validated using experimental data for corn, sorghum, winter wheat, alfalfa, soybean, sunflower, and fallow. Individual fields or groups of fields can be divided into the following ways: 100%; 50%-50%; 25%-75%; 33%-33%-33%; 25%-25%-%50; 25%-25%-25%-25%. Using current or expected crop and fuel prices, and production costs, several irrigation scenarios can be compared side-by-side with various cropping combinations in an iterative process. Each scenario generates net returns and the program arranges them from the most profitable option to the least (Fig. 3.11). This program benefits not only producers, but also economists, water resource managers, and policy makers in making water conservation decisions.

The Crop Yield Predictor (CYP) was conceptualized as an important decision tool after the crops have been planted. The CYP is an in-season adjustment tool for scheduling irrigation events when water is limited by projecting the yield at the end of the season for a particular irrigation strategy. As another decision support tool, KanSched is an ET-based irrigation scheduling tool that utilizes the daily ET estimates to calculate the water balance in the soil profile.

7 Conclusion

Groundwater levels in the Ogallala aquifer are declining and many well capacities are now unable to meet full crop ET. Deficit irrigation management offers opportunities for maintaining productivity of major irrigated crops such as corn under declining ground water supplies. Research from Kansas indicates moderately reducing irrigation water for corn by 20 % results in yield reductions ranging from 7 to 16 % depending on seasonal rainfall. Given the uncertain future of groundwater these modest yield reductions may be acceptable to some producers. As well capacities continue to decline, other strategies may be to combine proven dryland land soil water conservation techniques (e.g., reduced tillage and residue

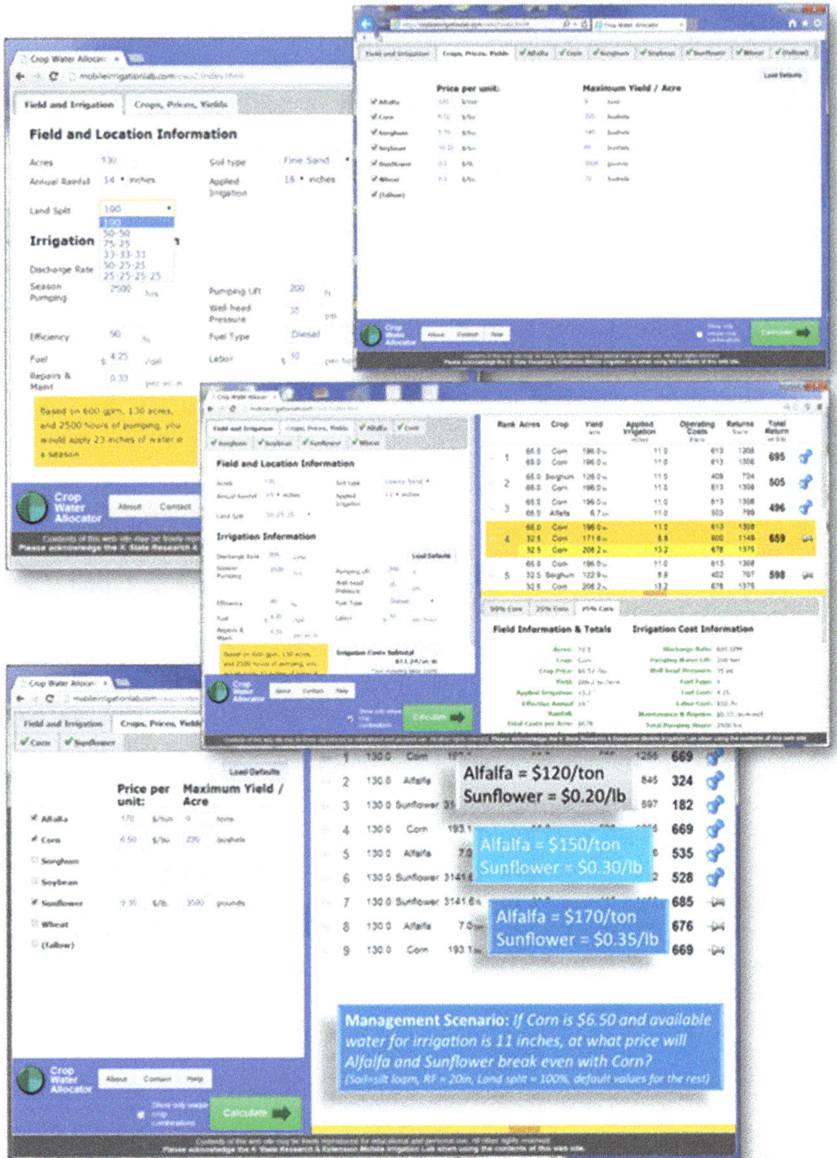

Fig. 3.11 Screenshots of the online version of Crop Water Allocator. Features on this program include customizable input, unique crop combinations button, tabbed format, hidden and resizable panes, and pinning option for comparison of results

management, ensuring sufficient soil water planting, selecting varieties that tolerate drought, managing crop microclimate to maximize ET) with deficit irrigation strategies, thus reducing risk and maintaining grain yields under water limited environments. Decision support tools are available to help producers assess production options with limited water.

References

Buchanan RC, Buddmeier RR, Wilson BB (2009) Public information circular 18: the high plains aquifer. Kansas Geological Survey, Lawrence, KS

Debaeke P, Aboudrare A (2004) Adaptation of crop management to water-limited environments. Eur J Agron 21:433–446

Erhart A (1969) Early Kansas irrigation. One of three article series, "A page out of Irrigation History". Irrigation Age Magazine. Feb 1969. 4pp

Gurdak JJ, Roe CD (2009) Recharge rates and chemistry beneath playas of the High Plains aquifer a literature review and synthesis. U.S. Geological Survey Circular, 2009

Howell TA, Copeland KS, Schneider AD, Dusek DA (1989) Sprinkler irrigation management for corn – Southern Great Plains. Trans ASAE 32(1):147, 154, 160

Kansas Water Office (2015) A long term vision for the future of water supply in Kansas. http://www.kwo.org/50_year_vision/Kansas%20Water%20Vision%20Draft%20II%2011.12.2014Final.pdf. Accessed 20 April 2016

Kisekka I, Aguilar JP, Rogers DH, Holman J, O'Brien DM, Klocke N (2015) Assessing deficit irrigation strategies for corn using simulation. Trans ASABE In Press

Klocke NL, Currie RS, Aiken RM (2009) Soil water evaporation and crop residues. Trans ASABE 52(1):103–110

Klocke NL, Stone LR, Clark GA, Dumler TJ, Briggeman S (2006) Water allocation model for limited irrigation. J App Eng In Agric 22(3):381–389

Lamm FR, Nelson ME, Rogers DH (1993) Resource allocation in corn production with water resource constraints. App Engr Iin Agric 9(4):379–385

Lamm FR, Rogers DH, Clark GA (1996) Irrigation scheduling for corn: macromanagement. In Proceedings of the Evapotranspiration and Irrigation Scheduling Conference, San Antonio, TX, Nov. 3–6 1996. Published by ASAE. p 741–748

Lamm FR, Rogers DH, Aguilar J, Kisekka I (2014) Deficit irrigation of grain and oilseed crops. In Proceedings of the 2014 Irrigation Association Technical Conference, Phoenix, Arizona, 19–20 November, Available from the Irrigation Association, Falls Church, Virginia

Miller JA, Appel CL (1997) Ground Water Atlas of the United States: Kansas, Missouri, and Nebraska. Washington DC: United States Geological Survey Number HA 730-D

Nielsen DC, Vigil MF, Anderson RL, Bowman RA, Benjamin JG, Halvorson AD (2002) Cropping system influence on planting water content and yield of winter wheat. Agron J 94:962–967

Norwood CA (1992) Tillage and cropping system effects on winter wheat and grain sorghum. J Prod Agric 5:120–126

Opie J (2000) Ogallala: water for a dry land, 2nd edn. University of Nebraska Press, Lincoln

Prihar SS, Stewart BA (1990) Using upper-bound slope through origin to estimate genetic harvest index. Agron. J. 82:1160–1165

Ritchie JT, Burnett E (1971) Dryland evaporative flux in a subhumid climate: I. micrometeorological influences. Agron J 63:56–62

Rogers DH, Lamm F (2012) Kansas irrigation trends. Proceedings of the 24th Annual Central Plains Irrigation Conference, Colby, Kansas, February 21–22, 2012 Available from CPIA, 760 N. Thompson, Colby, Kansa

Rogers DH, Alam M, Shaw LK (2008) Kansas irrigation trends. Kansas State University Agricultural Experiment Station and Cooperative Extension Service, MF2849

Schlegel AJ, Stone LR, Dumler TJ, Lamm FR (2012) A return look at dormant season irrigation strategies. In: Proc. 24th annual Central Plains Irrigation Conference Feb. 21–22, 2012, Colby, Kansas. Available from CPIA, 760 N. Thompson, Colby, Kansas. p 51–58

Schneekloth J, Bauder T, Hansen N (2003) Limited irrigation management: principles and practices. 20142003. CSU, Extension Fact Sheet No.4.720

Sinclair TR (2009) Taking measure of biofuel limits. Am Sci 97:400–407. doi:10.1511/2009.80.1

Sinclair TR, Weiss A (2010) Principles of ecology in plant production, 2nd edn. CAB Int, Cambridge, MA

Smika DE, Wicks GA (1968) Soil water storage during fallow in the central Great Plains as influenced by tillage and herbicide treatments. Soil Sci Soc Am Proc 32:591–595

Sophocleous M (2012) Conserving and extending the useful life of the largest aquifer in North America: the future of the High Plains/Ogallala Aquifer. Ground Water 50:831–839. doi:10.1111/j.1745-6584.2012.00965.x

Stewart BA, Peterson GA (2014) Managing green water in dryland agriculture. Agron J 106:1–10. doi:10.2134/agronj14.0038

Steiner JL, Schomberg HH, and Morrison Jr. JE (1994) Measuring surface residue and calculating losses from decomposition and redistribution. In: Crop residue management to reduce erosion and improve soil quality: Southern Great Plains. Conservation Res. Rep. 37. Agricultural Research Service, USDA, Washington, DC, pp 21–29

Tanner CB, Sinclair TR (1983) Efficient water use in crop production: research or re-search. In: Taylor HM, Jordan WR, Sinclair TR (eds) Limitations to efficient water use in crop production. American Society of Agronomy, Madison, pp 1–27

Chapter 4
Hydraulic Fracturing and Its Potential Impact on Shallow Groundwater

Ripendra Awal and Ali Fares

Abstract Unconventional natural gas extraction from impermeable geologic formations is getting momentum in recent years due to advances in horizontal drilling and hydraulic fracturing. By 2040, shale resources are projected to account for 53 % of all natural gas production in the U.S. However, the development of unconventional oil/gas production from hydraulic fracturing has raised serious concerns about its potential impact on the quantity and quality of water resources and the environment due to the large volume of water needed and the use of toxic substances in hydraulic fracturing fluids. This paper gives an overview of the hydraulic fracturing used to extract shale gas, its potential impact on water resources, provides an overview of modeling studies and tools used to assess its potential impacts, and regulation issues related to it. The most significant risks resulting from hydraulic fracturing and shale gas development are (1) the excessive withdrawal of water, (2) gas migration and groundwater contamination due to faulty well construction, blowouts, (3) contamination by wastewater disposal, and (4) accidental leaks and spill of wastewater and chemicals used during drilling and hydraulic fracturing process.

1 Introduction

Unconventional natural gas extraction from impermeable geologic formations has been gaining momentum in recent years due to advances in horizontal drilling and hydraulic fracturing. The top ten countries holding the largest shale gas resources based on an assessment by the U.S. Energy Information Administration (USEIA 2014) in 41 countries are China, Argentina, Algeria, U.S., Canada, Mexico, Australia, South Africa, Russia, and Brazil. Updated estimate of global shale gas resources indicate availability of technically recoverable 55×10^9 m^3 of world shale oil resources and 207×10^{12} m^3 shale gas resources (USEIA 2013). The U.S. has a number of natural gas shale basins located in different states. The

R. Awal (✉) • A. Fares
College of Agriculture and Human Sciences, Prairie View A&M University, Prairie View, TX 77446, USA
e-mail: riawal@PVAMU.EDU

A. Fares (ed.), *Emerging Issues in Groundwater Resources*, Advances in Water Security, DOI 10.1007/978-3-319-32008-3_4

major shale gas basins in the U.S. are Barnett, Haynesville, Eagle Ford, Fayetteville, Woodford, Utica, and Marcellus shale formations. By 2040, shale resources are projected to account for 53 % of all U.S. natural gas production (USEIA 2014).

The development of unconventional oil/gas has raised serious concerns about its potential impact on the quantities and quality of local water resources due to the large water volume withdrawals it uses for hydraulic fracturing and the use of toxic substances in hydraulic fracturing fluids. These toxic substances also cause threats to the environment. This paper explores potential impacts of hydraulic fracturing used for the production of shale gas on groundwater resources. It begins with a brief introduction on the unconventional oil and natural gas source; then it provides an overview of the hydraulic fracturing process. The following section contains an overview of potential impacts of hydraulic fracturing and shale gas development on groundwater resources. A brief review of modeling studies and tools used to assess potential impacts of hydraulic fracturing is then presented. The last section before the summary section of this paper discusses the current regulations related to hydraulic fracturing.

2 Unconventional Oil and Natural Gas Production: Shale, Coalbed Methane, and Tight Fracturing

Unconventional oil and gas resources are trapped in very tight or low permeability rock; thus, the effort required to extract them is greater than for conventional resources. Hydraulic fracturing is often used to stimulate the production of hydrocarbons by creating pathways for the natural gas to flow out of the source rock and into the well from unconventional oil and gas reservoirs to make oil and gas production cost-effective (USEPA 2011a). Hydraulic fracturing is also used to increase the gas flow in wells that are considered conventional reservoirs and make them even more economically viable (Martin and Valkó 2007). The three main categories of unconventional oil and natural gas resources are—shales, coalbeds, and tight sands. Figure 4.1 represents a schematic of geologic nature of the most major sources of natural gas in the United States and illustrates both conventional and unconventional gas resources. Gas-rich shale is the source rock for many natural gas resources. The application of hydraulic fracturing and horizontal drilling technologies have made shale gas an economically viable alternative to conventional gas resources. Conventional gas accumulations occur when gas migrates from gas-rich shale into an overlying sandstone formation, and then becomes trapped by an overlying impermeable formation called the seal (https://www.eia.gov/todayinenergy/detail.cfm?id=110). Associated gas accumulates in conjunction with oil, either dissolved in the oil or as a free gas cap above the oil in the reservoir, while non-associated gas does not accumulate with oil, but it is found in reservoirs containing no oil (dry wells). Tight sand gas accumulations occur in a variety of geologic settings where gas migrates from a source rock into a

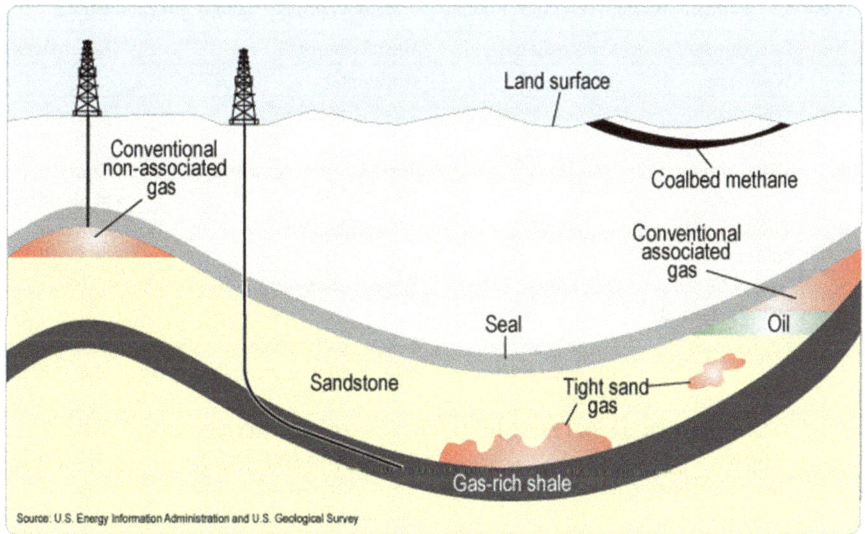

Fig. 4.1 Schematic geology of natural gas resources (http://www.eia.gov/todayinenergy/detail.cfm?id=110)

sandstone formation, but is limited in its ability to migrate upward due to reduced permeability in the sandstone. Coalbed methane does not migrate from shale, but is generated during the transformation of organic material to coal.

The conventional natural gas production differs from unconventional shale gas in several ways (Table 4.1).

The three types of unconventional gas reservoirs are discussed here.

Shale gas: Shale rock formations are present in many locations across the U.S.; they have become an important source of natural gas in the U.S. (USEPA 2011a). The distribution of shale plays in the contiguous U.S. is illustrated in Fig. 4.2. Depths for shale gas formations can range from 150 to 4110 m below the earth's surface (GWPC and ALL Consulting 2009). At the end of 2014, the six most productive shale gas fields in the country—the Marcellus, Eagle Ford, Barnett, Haynesville, Fayetteville and Woodford Shales—were producing 896×10^6 m^3 of natural gas per day according to EIA's official shale gas estimates in November, 2015 (Fig. 4.3). According to recent report by EIA, shale gas constituted 40 % of the total U.S. natural gas supply in 2012, and will make up 53 % of the U.S. gas supply in 2040 if current trends and policies persist (USEIA 2014).

Use of hydraulic fracturing and directional drilling technologies has also contributed to the increase in oil production in the U.S. over the last few years. Although oil production from shales is increasing in the Eagle Ford Shale in Texas, the Niobrara Shale in Colorado, Nebraska, and Wyoming, and the Utica Shale in Ohio, it has been concentrated primarily in the Williston Basin in North Dakota (USEIA 2010; USEPA 2011a). Between January 2008 and May 2014, U.S. monthly crude oil production rose by 3.2 million barrels per day, with about

Table 4.1 Comparison between conventional and unconventional natural gas productions

No.	Description	Conventional	Unconventional	References
1.	Well type	Vertical	Horizontal	
2.	Well pad footprint	>0.4–1.2 × 10^4 m^2	1.2–2.4 × 10^4 m^2	Smrecak et al. (2012)
3.	Water required	70–300 m^3	7.5–34 × 10^3 m^3; average 15 × 10^3 m^3	Smrecak et al. (2012)
4.	Time to drill well	~1 month	~3 months	Smrecak et al. (2012)
5.	Hydraulic fracturing required	Sometimes	Almost always	
6.	Source rock	Large pocket of resource; easy to extract	Resource scattered throughout rock, hard to extract, expensive	Smrecak et al. (2012)
7.	Total production volumes	850–1130 × 10^6 m^3	Fraction of amount produced by conventional (see Sect. 3.4)	Lake et al. (2013)
8.	Rate of decline in production volumes	Low rate of decline	Very steep rate of decline in the initial production period (see Sect. 3.4)	Lake et al. (2013)

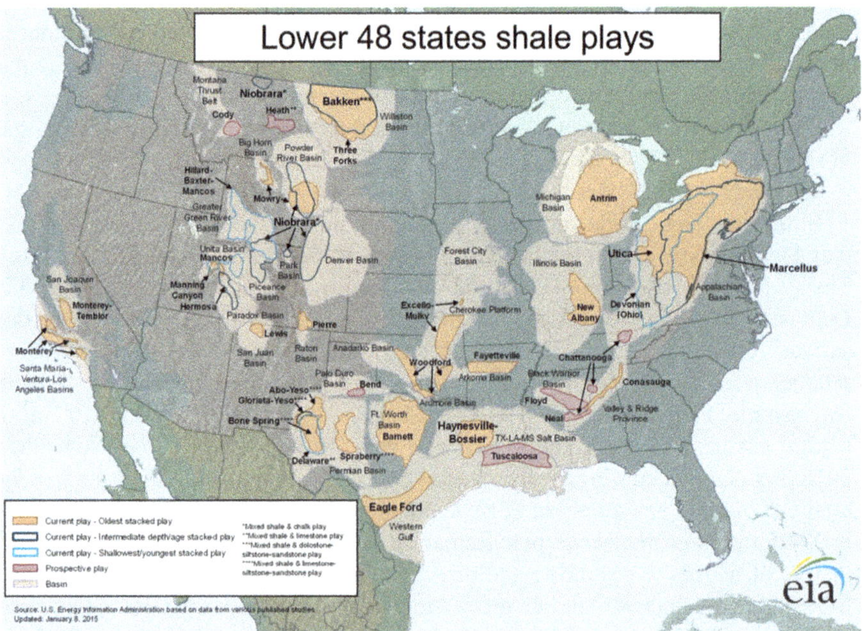

Fig. 4.2 Shale gas plays in the contiguous U.S. (http://www.eia.gov/oil_gas/rpd/shale_gas.jpg)

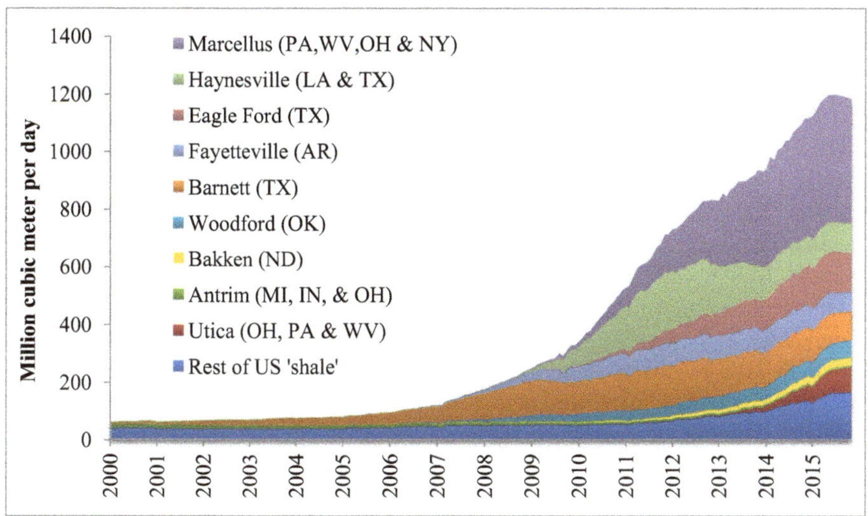

Fig. 4.3 Monthly dry shale gas production in the U.S. (Data source: U.S. EIA, http://www.eia.gov/naturalgas/weekly/)

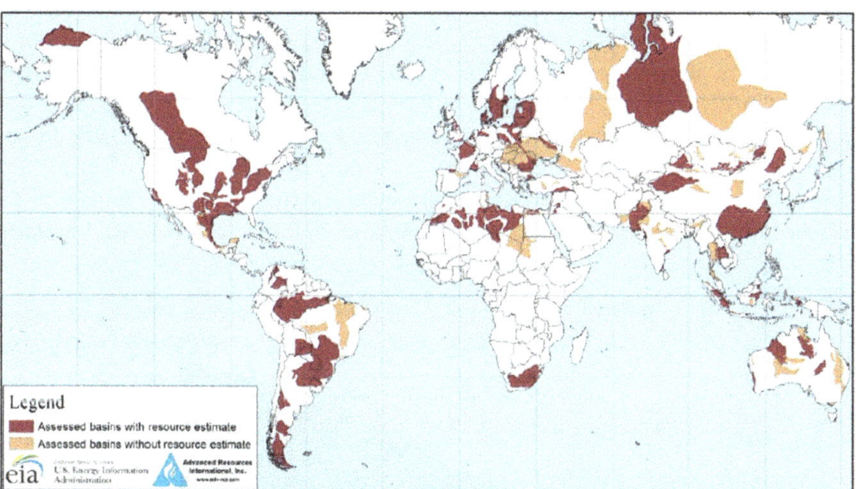

Source: United States basins from U.S. Energy Information Administration and United States Geological Survey; other basins from ARI based on data from various published studies.

Fig. 4.4 Map of basins with assessed shale oil and shale gas formations, as of May 2013

85 % of the increase coming from shale and related tight oil formations in Texas and North Dakota (Ratner and Tiemann 2014).

The new global shale gas resource estimate based on 137 shale formations in 41 countries outside the U.S. and EIA's own assessment of resources within the U.S. indicate technically recoverable resources of 55×10^9 m^3 of world shale oil resources and 207×10^{12} m^3 of world shale gas resources (USEIA 2013). The basins with assessed shale oil and shale gas formations are illustrated in Fig. 4.4.

Coalbed methane: Coalbed methane is formed as part of the geological process of coal generation and is contained in varying quantities within all coal (USEPA 2011a). Depths of coalbed methane formations range from 140 m to greater than 3050 m (Rogers et al. 2007; National Research Council 2010). The permeability decreases at greater depths lowering production; thus, efficient production of coalbed methane can be challenging from a cost-effectiveness perspective at depths greater than 2130 m (Rogers et al. 2007). In 1984, there were very few coalbed methane wells in the U.S.; by 1990, there were almost 8000, and in 2000, there were almost 14,000 (USEPA 2004). In 2012, natural gas production from coalbed methane reservoirs made up 5 % of the total U.S. natural gas production; this percentage is expected to 7 % in 2040 (USEIA 2014). Production of gas from coalbeds almost always requires hydraulic fracturing (USEPA 2004), and many existing coalbed methane wells that have not been fractured are now being considered for hydraulic fracturing (USEPA 2011a).

Tight sands: Tight sands (gas-bearing, fine-grained sandstones or carbonates with a low permeability) accounted for 20 % of total gas production in the U.S. in 2012 (USEIA 2014), but may account for as much as 35 % of the nation's recoverable gas reserves (Oil and Gas Investor 2005). Depths of tight sand formations range from 370 to 6100 m across the U.S. (Prouty 2001). Almost all tight sand reservoirs require hydraulic fracturing to release gas unless natural fractures are present (USEPA 2011a).

There is a 56 % projected increase in total natural gas production by 2040 from its level in 2012; this will result from increased development of shale gas, tight gas, and offshore natural gas resources (Fig. 4.5). Shale gas production is the largest contributor, growing by more than 280×10^9 m^3, from 270×10^9 m^3 in 2012 to 560×10^9 m^3 in 2040 (USEIA 2014). The shale gas share of total U.S. natural gas production is projected to increase from 40 % in 2012 to 53 % in 2040. Similarly,

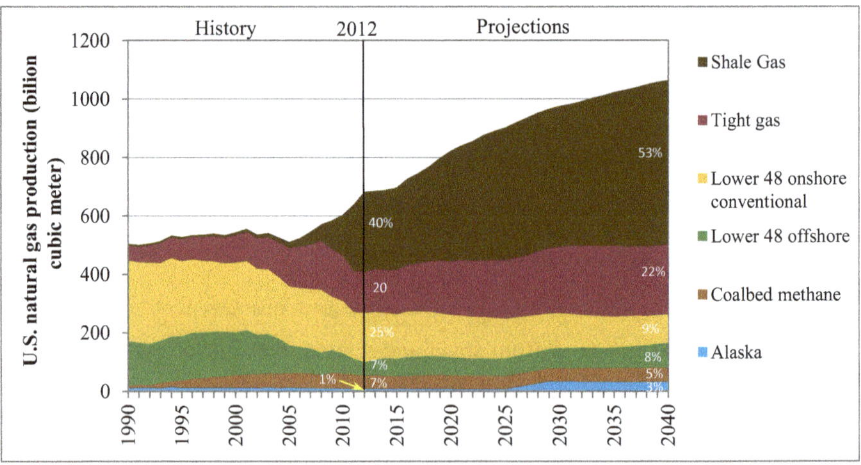

Fig. 4.5 U.S. natural gas production by source in the Reference case, 1990–2040 (Data source: USEIA 2014)

tight gas and offshore gas production shares of total U.S. natural gas production will increase by 2 % and 1 %, respectively, from 2012 to 2040. Their shares of total production remain relatively constant.

3 Overview of Hydraulic Fracturing Process

The following sections provide an overview of unconventional natural gas production based on the USEPA (2011a), which includes site selection and preparation, well construction and development, hydraulic fracturing, and natural gas production and closure.

3.1 Site Selection and Preparation

Site selection is the first process in any hydraulic fracturing project; it starts with exploring possible well sites then one of these sites will be selected for production; soon after that site preparation will follow. Selecting an appropriate site is very important for production of substantial quantities of natural gas at a minimum cost. Several factors which may be considered in the selection process include the site location with respect to adjacent buildings and other infrastructure, geologic formations, distance to natural gas pipelines or feasibility of installing new pipelines (Chesapeake Energy 2009), local ecology, water availability and disposal options, seasonal restrictions related to climate or wildlife concerns (IEA 2012) and laws and regulations (USEPA 2011a).

A well pad of sufficient area to accommodate the needed equipment during the drilling and production phases is first constructed prior to the start of any drilling work. The selected area should be large enough to host multiple wellheads; water storage tanks and/or pits. The well pad should also accommodate additional above ground storage capabilities needed to store drilling fluids and waste. The well pad should have enough space for trucks and other equipment. On average, a typical shale gas production site occupies between 1.2 and 2.4×10^4 m^2 in addition to access roads for transporting materials to and from the well site (USEPA 2011a).

3.2 Well Construction and Development

3.2.1 Types of Wells

Vertical, horizontal and directional wells with at least one section which has a curved axis (J-shaped and S-shaped) represent the current practices in drilling for natural gas. In the U.S., 63 % of wells drilled were horizontal, 11 % were

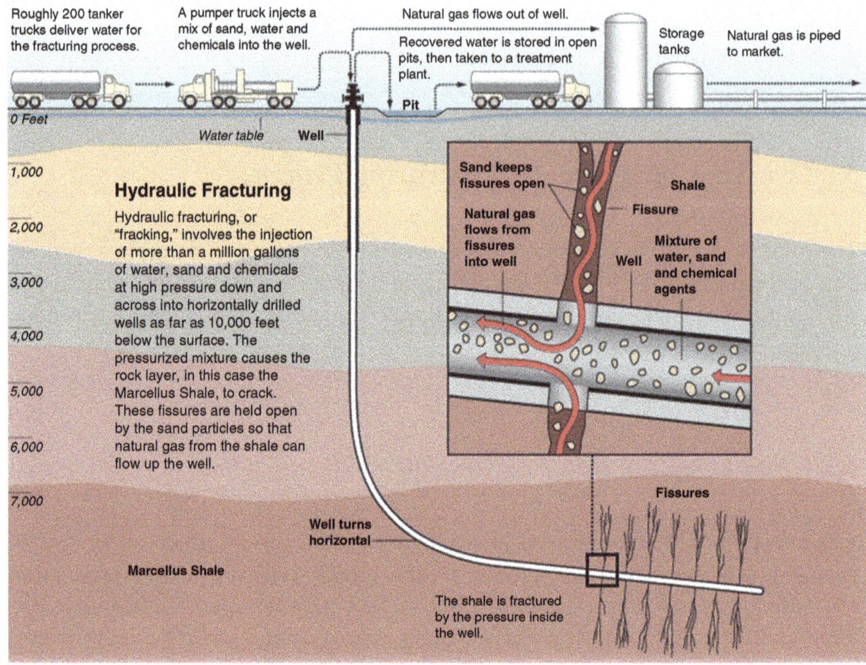

Fig. 4.6 Unconventional gas production method (Graphic by Greenberg/ProPublica) (http://www.propublica.org/special/hydraulic-fracturing-national)

directional, and 26 % were vertical by the end of 2012 (Amer et al. 2013). A schematic representation of a typical well completion at deep shale gas-bearing formation is shown in Fig. 4.6; such formations contain rock formations of several hundred of meters separating shallow groundwater aquifers. The well shown in Fig. 4.6 is a horizontal well composed of vertical and horizontal legs; its dimensions (length and depth) are location specific. They are a function of the location and the properties of the exploited geological formations. A production well may have multiple layers of cemented casing (i.e., conductor, surface, intermediate, and production) isolating it from its adjacent geological formations. The surface casing extends below the base of the shallow groundwater aquifers. It isolates and protects the shallow groundwater aquifers from potential contamination from the production well. The production casing, which is the innermost casing, usually runs all the way to the reservoir. A typical well can extend more than a kilometer below the ground surface, while the "toe" of its horizontal leg can extend more than 3 km from the vertical leg (Zoback et al. 2010). The exposure to gas-containing formation in horizontal drilling is comparatively higher than in a vertical well. Thus, production is more economical in horizontal drilling (USEPA 2011a). It may also have the advantage of limiting the amount of disruption and thereby the overall environmental impact of well construction on the surface because fewer wells are needed to access natural gas resources in a given area (GWPC and ALL Consulting 2009).

Multilateral wells are becoming more popular in shale gas production in the Marcellus Shale region, eastern North America (Kargbo et al. 2010) and other regions. For a multilateral well, at least two horizontal production holes are drilled from a single surface location (Ruszka 2007) to create an arrangement similar to an upside-down tree, with the vertical portion of the well as the "trunk" and multiple "branches" extending out from it in different directions and at different depths (USEPA 2011a).

3.2.2 Design and Construction of Production Wells

An appropriate well design and construction should ensure no negative impact on the environment; it also must assure a safe production of hydrocarbons by avoiding leaks outside the well, protecting adjacent groundwater aquifers, separating the production formations; and by proper implementation of hydraulic fractures and other stimulation operations (API 2009a). Appropriate well construction is vital for protecting shallow groundwater resources from any leak originating from the production zone and comprises; (1) drilling a hole, (2) casing it (installing a steel pipe), and (3) cementing the pipe in place.

Drilling of a Production Well: The drilling string used to drill a well consists of a drill bit, drill collars (heavy weight pipe to put weight on the bit), and a drill pipe. A drilling fluid, e.g., oil-based liquid, compressed air, or water, is injected down the drilling string. Water-based liquids are a mixture of water, barite, clay, and chemical additives (http://www.oilgasglossary.com/drilling_fluid.html). Drilling fluids are essential to cool the drill bit, lubricate the drilling assembly, remove the formation cuttings drilled, maintain the pressure control of the well, and protect the integrity of the well being drilled (API 2009b). Drill cuttings and drilling liquids are treated, recycled or disposed of after their removal from the wellbore.

Casing: The borehole is lined with casings, made of steel pipes, whose purpose is to isolate the geologic formation from the materials and equipment in the well. This preserves the borehole structure, limits the injected/produced fluid to the wellbore and the production zone, and offers a method of pressure control. The casing is subject to tremendous pressure within and from outside the borehole during installation, well cementing, fracturing of production formations, and well operation; consequently, it must be very strong to endure all stresses related to these activities. Containing the fluid within the casing substantially reduces the chances of contaminating the zones adjacent to the well. Some operators may decide not to case wells dug in strong geological formations which are relatively immune from collapsing on themselves. There are several types of casing used in well construction including conductor, surface, intermediate, and production which serve different purposes. The surface casing is intended to protect the shallow groundwater aquifers from contamination during the production, drilling, and completion of the well. This casing is cemented into place; it starts from the surface all the way to below the deepest surface groundwater aquifer. A production well may have multiple layers of cemented casing isolating it from its adjacent geological formations. Additional and more detailed information about this topic can be found in the literature, e.g., USEPA (2011a), USEPA (2015c), and API (2009b).

Cementing: Cement plays a vital role in the proper functioning of production wells; mechanically, it supports the casing. It also prevents the spread of the fluids outside the wellbore. The type of cement and the way it is placed in the well are crucial in the success of this phase (USEPA 2011a). Having adequate layer and uniform layer of cement around each casing combined with ability of the cement to confine fluid movement outside the well are crucial in establishing and maintaining the mechanical integrity of the well, while an adequately constructed production well can fail as a result of corrosion and the downhole stresses (Bellabarba et al. 2008). Most wells have cement behind the surface casing, however, the length and location of cement behind intermediate and production casings can vary based on the presence and locations of over-pressured formations, formations containing fluids, or geologically weak formations. Additional details on this topic can be found in literature, e.g., USEPA (2011a), USEPA (2015c), and API (2009b).

3.3 Hydraulic Fracturing of Vertical, Horizontal and Multilateral Wells

Once a well is constructed, the production geological formation, i.e., tight sands, shale or coalbed, is hydraulically fractured to enhance natural gas production. The hydraulic fracturing process needs large volumes of readily available water at the well site (Fig. 4.6). The water is mixed with different chemicals including propping agents (called proppants which are solid materials) that prevent the fractures from closing after pressure is reduced in the well (Palisch et al. 2008). Sand is the most common proppant (Carter et al. 1996); however, bauxite, ceramics, and resin-coated sand are also commonly used (Arthur et al. 2008; Palisch et al. 2008). Almost all water-based fracturing techniques use proppants; however, it's worth noticing that some fracturing techniques do not use any proppants, e.g., nitrogen gas is commonly used to fracture coalbeds and does not require the use of proppants (Rowan 2009). The production casing is perforated by explosive charges introduced into the well; then, the geological formation is fractured when hydraulic fracturing fluid is injected down the well under high pressure (USEPA 2011a). The fractures of production formations are enhanced with proppant carried by the hydraulic fracturing fluid (Fig. 4.6).

The fluid that returns to the surface can be referred to as either flowback or produced water, and may contain both hydraulic fracturing fluid and natural water. During the flowback period, which usually lasts up to 2 weeks, approximately 10–40 % of the fracturing fluid returns to the surface (Gregory et al. 2011). After the well is placed into production, aqueous and nonaqueous liquid continues to be produced at the surface in much smaller volumes (2–8 $m^3 d^{-1}$) over the lifetime of the well (Gregory et al. 2011). Flowback and produced water are collectively referred to as hydraulic fracturing wastewaters (USEPA 2011a). These wastewaters are typically stored on-site in tanks or pits before being transported for treatment, disposal, land application, and/or discharge.

3.4 *Well Production and Closure*

The production rates of natural gas can significantly vary between basins and within regions of each basin, depending on a number of factors mainly geologic factors and completion techniques. The mean well production rates for coalbed methane formations range from 1420 to 14,160 $m^3 d^{-1}$ across the U.S., with maximum production rates reaching 566,340 $m^3 d^{-1}$ in the San Juan Basin and 28,320 $m^3 d^{-1}$ in the Raton Basin (Rogers et al. 2007). The industry estimates that a typical well in the Marcellus Shale initially produces 79,290 $m^3 d^{-1}$; the production rate decreases to 15,570 $m^3 d^{-1}$ after 5 years and 6370 $m^3 d^{-1}$ after 10 years, after which it drops approximately 3 % a year (NYSDEC 2011). The average actual production rate of a well in the Barnett Shale is about $22.7 \times 10^6 m^3$ during its lifetime, which averages about 7.5 years (Berman 2009). It is not uncommon to refracture an oil or gas well when it is no longer cost-effectively producing hydrocarbons from it (USEPA 2011a). Wells may be refractured after roughly 5 years of service (NYSDEC 2011).

Wells that are no longer producing gas economically must be properly plugged to prevent possible fluid migration that could contaminate soils or waters; the primary environmental concerns are protecting freshwater aquifers and shallow groundwater resources, and isolating downhole formations that contain hydrocarbons (API 2009a).

Fluid may flow down the wellbore or up the well toward ground or surface waters causing serious contamination of water resources and the environment (API 2009a). A secure surface plug is essential to prevent surface water from seeping into the wellbore and migrating into groundwater resources (API 2009a); setting cement plugs is highly essential to isolate hydrocarbon and injection/disposal intervals, in addition to setting a plug at the base of the lowermost shallow groundwater resources present in the formation (API 2009a).

3.5 *Cost of Production*

The major components of cost in the drilling and completion of a shale gas well are the rig, associated drilling services, and the hydraulic fracturing stage of well completion. Well construction costs are primarily influenced by a number of factors, mainly the geographical location, the well depth, to some extent reservoir pressure, and by the market and infrastructure conditions in the country or region under consideration (IEA 2012). Operational costs vary with local conditions: for example, just as for drilling, operating costs in Europe are expected to be 30–50 % higher than those in the United States for a similar shale gas operation (IEA 2012).

The International Energy Agency's analysis of a typical shale gas well that was regularly drilled in 2011 into deep shale reservoirs (such as the Haynesville and Eagle Ford shale plays) with a depth of the order of 3000 m, having a horizontal section of around 12,000 m and completed with 20 facture stages using a total of 2000 tons of proppant and 15,000 m^3 of water is about $8 million (IEA 2012). Drilling costs change from month and also vary by location. The cost of drilling and completing a well in the Woodford shale (Oklahoma), where the shale is 1800–2400 m deep is estimated to be $6.7 million, whereas the cost of drilling and completing a new well in the Haynesville shale (TX), where the shale is 3000–4200 m deep is estimated to be $9.5 million (Lake et al. 2013). The hydraulic fracturing process accounts for around 40 % of the total well cost—around twice as much as the second most expensive item, the rig itself (IEA 2012). The cost of drilling the well is approximately 60 % of the total and completion costs make up the remaining 40 % (Lake et al. 2013).

The International Energy Agency developed a set of "Golden Rules", suggesting principles that can allow policymakers, regulators, operators and others to address environmental and social impacts (IEA 2012). According to IEA's estimate, applying the Golden Rules could increase the overall financial cost of development of a typical shale-gas well by an estimated 7 %. However, for a larger development project with multiple wells, additional investment in measures to mitigate environmental impacts may be offset by lower operating costs (IEA 2012).

4 Potential Impact of Hydraulic Fracturing and Shale Gas Development on Shallow Groundwater

Groundwater is an important resource that provides drinking water for nearly half the people in the U.S. (Nolan and Hitt 2006). Unfortunately, the groundwater resource is susceptible to contamination due to multiple activities associated with shale gas development. These impacts are caused by activities and products related to hydraulic fracturing processes (e.g. fracturing chemicals, and waste generated by hydraulic fracturing operations). The major issues that are related to hydraulic fracturing and shale gas development include groundwater (aquifer) contamination by hydraulic fracturing chemicals, accidental chemical spills, waste disposal, air quality, the land footprint of drilling activities, pipeline placement and safety, and the amount of water used (Rahm 2011). The most significant environmental risks associated with the development of shale gas are similar to those associated with conventional onshore gas, including gas migration and groundwater contamination due to faulty well construction, blowouts, and above-ground leaks and spill of wastewater and chemicals used during drilling and hydraulic fracturing (Zoback et al. 2010). A lack of sound scientific hydrogeological field observations and a scarcity of published peer-reviewed articles on the effects of both conventional and unconventional oil and gas activities on shallow groundwater make it difficult to address these issues (Jackson et al. 2013). This section will discuss potential impact of shale gas development and hydraulic fracturing on shallow groundwater.

4.1 Water Use in Hydraulic Fracturing

Large volumes of fresh water from surface or subsurface sources are required for fracturing operations. Most sites in the U.S. need between 10 and 15×10^3 m^3 of water per well (Table 4.2). Plays in more humid regions generally use surface water (e.g. Marcellus Shale play), whereas limited surface water availability in more semiarid regions may result in more groundwater (e.g. Eagle Ford play) use (Nicot et al. 2014). It was estimated that about 35,000 wells were hydraulically fractured in 2006 in the U.S. (White Paper 2008). The annual national water requirement may range from 265 to 530×10^6 m^3. This amount of water is equivalent to annual water consumption of five million people (USEPA 2011a). Nearly half of the wells hydraulically fractured since 2011 were in regions with high or extremely high water stress and over 55 % were in areas experiencing drought (Freyman 2014). Total water consumption for shale gas development will grow with the increase in number of wells and shale gas production. Almost all of water injected for hydraulic fracturing will be non-usable permanently unless the flowback is treated and re-introduced back into the hydrologic cycle (Donaldson et al. 2013). Water use estimates for hydraulic fracturing represent a small fraction of water used in each state (~0.1 % in Colorado, ~0.5 % in Texas, and <0.5 % in Oklahoma), the volume may be significant locally, depending on competition with other sectors, especially in areas where water is scarce (Nicot et al. 2014). The large amount of water required for hydraulic fracturing in a very short time can have a significant adverse impact on the local water system. However, so far, it is not clear how the large volume water withdrawals impact the local water system (Kellman and Schneider

Table 4.2 Water needs per-well from shale gas plays in the United States (Sakmar 2011)

Shale gas play	Volume of drilling water per well (m^3)	Volume of fracturing water per well (m^3)	Total volume of water (m^3)
Barnett shale	1510	8710	10,220
Fayetteville shale	230	10,980	11,580
Haynesville shale	3790	10,220	14,010
Marcellus shale	300	14,380	14,690

2010; Litovitz et al. 2013). For aquifers with low permeability, the radial distance from the wellbore that is affected by the water withdrawal can be small but the ecological impact on such small distances can be severe (Donaldson et al. 2013). This can also lead to lowering of the water table or dewatering of drinking water aquifers, decreased groundwater contribution to stream flows, and reduced volumes of water in surface water reservoirs. Lowering the water table in coastal regions can cause salt water intrusions into the groundwater aquifer. Impacts that can produce long-term alterations of the aquifer can include the introduction of oxygen in water and recharging the subsurface aquifer, which in turn can generate geochemical reactions (Donaldson et al. 2013). Recharged water trickling into the subsurface aquifer can introduce contaminants and excessive withdrawal of water can cause compaction of the aquifer, resulting in subsidence and permanent loss of water-bearing capacity (Donaldson et al. 2013). Since surface and subsurface water sources are hydraulically connected, changes of the quality and quantity of the surface water can produce adverse effects on shallow subsurface aquifers and vice versa (Winter et al. 1998).

One of the alternatives to offset the large water requirements for hydraulic fracturing is to treat and recycle the flowback and produced water in the hydraulic fracturing process (Pickett 2009; Veil 2010). The produced water may be treated and reused to compose a new fracturing solution by adding additional chemicals as well as freshwater. There are, however, challenges associated with reusing flowback due to the high concentrations of total dissolved solids (TDS) and other dissolved constituents found in flowback (Bryant et al. 2010; Ferrer and Thurman 2015). Constituents such as specific cations (e.g., calcium, magnesium, iron, barium, and strontium) and anions (e.g., chloride, bicarbonate, phosphate, and sulfate) can interfere with hydraulic fracturing fluid performance by producing scale or by interfering with chemical additives in the fluids (Godsey 2011). There are also some challenges in constructing a treatment system on-site to quickly treat a large volume of water (20–75 % of the injected hydraulic fracturing fluid) on a temporary basis and then moving the equipment and setting it up at a new site in few weeks. Another alternative of reducing the demand for fresh water on local resources is to use non-potable groundwater, but non-potable water may require treatment for removal of salts that could cause precipitation when chemicals necessary for the hydraulic fracturing process are added (Donaldson et al. 2013). Collection and store of surface water in impoundments during wet seasons to provide water for the hydraulic fracturing process during dry seasons (Donaldson et al. 2013) is an another alternative in reducing water demand on groundwater or surface water.

4.2 *Well Drilling and Production*

Groundwater contamination from shale gas operations can occur as a result of a several malfunctioning and or accidents e.g., leakage from a poorly cemented well annulus; leakage from the faulty well casings (e.g., poorly joined or corroded casings); migration of hydrocarbon gas along natural deformation features (e.g., faults, joints, or fractures) or those initiated by drilling (e.g., faults or fractures created, reopened, or intersected by drilling or hydraulic fracturing activities); migration of gases through abandoned or legacy wells (Darrah et al. 2014); surface spills and leaks from storage and production facilities etc. Failure of the cement or casing surrounding the wellbore poses a greater risk to shallow groundwater resources. If the annulus is improperly sealed, natural gas, hydraulic fracturing fluids, and formation water containing high concentrations of dissolved solids may be communicated directly along the outside of the wellbore among the target formation, shallow groundwater aquifers, and layers of rock in between. Although there are state requirements for well casing and integrity, accidents and failures still occur, as was demonstrated by an explosion in Dimock Township, Pennsylvania (Cooley and Donnelly 2014). In 2009, a residential water well in Dimock exploded as a result of methane buildup in the well. Further investigation found methane gas in other drinking water wells, which were located near drilling wells with improper or insufficient casings.

A study in the states of New York and Pennsylvania found that methane levels in shallow wells in active gas production areas were 17 times higher than in those outside of active gas production areas. Isotopic compositions of methane and higher molecular weight hydrocarbons such as ethane and propane and methane-to-ethane ratios can help determine whether the gas is thermogenic or biogenic in origin and whether it is derived from shale or other formations (Gorody 2012; Barker and Fritz 1981). However, determining the source of methane does not necessarily establish the migration pathway. An isotopic analysis of the methane suggests that the methane in the active gas production areas originated from deep underground and is consistent with deeper thermogenic methane sources and matched gas geochemistry from gas wells nearby (Osborn et al. 2011). Jackson et al. (2013) analyzed 141 drinking water wells in the Marcellus Shale region of northeastern Pennsylvania, examining natural gas concentrations and isotopic signatures with proximity to shale gas wells. Methane was detected in 82 % of drinking water samples, with average concentrations six times higher for homes <1 km from natural gas wells. Ethane was 23 times higher in homes <1 km from gas wells; propane was detected in ten water wells, all within approximately 1 km distance. They also found that the concentration of dissolved methane in drinking water well samples correlated well with the distance from hydraulic fracturing wells. Overall, their data suggest that some homeowners living within 1 km from gas wells have drinking water contaminated with stray gases. Molofsky et al. (2013) evaluated data from 1701 water wells in Suquehanna County, northern Pennsylvania. Based on statistical analysis they determined that there is not a significantly significant difference between the methane concentrations in production and nonproduction areas. They found that higher methane concentrations in groundwater more closely correlated with topography and elevation. Their findings also show that isotopic signatures are not indicative of thermogenic methane resulting from stray gas related to Marcellus Shale production. However, this study indicates that the interpretation of isotopic data is complex because multiple thermogenic gas sources may exhibit subtle variations that are difficult to detect and evaluate. These conclusions are not consistent with those of Jackson et al. (2013).

Darrah et al. (2014) analyzed 113 and 20 samples from drinking-water wells overlying the Marcellus and Barnett Shales, respectively; they examined hydrocarbon abundance and isotopic compositions and provided the first comprehensive analyses of noble gases and their isotopes in groundwater near shale-gas wells. Using noble gas and hydrocarbon tracers, they distinguished natural sources of methane from anthropogenic contamination and evaluate the mechanisms that cause elevated hydrocarbon concentrations in drinking water near natural-gas wells. They documented fugitive gases in eight clusters of domestic water wells overlying the Marcellus and Barnett Shales, including declining water quality through time over the Barnett. Gas geochemistry data implicate leaks through annulus cement (four cases), production casings (three cases), and underground well failure (one case) rather than gas migration induced by hydraulic fracturing deep underground. Optimizing well integrity is a critical, feasible, and cost-effective way to reduce problems with shallow groundwater contamination (Darrah et al. 2014).

Old, abandoned and improperly plugged wells can also potentially serve as migration pathways (USEPA 2011a) for contaminants to enter groundwater systems. Natural underground fractures, as well as those potentially created during the fracturing process, could also serve as conduits for groundwater contamination (Myers 2012). Based on modeling studies, Myers (2012) suggested the possibility of significant hydraulic fracturing fluid migration even within decades. However, many of the subsequent modeling studies suggest that fluid migration over vertical distances of the order of 1–2 km will deliver very low amounts of fracturing fluid (Kissinger et al. 2013), or are significant only over 1000 year time scales (Gassiat et al. 2013), even in the presence of a permeable pathway. Results from various modeling studies proved that the probability of hydraulic fracturing fluid migration over the large vertical distances that separate the upper limit of induced hydraulic fractures and deepest groundwater aquifers is very small (Birdsell et al. 2015).

4.3 Hydraulic Fracturing Fluid

Hydraulic fracturing utilizes up to 10 and 15×10^3 m^3 of hydraulic fracturing fluid per hydraulic fracturing event (Sakmar 2011). Sand and other inert solids, such as ceramic beads, are injected into the formation to provide a support, or "proppant", which prevents the fractures from closing once the well pressure is released. In addition to proppant, other chemicals are added to the injected hydraulic fracturing fluids. These chemicals serve various functions as illustrated in Table 4.3. The U.S. EPA has identified hundreds of chemicals that are used as additives in hydraulic fracturing fluids (USEPA 2011a). Stringfellow et al. (2014) analyzed physical, chemical, and biological characteristics of 81 common hydraulic fracturing chemical additives. The analytical characterization of such additives is

Table 4.3 Volumetric composition and purposes of the typical constituents of hydraulic fracturing fluid (Gregory et al. 2011)

Constituent	Composition by volume (%)	Example	Purpose
Water and sand	99.500	Sand suspension	"Proppant" sand grains hold microfractures open
Acid	0.123	Hydrochloric or muriatic acid	Dissolves minerals and initiates cracks in the rock
Friction reducer	0.088	Polyacrylamide or mineral oil	Minimizes friction between the fluid and the pipe
Surfactant	0.085	Isopropanol	Increases the viscosity of the fracture fluid
Salt	0.060	Potassium chloride	Creates a brine carrier fluid
Scale inhibitor	0.043	Ethylene glycol	Prevents scale deposits in pipes
pH-adjusting agent	0.011	Sodium or potassium carbonate	Maintains effectiveness of chemical additives
Iron control	0.004	Citric acid	Prevents precipitation of metal oxides
Corrosion inhibitor	0.002	*n,n*-dimethyl formamide	Prevents pipe corrosion
Biocide	0.001	Glutaraldehyde	Minimizes growth of bacteria that produce corrosive and toxic by-products

important to understand the transport, environmental fate and ultimate potential health impact in various water compartments associated with hydraulic fracturing (Ferrer and Thurman 2015). Many additives used in hydraulic fracturing fluids are known to be toxic to human and wildlife. The potential for contamination of aquifers by the residual hydraulic fracturing fluids that remain underground must be considered (Finkel et al. 2013). Hydraulic fracturing fluid could migrate along abandoned and improperly plugged oil and gas wells through an inadequately sealed annulus between the wellbore and casing or through natural or induced fractures outside the target formation (Vidic et al. 2013). The likelihood of spills throughout the entire lifecycle of shale gas development, blowouts (uncontrolled release of natural gas from a gas well after pressure control systems have failed), leakage from pits or tanks that store the hydraulic fracturing fluids may raise the serious risk of groundwater and surface water contamination. Even small quantities of the toxic hydraulic fracturing fluids can contaminate shallow aquifers with hydrocarbons, toxic chemicals, heavy metals, and radioactive materials (Finkel et al. 2013). According to draft reports of EPA released in December 2011 (USEPA 2011b) and September 2012 (USEPA 2012), EPA testing detected the presence of chemicals commonly associated with hydraulic fracturing in drinking water wells in Pavillion, Wyoming. Encana Oil and Gas Inc., the company responsible for the natural gas wells, disputed the findings of the study criticizing the EPA's testing methods and assumptions as well as the processes used to construct the monitoring wells and analyze the results (Cooley and Donnelly 2014). The organic contaminants, likely derived from drilling or hydraulic fracturing fluids, were detected in initially potable groundwater used by several households in southeastern Bradford County, Pennsylvania (Llewellyn et al. 2015). In the absence of baseline data, it is difficult to confirm or deny reports of groundwater contamination. Collecting baseline groundwater quality data prior to initiating hydraulic fracturing would likely have been an effective way to evaluate potential impacts (Stephens 2014).

4.4 Wastewater

Partially returned hydraulic fracturing fluid, injected for hydraulic fracturing process, is called flowback water. Flowback of the hydraulic fracturing fluid occurs over a few days to a few weeks following hydraulic fracturing, depending on the geology and geomechanics of the formation (Gregory et al. 2011). The highest rate of flowback occurs on the first day, and the rate diminishes over time; the typical initial rate may be as high as 1000 m^3 d^{-1} (GWPC and ALL Consulting 2009). Flowback water contains some of the chemicals used in the hydraulic fracturing process, together with metals, minerals and hydrocarbons leached from the reservoir rock. High levels of salinity are quite common and, in some reservoirs, the leached minerals can be weakly radioactive, requiring specific precautions at the surface (IEA: World Energy Outlook 2012).

Wastewater management and disposal laws forbid operators from directly discharging wastewater associated with shale-gas production into waterways (Donaldson et al. 2013). Flowback water as well as water that is returned to the surface over the life of the gas well (produced water) must be stored and then treated and/or properly disposed of. There are different options available for dealing with wastewater from hydraulic fracturing:

- Underground injection
- Wastewater treatment at municipal and industrial wastewater treatment plants
- Recycling/reuse
- Other industrial uses

The choice of treatment of flowback mainly depends on the quality of wastewater and the intended use of the treated effluent (Lester et al. 2015). Flowback fluids usually require on-site storage followed by recycling, reinjection, or deep-well disposal into a saline aquifer. Flowback fluids may contaminate shallow groundwater due to release from lagoons during extreme precipitation events or failure of the impoundments (Jackson et al. 2013). Inadequate treatment of wastewaters and subsequent discharge into surface or groundwater could potentially lead to water quantity and quality impacts (Rahm and Riha 2012). Most wastewater generated from oil and gas production in the United States is disposed of through deep underground injection (Clark and Veil 2009). When underground injection is utilized, such operations are performed using Class II (disposal) underground injection control wells as defined by the U.S. Environmental Protection Agency (Veil et al. 2004). However, all regions have not adequate deep-well disposal capacity for shale gas development. In Texas, there were over 11,000 Class II disposal wells in 2008, or slightly more than one disposal well per gas-producing well in the Barnett Shale (Tintera 2008). In contrast, the whole state of Pennsylvania has only seven Class II disposal wells available for receiving flowback water (Gregory et al. 2011). Thus, wastewater reinjection is not generally an attractive option, as injection well facilities are usually located long distances from well pads, requiring significant transportation and incurring high costs (Veil 2010). There is a possibility of groundwater contamination if regulations and guidelines for construction and operation of deep injection wells are not strictly followed. The subsurface formation selected for injection of wastewater should have high porosity and permeability, and there should be impermeable zones above the injection zone to offer protection to any shallow water aquifers (Cooley and Donnelly 2014). Future research efforts should focus on wastewater disposal and on the efficacy of contaminant removal by industrial and municipal wastewater treatment facilities (Rozell and Reaven 2012).

4.5 *Surface Spills and Leaks*

Hydraulic fracturing operations are intensive over a short period of time, usually a few weeks, and are equipment intensive and require large amounts of chemicals, such as: pumps, proppants, vehicles, and other equipment that can result in unintentional spills on the surface (Donaldson et al. 2013). Spills and leaks from storage and production facilities may occur at well pads or result from accidents during transport of chemicals and wastes leading to potential impact on surface water and groundwater. Most of the water quality issues in the U.S. associated with hydraulic fracturing activities are the result of surface spills or leakage into the shallow water formations (Metzger 2011). Spills and leaks are an important route of potential groundwater contamination from hydraulic fracturing activities (Gross et al. 2013). Fluid released onto the ground from spills can seep into shallow groundwater aquifers, flow into surface waters, evaporate into the air, or stay on the surface soil depending upon the type of spill and its location. Between July 2010 and July 2011, Gross et al. (2013) noted 77 reported surface spills (about 0.5 % of the active wells) impacting the groundwater in Weld County, Colorado. Their analysis provides scientific evidence that benzene can contaminate groundwater sources following surface spills at active well sites. Surface spills can occur as a result of a host of unpredictable accidents, such as: tank ruptures, equipment or surface impoundment failures, overfills, accidents, ground fires, or improper operations. Vandalism and other illegal activities can also result in spills and improper wastewater disposal (Cooley and Donnelly 2014). In 2012, a spill of 76 m^3 of hydraulic fracturing wastewater is being investigated as criminal mischief in

Canton Township, Pennsylvania (Clarke 2012). Accidents of truck carrying hydraulic fracturing wastewater can also lead to chemical and wastewater spills. A truck accident in Mifflin Township, Pennsylvania, released hydraulic fracturing wastewater into a nearby creek in December 2011 (Reppert 2011). The pathway by which spilled chemicals may migrate to groundwater, surface water, and air depends on many factors, including the site, type of chemicals, and/or fluid properties. Some of the chemicals used in hydraulic fracturing are hazardous to human health (Stringfellow et al. 2014). Thus, these chemicals should be managed properly by making sure that they will not spill and contaminate the environment.

Water impoundments are used to store water for preparing hydraulic fracturing fluids. They also can be used to store flowback water before and after treatment. The failure of impoundments can release waste. In fall 2011, some wastewater ponds in Pennsylvania overflowed because of excessive rainfall from tropical storm Lee (Donaldson et al. 2013).

Surface spills and their effects can be minimized by: (1) providing adequate training for the crew handling equipment and chemicals; (2) using chemicals that are not toxic to the environment and are bio-degradable; (3) using appropriate liners to contain spills; (4) using double-walled tanks to minimize accidents related to rupture of single walls; (5) having site-specific spill prevention control, countermeasure plans, and the associated equipment/chemicals necessary to neutralize any spill; and (6) proper housekeeping (Donaldson et al. 2013).

The most commonly reported remediation activity, to clean up a spill and its affected environmental media, mentioned in approximately half of the hydraulic fracturing-related spill records evaluated by the U.S. EPA, was removal of spilled fluid and/or affected media, typically soil (US EPA 2015b). Other remediation methods reported by the U.S. EPA (USEPA 2015b) included the use of absorbent material, vacuum trucks, flushing the affected area with water, and neutralizing the spilled material.

5 Modeling Studies and Tools

Hydraulic fracturing of the shale formations have raised concerns of environmental risks to groundwater quality of aquifers, particularly due to gas migration and contaminant transport through induced and natural fractures (Vidic et al. 2013). Many studies, which have found thermogenic gas in water wells found more gas near fracture zones (DiGiulio et al. 2011; Osborn et al. 2011; Breen et al. 2007), suggesting that fractures are pathways for gas transport (Myers, 2012). Hydraulic fracturing fluid has been found in aquifers (DiGiulio et al. 2011; USEPA 1987), although the exact source and pathways was not known. It is difficult to determine whether hydraulic fracturing of the shale formations has affected groundwater quality, because it requires baseline conditions and detailed information of well drilling and casing. Most importantly, complex flow and solute transport through induced and natural fractures, which are generally from over 1000 m below overlying aquifers, requires a long-term and extensive monitoring program, which is often unavailable (Cai and Ofterdinger 2014). With so little data concerning the movement of contaminants along pathways from depth, either from wellbores or from deep formations, to aquifers, conceptual analyses are an alternative means to consider the risks (Myers 2012). The different modeling tools used to analyze flow and transport in the fractured rock system are summarized in Table 4.4.

A recent study applied a groundwater flow model to estimate the risk of groundwater contamination from hydraulically fractured shale by using particle tracking method (Myers 2012). The study concluded that hydraulic fracturing could create high upward advective flow in the fracture and advective travel time between the shale formations to overlying aquifers could be less than 10 years. However, there were concerns on the modeling framework of Myers (2012) that it neglects critical hydrologic processes and misrepresents physical flow boundary conditions, which could in turn severely compromise its conclusions (Cai and Ofterdinger 2014; Cohen et al. 2013; Saiers and Barth 2012).

Kissinger et al. (2013) studied different qualitative scenarios dealing with the flow/leakage paths; (1) Flow of fracturing fluid and/or brine through natural fault zones, (2) Flow of methane through natural fault zones and (3) Flow of methane through the rock using the simulation software DuMu^x. Their results show that a significant fluid migration is only possible if a combination of several conservative assumptions is met by a scenario.

Gassiat et al. (2013) studied the potential for slow contamination of shallow groundwater due to hydraulic fracturing at depth via fluid migration along conductive faults. They used two-dimensional, single-phase, multispecies, density-dependent, finite-element numerical groundwater flow and mass transport model (SUTRA-MS). A sensitivity analysis of contaminant migration along the fault considered basin show that specific conditions are needed for the slow contamination of a shallow aquifer: a high permeability fault, high overpressure in the shale unit, and hydrofracturing in the upper portion of the shale near the fault. Under such conditions, contaminants from the shale unit reach the shallow aquifer in less than 1000 years following hydraulic fracturing, at concentrations of solutes up to 90 % of their initial concentration in the shale, indicating that the impact on groundwater quality could be significant.

Cai and Ofterdinger (2014) conducted realizations of numerical modeling simulations to assess fluid flow and chloride transport from a synthetic Bowland Shale (UK) over a period of 11,000 years. This study investigated fluid and chloride mass fluxes before, during, and after hydraulic fracturing of the Bowland Shale using HydroGeoSphere model. This modeling revealed that the hydraulically fractured Bowland Shale is unlikely to pose a risk to its overlying groundwater quality when the induced fracture aperture is 200 μm. In the extremely unlikely event of the upward fracture growth directly connecting the shale formation to the overlying Sherwood Sandstone aquifer with the fracture aperture $\geq$1000 μm, the upward chloride mass flux could potentially pose risks to the overlying aquifer in 100 years. The model study also revealed that the upward mass flux is significantly intercepted by the horizontal mass flux within a high permeable layer between the Bowland

Table 4.4 Summary of numerical models used to investigate different possible mechanisms that could lead to upward migration of fluids/gases from a shale gas reservoir

S. No.	Model	Model description	Model references	Application references
1	MODFLOW-2000	MODFLOW is a computer program that simulates three-dimensional groundwater flow through a porous medium using a finite-difference method	Harbaugh et al. (2000)	Myers (2012)
2	DuMu^x	DuMu^x is a free and open-source simulator for flow and transport processes in porous media, based on the Distributed and Unified Numerics Environment (DUNE)	Flemisch et al. (2011)	Kissinger et al. (2013)
3	SUTRA-MS	U.S. Geological Survey numerical variable-density simulator of groundwater flow as well as the transport of multiple-solutes (A modified version of SUTRA)	Hughes and Sanford (2005)	Gassiat et al. (2013)
4	HydroGeoSphere	HydroGeoSphere is a fully integrated surface and subsurface flow and transport numerical model. The subsurface module is based on a three-dimensional (3-D) subsurface flow and transport code FRAC3DVS which is an efficient and robust numerical model that solves the three-dimensional variably saturated subsurface flow and solute transport equations in nonfractured or fractured media	Therrien et al. (2004), Therrien and Sudicky (1996)	Cai and Ofterdinger. (2014)

(continued)

Table 4.4 (continued)

S. No.	Model	Model description	Model references	Application references
5	TOUGH+	The TOUGH+ code, developed at the Lawrence Berkeley National Laboratory, includes equation-of-state modules that describe the non-isothermal flow of real gas mixtures, water, and solutes through fractured porous media and accounts for all processes involved in flow through tight and shale gas reservoirs (i.e., gas-specific Knudsen diffusion, gas and solute sorption onto the media, non-Darcy flow, salt precipitation as temperature and pressure drop in the ascending reservoir, etc.)	Freeman (2010), Freeman et al. (2011), Freeman et al. (2012), Freeman et al. (2013), Moridis et al. (2010), Olorode (2011)	USEPA (2015a)
	TOUGH+Rgas	Describes the coupled flow of a real gas mixture and heat in geologic media	Moridis and Freeman (2012)	
	TOUGH +RgasH_2O	Describes the non-isothermal two-phase flow of a real gas mixture and water and the transport of heat in a gas reservoir, including tight/shale gas reservoirs	Moridis and Freeman (2012)	
	TOUGH +RGasH_2OCont	Describes physics and chemistry of flow and transport of heat, water, gases, and dissolved contaminants in porous/fractured media	Moridis and Webb (2014)	
	ROCMECH	Simulates geomechanical behavior of multiple porosity/permeability continuum systems and can accurately simulate the evolution and propagation of fractures in a formation following hydraulic fracturing	Kim and Moridis (2012a, b, c), Kim and Moridis (2013)	

Shale and its overlying aquifers, reducing further upward flux toward the overlying aquifers.

Lawrence Berkeley National Laboratory (LBNL), in coordination with the U.S. EPA, used numerical simulations (Table 4.4) to investigate six possible mechanisms that could lead to upward migration of fluids, including gases, from a shale gas reservoir and the conditions under which such hypothetical scenarios may be possible (USEPA 2015a). Six possible mechanisms included in EPA's subsurface migration modelling are:

- Scenario A: Defective or insufficient well construction coupled with excessive pressure during hydraulic fracturing operations results in damage to well integrity during the stimulation process. A migration pathway is then established through which fluids could travel through the cement or area near the wellbore into overlying aquifers. In this scenario, the overburden is not necessarily fractured.
- Scenario B1: Fracturing of the overburden because inadequate design of the hydraulic fracturing operation results in fractures allowing fluid communication, either directly or indirectly, between shale gas reservoirs and aquifers above them. Indirect communication would occur if fractures intercept a permeable formation between the shale gas formation and the aquifer. Generally, the aquifer would be located at a more shallow depth than the permeable formation.
- Scenario B2: Similar to Scenario B1, fracturing of the overburden allows in direct fluid communication between the shale gas reservoir and the aquifers after intercepting conventional hydrocarbon reservoirs, which may create a dual source of contamination for the aquifer.
- Scenario C: Sealed/dormant fractures and faults are activated by the hydraulic fracturing operation, creating pathways for upward migration of hydrocarbons and other contaminants.
- Scenario D1: Fracturing of the overburden creates pathways for movement of hydrocarbons and other contaminants into offset wells (or their vicinity) in conventional reservoirs with deteriorating cement. The offset wells may intersect and communicate with aquifers, and inadequate or failing completions/cement can create pathways for contaminants to reach the groundwater aquifer.
- Scenario D2: Similar to Scenario D1, fracturing of the overburden results in movement of hydrocarbons and other contaminants into improperly closed offset wells (or their vicinity) with compromised casing in conventional reservoirs. The offset well could provide a low-resistance pathway connecting the shale gas reservoir with the groundwater aquifer

The EPA's subsurface migration modeling project is proceeding along two main tracks (USEPA 2015a). The first addresses the geomechanical reality of the mechanisms and seeks to determine whether it is physically possible (as determined and constrained by the laws of physics and the operational quantities and limitations involved in hydraulic fracturing operations) for the six migration mechanisms (Scenarios A to D2) to occur. The second axis focuses on contaminant transport, assuming that a subsurface migration has occurred as described in the six scenarios, and attempts to determine a timeframe for contaminants (liquid or gas phase) escaping from a shale gas reservoir to reach the groundwater aquifer. There is currently no single numerical model that includes all the processes in order to accurately describe the hypothetical scenario conditions; thus, the LBNL chose the Transport of Unsaturated Groundwater and Heat (TOUGH) family codes in combination with existing modules to create a model that better simulates the subsurface flow and geomechanical conditions in the migration scenarios (USEPA 2015a).

Gaps still remain in the knowledge of the hydraulic fracturing process. Future work needs reliable field data to calibrate and validate numerical models (Kissinger et al. 2013). The role of fault zones may have a considerable influence on the migration of fracturing fluid or brine; thus, possible connections between shallow and deep groundwater should be analyzed and quantified through isotope and or chemical analysis (Kissinger et al. 2013). Numerical modeling tools used to analyze flow and transport in the fractured rock system should be improved by incorporating all major processes involved in flow and transport in the fractured rock system. The required input parameters for numerical modeling are very hard to obtain, and a validation with field data requires an intensive, very costly and a time-consuming monitoring program (Kissinger et al. 2013).

6 Regulations

Managing and regulating the development of shale gas resources is a growing global interest and challenge because of concerns for potential environmental impacts due to multiple activities associated with shale gas development, coupled with the broad occurrence of shale gas throughout the world (Rahm and Riha 2012). The development of shale gas raises several legal issues related to potential impacts on the environment. In the United States, federal, state, and local governments have responded with a large number of new regulations to address these issues, and there continue to be frequent developments relating to the regulation of hydraulic fracturing (Hall 2014). There are several environmental statutes that have implications for the regulation of hydraulic fracturing by the federal government and states (Vann et al. 2013). Different laws and regulations are associated with varying phases and protocols of shale gas development (Table 4.5).

In recent years, different state enacted regulations requiring the mandatory disclosure of hydraulic fracturing fluid composition. As of August 2013, about 19 states had enacted mandatory disclosure regulations (Hall 2014). Though, the mandatory disclosure regulations enacted by the various states have important similarities, but it is different in some ways (Hall 2014). Currently, 16 states use FracFocus, managed by the Ground Water Protection Council and Interstate Oil and Gas Compact Commission, as a means of official state chemical disclosure (https://fracfocus.org/). The site was created to provide the public access to reported chemicals used for hydraulic fracturing within their area. A few states have enacted

Table 4.5 An overview of relevant legislation (Modified from Holloway and Rudd 2013)

S. no.	Laws and regulations	Applicable phases and protocols
1	Clean Water Act (CWA)	Oil and gas operators must obtain a storm water permit under the CWA for the construction and operation of a well pad and access road that is one acre or greater. This act also prohibits the discharge of any pollutant into U.S. waters without a permit.
2	Clean Air Act (CAA)	The new regulations issued by the Environmental Protection Agency (EPA) under CAA are intended for operators to control emissions of volatile organic compounds (VOCs) from flowback during the hydraulic fracturing process by adopting volatile organic compound capture techniques called "green completion" or "reduced emissions completions".
3	Endangered Species Act (ESA)	Operators must consult with the U.S. Fish and Wildlife Service and potentially obtain an incidental "take" permit if their operations may affect endangered or threatened species by oil and gas development or the withdrawal of water from streams
4	Migratory Bird Treaty Act (MBTA)	Operators are held strictly liable for any harm to migratory birds, and must ensure that maintenance of surface pits or use of rigs does not attract and harm these birds
5	Emergency Planning and Community Right-to-Know Act (EPCRA) and Occupational Safety and Health Act (OSHA)	Operators must meet safety requirements in a myriad of work processes such as working at heights, tank entry, excavation, medical surveillance, first aid, and chemical storage. Operators must also maintain Material Safety Data Sheets (MSDS) for certain hazardous chemicals that are stored on site in threshold quantities
6	Comprehensive Environmental Responsibility, Compensation, and Liability Act (CERCLA)	Operators must report releases of hazardous chemicals of threshold quantities and may potentially be liable for cleaning up spills
7	Resource Conservation and Recovery Act (RCRA)	Most wastes from hydraulic fracturing and drilling are exempt from the hazardous waste disposal restrictions, meaning that states—not the federal government—have responsibility for disposal procedures for the waste
8	Safe Drinking Water Act (SDWA)	Hydraulic fracturing operators are exempt from the SDWA, which requires that entities that inject substances underground prevent underground water pollution. The SDWA applies only to waste from hydraulic fracturing and drilling that is disposed of in underground injection control wells. If operators use diesel fuel in hydraulic fracturing, however, they are not exempt from SDWA

provisions that either require or encourage baseline testing of water samples from water wells within a specified radius from a proposed oil and gas well before an oil or gas well is drilled or fractured (Hall 2014). Stronger and fully-enforced government regulations are needed in many states to ensure sufficient environmental protection from rapid shale gas developments (Zoback et al. 2010).

7 Summary

This work provides an overview of hydraulic fracturing used to extract shale gas, potential impact of hydraulic fracturing and shale gas development on groundwater, an overview of modeling studies and tools to access potential impacts, and regulations related to hydraulic fracturing. The most significant risks related to groundwater contamination from the development of shale gas are; (1) the excessive withdrawal of water, (2) gas migration and groundwater contamination due to faulty well construction, blowouts, (3) contamination by wastewater disposal, and (4) accidental leaks and spill of wastewater and chemicals used during drilling and hydraulic fracturing.

- Unconventional natural gas extraction from impermeable geologic formations is getting momentum in recent years due to advances in horizontal drilling and hydraulic fracturing.
- The development of shale gas has raised a serious concern about the potential impact on hydrological cycle due to water withdrawal for hydraulic fracturing and potential impact on water resources and environments due to use of toxic substances in hydraulic fracturing fluids.
- In places with limited available water, scientific water balance studies are necessary to insure water supplies for other needs can be maintained. Different alternatives of reducing the demand for fresh water for hydraulic fracturing on local resources is to treat and recycle the flowback and produced water in the fracturing process, use non-potable groundwater, collect and store of surface water in impoundments during wet seasons to provide water for the fracturing operation during dry seasons.
- Maintaining well integrity during both the drilling and operating phases of shale gas development is a critical, feasible, and cost-effective way to reduce possible contamination of shallow groundwater.
- Collecting baseline conditions and detailed information of groundwater quality data prior to initiating hydraulic fracturing is an effective way to evaluate potential impacts of hydraulic fracturing on groundwater.
- Proper handling of flowback fluids, on-site storage, recycling, reinjection, or deep-well disposal into a saline aquifer may protect possible contamination of water resources.

- Different chemicals used in the hydraulic fracturing fluid should be managed properly, making sure that it does not spill, leak and contaminate the groundwater and environment.
- Numerical modeling tools used to analyze flow and transport in the fractured rock system should be improved by incorporating all major processes involved in flow and transport in the fractured rock system.
- Stronger, fully-enforced government regulations are needed in many states to provide sufficient protection against potential impact of hydraulic fracturing and shale gas development to the environment and shallow groundwater aquifer.

References

Amer A, Chinellato F, Collins S, Denichou JM, Dubourg I, Griffiths R, Koepsell R, Lyngra S, Marza P, Roberts IB (2013) Structural steering—a path to productivity. Oilfield Rev 25 (1):14–31

API (American Petroleum Institute) (2009a) Environmental protection for onshore oil and gas production operations and leases. API recommended practice 51R, 1st edn. American Petroleum Institute, Washington, DC. http://www.api.org/~/media/files/policy/exploration/api_rp_51r.pdf. Accessed 22 Feb 2015

API (American Petroleum Institute) (2009b) Hydraulic fracturing operations—well construction and integrity guidelines. API guidance document HF1. American Petroleum Institute, Washington, DC

Arthur JD, Bohm B, Coughlin BJ, Layne M (2008) Evaluating the environmental implications of hydraulic fracturing in shale gas reservoirs. ALL consulting. http://www.all-llc.com/publicdownloads/ArthurHydrFracPaperFINAL.pdf. Accessed 22 Feb 2015

Barker JF, Fritz P (1981) Carbon isotope fractionation during microbial methane oxidation. Nature 293:289–291

Bellabarba M, Bulte-Loyer H, Froelich B, Le Roy-Delage S, Kujik R, Zerouy S, Guillot D, Meroni N, Pastor S, Zanchi A (2008) Ensuring zonal isolation beyond the life of the well. Oilfield Review 20(1):18–31

Berman A (2009) Lessons from the Barnett Shale suggest caution in other shale plays. World Oil 230(8)

Birdsell DT, Rajaram H, Dempsey D, Viswanthan HS (2015) Hydraulic fracturing fluid migration in the subsurface: a review and expanded modeling results. Water Resour Res 51:7159–7188

Breen KJ, Revesz K, Baldassare FJ, McAuley SD (2007) Natural gases in ground water near Tioga Junction, Tioga County, north-central Pennsylvania—Occurrence and use of isotopes to determine origins. Scientific Investigations Report Series 2007-5085. U.S. Geological Survey, Reston, Virginia

Bryant J, Welton T, Haggstrom J (2010) Will flowback or produced water do? http://www.halliburton.com/public/pe/contents/Papers_and_Articles/web/A_through_P/0910HEP_Sand_HAL.PDF. Accessed 22 Feb 2015

Cai Z, Ofterdinger U (2014) Numerical assessment of potential impacts of hydraulically fractured Bowland Shale on overlying aquifers. Water Resour Res 50(7):6236–6259

Carter RH, Holditch SA, Wolhart SL (1996) Results of a 1995 hydraulic fracturing survey and a comparison of 1995 and 1990 industry practices. Society of Petroleum Engineers Annual Technical Conference, Denver, CO

Chesapeake Energy (2009) Barnett Shale—natural gas production. http://www.askchesapeake.com/Barnett-Shale/Production/Pages/information.aspx

Clark CE, Veil JA (2009) Produced Water Volumes and Management Practices in the United States. United States Department of Energy, Argonne National Laboratory ANL/EVS/R-09/1
Clarke CR (2012) Police: spill at gas well site may have been vandalism. http://www.sungazette.com/page/content.detail/id/573278/Police--Spill-at-gas-well-site-may-have-been-vandalism.html?nav=5176. Accessed 22 Feb 2015
Cohen HA, Parratt T, Andrews CB (2013) Potential contaminant pathways from hydraulically fractured shale to aquifers. Ground Water 51(3):317–319
Cooley H, Donnelly K (2014) Hydraulic fracturing and water resources: what do we know and need to know? In: Gleick et al (ed) The World's water volume 8: the biennial report on freshwater resources, p 63–81. Accessed 22 Feb 2015
Darrah TH, Vengosha A, Jacksona RB, Warnera NR, Poredae RJ (2014) Noble gases identify the mechanisms of fugitive gas contamination in drinking-water wells overlying the Marcellus and Barnett Shales. Proc Natl Acad Sci U S A 111(39):14076–14081
DiGiulio DC, Wilkin RT, Miller C, Oberly G (2011) DRAFT: investigation of ground water contamination near Pavillion, Wyoming. U.S. Environmental Protection Agency, Office of Research and Development, Ada, Oklahoma
Donaldson EC, Alam W, Begum N (2013) Hydraulic fracturing explained: evaluation, implementation and challenges. Gulf Publishing Company, Texas
Ferrer I, Thurman EM (2015) Chemical constituents and analytical approaches for hydraulic fracturing waters. Trends Environ Analytic Chem 5:18–25
Finkel ML, Hays J, Law A (2013) Modern natural gas development and harm to health: the need for proactive public health policies. ISRN Public Health 2013:5. doi:10.1155/2013/408658
Flemisch B, Darcis M, Erbertseder K, Faigle B, Mosthaf K, Müthing S, Nuske P, Tatomir A, Wolf M, Helmig R (2011) DuMux: DUNE for multi-phase, component, scale, physics, flow and transport in porous media. Adv Water Resour 34(9):1102–1112. doi:10.1016/j.advwatres.2011.03.007
Freeman CM (2010) Study of flow regimes in multiply-fractured horizontal wells in tight gas and shale gas reservoir systems. MS Thesis. Texas A & M University, College Station, Texas
Freeman CM, Moridis GJ, Blasingame TA (2011) A numerical study of microscale flow behavior in tight gas and shale gas reservoir systems. Transport Porous Med 90(1):253–268
Freeman CM, Moridis GJ, Michael GE, Blasingame TA (2012) Measurement, modeling, and diagnostics of flowing gas composition changes in shale gas wells. SPE Latin American and Caribbean Petroleum Engineering Conference, Mexico City, Mexico
Freeman CM, Moridis GJ, Ilk D, Blasingame TA (2013) A numerical study of performance for tight gas and shale gas reservoir systems. J Petrol Sci Eng 108:22–39
Freyman M (2014) Hydraulic fracturing & water stress: water demand by the numbers—shareholder, Lender & Operator Guide to Water Sourcing, Ceres. http://www.ceres.org/resources/reports/hydraulic-fracturing-water-stress-water-demand-by-the-numbers. Accessed 28 Nov 2014
Gassiat C, Gleeson T, Lefebvre R, McKenzie J (2013) Hydraulic fracturing in faulted sedimentary basins: numerical simulation of potential contamination of shallow aquifers over long time scales. Water Resour Res 49:8310–8327. doi:10.1002/2013WR014287
Godsey WE (2011) Fresh, brackish, or saline water for hydraulic fracs: what are the options? EPA's Hydraulic Fracturing Technical Workshop 4. Washington, DC 2011
Gorody AW (2012) Factors affecting the variability of stray gas concentration and composition in groundwater. Environ Geosci 19:17–31
Gregory KB, Vidic RD, Dzombak DA (2011) Water management challenges associated with the production of shale gas by hydraulic fracturing. Elements 7:181–186
Gross SA, Avens HJ, Banducci AM, Sahmel J, Panko JM, Tvermoes BE (2013) Analysis of BTEX groundwater concentrations from surface spills associated with hydraulic fracturing operations. J Air Waste Manage Assoc 63(4):424–432
GWPC (Ground Water Protection Council) & ALL Consulting (2009) Modern shale gas development in the US: A primer. Contract DE-FG26-04NT15455. Washington, DC: US Department of

Energy, Office of Fossil Energy and National Energy Technology Laboratory. www.netl.doe.gov/File%20Library/Research/Oil-Gas/Shale_Gas_Primer_2009.pdf. Accessed 27 Feb 2014
Hall KB (2014) Recent developments in hydraulic fracturing regulation and litigation. Journal Articles. Paper 129. http://digitalcommons.law.lsu.edu/faculty_scholarship/129
Harbaugh AW, Banta ER, Hill MC, McDonald MG (2000). Modflow-2000, the U.S. Geological Survey modular ground-water model—User guide to modularization, concepts and the ground-water flow process. Open-File Report 00-92. U.S. Geological Survey, Reston, Virginia
Holloway MD, Rudd O (2013) Fracking: the operations and environmental consequences of hydraulic fracturing, Scrivener Publishing LLC
Hughes J, Sanford W (2005) SUTRA-MS a version of SUTRA modified to simulate heat and multiple-solute transport, Report, 141 pp, U.S. Geol. Surv. Open-File Rep. 2004-1207, Reston, Virginia
International Energy Agency (IEA) (2012) Golden rules for a golden age of gas. World Energy Outlook Report—Special Report on Unconventional Gas; IEA: Paris, 2012
Jackson RB, Vengosh A, Darrah TH, Warner NR, Down A, Poreda RJ, Osborn SG, Zhao K, Karr JD (2013) Increased stray gas abundance in a subset of drinking water wells near Marcellus shale gas extraction. Proc Natl Acad Sci U S A 110(28):11250–11255
Kargbo DM, Wilhelm RG, Campbell DJ (2010) Natural gas plays in the Marcellus Shale: challenges and potential opportunities. Environ Sci Technol 44(15):5679–5684
Kellman S, Schneider K (2010) Water demand is flash point in Dakota oil boom. Circle of Blue Waternews. http://www.circleofblue.org/waternews/2010/world/scarce-water-is-no-limit-yet-to-north-dakota-oil-shale-boom/. Accessed 5 Jan 2015
Kim J, Moridis GJ (2012a) Gas flow tightly coupled to elastoplastic geomechanics for tight and shale gas reservoirs: material failure and enhanced permeability. In: SPE Unconventional Resources Conference, Pittsburgh, PA, spe 155640
Kim J, Moridis GJ (2012b) Numerical geomechanical analyses on hydraulic fracturing in TightGas and shale gas reservoirs. Presented at 2012 SPE Annual Technical Conference and Exhibition, San Antonio, Texas
Kim J, Moridis GJ (2012c) Numerical studies for naturally fractured shale gas reservoirs: coupled flow and geomechanics in multiple porosity/permeability materials, 46th U.S. Rock Mechanics/Geomechanics Symposium, Chicago, Illinois, 24–27 June 2012
Kim J, Moridis GJ (2013) Development of the T+M coupled flow–geomechanical simulator to describe fracture propagation and coupled flow–thermal–geomechanical processes in tight/shale gas systems. Comput Geosci 60:184–198
Kissinger A, Helmig R, Ebigbo A, Class H, Lange T, Sauter M, Heitfeld M, Klünker J, Jahnke W (2013) Hydraulic fracturing in unconventional gas reservoirs: risks in the geological system, part 2. Modelling the transport of fracturing fluids, brine and methane. Environ Earth Sci 70:3855–3873. doi:10.1007/s12665-013-2578-6
Lake LW, Martin J, Ramsey JD, Titman S (2013) A primer on the economics of shale gas production just how cheap is shale gas? J Appl Corp Finance 25(4):87–96
Lester Y, Ferrer I, Thurman EM, Sitterley KA, Korak JA, Aiken G, Linden KG (2015) Characterization of hydraulic fracturing flowback water in Colorado: implications for water treatment. Sci Total Environ 512–513:637–644
Litovitz A, Curtright A, Abramzon S, Burger N, Samaras C (2013) Estimation of regional air-quality damages from Marcellus Shale natural gas extraction in Pennsylvania. Environ Res Lett 8:014017. doi:10.1088/1748-9326/8/1/014017
Llewellyn GT, Dorman F, Westland JL, Yoxtheimer D, Grieve P, Sowers T, Humston-Fulmer E, Brantley SL (2015) Evaluating a groundwater supply contamination incident attributed to Marcellus Shale gas development. PANAS 112(20):6325–6330
Martin T, Valkó P (2007) Hydraulic fracture design for production enhancement. In: Economides MJ, Martin T (eds) Modern fracturing: enhancing natural gas production. ET Publishing, Houston, TX, p 95
Metzger M (2011) Hydrofracturing and the environment. Water Qual Prod 16(11):14–15

Molofsky LJ, Connor JA, Wylie AS, Wagner T, Farhat SK (2013) Evaluation of methane sources in groundwater in Northeastern Pennsylvania. Ground Water 51(3):333–349
Moridis GJ, Freeman CM (2012) The RGasH2OCont module of the TOUGH+ code for simulation of coupled fluid and heat flow, and contaminant transport, in tight/shale gas systems. In preparation for journal submission. Lawrence Berkeley National Laboratory, Berkeley, California
Moridis F, Webb F (2014) The RealGas and RealGasH2O options of the TOUGH+ code for the simulation of coupled fluid and heat flow in tight/shale gas systems. Comput Geosci 65:56–71
Moridis GJ, Blasingame T, Freeman CM (2010) Analysis of mechanisms of flow in fractured tight-gas and shale-gas reservoirs. SPE paper 139250 presented at the SPE Latin American & Caribbean Petroleum Engineering Conference, 1–3 December, Lima, Peru
Myers T (2012) Potential contaminant pathways from hydraulically fractured shale to aquifers. Ground Water 50(6):872–882
National Research Council (2010) Management and effects of coalbed methane produced water in the western US. National Academies Press, Washington, DC
Nicot JP, Scanlon BR, Reedy RC, Costley RA (2014) Source and fate of hydraulic fracturing water in the barnett shale: a historical perspective. Environ Sci Technol 48(4):2464–2471. doi:10.1021/es404050r
Nolan BT, Hitt KJ (2006) Vulnerability of shallow groundwater and drinking-water wells to nitrate in the United States. Environ Sci Technol 40:7834–7840
NYSDEC (New York State Department of Environmental Conservation) (2011) Supplemental generic environmental impact statement on the oil, gas and solution mining regulatory program (revised draft). Well permit issuance for horizontal drilling and high-volume hydraulic fracturing to develop the Marcellus Shale and other low-permeability gas reservoirs. Albany, NY: New York State Department of Environmental Conservation. ftp://ftp.dec.state.ny.us/dmn/download/OGdSGEISFull.pdf. Accessed 27 Feb 2014
Oil and Gas Investor (2005) Tight Gas (special supplement). Houston, TX: Oil and Gas Investor/Hart Energy Publishing LP. Retrieved August 9, 2010, from http://www.oilandgasinvestor.com/pdf/Tight%20Gas.pdf
Olorode OM (2011) Numerical modeling of fractured shale-gas and tight-gas reservoirs using unstructured grids. MS Degree. Texas A&M University, College Station, TX
Osborn SG, Vengosh A, Warner NR, Jackson RB (2011) Methane contamination of drinking water accompanying gas-well drilling and hydraulic fracturing. Proc Natl Acad Sci 108 (20):8172–8176
Palisch TT, Vincent MC, Handren PJ (2008) Slickwater fracturing—food for thought. Society of Petroleum Engineers Annual Technical Conference, Denver, CO, 21–24 September 2008
Pickett A (2009) New solutions emerging to treat and recycle water used in hydraulic fracs 2009
Prouty JL (2001) Tight gas in the spotlight. Gas Technol Institute GasTIPS 7(2):4–10
Rahm D (2011) Regulating hydraulic fracturing in shale gas plays: the case of Texas. Energy Policy 39(5):2974–2981
Rahm BG, Riha SJ (2012) Toward strategic management of shale gas development: regional, collective impacts on water resources. Environ Sci Policy 17:12–23
Ratner M, Tiemann M (2014) An overview of unconventional oil and natural gas: resources and federal actions. Congressional Research Service
Reppert J (2011) Collision spills fracking fluid on state route. http://www.sungazette.com/page/content.detail/id/572658/Collision-spills-fracking-fluid-on-state-route.html?nav=5011. Accessed 7 Nov 2014
Rogers RE, Ramurthy M, Rodvelt G, Mullen M (2007) Coalbed methane: principles and practices. 3rd edn. Oktibbeha Publishing Co, Starkville, MS. http://www.halliburton.com/public/pe/contents/Books_and_Catalogs/web/CBM/CBM_Book_Intro.pdf. Accessed 27 Feb 2014
Rowan TM (2009). Spurring the Devonian: methods of fracturing the lower Huron in southern West Virginia and eastern Kentucky. Society for Petroleum Engineers Eastern Regional Meeting, Charleston, WV, 23–25 September 2009

Rozell DJ, Reaven SJ (2012) Water pollution risk associated with natural gas extraction from the Marcellus Shale. Risk Anal 32:1382–1393

Ruszka J (2007) Global challenges drive multilateral drilling. E&P. http://www.epmag.com/archives/features/583.htm. Accessed 27 Feb 2014

Saiers J, Barth E (2012) Potential contaminant pathways from hydraulically fractured shale aquifers. Ground Water 50(6):826–828

Sakmar SL (2011) The global shale gas initiative: will the United States be the role model for the development of shale gas around the world? Houst J Int Law 33:369–417

Smrecak TA, Marcellus Shale Team PRI (2012) Understanding drilling technology. Marcellus Shale 6:1–9

Stephens DB (2014) Analysis of the groundwater monitoring controversy at the Pavillion, Wyoming Natural Gas Field. Ground Water 53(1):29–37. doi:10.1111/gwat.12272

Stringfellow WT, Domen JK, Camarillo MK, Sandelin WL, Borglin S (2014) Physical, chemical, and biological characteristics of compounds used in hydraulic fracturing. J Hazard Mater 275:37–54

Therrien R, Sudicky EA (1996) Three-dimensional analysis of variably-saturated flow and solute transport in discretely-fractured porous media. J Contam Hydrol 23(1–2):1–44

Therrien R, McLaren RG, Sudicky EA, Panday SM (2004) HydroSphere: a three-dimensional numerical model describing fully integrated subsurface and surface flow and solute transport, University of Waterloo

Tintera J (2008) The regulatory framework of saltwater disposal 2008. Fort Worth Business Press Barnett Shale Symposium, Fort Worth, TX, 29 Feb 2008

USEIA (US Energy Information Administration) (2010) Annual energy outlook 2011: early release overview. US Department of Energy, Washington, DC

USEIA (2013) Technically recoverable shale oil and shale gas resources: an assessment of 137 shale formations in 41 countries outside the United States

USEIA (2014) Annual Energy Outlook 2014, U.S. Energy Information Administration

USEPA (US Environmental Protection Agency) (1987) Report to congress, management of wastes from the exploration, development, and production of crude oil, natural gas, and geothermal energy, volume 1 of 3, oil and gas. Washington, DC: EPA

USEPA (US Environmental Protection Agency) (2004) Evaluation of impacts to underground sources of drinking water by hydraulic fracturing of coalbed methane reservoirs. No. EPA/816/R-04/003. Washington, DC: US Environmental Protection Agency, Office of Water. http://water.epa.gov/type/groundwater/uic/class2/hydraulicfracturing/wells_coalbedmethanestudy.cfm. Accessed 27 Feb 2014

USEPA (US Environmental Protection Agency) (2011a) Plan to study the potential impacts of hydraulic fracturing on drinking water resources office of Research and Development U.S. Environmental Protection Agency, Washington, D.C. November 2011

USEPA (US Environmental Protection Agency) (2011b) Investigation of ground water contamination near Pavillion, Wyoming Draft. Office of Research and Development. http://www.epa.gov/region8/superfund/wy/pavillion/EPA_ReportOnPavillion_Dec-8-2011.pdf

USEPA (US Environmental Protection Agency) (2012) Investigation of ground water contamination near Pavillion, Wyoming, Phase V Sampling Event: Summary of Methods and Results. Denver, CO: US Environmental Protection Agency, Region 8. http://www.epa.gov/region8/superfund/wy/pavillion/phase5/PavillionSeptember2012Narrative.pdf

USEPA (US Environmental Protection Agency) (2015a) Study of the potential impacts of hydraulic fracturing on drinking water resources: progress report. In: McBroom (ed) The effects of induced hydraulic fracturing on the environment, CRC Press

USEPA (US Environmental Protection Agency) (2015b) Review of state and industry spill data: characterization of hydraulic fracturing-related spills [EPA Report]. (EPA/601/R-14/001). Washington, D.C.: Office of Research and Development, U.S. Environmental Protection Agency

USEPA (US Environmental Protection Agency) (2015c) Review of well operator files for hydraulically fractured oil and gas production wells: Well design and construction [EPA Report]. (EPA/601/R-14/002). Washington, D.C.: Office of Research and Development, U.S. Environmental Protection Agency

Vann A, Murrill BJ, Tiemann M (2013) Hydraulic fracturing: selected legal issues. Congressional Research Service, Washington, DC

Veil JA (2010) Water management technologies used by Marcellus Shale Gas Producers. Argonne National Laboratory, Argonne, IL Prepared for the U.S. Department of Energy, National Energy Technology Laboratory. http://www.marcellus.psu.edu/resources/PDFs/argonnelab.pdf. Accessed 28 Nov 2014

Veil JA, Puder MG, Elcock D, Redweik RJ (2004) A white paper describing produced water from production of crude oil, natural gas, and coal bed methane. Prepared for the US Department of Energy, National Energy Technology Laboratory. Argonne National Laboratory, Argonne, IL

Vidic RD, Brantley SL, Vandenbossche JM, Yoxtheimer D, Abad JD (2013) Impact of Shale gas development on regional water quality. Science 340:1235009. doi:10.1126/science.1235009

Winter TC, Harvey JW, Franke QL, Alley WM (1998) Ground water and surface water: a single resource. US Geological Survey Circular 1139

White Paper, US shale gas—an unconventional resource, unconventional challenge. Halliburton; 2008

Zoback M, Kitaseib S, Copithornec B (2010) Addressing the environmental risks from shale gas development, Briefing Paper 1, Worldwatch institute

Chapter 5
Pharmaceuticals and Groundwater Resources

Matteo D'Alessio and Chittaranjan Ray

Abstract Due to advances in analytical techniques and increasing concerns related to the potential impact on aquatic and terrestrial organisms, the reported occurrence of pharmaceutical compounds in groundwater has significantly increased during the past two decades. This chapter provides an overview of the detection of pharmaceutical, life-style, and endocrine disruptor compounds in groundwater from five geographical areas (Africa, Asia, Central and South America, Europe, and North America). The occurrence of these compounds has been linked to the four major sources of contamination: agricultural practices/wastes, landfill, septic tanks, and wastewater. The concentration of the detected compounds ranged between ng/L and µg/L. Pharmaceutical compounds, in particular antibiotic and analgesic/anti-inflammatory, represent the most common group detected in groundwater, regardless of the geographic location. Carbamazepine (anticonvulsant), sulfamethoxazole (antibiotic), and caffeine (life-style) represent the three most common compounds detected in groundwater. The occurrence of the detected compounds in groundwater is primarily related to wastewater and/or agricultural practices/wastes. None of the detected compounds has been linked to all four major sources of contamination. However, five compounds (ibuprofen, paracetamol, triclosan, caffeine and cotinine) have been linked to three sources of contamination such as wastewater, landfill, and septic tanks.

1 Introduction

During the last few decades, the reported presence of organic-micropollutants—such as personal care products, household chemicals, and pharmaceutically active compounds—dramatically increased in the environment (Ternes 1998; Kolpin et al. 2002; Focazio et al. 2008; Snyder et al. 2008; Benotti et al. 2009; Luo et al. 2014). The increasing concern associated with these micropollutants was due to their potential impact on aquatic and terrestrial organisms (Thiele-Bruhn and

M. D'Alessio (✉) • C. Ray
Nebraska Water Center, Nebraska Innovation Campus, 2021 Transformation Drive, Lincoln, Nebraska 68588-6204, USA
e-mail: mdalessio2@unl.edu

A. Fares (ed.), *Emerging Issues in Groundwater Resources*, Advances in Water Security, DOI 10.1007/978-3-319-32008-3_5

Beck 2005; Madureira et al. 2011) and human health (Boxall 2004; Fent et al. 2006; Bolong et al. 2009). Most of these studies focused on surface waters, in which the occurrence of pharmaceutical compounds was more frequent, and occurred at higher concentrations compared to groundwater (Ternes 1998; Kolpin et al. 2002; Ashton et al. 2004; Bartelt-Hunt et al. 2009). However, advances in analytical instrumentation enhanced the ability to detect pharmaceutical compounds at consistently lower concentrations (Richardson and Ternes 2014). As a consequence of these improvements and public interest over these compounds, the number of studies investigating the occurrence of these compounds in groundwater drastically increased over the last two decades (Barnes et al. 2008; Teijon et al. 2010; Estévez et al. 2012; Lapworth et al. 2012; Meffe and de Bustamante 2014; de Jesus Gaffney et al. 2015).

The focus of this chapter is to examine the occurrence of pharmaceutical compounds, life style compounds, and endocrine disruptors exclusively in groundwater. In contrast with other studies, we investigated not only the occurrence of these compounds in Europe and the USA but also in Africa, Asia, and Central and South America. For the purpose of this chapter, pesticides, industrial products, and industrial by-products are not investigated due to the extensive literature already available regarding their occurrence in surface and groundwater (Spalding and Exner 1993; Squillace et al. 1996; Kolpin et al. 1998; Höhener et al. 2003; Verstraeten et al. 2003; Gilliom 2007). In this chapter, we first examine the legal framework related to groundwater protection and prevention of possible contaminants such as pharmaceutical compounds; second, we investigate the possible sources of contaminations and the pathways leading to the occurrence of these compounds in groundwater based on approximately 100 studies published between 1990 and 2014; third, we identify the most common pharmaceutical compounds detected in groundwater across the different geographical areas and their potential source.

2 Legal Framework

Monitoring and protection of groundwater, especially in terms of pharmaceutical compounds, shows significant discrepancy between the different countries. In the United States, the Environmental Protection Agency (EPA) promulgated the final Ground Water Rule (GWR) in 2006 in an attempt to minimize the risk of exposure to fecal contamination that may occur in public water systems that use ground water sources (815-F-06-003, USEPA 2006). In addition to bacterial contamination, the EPA defined the primary standards (NPDRWs) to protect public health by limiting the levels of contaminants in drinking water. The maximum contaminant level goal (MCLG, a non-enforceable public health goal) and maximum contaminant level (MCL, an enforceable standards) are defined for disinfectants (3), disinfection by-products (4), inorganic chemicals (16), organic chemicals (53), radionuclides

(4), and microorganisms (7) (http://water.epa.gov/drink/contaminants/#Primary). In addition to the NPDRWs standard, USEPA has also defined a contaminant candidate list (CCL), which are known or anticipated to occur in public water system and may require additional regulations. There are 116 candidates in the latest CCL, CCL3 2009, including pesticides, disinfection by-products, chemical used in commerce, waterborne pathogens, pharmaceuticals and biological toxins (http://www2.epa.gov/ccl/contaminant-candidate-list-3-ccl-3). Among the pharmaceutical compounds, nine hormones (17α-estradiol, Equilenin, Equilin, 17-β-estradiol, estriol, estrone, 17α-ethynyl estradiol, Mestranol, and 19-Norethisterone) and one antibiotic (erythromycin) are included in CCL3.

All the European Union Countries are fully involved in the process of implementing the principles of the European Water Framework Directive (2000; 2000/60/EC) and the Groundwater Daughter Directive (2006; 2006/118/EC). The overall goals of these laws are to improve groundwater protection and knowledge, and achieve effective management of this resource. In addition, the Drinking Water Directive (1998) defines the limits for a small number of organic micropollutants including aromatic hydrocarbons, chlorinated solvents and disinfection by-products. The European Commission (EC 2008) defines the priority substances directive, while the EC 2011 proposed revision and addition of pharmaceutical compounds such as ibuprofen, diclofenac, a-ethinyloestradiol, and perfluorooctane sulfonate.

In Japan, the environmental quality standards (EQS) for groundwater pollution were issued in 1997 and were further revised during the last decade (Findikakis and Sato 2011). In the latest version of 2012, 28 hazardous substances including heavy metals, chlorinated organic compounds and volatile organic carbons, are regulated (http://www.env.go.jp/en/water/gw/gwp.html). However, pharmaceutical compounds and their potential degradation by-products are not included.

Similar to Japan, Mexico and China do not include pharmaceutical compounds in their list of targeted compounds for groundwater protection. In Mexico, surface water and groundwater are required to comply with the 44 physical, chemical and biological quality parameters established by Mexican law in order to be considered suitable for human consumption (Félix-Cañedo et al. 2013). In China, the Environmental Quality Standard for Groundwater, established in 1993, focused on a limited number of inorganic salts and heavy metals, two pesticides (BHC and DDT) and biological parameters (Findikakis and Sato 2011).

3 Sources of Contamination and Pathways

Sources of pharmaceutical compounds in the environment can be classified as: point-sources and diffuse sources. Figure 5.1 shows the possible sources of contamination, the pathways and the final receptors of pharmaceutical compounds in the environment.

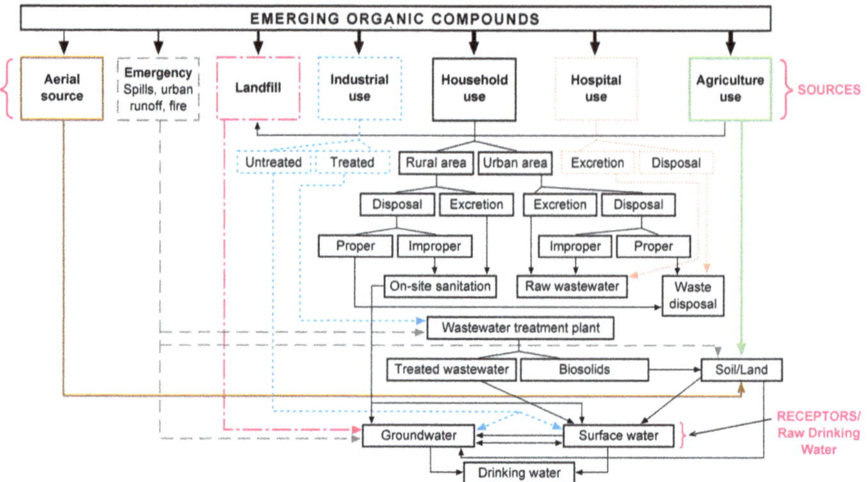

Fig. 5.1 Sources, pathways and final receptors for pharmaceutical compounds, life-style compounds and endocrine disruptor compounds (Redrawn, sources: Lapworth et al. 2012; Stuart et al. 2012; Deo and Halden 2013)

Wastewater treatment plants (Ternes and Hirsch 2000; Nakada et al. 2008; Vulliet and Cren-Olivé 2011; López-Serna et al. 2013; Bradley et al. 2014), potential leakage from landfills (Holm et al. 1995; Barnes et al. 2004; Buszka et al. 2009) and septic tanks (Verstraeten et al. 2005; Carrara et al. 2008; Schaider et al. 2014) represent the most common point-sources, while application of manure and bio-solids from wastewater treatment plants (Kolodziej et al. 2004; Bartelt-Hunt et al. 2011; Pinheiro et al. 2013; Zhou et al. 2013), spills (Reddersen et al. 2002), and urban runoff represent the most common diffuse sources of pharmaceutical compounds in the environment. Aerial source represents a new possible diffusive-source of pharmaceutical compounds. According to Hamscher and Hartung (2008), dust can represent a new route for veterinary compounds to enter the environment.

The occurrence of pharmaceutical compounds in groundwater due to wastewater reuse or disposal has been well documented in developed countries (i.e., European Union, United States, and Japan) as well as in developing countries (i.e. China) (Clara et al. 2004; Vulliet et al. 2008; Kuroda et al. 2012; Zhou et al. 2013). On the other hand, the presence of these compounds in groundwater due to leakages from septic tanks and landfills has been reported only in Europe (Holm et al. 1995) and the United States (Eckel et al. 1993; Swartz et al. 2006; Carrara et al. 2008; Buszka et al. 2009; Schaider et al. 2014). Even if limited, the studies investigating the occurrence of pharmaceutical compounds in groundwater due to agricultural practices are well distributed across the world (Krapac et al. 2004; Hu et al. 2010; Bartelt-Hunt et al. 2011; Pinheiro et al. 2013).

The origin of point-source pollution can be easily identified and the spatial distribution can be confined. Point-source pollution is characterized by a high frequency of occurrence, high concentration of the pharmaceutical compounds, and limited natural attenuation through the soil and subsurface due to reduced contact time. On the other hand, diffuse pollution originates from less defined sources and can disperse over broad regions. The concentrations of pharmaceutical compounds associated with diffuse contamination are usually low but the natural attenuation through soil and subsurface is usually enhanced by the longer contact time.

This chapter reviews 99 studies that have been published between 1992 and 2014 showing the occurrence of pharmaceutical compounds in groundwater around the world. Among the 99 studies investigated, 91 (93 %) have been published during the last 14 years and 43 (44 %) in the last 4 years. For simplicity, the data presented are grouped based on the geographical area in which the studies have been conducted and the type of study. Five distinct geographical areas have been identified: Africa, Asia, Central and South America, Europe, and North America (Canada and USA). There is a significant disparity in terms of the number of studies conducted within each geographical area. Of the 99 studies investigated, a majority (78) have been conducted in Europe (52) and North America (36), while only 11 studies in the remaining three areas (7 in Asia, 2 in Central and South America, and 2 in Africa) (Fig. 5.2). Among the European countries, most of the studies have been conducted in Germany (28), France (7), Spain (5) and the UK (4). The studies have been further divided as research studies (75) and survey-reviews (23).

The research studies have been subsequently divided into four groups based on the potential source of contamination. Three point-source contaminations (landfill, septic tanks, and wastewater), and one diffuse source (agricultural practices and wastes) have been identified. A total of 56 studies have been related to wastewater, 12 to agricultural practices, 4 to landfills and 4 to septic tanks. The frequency of the four most common sources of contamination varies among the different regions (Fig. 5.3).

In Africa, the only two studies available are a survey conducted in Zambia (Sorensen et al. 2014) and a field study conducted in Tunisia (Cary et al. 2013); of the 2 studies available for Central and South America, one is a survey conducted in Mexico (Félix-Cañedo et al. 2013) while the other was conducted in Brazil and has been related to the detection of veterinary antibiotics and hormones in groundwater from the application of a pig slurry (Pinheiro et al. 2013). Among the 7 studies investigating the occurrence of pharmaceuticals in Asia, 5 are related to wastewater and 2 to agricultural practices. Wastewater has been the primary source of contamination investigated in Europe. Among the 52 studies conducted in Europe, 38 focused on wastewater, 13 on survey-review, and only one is related to leakage from a landfill. In North America, the distribution of 36 for the four major sources of contamination include 12 relating to wastewater, 9 to agricultural practices and wastes, 8 to surveys-reviews, 4 to the investigation of leaching from septic tanks and 3 to leaching from a landfill.

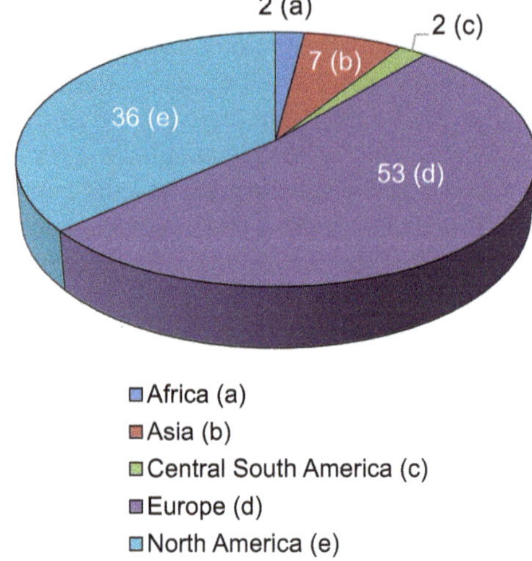

Fig. 5.2 Distribution of the 99 studies reporting the presence of pharmaceutical compounds, life-style compounds and endocrine disruptor compounds among the five geographical areas identified

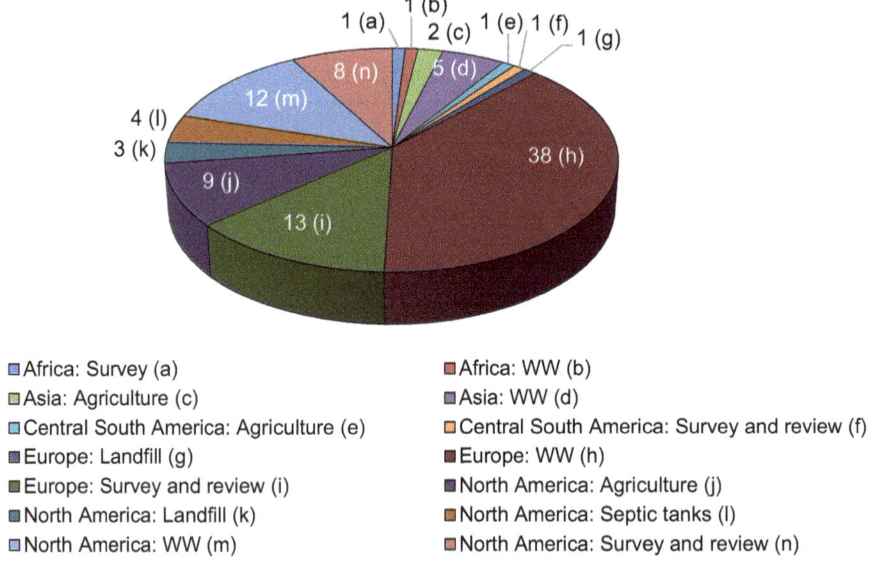

Fig. 5.3 Distribution of studies among the five geographical areas identified based on the potential source of contamination [*WW* = *wastewater*]

4 Pharmaceutical Compounds Investigated and Their Occurrence Across the World

In the context of this chapter, pharmaceuticals have been divided into three groups: pharmaceutical, life-style and endocrine disruptor compounds. Due to the extensive literature available, pesticides and industrial additives and by-products are not included.

Pharmaceutical compounds include a large group of compounds such as antibiotic, analgesic and anti-inflammatory, anticonvulsant, cardiovascular drug, and insect repellent. These compounds have been detected in groundwater at concentrations up to a few μg/L (Barnes et al. 2008 Watanabe et al. 2008; Fang et al. 2012; Jiang et al. 2014).

The life-style group primarily includes compounds acting on the central nervous system (CNS stimulant) and artificial sweeteners. Caffeine and cotinine, breakdown product of nicotine, represent the most common life-style compounds and they have been frequently detected in wastewater effluents, and consequently in groundwater (Seiler et al. 1999; Nakada et al. 2008; Fram and Belitz 2011; Reh et al. 2013; Stuart et al. 2014). Artificial sweeteners, common in developed countries, can be used as a tracer of human wastewater (Engelhardt et al. 2011; Wolf et al. 2012), including in urban settings with complex hydrology (Van Stempvoort et al. 2011).

The endocrine disruptor group occurs naturally and are synthetically produced for use in oral contraceptives and hormonal therapy as well as human antibiotics and in veterinary treatment of domestic and farm animals (Wise et al. 2011). Estrone, the breakdown product of 17β–estradiol, has been frequently identified (Kolodziej et al. 2004; Loos et al. 2010; Vulliet and Cren-Olivé 2011).

The pharmaceutical group represents the largest group of compounds detected in groundwater around the world, followed by life-style and endocrine disruptors (Fig. 5.4).

4.1 *Africa*

As shown in Fig. 5.2 there is a lack of studies investigating the occurrence of pharmaceutical compounds in groundwater in Africa. Few studies have investigated the occurrence of trace elements (Chinyem 2013), bacteria (Bello et al. 2013a, b) and pesticides (Achagra et al. 2013; Fekkoul et al. 2013) in groundwater. The only studies investigating the occurrence of emerging contaminants have been conducted in a monitoring study in Zambia (Sorensen et al. 2014) and in a field study in Tunisia (Cary et al. 2013). Two pharmaceutical compounds, triclosan (antiseptic) and *N*,*N*-Diethyl-*m*-toluamide (DEET) (insect repellent), and one life-style compounds (caffeine) were identified in Zambia (Sorensen et al. 2014), while carbamazepine was identified in Tunisia (Cary et al. 2013) (Fig. 5.5).

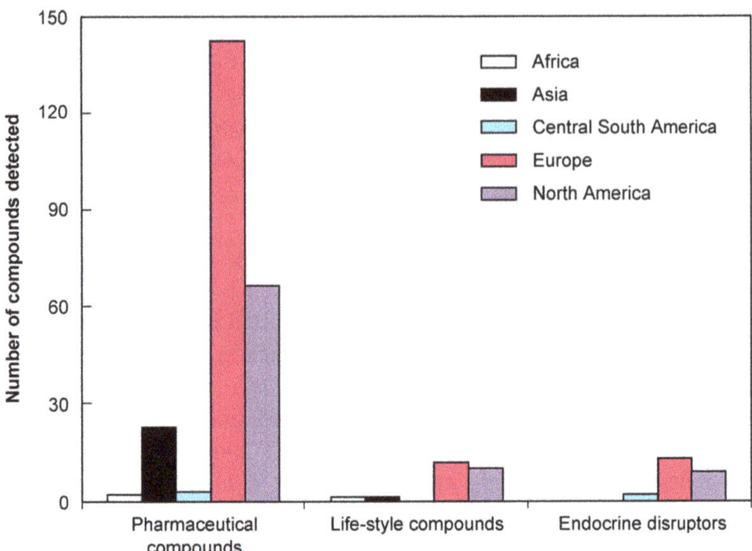

Fig. 5.4 Distribution of pharmaceutical compounds, life-style compounds, and endocrine disruptors in ground water samples in the five geographical areas investigated

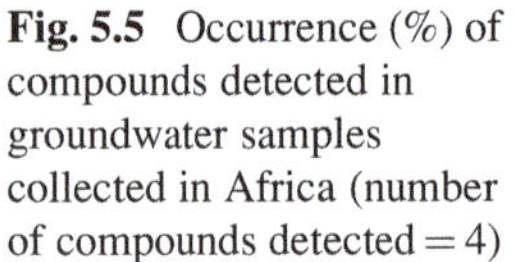

Fig. 5.5 Occurrence (%) of compounds detected in groundwater samples collected in Africa (number of compounds detected = 4)

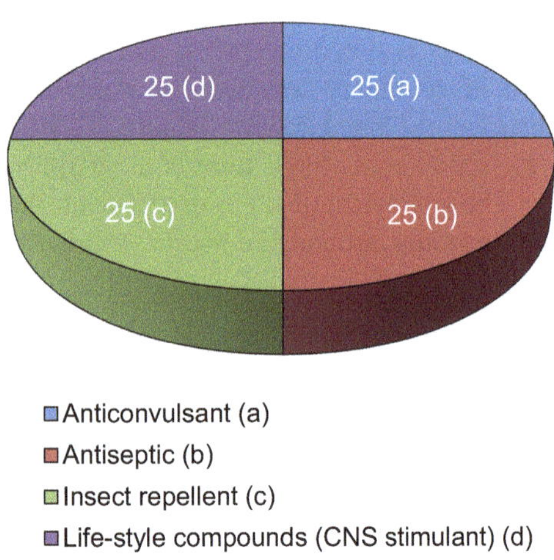

Among the three compounds identified in Zambia, DEET has been detected more frequently (37 times) and at higher concentration (≤1.8 ng/L) compared to triclosan (6 times, ≤0.03 ng/L) and caffeine (once, 0.17 ng/L). This study has also revealed the absence of other life-style compounds, and pharmaceutical compounds that are commonly detected in developing countries (Sorensen et al. 2014). In Tunisia, the occurrence and variability of carbamazepine in groundwater (ranging between below the analytical detection limit to 910 ng/L) was related to the quality of the treated wastewater impacting the field site (Cary et al. 2013).

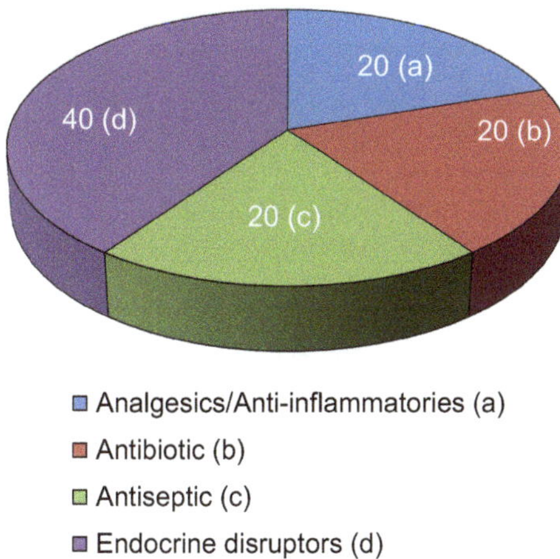

Fig. 5.6 Occurrence (%) of compounds detected in groundwater samples collected in Central and South America (number of compounds identified = 5)

4.2 Central and South America

Similar to Africa, a limited number of studies are also available for Central and South America. Five compounds (three pharmaceutical and two endocrine disruptors) have been identified in the two studies conducted in Mexico (Félix-Cañedo et al. 2013) and in Brazil (Pinheiro et al. 2013) (Fig. 5.6; Table 5.1).

Salicyclic acid (analgesic/anti-inflammatory), sulfadimidine (antibiotic), and triclosan (antibiotic) are the three pharmaceutical compounds detected (Félix-Cañedo et al. 2013; Pinheiro et al. 2013), while estrone and 17β-estradiol are the two endocrine disruptors identified (Pinheiro et al. 2013) (Table 5.1). The occurrence of the two endocrine disruptors as well as of sulfadimidine has been related to the application of pig slurry to the soil (Pinheiro et al. 2013). Salicylic acid and triclosan have been commonly detected at concentrations ranging between 1 and 464 ng/L in a survey conducted in Mexico City (Félix-Cañedo et al. 2013).

4.3 Asia

Pharmaceutical compounds were reported only in the groundwater of the Asian countries China, Japan, and Israel. Twenty-four compounds were identified, which include 23 pharmaceutical and one life-style compound (caffeine). The current literature does not report the occurrence of endocrine disruptor compounds (Fig. 5.7).

Antibiotic (16) and analgesic/anti-inflammatory (3) represent the two most common detected classes of pharmaceutical compounds. In particular, the most common compounds identified include (Fig. 5.8) two antibiotics, sulfamethoxazole (Hu et al. 2010; Zhou et al. 2013; Jiang et al. 2014), and tetracycline (Hu et al. 2010;

Table 5.1 Compounds, with relative pharmacological class, source, number of studies conducted and reference, detected in groundwater in Central and South America

Compound	Class	References	Source	Number of studies
Pharmaceuticals				
Salicylic acid	Analgesics/anti-inflammatories byproducts	Félix-Cañedo et al. (2013)	Review	1
Sulfadimidine	Antibiotic	Pinheiro et al. (2013)	Manure	1
Triclosan	Antiseptic	Félix-Cañedo et al. (2013)	Review	1
Endocrine disruptors				
Estrone		Pinheiro et al. (2013)	Manure	1
17b-Estradiol		Pinheiro et al. (2013)	Manure	1

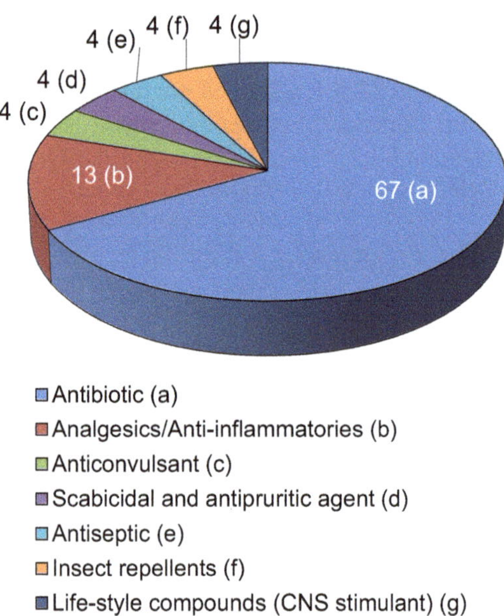

Fig. 5.7 Occurrence (%) of compounds detected in groundwater samples collected in Asia (number of compounds identified = 24)

Chen et al. 2011; Jiang et al. 2014), and one anticonvulsant, carbamazepine (Nakada et al. 2008; Gasser et al. 2010; Kuroda et al. 2012).

The majority of the compounds (14) have been reported in a single study. Wastewater and agricultural practices are the only two sources of contamination investigated. The occurrence of seven antibiotics has been related to agricultural practices and wastewater, while the occurrence of the remaining compounds has only been related to wastewater (Table 5.2). Tetracyclines, including oxytetracycline, tetracycline, and chlortetracycline, and crotamiton have showed the highest

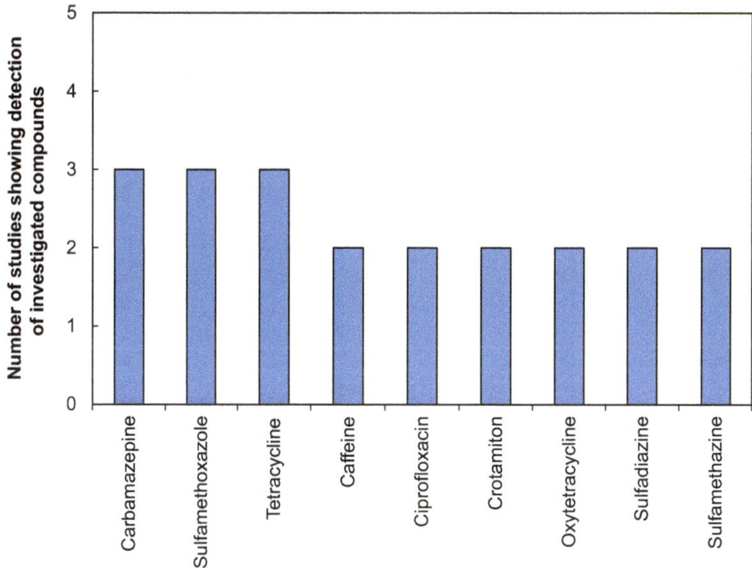

Fig. 5.8 Most common compounds detected in groundwater in Asia

concentrations in groundwater (ranging between 1 and 47 μg/L) (Kuroda et al. 2012; Jiang et al. 2014). The remaining compounds have been detected at lower concentrations (<200 ng/L) (Nakada et al. 2008; Chen et al. 2011; Kuroda et al. 2012; Zhou et al. 2013; Jiang et al. 2014).

4.4 North America

In North America, 86 compounds have been reported. Pharmaceutical compounds are the most abundant (67), followed by life-style compounds (10), and endocrine disruptors (9). Among the pharmaceutical compounds, antibiotic (18), analgesic/anti-inflammatory (12), anticonvulsant (12), and compounds acting as central nervous system depressant (8) represent the most common groups (Fig. 5.9).

Sulfamethoxazole (antibiotic) has been detected ten times, carbamazepine (anticonvulsant), and caffeine (life-style compound) have been detected nine times, and represent the most common compounds detected in groundwater in North America (Fig. 5.10).

Wastewater and agricultural practices represent the two most common sources of groundwater contamination in North America. Among the 86 compounds detected, 5 (ibuprofen, paracetamol, triclosan, caffeine and cotinine) share 3 sources of contaminations, 14 share two sources of contamination, and the remaining 67 has been related to one single source of contamination (Table 5.3). None of the compounds detected share the four sources of contaminations.

Table 5.2 Compounds, with relative pharmacological class, source, number of studies conducted and reference, detected in groundwater in Asia

Compound	Class	References	Source	Number of studies
Pharmaceuticals				
Sulfamethoxazole	Antibiotic	Hu et al. (2010), Jiang et al. (2014), Zhou et al. (2013)	Manure, WW	3
Tetracycline	Antibiotic	Hu et al. (2010), Chen et al. (2011), Jiang et al. (2014)	Manure, WW	3
Trimethoprim	Antibiotic	Chen et al. (2011), Jiang et al. (2014)	Manure, WW	2
Ciprofloxacin	Antibiotic	Hu et al. (2010), Jiang et al. (2014)	Manure, WW	2
Oxytetracycline	Antibiotic	Chen et al. (2011), Jiang et al. (2014)	Manure, WW	2
Sulfadiazine	Antibiotic	Chen et al. (2011), Jiang et al. (2014)	Manure, WW	2
Sulfamethazine	Antibiotic	Chen et al. (2011), Zhou et al. (2013)	Manure, WW	2
Norfloxacin	Antibiotic	Jiang et al. (2014)	WW	1
Chloramphenicol	Antibiotic	Hu et al. (2010)	Manure	1
Chlortetracycline	Antibiotic	Jiang et al. (2014)	WW	1
Enrofloxacin	Antibiotic	Jiang et al. (2014)	WW	1
Lincomycin	Antibiotic	Hu et al. (2010)	Manure	1
Ofloxacin	Antibiotic	Jiang et al. (2014)	WW	1
Roxithromycin	Antibiotic	Jiang et al. (2014)	WW	1
Sulfamonomethoxine	Antibiotic	Zhou et al. (2013)	Manure	1
Sulfadoxine	Antibiotic	Hu et al. (2010)	Manure	1
Propyphenazone	Analgesics/anti-inflammatories	Kuroda et al. (2012)	WW	1
Ethenzamide	Analgesics/anti-inflammatories	Kuroda et al. (2012)	WW	1
Salicylic acid	Analgesics/anti-inflammatories byproducts	Chen et al. (2011)	WW	1
Carbamazepine	Anticonvulsant	Nakada et al. (2008), Kuroda et al. (2012), Gasser et al. (2010)	WW	3
Crotamiton	Scabicidal and antipruritic agent	Nakada et al. (2008), Kuroda et al. (2012)	WW	2
Triclosan	Antiseptic	Chen et al. (2011)	WW	1
Diethyltoluamide	Insect repellents	Kuroda et al. (2012)	WW	1
Life-style compounds				
Caffeine	CNS stimulant	Nakada et al. (2008), Kuroda et al. (2012)	WW	2

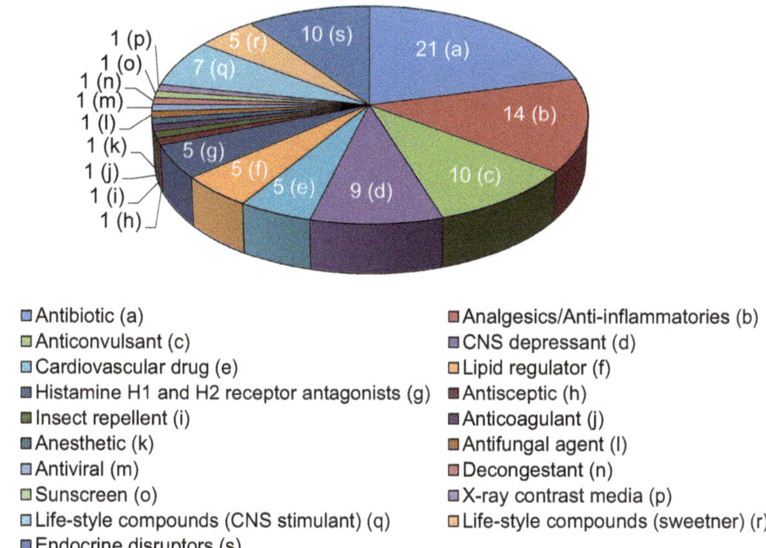

Fig. 5.9 Occurrence (%) of compounds detected in groundwater samples collected in North America (number of compounds identified = 86)

Concentrations of detected compounds range between ng/L and μg/L (Snyder et al. 2004; Barnes et al. 2008; Gottschall et al. 2011; Fang et al. 2012). Gemfibrozil, ibuprofen, sulfamethazine, lincomycin, and acetaminophen show the highest concentrations among the identified compounds ranging between 1.9 and 6.8 μg/L (Barnes et al. 2008; Watanabe et al. 2010; Fang et al. 2012; Deo and Halden 2013).

4.5 *Europe*

Europe provides the largest number of compounds detected in groundwater. Among the 169 compounds detected, pharmaceuticals are the most abundant (144), followed by endocrine disruptors (13), and life-style (12). Among the pharmaceutical compounds, antibiotic (37), analgesic/anti-inflammatory (24), anticonvulsant (18), and the cardiovascular drug (17) represents the most common groups (Fig. 5.11).

An anticonvulsant (carbamazepine), an antibiotic (sulfamethoxazole), and two analgesic/anti-inflammatory drug (diclofenac and ibuprofen) were the most commonly detected compounds in groundwater in Europe (Fig. 5.12).

Wastewater represented the major source of contamination in Europe. Among the 52 studies investigated, only one (Holm et al. 1995) studied the occurrence of these compounds in groundwater due to the possible leaching from a landfill (Table 5.4).

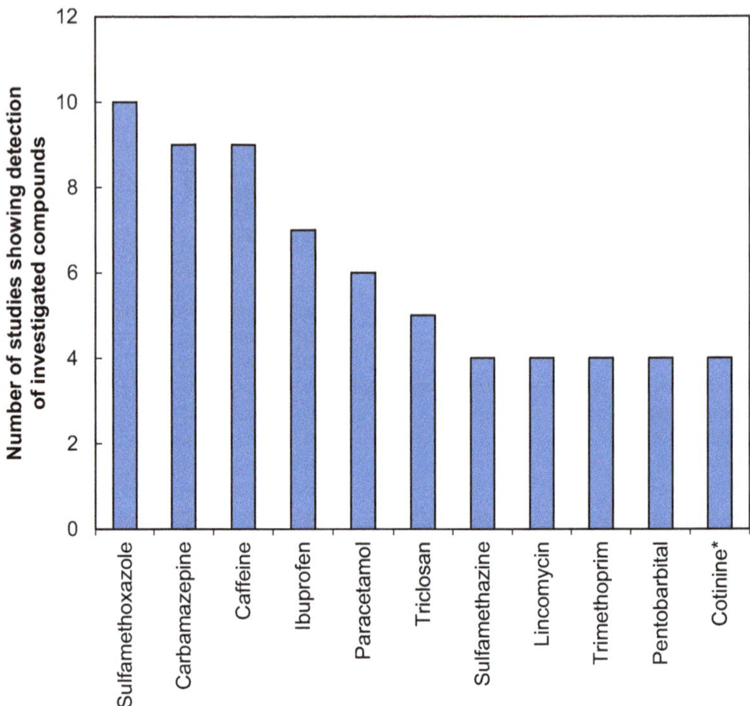

Fig. 5.10 Most common compounds detected in groundwater in North America [*: breakdown product]

In most of the studies, the concentration of the pharmaceutical compounds, as well as life-style and endocrine disruptor, were found to be below a few ng/L (Rabiet et al. 2006; Maskaoui and Zhou 2010; Wolf et al. 2012; López-Serna et al. 2013). However, concentrations higher than 1 μg/L were often detected in groundwater (Kreuzinger et al. 2004; Wolf et al. 2012; Reh et al. 2013; Meffe and de Bustamante 2014).

5 Conclusion

The occurrence of pharmaceutical life-style, and endocrine disruptor compounds has been reported during the past two decades around the world. There is a clear discrepancy regarding the number of studies conducted to investigate the presence of these compounds in groundwater among the five different geographical regions as well as the potential sources of contamination. The majority of the studies have been conducted in Europe and only relate to wastewater. In North America, the studies are more distributed among the four potential sources of contamination. In Asia, the only studies available are conducted in China, Japan, and Israel, and are

Table 5.3 Compounds, with relative pharmacological class, source, number of studies conducted and reference, detected in groundwater in North America

Compound	Class	References	Source	Number of studies
Pharmaceuticals				
Sulfamethoxazole	Antibiotic	Lindsey et al. (2001), Hinkle et al. (2005), Verstraeten et al. (2005), Miller and Meek (2006), Barnes et al. (2008), Godfrey et al. (2009), Watanabe et al. (2010), Fram and Belitz (2011), Bradley et al. (2014), Schaider et al. (2014)	Review, WW, septic tanks	10
Sulfamethazine	Antibiotic	Batt et al. (2006), Barnes et al. (2008), Watanabe et al. (2010), Bartelt-Hunt et al. (2011)	Manure, review	4
Lincomycin	Antibiotic	Barnes et al. (2004), Barnes et al. (2008), Watanabe et al. (2010), Bartelt-Hunt et al. (2011)	Landfill, review, manure	4
Trimethoprim	Antibiotic	Verstraeten et al. (2005), Fram and Belitz (2011), Bradley et al. (2014), Schaider et al. (2014)	Septic tanks, WW, review	4
Erythromycin	Antibiotic	Verstraeten et al. (2005), Bartelt-Hunt et al. (2011)	Manure, septic tanks	2
Monensin	Antibiotic	Watanabe et al. (2008), Bartelt-Hunt et al. (2011)	Manure	2
Oxytetracycline	Antibiotic	Krapac et al. (2004), Mackie et al. (2006)	Manure	2
Sulfadimethoxine	Antibiotic	Batt et al. (2006), Watanabe et al. (2010)	Manure	2
Tetracycline	Antibiotic	Krapac et al. (2004), Monteiro and Boxall (2010)	Manure, review	2
Anhydrotetracycline	Antibiotic	Krapac et al. (2004)	Manure	1
Anhydrochlortetracycline	Antibiotic	Krapac et al. (2004)	Manure	1
Beta-Apooxytetracycline	Antibiotic	Krapac et al. (2004)	Manure	1
Ciprofloxacin	Antibiotic	Karnjanapiboonwong et al. (2011)	WW	1
Sulfamerazine	Antibiotic	Bartelt-Hunt et al. (2011)	Manure	1
Sulfamethizole	Antibiotic	Schaider et al. (2014)	Septic tanks	1
Sulfathiazole	Antibiotic	Bartelt-Hunt et al. (2011)	Manure	1

(continued)

Table 5.3 (continued)

Compound	Class	References	Source	Number of studies
Tylosin	Antibiotic	Watanabe et al. (2010)	Manure	1
Triclocarban	Antibiotic	Gottschall et al. (2011)	Review	1
Ibuprofen	Analgesics/anti-inflammatories	Drewes et al. (2003), Verstraeten et al. (2005), Miller and Meek (2006), Carrara et al. (2008), Barnes et al. (2008), Buszka et al. (2009), Gottschall et al. (2011)	Septic tanks, WW, review, landfill	7
Paracetamol	Analgesics/anti-inflammatories	Snyder et al. (2004), Hinkle et al. (2005), Verstraeten et al. (2005), Barnes et al. (2008), Buszka et al. (2009), Katz et al. (2009)	Review, WW, septic tanks, landfill	6
Diclofenac	Analgesics/anti-inflammatories	Drewes et al. (2002), Miller and Meek (2006), Carrara et al. (2008)	Septic tanks, WW, review	3
Compound	*Class*	*References*	*Source*	*Number of studies*
Naproxen	Analgesics/anti-inflammatories	Drewes et al. (2003), Miller and Meek (2006), Carrara et al. (2008)	WW, septic tanks, review	3
Antipyrine	Analgesics/anti-inflammatories	Bradley et al. (2014), Schaider et al. (2014)	Septic tanks, WW	2
Acetaminophen	Analgesics/anti-inflammatories	Fram and Belitz (2011), Bradley et al. (2014)	Review, WW	2
Codeine	Analgesics/anti-inflammatories	Verstraeten et al. (2005), Fram and Belitz (2011)	Review, septic tanks	2
Indomethacin	Analgesics/anti-inflammatories	Carrara et al. (2008)	Septic tanks	1
Ketoprofen	Analgesics/anti-inflammatories	Carrara et al. (2008)	Septic tanks	1
Propyphenazone	Analgesics/anti-inflammatories	Drewes et al. (2003)	WW	1
Salicylic acid	Analgesics/anti-inflammatories byproducts	Carrara et al. (2008)	Septic tanks	1
Tramadol	Analgesics/anti-inflammatories	Bradley et al. (2014)	WW	1

(continued)

Table 5.3 (continued)

Compound	Class	References	Source	Number of studies
Carbamazepine	Anticonvulsant	Seiler et al. (1999), Drewes et al. (2003), Snyder et al. (2004), Miller and Meek (2006), Katz et al. (2009), Godfrey et al. (2009), Fram and Belitz (2011), Bradley et al. (2014), Schaider et al. (2014)	WW; review; septic tanks	9
Phenytoin	Anticonvulsalnt	Miller and Meek (2006), Bradley et al. (2014), Schaider et al. (2014)	Review, WW, septic tanks	3
Fluoxetine	Anticonvulsant	Barnes et al. (2008), Katz et al. (2009)	WW, review	2
Bupropion	Anticonvulsant	Bradley et al. (2014)	WW	1
Citalopram	Anticonvulsant	Bradley et al. (2014)	WW	1
Desvenlafaxine	Anticonvulsant byproduct	Bradley et al. (2014)	WW	1
Diazepam	Anticonvulsant	Bradley et al. (2014)	WW	1
O-desmethyl venlafaxine	Anticonvulsant	Gottschall et al. (2011)		1
Venlafaxine	Anticonvulsant	Bradley et al. (2014)	WW	1
Carisoprodol	CNS depressant	Bradley et al. (2014)	WW	1
Meprobamate	CNS depressant	Snyder et al. (2004), Barnes et al. (2008), Schaider et al. (2014), Bradley et al. (2014)	Review; septic tanks	3
Methocarbamol	CNS depressant	Bradley et al. (2014)	WW	1
Metaxalone	CNS depressant	Bradley et al. (2014)	WW	1
Pentobarbital	CNS depressant	Eckel et al. (1993)	Landfill	4
Primidone	CNS depressant	Drewes et al. (2003), Hinkle et al. (2005), Zhao et al. (2011)	WW	3
Promethazine	CNS depressant	Bradley et al. (2014)	WW	1
Temazepam	CNS depressant	Bradley et al. (2014)	WW	1
Atenolol	Cardiovascular drug	Bradley et al. (2014), Schaider et al. (2014)	Septic tanks, WW	2
Compound	*Class*	*References*	*Source*	*Number of studies*
Dehydronifedipine	Cardiovascular drug	Verstraeten et al. (2005), Barnes et al. (2008)	Septic tanks; review	1

(continued)

Table 5.3 (continued)

Compound	Class	References	Source	Number of studies
Diltiazem	Cardiovascular drug	Barnes et al. (2008)	Review	1
Metoprolol	Cardiovascular drug	Bradley et al. (2014)	WW	1
Gemfibrozil	Lipid regulating	Carrara et al. (2008), Fang et al. (2012), Schaider et al. (2014)	Septic tanks; review	3
Glyburide	Lipid regulator	Bradley et al. (2014)	WW	1
Metformin	Lipid regulator	Bradley et al. (2014)	WW	1
Clofibric acid	Lipid regulator	Drewes et al. (2002)	WW	1
Ranitidine	Histamine H1 and H2 receptor antagonists	Bradley et al. (2014)	WW	1
Diphenhydramine	Histamine H1 and H2 receptor antagonists	Bradley et al. (2014)	WW	1
Famotidine	Histamine H1 and H2 receptor antagonists	Bradley et al. (2014)	WW	1
Fexofenadine	Histamine H1 and H2 receptor antagonists	Bradley et al. (2014)	WW	1
Triclosan	Antiseptic	Barnes et al. (2004), Godfrey et al. (2009), Carrara et al. (2008), Fram and Belitz (2011), Karnjanapiboonwong et al. (2011)	Septic tanks, landfill, WW, review	5
N,N-Diethyl-*m*-toluamide (DEET)	Insect repellent	Hinkle et al. (2005), Barnes et al. (2008)	WW, review	2
Warfarin	Anticoagulant	Verstraeten et al. (2005), Bradley et al. (2014)	Septic tanks, WW	2
Lidocaine	Anesthetic	Bradley et al. (2014)	WW	1
Fluconazole	Antifungal agent	Bradley et al. (2014)	WW	1
Acyclovir	Antiviral	Bradley et al. (2014)	WW	1
Pseudoephedrine	Decongestant	Bradley et al. (2014)	WW	1
Oxybenzone	Sunscreen	Snyder et al. (2004), Miller and Meek (2006)	Review	2
Life-style compounds				
Caffeine	CNS stimulant	Seiler et al. (1999), Hinkle et al. (2005), Verstraeten et al. (2005), Swartz et al. (2006), Barnes et al. (2008), Buszka et al. (2009), Katz et al. (2009), Fram and Belitz (2011), Bradley et al. (2014)	WW, landfill, review, septic tanks	9

(continued)

Table 5.3 (continued)

Compound	Class	References	Source	Number of studies
Cotinine	CNS stimulant	Barnes et al. (2004), Verstraeten et al. (2005), Buszka et al. (2009), Bradley et al. (2014)	Landfill, septic tanks, WW	4
1,7-dimethylxanthine	CNS stimulant byproducts	Verstraeten et al. (2005), Barnes et al. (2008), Buszka et al. (2009)	Septic tanks, review, landfill	3
Nicotine	CNS stimulant	Godfrey et al. (2009), Bradley et al. (2014)	WW	2
Paraxanthine	CNS stimulant	Swartz et al. (2006)	Septic tanks	1
p-Xanthine	CNS stimulant byproduct	Fram and Belitz (2011)	Review	1
Compound	*Class*	*References*	*Source*	*Number of studies*
Acesulfame	Sweetener	Van Stempvoort et al. (2011)	WW	1
Saccharin	Sweetener	Van Stempvoort et al. (2011)	WW	1
Sucralose	Sweetener	Van Stempvoort et al. (2011)	WW	1
Cyclamate	Sweetner	Van Stempvoort et al. (2011)	WW	1
Endocrine disruptors				
Estrone		Kolodziej et al. (2004), Swartz et al. (2006)	Septic tanks, manure	2
17b-Estradiol		Mansell and Drewes (2004), Swartz et al. (2006)	WW, septic tanks	2
Cholesterol		Barnes et al. (2004), Barnes et al. (2008)	Landfill, review	2
Coprostanol		Barnes et al. (2004), Barnes et al. (2008)	Landfill, review	2
Oestriol		Miller and Meek (2006)	Review	1
Oestrone		Miller and Meek (2006)	Review	1
Testosterone		Kolodziej et al. (2004)	Manure	1
Stigmastanol		Barnes et al. (2004)	Landfill	1
17-α-ethinylestradiol		Karnjanapiboonwong et al. (2011)	WW	1

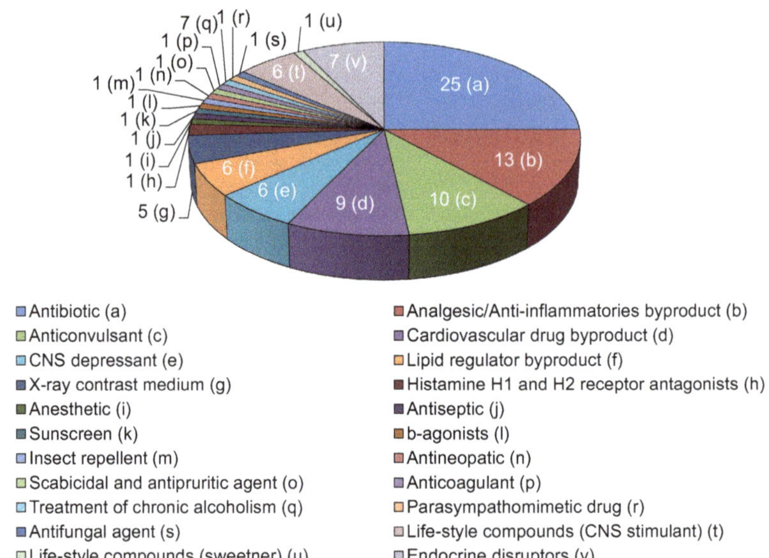

Fig. 5.11 Occurrence (%) of compounds detected in groundwater samples collected in Europe (number of compounds identified = 168)

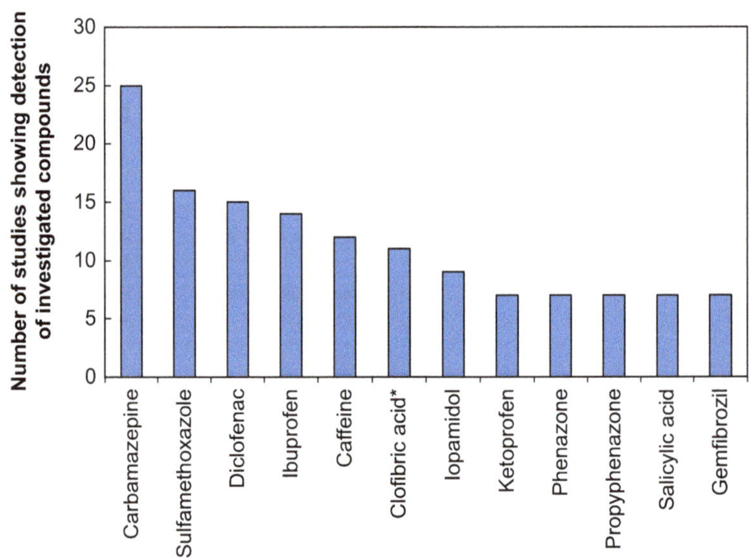

Fig. 5.12 Most common compounds detected in groundwater in Europe [*: breakdown product]

focused on wastewater and agricultural practices. No studies, and consequently no data are available for many developing Asian countries, in which poor waste disposal practices occur (AECOM 2010), leading to a high risk of groundwater

Table 5.4 Compounds, with relative pharmacological class, source, number of studies conducted and reference, detected in groundwater in Europe

Compound	Class	References	Source	Number of studies
Pharmaceuticals				
Sulfamethoxazole	Antibiotic	Hirsch et al. (1999), Sacher et al. (2001), Schmidt et al. (2003), Bruchet et al. (2005), Ternes et al. (2007), Tracol and Duchemin (2009), Loos et al. (2010), NAQUA (2009), Maskaoui and Zhou (2010), Teijon et al. (2010), Vulliet and Cren Olivé (2011), de Jesus Gaffney et al. (2015), Cabeza et al. (2012), Reh et al. (2013), López-Serna et al. (2013), Meffe and de Bustamante (2014)	WW, review	16
Sulfamethazine	Antibiotic	Hirsch et al. (1999), NAQUA (2009), Teijon et al. (2010), Cabeza et al. (2012), López-Serna et al. (2013), de Jesus Gaffney et al. (2015)	Review, WW	6
Erythromycin	Antibiotic	Sacher et al. (2001), Cabeza et al. (2012), López-Serna et al. (2013), Reh et al. (2013), de Jesus Gaffney et al. (2015)	Review, WW	5

(continued)

Table 5.4 (continued)

Compound	Class	References	Source	Number of studies
Ofloxacin	Antibiotic	Teijon et al. (2010), Cabeza et al. (2012), López-Serna et al. (2013)	WW	3
Roxithromycin	Antibiotic	Vulliet and Cren Olivé (2011), López-Serna et al. (2013), Reh et al. (2013)	WW	3
Sulfapyridine	Antibiotic	Teijon et al. (2010), Cabeza et al. (2012), de Jesus Gaffney et al. (2015)	WW, review	3
Trimethoprim	Antibiotic	Vulliet and Cren Olivé (2011), López-Serna et al. (2013), Reh et al. (2013)	WW	3
Ciprofloxacin	Antibiotic	Cabeza et al. (2012), López-Serna et al. (2013)	WW	2
Clarithromycin	Antibiotic	López-Serna et al. (2013), Reh et al. (2013)	WW	2
Doxycycline	Antibiotic	Faille (2010), López-Serna et al. (2013)	Review, WW	2
Enoxacin	Antibiotic	NAQUA (2009), López-Serna et al. (2013)	WW	2
Josamycin	Antibiotic	López-Serna et al. (2013), Meffe and de Bustamante (2014)	WW, review	2
Tylosin	Antibiotic	López-Serna et al. (2013), Meffe and de Bustamante (2014)	WW	2
Anhydroerythromycin	Antibiotic byproduct	López-Serna et al. (2013)	WW	1
Azithromycin	Antibiotics	López-Serna et al. (2013)	WW	1

(continued)

Table 5.4 (continued)

Compound	Class	References	Source	Number of studies
Chlortetracycline	Antibiotic	López-Serna et al. (2013)	WW	1
Danofloxacin	Antibiotic	López-Serna et al. (2013)	WW	1
Dimetridazole	Antibiotic	Stuart et al. (2011)	Review	1
N-acetyl-sulfadiazine	Antibiotic byproduct	López-Serna et al. (2013)	WW	1
Desamino-SMX	Antibiotic	Reh et al. (2013)	WW	1
Enrofloxacin	Antibiotic	López-Serna et al. (2013)	WW	1
Flumequine	Antibiotic	López-Serna et al. (2013)	WW	1
Metronidazole	Antibiotic	López-Serna et al. (2013)	WW	1
Nifuroxazide	Antibiotic	Estevez et al. (2012)	WW	1
N-acetyl-sulfamethoxazole	Antibiotic byproduct	López-Serna et al. (2013)	WW	1
Norfloxacin	Antibiotic	NAQUA (2009), López-Serna et al. (2013)	WW	2
Oxytetracycline	Antibiotic	López-Serna et al. (2013)	WW	1
Albendazole	Antibiotic	Petrovíc et al. (2014)	Review	1
Compound	*Class*	*References*	*Source*	*Number of studies*
Spiramycin	Antibiotic	López-Serna et al. (2013)	WW'	1
Sulfadiazine	Sulfonamide antibiotic	NAQUA (2009), López-Serna et al. (2013), de Jesus Gaffney et al. (2015)	WW, review	3
Sulfadimethoxine	Antibiotic	Meffe and de Bustamante (2014)	WW	1
Sulfamethizole	Antibiotic	Meffe and de Bustamante (2014)	WW	1
Sulfanilamide	Antibiotic	Holm et al. (1995)	Landfill	1
Sulfanilic acid	Antibiotic	Holm et al. (1995)	Landfill	1
Tetracycline	Antibiotic	López-Serna et al. (2013)	WW	1

(continued)

Table 5.4 (continued)

Compound	Class	References	Source	Number of studies
Tilmicosin	Antibiotic	López-Serna et al. (2013)	WW	1
4-Nitro-SMX	Antibiotic	Reh et al. (2013)	WW	1
Diclofenac	Analgesics/anti-inflammatories	Sacher et al. (2001), Heberer (2002), Heberer et al. (2004), Ellis (2006), Rabiet et al. (2006), Einsiedl et al. (2010), Faille (2010), Loos et al. (2010), Maskaoui and Zhou (2010), Teijon et al. (2010), Vulliet and Cren Olivé (2011), Cabeza et al. (2012), Wolf et al. (2012), López-Serna et al. (2013), Reh et al. (2013)	Review, WW	15
Ibuprofen	Analgesics/anti-inflammatories	Heberer et al. (1997), Heberer (2002), Rabiet et al. (2006), Ellis (2006), Einsiedl et al. (2010), Faille (2010), Loos et al. (2010), Makris and Snyder (2010), Teijon et al. (2010), Stuart et al. (2011), Wolf et al. (2012), López-Serna et al. (2013), Reh et al. (2013), Petrovíc et al. (2014)	Review, WW	14

(continued)

Table 5.4 (continued)

Compound	Class	References	Source	Number of studies
Ketoprofen	Analgesics/anti-inflammatories	Heberer et al. (1997), Heberer (2002), Rabiet et al. (2006), Faille (2010), Loos et al. (2010), Vulliet and Cren Olivé (2011), López-Serna et al. (2013)	Review, WW	7
Phenazone	Analgesics/anti-inflammatories	Heberer et al. (1997), Sacher et al. (2001), Reddersen et al. (2002), Maeng et al. (2010), López-Serna et al. (2013), Reh et al. (2013), Petrovíc et al. (2014)	WW, Spill	7
Propyphenazone	Analgesics/anti-inflammatories	Heberer (2002), Reddersen et al. (2002), Heberer et al. (2004), Maeng et al. (2010), Cabeza et al. (2012), López-Serna et al. (2013), Petrovíc et al. (2014)	Spill, WW	7
Salicylic acid	Analgesics/anti-inflammatories byproducts	Heberer (2002), Togola and Budzinski (2007), Faille (2010), Vulliet and Cren Olivé (2011), Cabeza et al. (2012), López-Serna et al. (2013), Petrovíc et al. (2014)	Review; WW	7

(continued)

Table 5.4 (continued)

Compound	Class	References	Source	Number of studies
Naproxen	Analgesics/anti-inflammatories	Rabiet et al. (2006), Teijon et al. (2010), Vulliet and Cren Olivé (2011), Cabeza et al. (2012), López-Serna et al. (2013), Petrovíc et al. (2014)	Review, WW	6
Acetaminophen	Analgesic—anti-inflammatory	Vulliet and Cren Olivé (2011), López-Serna et al. (2013), de Jesus Gaffney et al. (2015)	Review; WW	3
AMDOPH	Analgesic—anti-inflammatory	Reddersen et al. (2002), Heberer et al. (2004), Maeng et al. (2010)	Spill	3
Paracetamol	Analgesics/anti-inflammatories	Rabiet et al. (2006), Faille (2010), Reh et al. (2013)	Review, WW	3
Acetylsalicylic acid	Analgesics/anti-inflammatories	Rabiet et al. (2006), Tracol and Duchemin (2009)	WW	2
Codeine	Analgesics/anti-inflammatories	Teijon et al. (2010), López-Serna et al. (2013)	WW	2
Indomethacin	Analgesics/anti-inflammatories	Maskaoui and Zhou (2010), López-Serna et al. (2013)	WW	2
Compound	*Class*	*References*	*Source*	*Number of studies*
N-methyl phenacetin	Analgesics/anti-inflammatories	Heberer et al. (1997), Heberer (2002)		2

(continued)

Table 5.4 (continued)

Compound	Class	References	Source	Number of studies
Antipyrine	Analgesics/anti-inflammatories	Cabeza et al. (2012)	WW	1
Atorvastatin	Analgesics/anti-inflammatories	López-Serna et al. (2013)	WW	1
Dimethylaminophenazone	Analgesics/anti-inflammatories	Reddersen et al. (2002)	Spill	1
Gabapentin	Analgesics/anti-inflammatories	Stuart et al. (2014)	Review	1
Mefenamic acid	Analgesics/anti-inflammatories	López-Serna et al. (2013)	WW	1
N-acetyl-4-amino-antipiryne	Analgesics/anti-inflammatories	Teijon et al. (2010), Cabeza et al. (2012)	WW	2
N-formyl-4-amino-antipiryne		Teijon et al. (2010), Cabeza et al. (2012)	WW	2
Phenylbutazone	Analgesics/anti-inflammatories	López-Serna et al. (2013)	WW	1
Pyrazole	Analgesic, anti-inflammatory	Tracol et Duchemin (2009)		1
4OH diclofenac	Analgesics/anti-inflammatories byproduct	López-Serna et al. (2013)	WW	1
Carbamazepine	Anticonvulsant	Sacher et al. (2001), Clara et al. (2004), Heberer et al. (2004), Kreuzinger et al. (2004), Bruchet et al. (2005), Rabiet et al. (2006), Osenbruck et al. (2007), Ternes et al. (2007), Togola and Budzinski (2007), Musolff et al. (2009), NAQUA (2009),	Review, WW	25

(continued)

Table 5.4 (continued)

Compound	Class	References	Source	Number of studies
		Tracol et Duchemin (2009), Faille (2010), Loos et al. (2010), Maskaoui and Zhou (2010), Musolff et al. (2010), Teijon et al. (2010), Stuart et al. (2011), Vulliet and Cren Olivé (2011); Cabeza et al. (2012), Wolf et al. (2012), López-Serna et al. (2013), Reh et al. (2013), Petrovíc et al. (2014), Stuart et al. (2014)		
Primidone	Anticonvulsant	Heberer (2002), Heberer et al. (2004), Cabeza et al. (2012), Hass et al. (2012), Reh et al. (2013)	WW, review	5
Diazepam	Anticonvulsant	Faille (2010), Cabeza et al. (2012), López-Serna et al. (2013), Meffe and de Bustamante (2014)	Review	4
Fluoxetine	Anticonvulsant	López-Serna et al. (2013), Makris and Snyder (2010), Reh et al. (2013)	WW	3

(continued)

Table 5.4 (continued)

Compound	Class	References	Source	Number of studies
Lorazepam	Anticonvulsant	Vulliet and Cren Olivé (2011), López-Serna et al. (2013)	WW	2
Velafaxine	Anticonvulsant	Teijon et al. (2010), Cabeza et al. (2012)	WW	2
Acridone	Anticonvulsant byproduct	López-Serna et al. (2013)	WW	1
Citalopram	Anticonvulsant	Reh et al. (2013)	WW	1
Mirtazapine	Anticonvulsant	Stuart et al. (2011)	Review	1
Desmethyl diazepam	Anticonvulsant byproduct	López-Serna et al. (2013)	WW	1
Compound	*Class*	*References*	*Source*	*Number of studies*
Oxazepam glucuronide	Anticonvulsant byproduct	López-Serna et al. (2013)	WW	1
Paraldehyde	Anticonvulsant	Stuart et al. (2011)	Review	1
Paroxetine	Anticonvulsant	López-Serna et al. (2013)	WW	1
Phenylethylmalonamide	Anticonvulsant drug metabolite	Hass et al. (2012)	WW	1
Trimipramine	Anticonvulsant	Stuart et al. (2011)	Review	1
2OH carbamazepine	Anticonvulsant byproduct	López-Serna et al. (2013)	WW	1
3OH carbamazepine	Anticonvulsant byproduct	López-Serna et al. (2013)	WW	1
Methaqualone	Anticonvulsant	Stuart et al. (2014)	Review	1
Atenolol	Cardiovascular drug	Teijon et al. (2010), Vulliet and Cren Olivé (2011), Cabeza et al. (2012), López-Serna et al. (2013), Reh et al. (2013), de Jesus Gaffney et al. (2015)	Review, WW	6

(continued)

Table 5.4 (continued)

Compound	Class	References	Source	Number of studies
Metoprolol	Cardiovascular drug	Faille (2010), Vulliet and Cren Olivé (2011), López-Serna et al. (2013), Reh et al. (2013)	Review; WW	4
Propranolol	Cardiovascular drug	Maskaoui and Zhou (2010), Vulliet and Cren Olivé (2011), de Jesus Gaffney et al. (2015), Petrovíc et al. (2014)	Review; WW	4
Hydrochlorothiazide	Cardiovascular drug	Teijon et al. (2010), Cabeza et al. (2012), López-Serna et al. (2013)	WW	3
Sotalol	Cardiovascular drug	Sacher et al. (2001), López-Serna et al. (2013), Reh et al. (2013)	WW	3
Carazolol	Cardiovascular drug	López-Serna et al. (2013), Petrovíc et al. (2014)	WW' review	2
Betaxolol	Cardiovascular drug	López-Serna et al. (2013)	WW	1
Disopyramide	Cardiovascular drug	Stuart et al. (2011)	Review	1
Enalapril	Cardiovascular drug	López-Serna et al. (2013)	WW	1
Enalaprilat	Cardiovascular drug	López-Serna et al. (2013)	WW	1
Furosemide	Cardiovascular drug	Tracol et Duchemin (2009), Teijon et al. (2010), Cabeza et al. (2012), López-Serna et al. (2013)	WW	3
Lisinopril	Cardiovascular drug	López-Serna et al. (2013)	WW	1
Timolol	Cardiovascular drug	López-Serna et al. (2013)	WW	1

(continued)

Table 5.4 (continued)

Compound	Class	References	Source	Number of studies
Nadolol	Cardiovascular drug	López-Serna et al. (2013)	WW	1
Pindolol	Cardiovascular drug	López-Serna et al. (2013)	WW	1
Propanolol	Cardiovascular drug	López-Serna et al. (2013)	WW	1
4OH propranolol	Cardiovascular drug byproduct	López-Serna et al. (2013)	WW	1
Chlorzoxazone	CNS depressant	Stuart et al. (2011)	Review	1
Oxazepam	CNS depressant	Togola and Budzinski (2007), Tracol and Duchemin (2009), Vulliet and Cren Olivé (2011)	WW	3
Phenobarbital	CNS depressant	Stuart et al. (2011), Hass et al. (2012), López-Serna et al. (2013)	WW, review	3
Butalbital	CNS depressant	Tracol et Duchemin (2009), López-Serna et al. (2013)	WW	2
Pentobarbital	CNS depressant	Stuart et al. (2011), López-Serna et al. (2013)	WW, review	2
Compound	*Class*	*References*	*Source*	*Number of studies*
Alprazolam	CNS depressant	Tracol et Duchemin (2009)		1
Amobarbital	CNS depressant	Stuart et al. (2011)	Review	1
Bromazepam	CNS depressant	Faille (2010)	Review	1
Haloperidol	CNS depressant	Reh et al. (2013)	WW	1
Meprobamate	CNS depressant	Makris and Snyder (2010)	WW	1
Zolpidem	CNS depressant	Faille (2010)	Review	1

(continued)

Table 5.4 (continued)

Compound	Class	References	Source	Number of studies
Clofibric acid	Lipid regulator	Stan and Linkerhaigner (1992), Stan et al. (1994), Heberer et al. (1997), Heberer (2002), Heberer et al. (2004), Ellis (2006), NAQUA (2009), Wolf et al. (2012), López-Serna et al. (2013), Reh et al. (2013), Meffe and de Bustamante (2014)	Review, WW	11
Gemfibrozil	Lipid regulating	Heberer et al. (1997), Faille (2010), Teijon et al. (2010), Cabeza et al. (2012), Wolf et al. (2012), López-Serna et al. (2013), Reh et al. (2013)	WW, Review	7
Bezafibrate	Lipid regulator	Faille (2010), Cabeza et al. (2012), Wolf et al. (2012), López-Serna et al. (2013), Reh et al. (2013)	Review; WW	5
Fenofibrate	Lipid regulator	Heberer et al. (1997), Heberer (2002), López-Serna et al. (2013)	WW	3
Fenofibric acid	Lipid regulator	Faille (2010), Vulliet and Cren Olivé (2011)	Review	2

(continued)

Table 5.4 (continued)

Compound	Class	References	Source	Number of studies
Glyburide	Lipid regulator	López-Serna et al. (2013)	WW	1
Metformin	Lipid regulator	Vulliet and Cren Olivé (2011)	WW	1
Acridin	Lipid regulator byproduct	López-Serna et al. (2013)	WW	1
Pravastatin	Lipid regulators	López-Serna et al. (2013)	WW	1
2OH atorvast in Acyl-b-D-glucuronide	Lipid regulator byproduct	López-Serna et al. (2013)	WW	1
Iopamidol	X-ray contrast medium	Ternes and Hirsch (2000), Sacher et al. (2001), Bruchet et al. (2005), Ellis (2006), Ternes et al. (2007), NAQUA (2009), Teijon et al. (2010), Cabeza et al. (2012), Wolf et al. (2012)	WW	9
Diatrizoate	X-ray contrast medium	Ternes and Hirsch (2000), Bruchet et al. (2005), Ternes et al. (2007), Engelhardt et al. (2011)	WW, review	4
Iopromide	X-ray contrast medium	Ternes et Hirsch (2000), Schultz et al. (2008), Teijon et al. (2010), Wolf et al. (2012)	WW	4
Amidotrizoic acid	X-ray contrast medium	Sacher et al. (2001), Schmidt et al. (2003), Wolf et al. (2012)	WW	3
Ioxithalamic acid	X-ray contrast medium	Ternes and Hirsch (2000), Bruchet et al. (2005), Wolf et al. (2012)	WW	3

(continued)

Table 5.4 (continued)

Compound	Class	References	Source	Number of studies
Iohexol	X-ray contrast medium	Wolf et al. (2012), Reh et al. (2013)	WW	2
Iotalamic acid	X-ray contrast medium	Ternes et Hirsch (2000), Wolf et al. (2012)	WW	2
Iomeprol	X-ray contrast medium	Wolf et al. (2012)	WW	1
Loratadine	Histamine H1 and H2 receptor antagonists	López-Serna et al. (2013), Reh et al. (2013)	WW	2
Ranitidine	Histamine H1 and H2 receptor antagonists	López-Serna et al. (2013)	WW	1
Compound	*Class*	*References*	*Source*	*Number of studies*
Cimetidine	Histamine H1 and H2 receptor antagonists	López-Serna et al. (2013)	WW	1
Famotidine	Histamine H1 and H2 receptor antagonists	López-Serna et al. (2013)	WW	1
Mepivacaine	Anesthetic	Teijon et al. (2010), Cabeza et al. (2012)	WW	2
Lidocaine	Anesthetic	Stuart et al. (2011)	Review	1
Triclosan	Antiseptic	Loos et al. (2010), Makris and Snyder (2010), Stuart et al. (2011)	Review	3
Benzalkonium chloride	Antiseptic	Estévez et al. (2012)	WW	1
Ethylhexyl methoxycinnamate	Sunscreen	Teijon et al. (2010), Cabeza et al. (2012)	WW	2
Oxybenzone	Sunscreen	Stuart et al. (2011)	Review	1
Albuterol	β-agonists	López-Serna et al. (2013)	WW	1
Clenbuterol	β-agonists	López-Serna et al. (2013)	WW	1
N,*N*-Diethyl-*m*-toluamide (DEET)	Insect repellent	Loos et al. (2010), Stuart et al. (2011), Stuart et al. (2014)	Review	3

(continued)

Table 5.4 (continued)

Compound	Class	References	Source	Number of studies
Tamoxifen	Antineoplatic	López-Serna et al. (2013), Reh et al. (2013)	WW	2
Crotamiton	Scabicidal and antipruritic agent	Stuart et al. (2011), Stuart et al. (2014)	Review	2
Coumadin	Anticoagulant	Stuart et al. (2011)	Review	1
Disulfiram	Treatment of chronic alcoholism	Stuart et al. (2011)	Review	1
Diethyl p-nitrophenyl phosphate (Paraoxon)	Parasympathomimetic drug	Stuart et al. (2011)	Review	1
Clotrimazole	Antifungal mediacation	Faille (2010)	Review	1
Life-style compounds				
Caffeine	CNS stimulant	Rabiet et al. (2006), Musolff et al. (2009), Loos et al. (2010), Makris and Snyder (2010), Musolff et al. (2010), Teijon et al. (2010), Stuart et al. (2011), Cabeza et al. (2012), Estévez et al. (2012), Reh et al. (2013), de Jesus Gaffney et al. (2015), Stuart et al. (2014)	WW, review	12
Nicotine	CNS stimulant	Teijon et al. (2010), Stuart et al. (2011), Cabeza et al. (2012), Estévez et al. (2012)	WW, review	4

(continued)

Table 5.4 (continued)

Compound	Class	References	Source	Number of studies
Cotinine	CNS stimulant	Stuart et al. (2011), Cabeza et al. (2012), Reh et al. (2013)	WW, review	3
AMPH	CNS stimulant	Reddersen et al. (2002), Maeng et al. (2010)	Spill	2
Paraxanthine	CNS stimulant byproduct	Teijon et al. (2010), Reh et al. (2013)	WW	2
Theobromine	CNS stimulant	Estévez et al. (2012), Reh et al. (2013)	WW	2
Benzoylecgonine	CNS stimulant, byproduct	Reh et al. (2013)	WW	1
Theophylline	CNS stimulant byproduct	Reh et al. (2013)	WW	1
3-Methylxanthine	CNS stimulant byproduct	Reh et al. (2013)	WW	1
1-Methylxanthine	CNS stimulant byproduct	Reh et al. (2013)	WW	1
Compound	*Class*	*References*	*Source*	*Number of studies*
Acesulfame	Sweetener	Buerge et al. (2009), Engelhardt et al. (2011), Wolf et al. (2012)	WW	3
Cyclamate	Sweetner	Wolf et al. (2012)	WW	1
Endocrine disruptors				
Estrone		Hohenblum et al. (2004), Zuehlke et al. (2004), Vulliet et al. (2008), Loos et al. (2010), Vulliet and Cren Olivé (2011)	Review, WW	5
17b-Estradiol		Hohenblum et al. (2004), Vulliet et al. (2008), Vulliet and Cren Olivé (2011)	Review, WW	3

(continued)

Table 5.4 (continued)

Compound	Class	References	Source	Number of studies
17a-Ethinyl estadiol		Adler et al. (2001), Hohenblum et al. (2004), Vulliet et al. (2008)	Review, WW	3
Ethinylestradiol		Vulliet and Cren Olivé (2011)	WW	1
Testosterone		Vulliet et al. (2008), Vulliet and Cren Olivé (2011)	WW	2
17a-Estradiol		Hohenblum et al. (2004), Vulliet et al. (2008), Vulliet and Cren Olivé (2011)	Review, WW	3
Hexestrol		Stuart et al. (2011)	Review	1
Levonorgestrel		Vulliet et al. (2008), Vulliet and Cren Olivé (2011)	WW	2
Norethindrone		Vulliet et al. (2008), Vulliet and Cren Olivé (2011)	WW	2
Progesterone		Vulliet et al. (2008), Vulliet and Cren Olivé (2011)	WW	2
Androstenedione		Vulliet et al. (2008), Vulliet and Cren Olivé (2011)	WW	2
Estrone-3-sulfate		Kuster et al. (2010)	WW	1
Estriol		Hohenblum et al. (2004)	Review	1

contamination. A similar scenario, a lack and/or absence of data regarding the occurrence of pharmaceutical compounds, was observed in Central and South America as well as in Africa. Therefore, more studies primarily focused on pharmaceutical and life-style compounds, which are different from those commonly detected in developed countries, should be conducted in these areas.

Even if the concentrations of detected compounds in the groundwater are usually lower than those observed in the surface water, monitoring studies should be conducted to evaluate the occurrence of these compounds especially in areas impacted by wastewater reuse, agricultural practices, and leakages from landfills or septic tanks.

References

Achagra L, El Messari JS, Draoui M (2013) Assessing the vulnerability to pollution in the aquifer's Charf El Akab (Tangier, Morocco). J Black Sea/Mediterr Environ 19(2):268–273

Adler P, Steger-Hartmann T, Kalbfus W (2001) Vorkommen natürlicher und synthetischer östrogener Steroide in Wässern des sü-und mitteldeutschen Raumes. Acta Hydrochimica et Hydrobiologica 29(4):227–241

AECOM International Development (2010) A rapid assessment of septage management in Asia. Policies and Practices in India, Indonesia, Malaysia, the Philippines, Sri Lanka, Thailand, and Vietnam. USAID Publisher, Bangkok, p 144

Ashton D, Hilton M, Thomas KV (2004) Investigating the environmental transport of human pharmaceuticals to stream in the United Kingdom. Sci Total Environ 333(1–3):167–184

Barnes KK, Christenson SC, Kolpin DW et al (2004) Pharmaceuticals and other organic waste water contaminants within a leachate plume downgradient of a municipal landfill. Ground Water Monit Remediat 24(2):119–126

Barnes KK, Kolpin DW, Furlong ET et al (2008) A national reconnaissance of pharmaceuticals and other organic wastewater contaminants in the United States—I) Groundwater. Sci Total Environ 402(2–3):192–200

Bartelt-Hunt SL, Snow DD, Damon T et al (2009) The occurrence of illicit and therapeutic pharmaceuticals in wastewater effluent and surface waters in Nebraska. Environ Pollut 157 (3):786–791

Bartelt-Hunt SL, Snow DD, Damon-Powell T et al (2011) Occurrence of steroid hormones and antibiotics in shallow groundwater impacted by livestock waste control facilities. J Contam Hydrol 123(3–4):94–103

Batt AL, Snow DD, Aga DS (2006) Occurrence of sulfonamide antimicrobials in private water wells in Washington County, Idaho, USA. Chemosphere 64(11):1963–1971

Bello OO, Osho A, Bello TK (2013a) Microbial quality and antibiotics susceptibility profiles of bacterial isolates from borehole water used by some schools in Ijebu-Ode, Southwestern Nigeria. Sch Acad J Biosci 1(1):4–13

Bello OT, Bello OO, Egberongbe HO et al (2013b) Antibiotics resistance profile of *Escherichia Coli* and *Enterobacter Aerogenes* isolated from well waters in Ago-Iwoye, Southwestern Nigeria. J Adv Biol 2(2):135–144

Benotti MJ, Trenholm RA, Vanderford BJ et al (2009) Pharmaceuticals and endocrine disrupting compounds in U.S. drinking water. Environ Sci Technol 43(3):597–603

Bolong N, Ismail AF, Salim MR et al (2009) A review of the effects of emerging contaminants in wastewater and options for their removal. Desalination 239(1–3):229–246

Boxall ABA (2004) The environmental side effects of medication. EMBO Rep 5(12):1110–1116

Bradley PM, Barber LB, Duris JW et al (2014) Riverbank filtration potential of pharmaceuticals in a wastewater-impacted stream. Environ Pollut 193:173–180

Bruchet A, Hochereau C, Picard C et al (2005) Analysis of drugs and personal care products in French source and drinking waters: the analytical challenge and examples of application. Water Sci Technol 52(8):53–61

Buerge IJ, Buser H-R, Kahle M et al (2009) Ubiquitous occurrence of artificial sweetener acesulfame in the aquatic environment: an ideal chemical marker of domestic wastewater in groundwater. Environ Sci Technol 43(12):4381–4385

Buszka PM, Yeskis DJ, Kolpin DW et al (2009) Waste-indicator and pharmaceutical compounds in landfill-leachate-affected ground water near Elkhart, Indiana, 2000–2002. Bull Environ Contam Toxicol 82(6):653–659

Cabeza Y, Candela L, Ronen D et al (2012) Monitoring the occurrence of emerging contaminants in treated wastewater and groundwater between 2008 and 2010. The Baix Llobregat (Barcelona, Spain). J Hazard Mater 239–240:32–39

Carrara C, Ptacek CJ, Robertson WD et al (2008) Fate of pharmaceutical and trace organic compounds in three septic system plumes, Ontario, Canada. Environ Sci Technol 42 (8):2805–2811

Cary L, Casanova J, Gaaloul N et al (2013) Combining boron isotopes and carbamazepine to trace sewage in salinized groundwater: a case study in Cap Bon, Tunisia. Appl Geochem 34:126–139

Chen F, Ying GG, Kong LX et al (2011) Distribution and accumulation of endocrine-disrupting chemicals and pharmaceuticals in wastewater irrigated soils in Hebei, China. Environ Pollut 159(6):1490–1498

Chinyem FI (2013) Concentrations of some trace elements in surface and groundwater resources in Agepanu and Environs, Edo State, Nigeria. Indian J Sci Technol 6(5):4459–4462

Clara M, Strenn B, Kreuzinger N (2004) Carbamazepine as a possible anthropogenic marker in the aquatic environment: investigations on the behaviour of carbamazepine in wastewater treatment and during groundwater infiltration. Water Res 38(4):947–954

de Jesus GV, Almeida CMM, Rodrigues A et al (2015) Occurrence of pharmaceuticals in a water supply system and related human health risk assessment. Water Res 72:199–208

Deo RP, Halden RU (2013) Pharmaceuticals in the built and natural water environment of the United States—A review. Water 5(3):1346–1365

Drewes JE, Heberer T, Reddersen K (2002) Fate of pharmaceuticals during indirect potable reuse. Water Sci Technol 46(3):73–80

Drewes JE, Heberer T, Rauch T et al (2003) Fate of pharmaceuticals during ground water recharge. Ground Water Monit Remediat 23(3):64–72

Eckel WP, Ross B, Isensee RK (1993) Pentobarbital found in ground water. Ground Water 31 (5):801–804

Einsiedl F, Radke M, Maloszewski P (2010) Occurrence and transport of pharmaceuticals in a karst groundwater system affected by domestic wastewater treatment plants. J Contam Hydrol 117(1–4):26–36

Ellis JB (2006) Pharmaceutical and personal care products (PPCPs) in urban receiving waters. Environ Pollut 144(1):184–189

Engelhardt I, Piepenbrink M, Trauth N et al (2011) Comparison of tracer methods to quantify hydrodynamic exchange within the hyporheic zone. J Hydrol 400(1–2):255–266

Estévez E, Cabrera MC, Molina-Díaz A et al (2012) Screening of emerging contaminants and priority substances (2008/105/EC) in reclaimed water for irrigation and groundwater in a volcanic aquifer (Gran Canaria, Canary Islands, Spain). Sci Total Environ 433:538–546

European Commission (1998) Directive 1998/83/EC of the European Parliament and of the Council of 3 November 1998 on the quality of water intended for human consumption

European Commission (2000) Directive 2000/60/EC of the European Parliament and of the Council of 23 October 2000 establishing a framework for Community action in the field of water policy. Official Journal of the European Union L 327/1

European Commission (2006) Directive 2006/118/EC of the European Parliament and of the Council of 12 October 2006 on the protection of groundwater against pollution and deterioration. Official Journal of the European Union L 372/19

European Commission (2008) Directive 2008/105/EC of the European Parliament and of the Council on environmental quality standards in the field of water policy

European Commission (2011) Review of priority substances under the Water Frame Directive
Faille J (2010) Vulnérabilités des nappes d'eu souterraine aux pollutions médicamenteuses: aquifer vulnerability to drug pollution (a literature review). Masters dissertation, Lille polytechnic. p 19
Fang Y, Karnjanapiboonwong A, Chase DA et al (2012) Occurrence, fate, and persistence of gemfibrozil in water and soil. Environ Toxicol Chem 31(3):550–555
Fekkoul A, Zarhloule Y, Bourghriba M et al (2013) Impact of anthropogenic activities on the groundwater resources of the unconfined aquifer of Triffa plain (Eastern Morocco). Arab J Geosci 6:4917–4924
Félix-Cañedo TE, Durán-Álvarez J-CB (2013) The occurrence and distribution of a group of organic micropollutants in Mexico City's water sources. Sci Total Environ 454–455:109–118
Fent K, Weston AA, Caminada D (2006) Ecotoxicology of human pharmaceuticals. Aquat Toxicol 76(2):122–159
Findikakis AN, Sato K (2011) Groundwater management practices. IAHR Monograph. Taylor & Francis Group, London, p 436
Focazio MJ, Kolpin DW, Barnes KK et al (2008) A national reconnaissance for pharmaceuticals and other organic wastewater contaminants in the United States—II) Untreated drinking water sources. Sci Total Environ 402(2–3):201–216
Fram MS, Belitz K (2011) Occurrence and concentrations of pharmaceutical compounds in groundwater used for public drinking-water supply in California. Sci Total Environ 409 (18):3409–3417
Gasser G, Rona M, Voloshenko A et al (2010) Quantitative evaluation of tracers for quantification of wastewater contamination of potable water sources. Environ Sci Technol 44(10):3919–3925
Gilliom RJ (2007) Pesticides in US streams and groundwater. Environ Sci Technol 41 (10):3408–3414
Godfrey E, Woessner WW, Benotti MJ (2009) Pharmaceuticals in on-site sewage effluent and ground water, Western Montana. Ground Water 45(3):263–271
Gottschall N, Topp E, Metcalfe C (2011) Pharmaceutical and personal care products in groundwater, subsurface drainage, soil, and wheat grain, following a high single application of municipal biosolids to a field. Chemosphere 87(2):194–203
Hamscher G, Hartung J (2008) Veterinary antibiotics in dust: sources, environmental concentrations, and possible health hazards. In: Kümmerer K (ed) Pharmaceuticals in the environment. Sources, fate, effects, and risks. Springer, Berlin, pp 95–100, p. 521
Hass U, Duennbier U, Massmann G (2012) Occurrence and distribution of psychoactive compounds and their metabolites in the urban water cycle of Berlin (Germany). Water Res 46 (18):6013–6022
Heberer T (2002) Tracking persistent pharmaceutical residues from municipal sewage to drinking water. J Hydrol 266(3–4):175–189
Heberer T, Dünnbier U, Reilich C et al (1997) Detection of drugs and drug metabolites in ground water samples of drinking water treatment plant. Fresen Environ Bull 6:438–443
Heberer T, Mechlinski A, Fanck B et al (2004) Field studies on the fate and transport of pharmaceutical residues in bank filtration. Ground Water Monit Remediat 24(2):70–77
Hinkle SR, Weick RJ, Johnson JM et al (2005) Organic wastewater compounds, pharmaceuticals, and coliphage in ground water receiving discharge from onsite wastewater treatment systems near La Pine, Oregon: occurrence and implications for transport. Scientific Investigations Report 2005–5055. p. 98
Hirsch R, Ternes T, Haberer K et al (1999) Occurrence of antibiotics in the aquatic environment. Sci Total Environ 225(1–2):109–118
Hohenblum P, Gans O, Moche W et al (2004) Monitoring of selected estrogenic hormones and industrial chemicals in groundwaters and surface waters in Austria. Sci Total Environ 333 (1–3):185–193
Höhener P, Werner D, Balsiger C et al (2003) Worldwide occurrence and fate of chlorofluorocarbons in groundwater. Crit Rev Environ Sci Technol 33(1):1–29

Holm JV, Rügge K, Bjerg PL et al (1995) Occurrence and distribution of pharmaceutical organic compounds in the groundwater downgradient of a landfill (Grindsted, Denmark). Environ Sci Technol 29(5):1415–1420

http://water.epa.gov/drink/contaminants/#Primary. Accessed 20 Dec 2014

http://www.env.go.jp/en/water/gw/gwp.html. Accessed 20 Dec 2014

http://www2.epa.gov/ccl/contaminant-candidate-list-3-ccl-3. Accessed 20 Dec 2014

Hu X, Zhou Q, Luo Y (2010) Occurrence and source analysis of typical veterinary antibiotics in manure, soil, vegetables and groundwater from organic vegetable bases, northern China. Environ Pollut 158(9):2992–2998

Jiang Y, Li M, Guo C et al (2014) Distribution and ecological risk of antibiotics in a typical effluent-receiving river (Wangyang River) in north China. Chemosphere 112:267–274

Karnjanapiboonwong A, Suski JG, Shah AA et al (2011) Occurrence of PPCPs at a wastewater treatment plant and in soil and groundwater at a land application site. Water Air Soil Pollut 216:257–273

Katz BG, Griffin DW, Davis JH (2009) Groundwater quality impacts from the land application of treated municipal wastewater in a large karstic spring basin: chemical and microbiological indicators. Sci Total Environ 407(8):2872–2886

Kolodziej EP, Harter T, Sedlak DL (2004) Dairy wastewater, aquaculture, and spawning fish as sources of steroid hormones in the aquatic environment. Environ Sci Technol 38 (23):6377–6384

Kolpin DW, Barbash JE, Gilliom RJ (1998) Occurrence of pesticides in shallow groundwater of the United States: initial results from the national water-quality assessment program. Environ Sci Technol 32(5):558–566

Kolpin DW, Furlong ET, Meyer MT et al (2002) Pharmaceuticals, hormones, and other organic wastewater contaminants in U.S. streams, 1999–2000: a national reconnaissance. Environ Sci Technol 36(6):1202–1211

Krapac IG, Koike S, Meyer MT et al (2004) Long-term monitoring of the occurrence of antibiotic residues and antibiotic resistance genes in groundwater near swine confinement facilities. Proceedings of the 4th international conference on pharmaceuticals and endocrine disrupting chemicals in water, Minneapolis, MN, National Ground Water Association, 13–15 Oct, p 158–172

Kreuzinger N, Clara M, Strenn B et al (2004) Investigation on the behavior of selected pharmaceuticals in the groundwater after infiltration of treated wastewater. Water Sci Technol 50 (2):221–228

Kuroda K, Murakami M, Oguma K et al (2012) Assessment of groundwater pollution in Tokyo using PPCPs as sewage markers. Environ Sci Technol 46(3):1455–1464

Kuster M, Díaz-Cruz S, Rosell M et al (2010) Fate of selected pesticides, estrogens, progestogens and volatile organic compounds during artificial recharge using surface waters. Chemosphere 79(8):880–886

Lapworth DJ, Baran N, Stuart ME et al (2012) Emerging organic contaminants in groundwater: a review of sources, fate and occurrence. Environ Pollut 163:287–303

Lindsey ME, Meyer M, Thurman EM (2001) Analysis of trace levels of sulfonamide and tetracycline antimicrobials in groundwater and surface water using solid-phase extraction and liquid chromatography/mass spectrometry. Anal Chem 73(3):4640–4646

Loos R, Locoro G, Comero S et al (2010) Pan-European survey on the occurrence of selected polar organic persistent pollutants in ground water. Water Res 44(14):4115–4126

López-Serna R, Jurado A, Vázquez-Suñé E et al (2013) Occurrence of 95 pharmaceuticals and transformation products in urban groundwaters underlying the metropolis of Barcelona, Spain. Environ Pollut 174:305–315

Luo Y, Guo W, Ngo HH et al (2014) A review on the occurrence of micropollutants in the aquatic environment and their fate and removal during wastewater treatment. Sci Total Environ 473–474:619–641

Mackie RI, Koike S, Krapac I et al (2006) Tetracycline residues and tetracycline resistance genes in groundwater impacted by swine production facilities. Anim Biotechnol 17(2):157–176

Madureira TV, Rocha MJ, Cruzeiro C et al (2011) The toxicity potential of pharmaceuticals found in Douro River estuary (Portugal): assessing impacts on gonadal maturation with a histopathological and stereological study of zebrafish ovary and testis after sub-acute exposures. Aquat Toxicol 105(3–4):292–299

Maeng SK, Ameda E, Sharma SK et al (2010) Organic micropollutant removal from wastewater effluent-impacted drinking water sources during bank filtration and artificial recharge. Water Res 44(14):4003–4014

Makris KC, Snyder SA (2010) Screening of pharmaceuticals and endocrine disrupting compounds in water supplies of Cyprus. Water Sci Technol 62(11):2720–2728

Mansell J, Drewes JE (2004) Fate of steroidal hormones during soil-aquifer treatment. Ground Water Monit Remediat 24(2):94–101

Maskaoui K, Zhou JL (2010) Colloids as a sink for certain pharmaceuticals in the aquatic environment. Environ Sci Pollut Res Int 17(8):898–907

Meffe R, de Bustamante I (2014) Emerging organic contaminants in surface water and groundwater: a first overview of the situation in Italy. Sci Total Environ 481:280–295

Miller KJ, Meek J (2006) Helena Valley ground water: pharmaceuticals, personal care products, endocrine disruptors (PPCPs), and microbial indicators of fecal contamination. Helena, MT, USA. Montana Department of Environmental Quality

Monteiro SC, Boxall ABA (2010) Occurrence and fate of human pharmaceuticals in the environment. In: Whitacre DM (ed) Reviews of environmental contamination and toxicology. Springer, Summerfield, pp 53–154

Musolff A, Leschik S, Möder M et al (2009) Temporal and spatial patterns of micropollutants in urban receiving waters. Environ Pollut 157(11):3069–3077

Musolff A, Leschik S, Schafmeister MT et al (2010) Evaluation of xenobiotic impact on urban receiving waters by means of statistical methods. Water Sci Technol 62(3):684–692

Nakada N, Kiri K, Shinohara H et al (2008) Evaluation of pharmaceuticals and personal care products as water-soluble molecular makers of sewage. Environ Sci Technol 42 (17):6347–6353

NAQUA (2009) Ergebnisse der Grundwassereobachtung Scweiz (NAQUA). Zustand und Entwicklung 2004–2006. Umwelt-Zustand Nr. 0903. Bundesamt für Umwelt, Bern. p 144

Osenbrück K, Gläser HR, Knöller K et al (2007) Sources and transport of selected organic micropollutants in urban groundwater underlying the city of Halle (Saale), Germany. Water Res 41(15):3259–3270

Petrovíc M, Škrbíc B, Živančev J et al (2014) Determination of 81 pharmaceutical drugs by high performance liquid chromatography coupled to mass spectrometry with hybrid triple quadrupole-linear ion trap in different types of water in Serbia. Sci Total Environ 468–469:415–428

Pinheiro A, Albano RMR, Alves TC et al (2013) Veterinary antibiotics and hormone in water from application of pig slurry to soil. Agric Water Manage 129:1–8

Rabiet M, Togola A, Brissaud F et al (2006) Consequences of treated water recycling as regards pharmaceuticals and drugs in surface and ground waters of a medium-sized Mediterranean catchment. Environ Sci Technol 40(17):5282–5288

Reddersen K, Heberer T, Dünnbier U (2002) Identification and significance of phenazone drugs and their metabolites in ground- and drinking water. Chemosphere 49(6):539–544

Reh R, Licha T, Geyer T et al (2013) Occurrence and spatial distribution of organic micropollutants in a complex hydrogeological karst system during low flow and high flow periods, results of a two-year study. Sci Total Environ 443:438–445

Richardson SD, Ternes TA (2014) Water analysis: emerging contaminants and current issues. Anal Chem 86(6):2813–2848

Sacher F, Lange FT, Brauch H-J et al (2001) Pharmaceuticals in groundwaters: analytical methods and results of a monitoring program in Baden-Württemberg, Germany. J Chromatogr A 938 (1–2):199–210
Schaider LA, Rudel RA, Ackerman JM et al (2014) Pharmaceuticals, perfluorosurfactants, and other organic wastewater compounds in public drinking water wells in a shallow sand and gravel aquifer. Sci Total Environ 468–469:384–393
Schmidt CK, Lange FT, Brauch HJ et al (2003) Experiences with riverbank filtration and infiltration in Germany. In: DVGW-Water Technology Center (TZW). Karlsruhe, Germany, p 17
Schulz M, Löffler D, Wagner M et al (2008) Transformation of the X-ray contrast medium iopromide in soil and biological wastewater treatment. Environ Sci Technol 42(19):7207–7217
Seiler RL, Zaugg SD, Thomas JM et al (1999) Caffeine and pharmaceuticals as indicators of waste water contamination in wells. Ground Water 37(3):405–410
Snyder SA, Leising J, Westerhoff P et al (2004) Biological and physical attenuation of endocrine disruptors and pharmaceuticals: implications for water reuse. Ground Water Monit Remediat 24(2):108–118
Snyder SA, Wert EC, Lei H et al (2008) Removal of EDCs and pharmaceuticals in drinking and reuse treatment processes. American Water Works Association, IWA Publishing. p 368
Sorensen JP, Lapworth DJ, Nkhuwa DC et al (2014) Emerging contaminants in urban groundwater sources in Africa. Water Res 72:51–63
Spalding RF, Exner ME (1993) Occurrence of nitrate in groundwater—a review. J Environ Qual 22(3):392–402
Squillace PJ, Zogorski PJ, Wilber WG et al (1996) Preliminary assessment of the occurrence and possible sources of MTBE in groundwater in the United States, 1993–1994. Environ Sci Technol 30(5):1721–1730
Stan H, Linkerhagner M (1992) Identification of 2-(4-Chlorphenoxy)-2-methyl-propions acid in groundwater with GC-AED and GC-MS. Vom Wasser 79:75–88
Stan H, Heberer T, Linkerhagner M (1994) Occurrence of clofibric acid in the aquatic system—is the use of human medical care the source of the contamination of surface, ground and drinking water? Vom Wasser 83:57–68
Stuart ME, Manamsa K, Talbot JC et al (2011) Emerging contaminants in groundwater. British Geological Survey. Groundwater Science Program Open Report OR/11/013. p 111
Stuart ME, Lapworth DJ, Crane E et al (2012) Review of risk from potential emerging contaminants in UK groundwater. Sci Total Environ 416:1–21
Stuart ME, Lapworth DJ, Thomas J et al (2014) Fingerprinting groundwater pollution in catchments with contrasting contaminant sources using microorganic compounds. Sci Total Environ 468–469:564–577
Swartz CH, Reddy S, Benotti MJ et al (2006) Steroid estrogens, nonylphenol ethoxylate metabolites, and other wastewater contaminants in groundwater affected by a residential septic system on Cape Cod, MA. Environ Sci Technol 40(16):4894–4902
Teijon G, Candela L, Tamoh K et al (2010) Occurrence of emerging contaminants, priority substances (2008/105/CE) and heavy metals in treated wastewater and groundwater at Depurbaix facility (Barcelona, Sapin). Sci Total Environ 408(17):3584–3595
Ternes TA (1998) Occurrence of drugs in German sewage treatment plants and rivers. Water Res 32(11):3245–3260
Ternes TA, Hirsch R (2000) Occurrence and behavior of X-ray contrast media in sewage facilities and the aquatic environment. Environ Sci Technol 34(13):2741–2748
Ternes TA, Bonerz M, Herrmann N et al (2007) Irrigation of treated wastewater in Braunschweig, Germany: an option to remove pharmaceuticals and musk fragrances. Chemosphere 66 (5):894–904
Thiele-Bruhn S, Beck I-C (2005) Effects of sulfonamide and tetracycline antibiotics on soil microbial activity and microbial biomass. Chemosphere 59(4):457–465

Togola A, Budzinski H (2007) Development of polar organic integrative samplers for analysis of pharmaceuticals in aquatic systems. Anal Chem 79:6734–6741

Tracol R, Duchemin J (2009) Evaluation de l'occurrence des residues de medicaments dans unéchantillon de nappes souterranies vulnerable du basin Seine Normandie utililées pour laproduction d'eau destinée á la consummation humaine. Service santé-environment and Agence de eau Seine-Normadie. p 42

U.S. Environmental Protection Agency (USEPA) (2006) Ground Water Rule 815-F-06-003

Van Stempvoort DR, Roy JW, Brown SJ et al (2011) Artificial sweeteners as potential tracers in groundwater in urban environments. J Hydrol 401(1–2):126–133

Verstraeten IM, Heberer T, Scheytt T (2003) Occurrence, characteristics, transport, and fate of pesticides, pharmaceuticals, industrial products, and personal care products at riverbank filtration sites. Water Trans 43:175–227

Verstraeten IM, Fetterman GS, Meyer MT et al (2005) Use of tracers and isotopes to evaluate vulnerability of water in domestic wells to septic waste. Ground Water Monit Remediat 25 (2):107–117

Vulliet E, Cren-Olivé C (2011) Screening of pharmaceuticals and hormones at the regional scale, in surface and groundwaters intended to human consumption. Environ Pollut 159 (10):2929–2934

Vulliet E, Wiest L, Baudot R et al (2008) Multi-residue analysis of steroids at sub-ng/L levels in surface and ground-waters using liquid chromatography coupled to tandem mass spectrometry. J Chromatogr A 1210(1):84–91

Watanabe N, Harter TH, Bergamaschi BA (2008) Environmental occurrence and shallow ground water detection of the antibiotic monensin from dairy farms. J Environ Qual 37(5 Suppl):S-78–S-85

Watanabe N, Bergamaschi BA, Loftin KA et al (2010) Use and environmental occurrence of antibiotics in freestall dairy farms with manured forage fields. Environ Sci Technol 44 (17):6591–6600

Wise A, O'Brien K, Woodruff T (2011) Are oral contraceptives a significant contributor to the estrogenicity of drinking water? Environ Sci Technol 45(1):51–60

Wolf L, Zwiener C, Zemann M (2012) Tracking artificial sweeteners and pharmaceuticals introduced into urban groundwater by leaking sewer networks. Sci Total Environ 430:8–19

Zhao S, Zhang PF, Crusius J et al (2011) Use of pharmaceuticals and pesticides to constrain nutrient sources in coastal groundwater of northwestern Long Island, New York, USA. J Environ Monit 13(5):1337–1343

Zhou L-J, Ying G-G, Liu S (2013) Excretion masses and environmental occurrence of antibiotics in typical swine and dairy cattle farms in China. Sci Total Environ 444:183–195

Zuehlke S, Duennbier U, Heberer T et al (2004) Analysis of endocrine disrupting steroids: investigation of their release into the environment and their behaviour during bank filtration. Ground Water Monit Remediat 24(2):78–85

Chapter 6
Remote Sensing Applications for Monitoring Water Resources in the UAE Using Lake Zakher as a Water Storage Gauge

Dawit T. Ghebreyesus, Marouane Temimi, Ali Fares, and Haimanote K. Bayabil

Abstract The potential of remote sensing has been fully demonstrated in large scale and regional hydrological studies where *in situ* observations are limited. However, the use of satellite imagery to monitor water resources in small watersheds remains challenging, mainly due to coarse resolution satellite data. In this study, we assessed the efficacy of remotely sensed data to investigate changes in water storage in Al Ain watershed, in the United Arab Emirates (UAE). Lake Zakher, in the watershed, was used in this study as a gauge indicating changes in water storage. The area of the lake was monitored using Landsat 7 and 8 images, which were then used with a 15 m-Digital Elevation Model (DEM) to calculate time-series lake volumes. In addition, water storage anomalies over the watershed were estimated using Gravity Recovery and Climate Experiment (GRACE) images. Changes in water storage estimated from Landsat and GRACE were in agreement with water consumption and wastewater treatment reported by local agencies in the Emirate of Abu Dhabi. Discharged water from the wastewater treatment plant and volume of water in Lake Zakher showed similar patterns. The results from this study confirmed the reliability of remotely sensed data in monitoring water resources in arid and remote watersheds where ground-based observations are scarce.

D.T. Ghebreyesus • M. Temimi (✉)
Institute Center for Water and Environment (iWATER), Masdar Institute of Science and Technology, P.O. Box 54224, Masdar City, Abu Dhabi, United Arab Emirates
e-mail: mtemimi@masdar.ac.ae

A. Fares • H.K. Bayabil
College of Agriculture and Human Sciences Prairie View, Prairie View A&M University, MS 2008, P.O. Box 519, Prairie View, TX 77446, USA

A. Fares (ed.), *Emerging Issues in Groundwater Resources*, Advances in Water Security, DOI 10.1007/978-3-319-32008-3_6

1 Introduction

The United Arab Emirates (UAE) has an arid climate with average annual precipitation around 100 mm, which makes it among the countries with the lowest records of renewable water resources (FAO 2003). Fresh water in the UAE comes from groundwater and unconventional sources, mainly desalination. Treated water is also used in landscaping. Groundwater is mostly used for agricultural production and accounts for almost three quarters of the total water consumption; whereas potable water is provided from desalination plants (Shahin and Salem 2015). Despite water scarcity issues, however, countries in the Gulf region including the UAE are among the top water consuming nations globally. The disparity between dwindling available water resources and growing water demand shows the immediate need for implementation of effective water resources management policies before these resources are fully depleted.

Planning management policies that allow efficient use of scarcely available water resources requires an understanding of spatial and temporal variability of available water resources with respect to the growing water demand. This will enable policy makers to develop effective and sustainable management strategies. However, this also requires reliable time series data on water resources. Most of the rivers and Wadis in the UAE are ephemeral, especially in the western part of the country, which makes it difficult to continuously monitor their flow. Moreover, most of the watersheds are ungauged with very limited data records. The earliest rainfall records go back to the 1970s in only few stations. The rest of the stations in the country are more recent. Available data is mostly from short field campaigns, which is not sufficient to perform trend analysis.

Satellite imagery is a reliable option to fill hydrological data gaps (Sultan et al. 2008). Therefore, in data scarce regions, like the UAE, remote sensing techniques could be utilized to generate spatial and temporal data for ungauged watersheds (Giardino et al. 2010). Such techniques could also potentially provide better monitoring of weather variables like rainfall, where the common areal interpolation methods are less effective (Grimes et al. 1999). Global precipitation products are often blended products of ground measurements, climate models, and satellite readings (Cheema and Bastiaanssen 2012). Tropical Rainfall Measuring Mission (TRMM), Global Precipitation Measurement (GPM), Climate Prediction Center MORPHing (CMORPH), and Climate Research Unit Database are few of the numerous blended global precipitation products available to the public.

Surface area of reservoirs can be retrieved from Landsat satellite imagery (Liebe et al. 2005) and satellite-based radars can also be used because of their non-sensitivity to atmosphere as an alternative in case of clouds presence (Horritt et al. 2001). Storage fluctuation in reservoirs can be used to estimate runoff in a catchment. Reservoir volume—surface area relationship is a good estimator of flow with high precision (Liebe et al. 2005). Recent remote sensing techniques provide accurate as well as high resolution spatial and temporal data for some climatic variable like rainfall and temperature (Khalaf and Donoghue 2012). Thus, remote sensing applications are becoming viable options to study ungauged watersheds. A number of hydraulic parameters such as height, width, and area of water bodies can be remotely captured, and used to estimate river discharge. Changes in river runoff could be directly estimated using microwave sensors (Brakenridge et al. 2007; Harris et al. 2007; Khan et al. 2012; Smith et al. 1996; Smith and Pavelsky 2008; Temimi et al. 2007, 2011).

Moreover, remote sensing techniques are also used to measure change in water storage by quantifying the change in total mass. Gravity Recovery And Climate Experiment (GRACE) products can be used to investigate temporal and spatial changes of surface water resources (Forootan et al. 2014). This technique is more effective in places where there is no other type of mass displaced by natural or anthropogenic mechanisms. One of the limitations of such easily available satellite products is the coarse spatial resolution ($1^0 \times 1^0$) that lacks detailed spatial and temporal information about hydrological processes. Nevertheless, there is an increasing interest to expand the use of remote sensing to study hydrological processes and water availability within small watersheds in arid regions, similar to the one studied here, where *in situ* records are scarce.

The objective of this study was to investigate the applicability of remote sensing tools to calculate changes in water storage in the Al Ain watershed in the UAE. The Al Ain watershed was chosen because it comprises a lake (Zakher Lake), which was used in this study as an indicator of water storage changes. A Two-dimensional Earth resistivity imaging was implemented to assess the potential and quantity of an aquifer in the North of the UAE in the areas where there are no monitoring wells. Geographic Information System (GIS) techniques were used to quantify water resources (Ebraheem et al. 2014). Similarly, it was assumed that changes in Lake Zakher's volume reflects changes in watershed water storage and can be effectively estimated using remotely sensed data. Unlike most sophisticated (data intensive) hydrological models, our approach uses only freely available satellite images and water demand records in Al Ain watershed. Findings from this study will aid future efforts towards quantification of the major water budget components of watersheds with arid climate and limited ground observations.

2 Methodology

2.1 Study Area

The Al Ain municipality is located in the southeastern region of Abu Dhabi Emirate, about 130 km south of Dubai and east of the capital Abu Dhabi. It is located at 24°12′27″N and 55°44′41″E geographical coordinates. The total area of the city is approximately 13,100 km^2 with a total population of 518,316 according to 2012 census. The area is rich in fresh ground water compared to other regions of the Emirate of Abu Dhabi where brackish water is dominant, except in the Liwa

Crescent area on the edge of the empty quarter (Elmahdy and Mohamed 2012). The city of Al Ain is known as the garden city in the UAE. Historically, the city holds a precious reserve of fresh water, which is subjected to a growing demand not only for residential and industrial uses, but also for agriculture given the increasing number of farms developed in the western part of Al Ain municipality on the road to Abu Dhabi. As a result, despite the region's arid climate, with average annual rainfall of 100 mm (Böer 1997), the city is attracting local and international agricultural investors due to its rich ground water resources. There are many livestock farms within the city and fruit and vegetable production covers a significant proportion of the local market. The city's industrial sector is also growing rapidly contributing to the increasing demand for water.

The Al Ain watershed was selected because previous reports showed that groundwater resources are being rapidly depleted for agricultural uses (Sherif et al. 2011). Calculating the water budget of Al Ain watershed is crucial for management and regulation purposes in the region, where renewable water resources are becoming increasingly scarce.

The city has a wastewater treatment plant very close to Lake Zakher that discharges treated water into the lake since 2010 (Fig. 6.1b). Since then, the lake area has been expanding, which indicates increases in the total volume of treated water. We assumed that there is a direct correlation between changes in lake area and changes in water demand and consumption in the watershed, which we investigated using remote sensing techniques. It is worth noting that the wastewater treatment plant is the oldest and largest one in the region. Therefore, we assumed that, although discharge from the wastewater treatment plant does not fully represent total water discharge from the city, it still shows the trend in water demand and therefore the overall water storage variation in Lake Zakher.

2.2 *Data Set*

2.2.1 Satellite Data Acquisition and Analysis

Information on water storage was extracted from two satellites, namely; the Gravity Recovery and Climate Experiment (GRACE) and Landsat (7 and 8). The former showed anomalies in total water storage within the study area and the latter allowed monitoring of time series areal extent of Lake Zakher, which were then converted to lake volumes and compared to water storage fluctuations from GRACE. GRACE is a mission by the United States National Aeronautics and Space Administration's (NASA's) Earth System Science Pathfinder and was launched in March 2002. It consists of two identical satellites about 220 km apart and orbiting 500 km above the earth surface. The main objective of NASA's GRACE mission is to monitor the change of earth mass by accurately measuring the change in distance between the satellites due to variation in gravitational force. This data helps in analyzing the deep current movement in oceans, surface, and ground water movement in

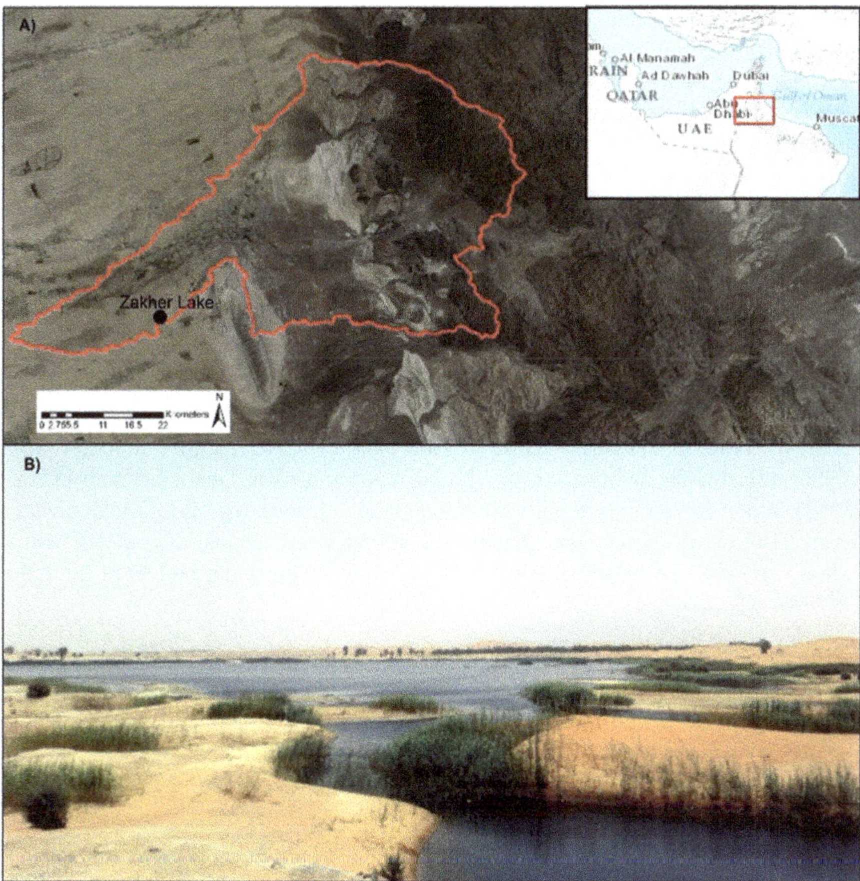

Fig. 6.1 The city of Al Ain Watershed and its location on the eastern United Arab Emirates boarders with Oman (**a**) and Lake Zakher (**b**)

terrestrial and the melting of ice sheets (Landerer and Swenson 2012; Swenson 2012). GRACE maps were corrected for glacial isostatic adjustment (GIA) according to the model by (Wahr et al. 1998). A de-striping filter has been applied to the data to minimize the effect of correlated errors whose telltale signal are N-S stripes in GRACE monthly maps. Gaussian filter was implemented to reduce the noise with 300 km wide window. In this study, level 3 gridded product from GRACE was used with 1° spatial and approximately 1 month temporal resolutions. The data was provided in anomaly relative to average values from January 2004 to December 2009 and was given in depth (cm) of water for each grid cell.

Landsat images (30 m spatial and 16 days temporal resolutions) were also used to calculate area changes of Lake Zakher using the morphological gradient detection. A total of 38 images were obtained from 2010 to 2014. A number of images were eliminated due to occasional interruptions as a result of cloud coverage. All processed images were classified and the lake boundary was visually inspected.

Changes in lateral extent of the lake were clearly visible. Lake area was calculated by delineating polygons using images from different dates. In addition, 15 m resolution DEM was used to determine the depth of the lake on different dates. The gentle topography in the desert area fostered a significant change in the surface area of the lake whenever the Lake volume changes. For each time period, volume of water in the lake was then calculated by multiplying lake area by lake depth obtained from the DEM.

2.2.2 Water Demand Data

Records of water demand, consumption (agricultural and domestic), recycled, and discharged from wastewater treatment plants were obtained from the Water and Energy reports of Statistics Centre of Abu Dhabi (SCAD), a governmental office responsible for the publication of official records in Abu Dhabi Emirate (SCAD 2013). Water and Energy reports contain annual consumption of desalinated water, annual inflow of wastewater, annual treated water, and annual quantity of reused water were used as input variables during computing the water budget of the Lake.

3 Results and Discussions

3.1 Surface Water Monitoring

Figure 6.2 shows area changes of Lake Zakher for selected dates detected using Landsat images. Lake area started to increase towards the end of 2009 and beginning of 2010. In May 2010, Landsat images showed that the lake area was just 0.35 km^2. The area of the lake started to increase rapidly and reached a maximum of 0.86 km^2 in September 2012. Afterwards, it started to diminish slowly reaching a plateau at 0.46 km^2 in July 2014.

The Landsat 07 images were affected due to the failure of the Scan Line Corrector (SLC) sensor. This affected the quality of the image by introducing a stripes of no value data on the image. Despite the missing values, the Landsat 07 images agree with the Landsat 07 images from 2013 to 2014 on the Lake's delineation as seen in the row 2 and row 3 in Fig. 6.2.

Figure 6.3 shows annual variation of the volume of Lake Zakher from 2010 to 2014, which was calculated using (15 m resolution) DEM. Overall, annual average volume of the lake, as shown by the area under the curve in Fig. 6.3, was 2.80 Mm^3. The lake volume was approximately 1 Mm^3 in January 2010. Then, it started to rise rapidly and peaked in 2012 at 4.5 Mm^3. Afterwards, the lake volume started to plummet until it reached 2 Mm^3 in 2013. This level was maintained with slight decrease in 2014. Changes in lake volume clearly indicated that there was a considerable influx of water into the lake that resulted in increased volume of Lake Zakher from its initial size (less than 1 Mm^3) in 2010.

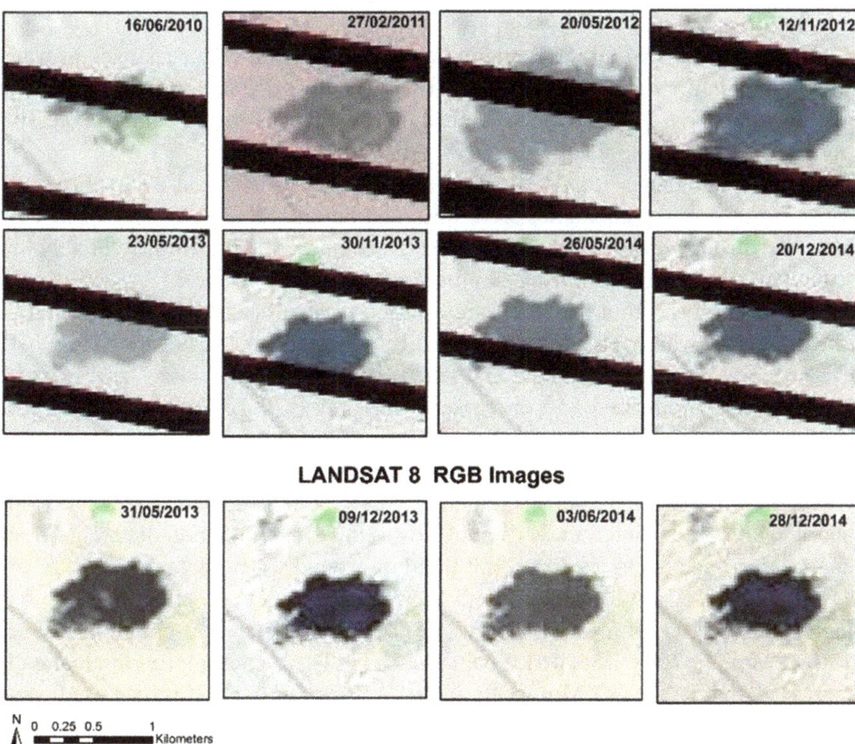

Fig. 6.2 Landsat 07 and 08 images of Lake Zakher at several periods, capturing the evolution of the Lake

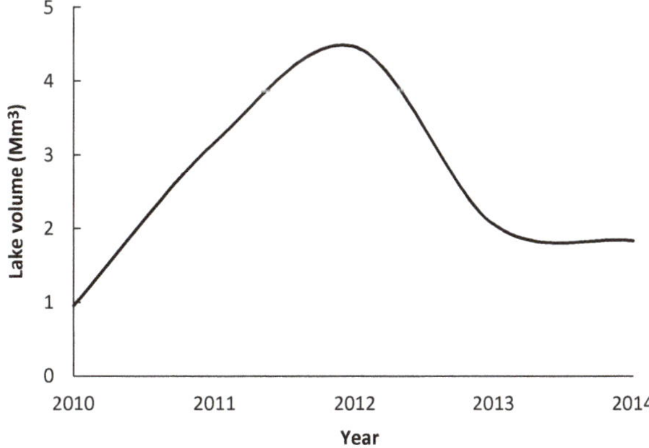

Fig. 6.3 Changes in annual volume of Lake Zakher calculated from Landsat images and DEM

3.2 Water Storage Monitoring

Time series results of GRACE anomalies were used to monitor changes in water storage in Al Ain watershed (Fig. 6.4). This was a valid assumption because there was no other activity within the watershed involving mass movement like excessive mining or oil production activities in Al Ain watershed.

GRACE anomalies for Al Ain city showed a clear biannual cycle (Fig. 6.4). Most of the positive anomalies were observed before 2008. Peak anomalies were observed during winter months of the year (December–March), while relatively low anomalies were recorded during summer months (June–October).

Water storage within the study area started to significantly decline from 2008 onwards as the values of GRACE anomalies were persistently negative (Fig. 6.4). The negative gradient continued and reached its lowest level (−2.7 cm) in 2013. This depletion coincided with increase in the area of Lake Zakher, which could result in loss of significant quantity of water to evaporation. This also indicates that ground water resources in the area were excessively exploited in the last decade. GRACE results showed a substantial decline in the rate of depletion as the monthly and annual values flatten in 2013 and 2014. However, further studies (e.g., trend analysis) are needed to verify overall trend of ground water depletion.

In line with observed depletion of water storage, water consumption in the city increased by 100 Mm^3 in 6 years (2009–2015). Water consumption reached 190.90 Mm^3 in 2009 from 161.2 Mm^3 in 2005. Then, consumption dramatically increased and hit the 293 Mm^3 mark in 2013. The main factor for increased water demand was the expansion of the city and the growth in industrial and agricultural sectors. Figure 6.4 shows the increase in the water consumption throughout the period of 2005–2013 versus the depletion in the ground water resource with GRACE results.

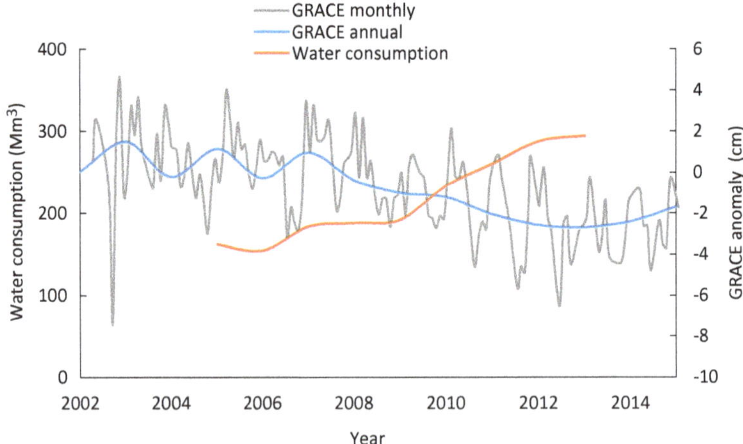

Fig. 6.4 Grace Anomaly of Al Ain and the water consumption for the city from Abu Dhabi Water and Electricity Authority, ADWEA

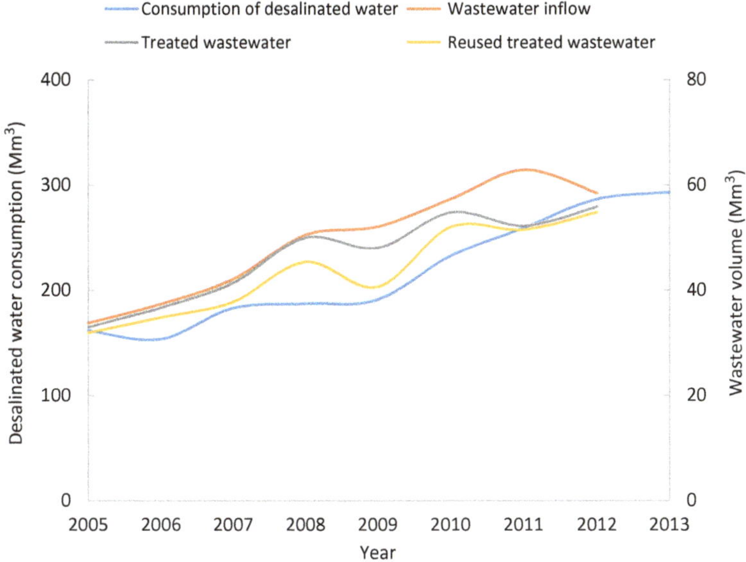

Fig. 6.5 Desalinated water and wastewater volumes as reported by the Statistics Centre of Abu Dhabi (SCAD 2013)

3.3 Water Consumption

Most of the water distributed in the city comes from desalinated seawater. In rural locations out of the city, some deep water pumping contributes to water supply as well but mostly to farming activities. Changes in water consumption in Al Ain is shown in Fig. 6.5. Water consumption increased with a relatively small amount between 2005 and 2009, only a 30 Mm^3 increase in a 4 year period (Fig. 6.5). However, water consumption started to soar and reached 300 Mm^3 in 2013, which corresponds to approximately 100 Mm^3 increase for the same duration (4 years).

Out of 160 Mm^3 consumed water in 2005, only 33.7 Mm^3 ended in the wastewater treatment plants. While in 2009, 52.1 Mm^3 water was collected by treatment plants out of the 190.9 Mm^3 water consumed. However, starting from 2010, wastewater inflow did not increase with the same rate as the water consumption. Wastewater inflow increased only by 12 % with the rise of more than 50 % of the consumption from 2009 to 2012. This means that increase of the wastewater inflow was less than 10 Mm^3, while water consumption soared by almost 100 Mm^3. The difference between wastewater inflow and treated water was relatively small from 2005 to 2008. Increased differences were observed after 2008 with highest difference of 10.53 Mm^3 observed in 2011.

In 2005, most of the treated water was reused by the public. Water reuse was mainly for landscaping in the city. Although the quantity of the treated water increased, the amount of the reused water did not increase at the same rate during

2008–2009. Greater reduction (5 Mm^3) in the amount of reused water was observed during the same period. The largest difference (7.5 Mm^3) between treated and reused water was observed in 2009. Then, in 2011, reused water rapidly increased and reached the same level with treated water. This could be mainly attributed to increased landscape irrigation in the city and increased efforts by the Government to raise public awareness on the benefits of water reuse.

3.4 Comparison Between Volume of Lake Zakher and Water Discharged from Wastewater Treatment Plants

Discharge water from treatment plants into the lake increased from 0.5 Mm^3 in 2008 to 4 Mm^3 in 2009 (Fig. 6.6). However, discharge decreased by about 50 % in the following year (2010), and then it soared and reached 10.53 Mm^3 in 2011. Such increment would likely saturate the water table in the aquifer as shown in a previous study by (Moustafa et al. 2014). In the following year (2012) discharge water significantly decreased and was only 2.5 Mm^3 (Fig. 6.6).

Based on the average annual volume obtained using remote sensing, the Lake expansion started in 2010 with an estimated volume of 1 Mm^3. Then, it increased and reached 3.17 Mm^3 in just 1 year. In just 2 years (2012), lake volume reached peak at 4.46 Mm^3. Afterwards, the volume decreased by almost half and reached approximately 2 Mm^3 in 2013.

Discharge water sharply increased from its level in 2010 and reached its highest level 10.53 Mm^3 in 2011, while the volume of the lake reached at peak (4.46 Mm^3) the following year (in 2012). Recession rates of discharge water and lake volume curves (Fig. 6.6) were almost identical. The peaks of the two curves had approximately a 1 year lag period. This suggests that it might take 1 year for discharge

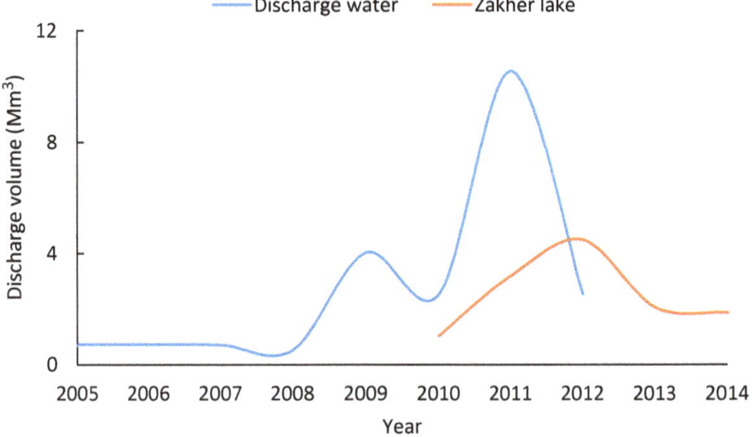

Fig. 6.6 Trends of annual average discharged water and volume of Lake Zakher

water from the wastewater treatment plant to reach the water table and eventually affect the volume of water in the Lake. Discharge water reduction was also marked by reduction in lake volume.

Remote sensing showed a potential to quantify the volume of the lake and calculate the water budget for the Al Ain city. Desalinated water, inflow to the lake, is the main input of the water budget for Lake Zakher. Rainfall input is assumed to be negligible as its volume seems to be relatively minimal in this arid desert region. It is worth noting here that most of the discharge water into Lake Zakher is lost through evaporation. During estimating lake volume, evaporation loss was not accounted, and only the stored water in the lake was calculated using Landsat images and DEM. Lake evaporation could be estimated using remote sensing imagery, and this would improve accuracy of the water budget. However, this was beyond the scope of this study and will be addressed in future studies.

An interesting agreement was found between the data from the GRACE satellite and changes in the volume of Zakher Lake. This finding corroborated the validity of the assumption that Zakher Lake could be considered as a water storage gauge, which could reflect the variability in total water budget. This was specifically valid in an arid watershed like the one studied here where Zakher Lake was a main outlet for the water system. The absence of perennial rivers and significant baseflow in the region (Alsharhan et al. 2001) made Zakher Lake an indicator of water storage variability.

The volume of the Lake was estimated to be 1 Mm^3 in 2010 and at that time the readings of GRACE anomaly was -1.27 cm. Then, the volume started to increase rapidly reaching its maximum in 2012 with a corresponding GRACE anomaly of -2.61 cm. Satellite readings declined slightly from 2012 to 2013 before starting to recover. This suggests that the expansion of the lake could lead to the depletion in the storage as shown by GRACE readings. Observed agreement between GRACE anomaly and change in lake volume supported the assumption that the lake could be considered as a water storage gauge that could be monitored from space.

4 Conclusion

The main goal of this study was to assess the application of remote sensing in water resource management for ungauged watersheds in an arid region. Results showed that water storage, especially groundwater, in Al Ain is at its lowest record level in the last 15 years. This was clearly demonstrated in this study through the analysis of GRACE readings. This finding was in agreement with previous study by (Murad et al. 2007), which indicated increased depletion of ground water across the UAE. Deep well pumping for agricultural purposes was the main cause of the declining groundwater resource. The study indicated that there is excessive depletion and leads to sea water intrusion which increased the salinity of ground water across the country. Murad et al. (2007) suggested that the only solution is to implement an integrated water resource management strategy.

Findings from this study demonstrated that remote sensing tools could be effectively used for planning water resources management. Water resources can be reasonably estimated using remote sensing techniques, which can also complement *in situ* measurements especially in arid and semi-arid areas where the main water input (gain) to the study area from desalination plant was closely monitored. Furthermore, the main output (loss) from the watershed was to the wastewater treatment plant, which was accurately measured. A significant fraction of the treated water was sent to Zakher Lake, which was monitored from space to infer changes in the water storage volumes.

Overall, this study showed that remote sensing could be a useful tool in monitoring of water resources in arid regions. This could be enhanced by using satellite imagery in the monitoring of irrigation and vegetation health. Therefore, satellite imagery should be an important component of any comprehensive water resource management strategy that should be implemented in the UAE to achieve water and food security.

References

Alsharhan AS, Rizk Z, Nairn AEM, Bakhit D, Alhajari S (2001) Hydrogeology of an arid region: the Arabian Gulf and adjoining areas. Elsevier, Amsterdam

Böer B (1997) An introduction to the climate of the United Arab Emirates. J Arid Environ 35 (1):3–16. doi:http://dx.doi.org/10.1006/jare.1996.0162

Brakenridge GR, Nghiem SV, Anderson E, Mic R (2007) Orbital microwave measurement of river discharge and ice status. Water Resour Res 43(4)

Cheema MJM, Bastiaanssen WG (2012) Local calibration of remotely sensed rainfall from the TRMM satellite for different periods and spatial scales in the Indus Basin. Int J Remote Sens 33 (8):2603–2627

Ebraheem AM, Al Mulla MM, Sherif MM, Awad O, Akram SF, Al Suweidi NB, Shetty A (2014) Mapping groundwater conditions in different geological environments in the northern area of UAE using 2D earth resistivity imaging survey. Environ Earth Sci 72(5):1599–1614. doi:10.1007/s12665-014-3064-5

Elmahdy SI, Mohamed MM (2012) Topographic attributes control groundwater flow and groundwater salinity of Al Ain, UAE: a prediction method using remote sensing and GIS. J Environ Earth Sci 2(8):1–13

FAO (2003). Review of world water resources by Country (N. R. M. a. E. Department, Trans.). Food and Agriculture Organization of the United Nations, Rome

Forootan E, Rietbroek R, Kusche J, Sharifi MA, Awange JL, Schmidt M, Omondi P, Famiglietti J (2014) Separation of large scale water storage patterns over Iran using GRACE, altimetry and hydrological data. Remote Sens Environ 140:580–595. doi:http://dx.doi.org/10.1016/j.rse.2013.09.025

Giardino C, Bresciani M, Villa P, Martinelli A (2010) Application of remote sensing in water resource management: the case study of Lake Trasimeno, Italy. Water Resour Manag 24 (14):3885–3899. doi:10.1007/s11269-010-9639-3

Grimes DIF, Pardo-Igúzquiza E, Bonifacio R (1999) Optimal areal rainfall estimation using raingauges and satellite data. J Hydrol 222(1–4):93–108. doi:http://dx.doi.org/10.1016/S0022-1694(99)00092-X

Harris A, Rahman S, Hossain F, Yarborough L, Bagtzoglou AC, Easson G (2007) Satellite-based flood modeling using TRMM-based rainfall products. Sensors 7(12):3416–3427

Horritt MS, Mason DC, Luckman AJ (2001) Flood boundary delineation from Synthetic Aperture Radar imagery using a statistical active contour model. Int J Remote Sens 22(13):2489–2507. doi:10.1080/01431160116902

Khalaf A, Donoghue D (2012) Estimating recharge distribution using remote sensing: a case study from the West Bank. J Hydrol 414–415:354–363. doi:10.1016/j.jhydrol.2011.11.006

Khan SI, Yang H, Vergara HJ, Gourley JJ, Brakenridge GR, De Groeve T, Flamig ZL, Policelli F, Bin Y (2012) Microwave satellite data for hydrologic modeling in ungauged basins. IEEE Geosci Remote Sens Lett 9(4):663–667. doi:10.1109/LGRS.2011.2177807

Landerer FW, Swenson SC (2012) Accuracy of scaled GRACE terrestrial water storage estimates. Water Resour Res 48:W04531. doi:10.1029/2011WR011453

Liebe J, van de Giesen N, Andreini M (2005) Estimation of small reservoir storage capacities in a semi-arid environment: a case study in the Upper East Region of Ghana. Phys Chem Earth, Parts A/B/C 30(6–7):448–454. doi:http://dx.doi.org/10.1016/j.pce.2005.06.011

Moustafa SSR, Alarifi N, Naeem M, Jafri MK (2014) An integrated technique for delineating groundwater contaminated zones using geophysical and remote sensing techniques: a case study of Al-Quway'iyah, central Saudi Arabia. Can J Earth Sci 51(8):797–808. doi:10.1139/cjes-2014-0008

Murad A, Al Nuaimi H, Al Hammadi M (2007) Comprehensive assessment of water resources in the United Arab Emirates (UAE). Water Resour Manag 21(9):1449–1463. doi:10.1007/s11269-006-9093-4

SCAD (2013) Energy and water in figures. Statistics Centre, Abu Dhabi

Shahin SM, Salem MA (2015) The challenges of water scarcity and the future of food security in the United Arab Emirates (UAE). Nat Resour Conserv 3(1):1–6

Sherif M, Mohamed M, Kacimov A, Shetty A (2011) Assessment of groundwater quality in the northeastern coastal area of UAE as precursor for desalination. Desalination 273(2):436–446. doi:10.1016/j.desal.2011.01.069

Smith LC, Pavelsky TM (2008) Estimation of river discharge, propagation speed, and hydraulic geometry from space: Lena River, Siberia. Water Resour Res 44(3):n/a–n/a. doi:10.1029/2007WR006133

Smith LC, Isacks BL, Bloom AL, Murray AB (1996) Estimation of discharge from three braided rivers using synthetic aperture radar satellite imagery: potential application to ungaged basins. Water Resour Res 32(7):2021–2034

Sultan M, Sturchio N, Al Sefry S, Milewski A, Becker R, Nasr I, Sagintayev Z (2008) Geochemical, isotopic, and remote sensing constraints on the origin and evolution of the Rub Al Khali aquifer system, Arabian Peninsula. J Hydrol 356(1–2):70–83. doi:http://dx.doi.org/10.1016/j.jhydrol.2008.04.001

Swenson SC (2012) GRACE monthly land water mass grids NETCDF RELEASE 5

Temimi M, Leconte R, Brissette F, Chaouch N (2007) Flood and soil wetness monitoring over the Mackenzie River Basin using AMSR-E 37 GHz brightness temperature. J Hydrol 333(2):317–328. doi:10.1016/j.jhydrol.2006.09.002

Temimi M, Lacava T, Lakhankar T, Tramutoli V, Ghedira H, Ata R, Khanbilvardi R (2011) A multi-temporal analysis of AMSR-E data for flood and discharge monitoring during the 2008 flood in Iowa. Hydrol Process 25(16):2623–2634. doi:10.1002/hyp.8020

Wahr J, Molenaar M, Bryan F (1998) Time variability of the Earth's gravity field: hydrological and oceanic effects and their possible detection using GRACE. J Geophys Res Solid Earth 103(B12):30205–30229. doi:10.1029/98JB02844

Chapter 7
Assessment of Groundwater Balance Terms Based on the Cross-Calibration of Two Different Independent Approaches

Giuseppe Passarella, Rita Masciale, Donato Sollitto, Maria Clementina Caputo, and Emanuele Barca

Abstract A reliable groundwater balance assessment is a fundamental tool for any effective resource exploitation plan. Nevertheless, some terms of the balance equation are, generally, very difficult to be estimated, even on average, especially when large and heterogeneous groundwater bodies are considered.

In this work, a methodology for mutually calibrating groundwater balances carried out by means of different methods is proposed, also capable of providing an average estimation of the specific yield of the considered aquifer, as by-product.

The method has been applied to the porous aquifer of the *Tavoliere di Puglia* located in Southern Italy. The plain is mostly exploited for agricultural uses and the aquifer represents the main source of the district water supply.

A long time series of groundwater balances has been calibrated by assessing the average value of the term "inflow/outflow", which had not been previously considered. Furthermore, the average value of the specific yield of the considered aquifer has been assessed.

1 Introduction

Groundwater is considered a valuable resource, commonly destined to drinking uses, particularly in areas affected by surface water scarcity. Nevertheless, the growing needs of the agricultural and manufacturing sectors led to an over-exploitation of the resource, resulting in many negative consequences, such as lowering of piezometric surfaces, depletion of groundwater wells and natural springs, growing subsidence phenomena and sea water intrusion, and general qualitative degradation (Masciopinto et al. 1994; Kehew et al. 1996; Caputo et al. 1998).

G. Passarella (✉) • R. Masciale • D. Sollitto • M.C. Caputo • E. Barca
Department of Bari, IRSA-CNR—Water Research Institute of the National Research Council, Viale De Blasio, 5, Bari 70132, Italy
e-mail: giuseppe.passarella@ba.irsa.cnr.it

A. Fares (ed.), *Emerging Issues in Groundwater Resources*, Advances in Water Security, DOI 10.1007/978-3-319-32008-3_7

A reliable assessment of the groundwater availability becomes more and more fundamental in order to develop sustainable exploitation policies by the local water resources authorities. The main managerial tool for addressing such issues is the hydrogeological balance. Scientific and technical literature provides a number of methods for groundwater balance assessment, mostly based on simple models taking into account all the positive and negative water volume contributions. In practice, some of the balance terms are complex or expensive to be assessed and sometimes they are coarsely approximated or not considered at all producing partial or uncertain results.

Quick and easy, but reliable, quantitative tools capable of calibrating groundwater balances and/or providing additional information related to groundwater resources availability, can support water authorities to better plan the resource exploitation.

In this work, a simple method for calibrating a couple of groundwater balance terms is presented, based on the knowledge of the average, yearly groundwater table, during a given period of time. Actually, given a time series of groundwater balances, in terms of water volumes (Mm^3) or water heights per unit area (mm/m^2) and the estimated groundwater tables (m.a.s.l.) for the same period, we propose a simple regressive model capable of assessing possible amounts of water volume neglected in the first series and the average specific yield of the considered aquifer, at the same time. The methodology has been applied to the aquifer of the *Tavoliere di Puglia* located in the Apulia region, South Italy.

2 Materials and Methods

2.1 Methodological Framework

In this work estimated yearly values of the hydrogeological balance assessed in the Water Protection Plan of the Apulia Region (AA.VV. 2009) and in a previous study, jointly carried out by the Apulian Water Basin Authority and the Water Research Institute of the Italian National Research Council (Portoghese et al. 2010), have been used and compared to the results of a different method proposed for the assessment of the groundwater balance, based on monitored values of groundwater levels.

In particular, for each Apulian groundwater body, the two aforementioned studies assessed the hydrogeological balance B_1', from 1950 to 2011 as follows:

$$GR = R + SI - ET - SR$$

$$B_1' = GR - GI - DI$$

where Groundwater Recharge $= GR$, Rainfall $= R$, Irrigation from Surface water $= SI$, real Evapotranspiration $= ET$, Surface Runoff $= SR$, Irrigation from

Groundwater = GI and withdrawals for Drinking and Industrial uses = DI. Unfortunately, these studies were not able to estimate the (positive/negative) contribution of the water volumes exchanged with neighbouring aquifers or discharged to the sea.

Furthermore, in the period between 2007 and 2011, the Regional Water Authority started a monitoring survey for the assessment of the quantitative and qualitative status of the regional aquifers. In this context, groundwater levels were monitored, allowing us to assess the average yearly groundwater table by means of geostatistical methods (ordinary cokriging).

Subtracting the estimated groundwater tables, year by year, the gross volumes B_2', between the two surfaces, have been calculated. Obviously, these volumes are total volumes, including solid matrix and water content.

The comparison between B_1' and B_2', allowed us to calibrate both the balances, in order to rectify the above mentioned lacks of information affecting each of them, and make some considerations about the resource availability.

In detail, the present study aims to:

(a) Establish a model connecting B_1' to B_2' over the period from 2008 to 2011;
(b) Estimate, by means of the previous model, a net balance concerning the water volumes exchanged with neighbouring aquifers or discharged to the sea (AD);
(c) Estimate, by means of the previous model, an approximate value of the average specific yield (S_y);
(d) Define significant and time-invariant relationships between rainfall and net balance;

The above listed goals have been achieved by means of the following multistep methodology:

(a) Yearly values of B_1' and B_2' have been compared by means of the following linear regression model:

$$B_1' = a^*B_2' + b \tag{7.1}$$

In principle, such two quantities should be equal:

$$B_1' = B_2' \tag{7.2}$$

where $a = 1$ e $b = 0$; however, this doesn't happen in practice because the first component B_1' is affected by the unknown quantity AD, while the second one B_2' doesn't take into account the specific yield coefficient.

(b) B_1' values have been transformed according to Eq. (7.3):

$$B_1 = B_1' - b \tag{7.3}$$

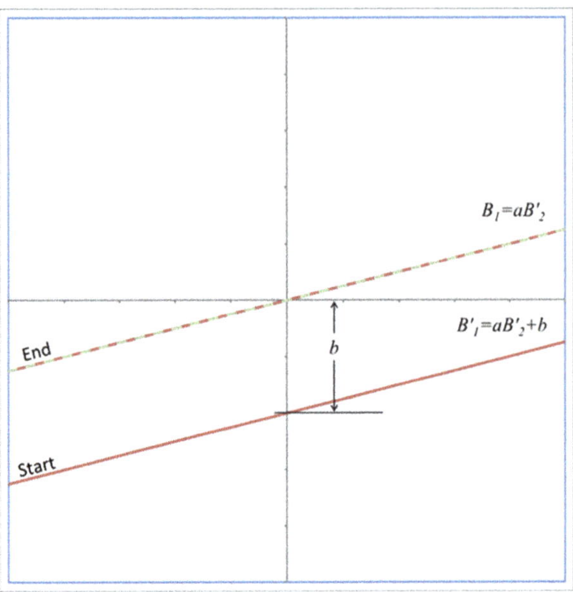

Fig. 7.1 Graphical representation of transformations (3)

Transformations (3) consists in a traslation and it is graphically shown in Fig. 7.1.

(c) $B_2^{'}$ values have been transformed according to equation (Eq. 7.4):

$$B_2 = a^* B_2^{'} \tag{7.4}$$

Transformations (4) consists in a rotation and it is graphically shown in Fig. 7.2.

(d) Once transformation (3) and (4) have been carried out, B_1 values of the whole time series (1950–2011) have been coupled with the corresponding yearly rainfall rates.

Transformations (3) and (4) can be viewed by a physical standpoint: in fact, concerning the traslation, the amount b can be physically considered as the average inflow/outflow through its boundary (AD). Consequently, the transformed variable B_1 can be viewed as the actual, yearly net balance of the considered aquifer from 1950 to 2011.

Similarly, the coefficient a of equation (Eq. 7.4) can be physically viewed as the average specific yield (S_y) of the considered aquifer. Consequently, the transformed variable B_2 can be viewed as the actual, yearly net balance of the considered aquifer from 2008 to 2011.

On the basis of the regression theory (Neter et al. 1996), it is possible to estimate the confidence limits associated to the coefficients a and b. This allowed us to define likelihood intervals, significant at 95 %, both for a and b and consequently to assess the uncertainty related to B_1 and B_2.

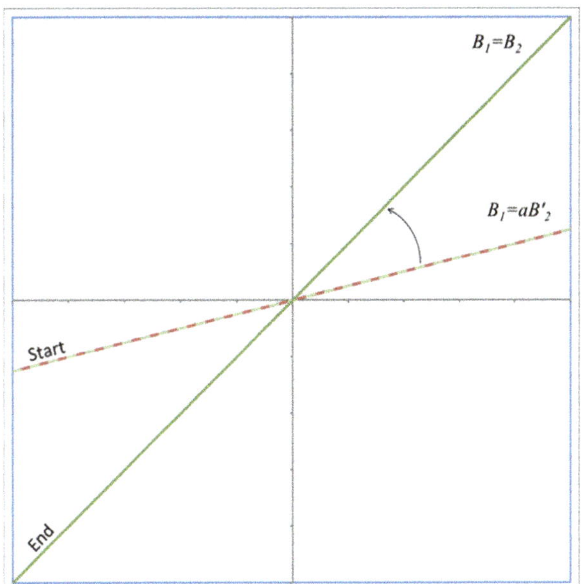

Fig. 7.2 Graphical representation of transformations (4)

As reported at point 4. above, the long time series of B_1 values has been coupled with the corresponding rainfall rates through a linear regression model. The two time series showed a good accordance, proved by a high coefficient of determination and a slim confidence band. Finally, this model has been used in order to assess the rainfall rate threshold interval, below which the considered aquifer can undergo to a stress status. In these cases, in fact, the recharge cannot be sufficient to restore the natural and man-induced water losses.

2.2 Multivariate Geostatistics

A critical stage of the applied methodology concerns the estimation of the groundwater tables at the aquifers' scale. Nevertheless, geostatistics supported us providing reliable tools able to address the issue.

Groundwater table can be considered as a regionalized variable according to the Matheron's theory (Matheron 1970). In practice, it can be viewed as a *Random Function* (*RF*), which represents the spatial law of distribution of the groundwater levels, each associated with a specific spatial position.

Geostatistics offers a great number of reliable techniques (e.g.: *ordinary kriging*) to reproduce *quantitatively* the spatial law of the considered property on the observed domain.

The estimation accuracy can be furtherly improved by taking advantage of secondary information collected at the same area, when available.

In fact, the resulting estimation accuracy furtherly improves, using multivariate geostatistics, when the secondary information is significantly related to the primary variable and more abundant in terms of dataset size (Castrignanò et al. 2000). The main multivariate geostatistical method is the cokriging (Passarella et al. 2003). If an unsampled location x_0 is considered for estimating the variable $z_{i_0}(x_0)$, the kriging predictor equation can be expressed in the following way:

$$z_{i_0}^*(x_0) = \sum_{i=1}^{n}\sum_{\alpha=1}^{n_i}\lambda_\alpha^i z_i(x_\alpha) \tag{7.5}$$

where i indicates the variable of interest among those observed, α is indicates a generic sampled location and λ is a weight value.

For the cokriging case, the previous equation takes the form of the following equations system:

$$\begin{cases} \sum_{j=1}^{n}\sum_{\beta=1}^{n_j}\lambda_\beta^i \gamma_{ij}(x_\alpha, x_\beta) + \mu_i = \gamma_{ii_0}(x_\alpha, x_0) \quad i = 1, \ldots, n \text{ and } \alpha = 1, \ldots, n_i \\ \ldots \\ \sum_{\beta=1}^{n_i}\lambda_\beta^i = \delta_{ii_0} \quad i = 1, \ldots, n \end{cases} \tag{7.6}$$

The system is composed by $\sum_{i=1}^{n} n_i + n$ linear equations and $\sum_{i=1}^{n} n_i$ unknowns; λ_β^i are weights, while μ_i are the n Lagrange multipliers and δ_{ii_0} represents the *Kronecker symbol* which can assume only the values 1 or 0 according to the case indices i and j coincide or not and γ is the variogram model. Cokriging variance is the following (Isaaks and Srivastava 1989):

$$\sigma_{CK}^2(x_0) = 2\sum_{i=1}^{n}\sum_{\alpha=1}^{n_i}\lambda_\alpha^i \gamma_{ii_0}(x_\alpha, x_0) - \sum_{i=1}^{n}\sum_{j=1}^{n}\sum_{\alpha=1}^{n_i}\sum_{\beta=1}^{n_i}\lambda_\alpha^i \lambda_\beta^j (x_\alpha, x_0) \tag{7.7}$$

3 Case Study

3.1 Study Area

The study area is located in the Tavoliere di Puglia, the largest alluvial plain of Southern Italy which extends for about 2830 km^2 in the Northern Apulia region (Fig. 7.3).

From the geological standpoint, the Tavoliere di Puglia represents the Northern sector of the Bradanic foredeep, which is bounded to the West by the Appenines

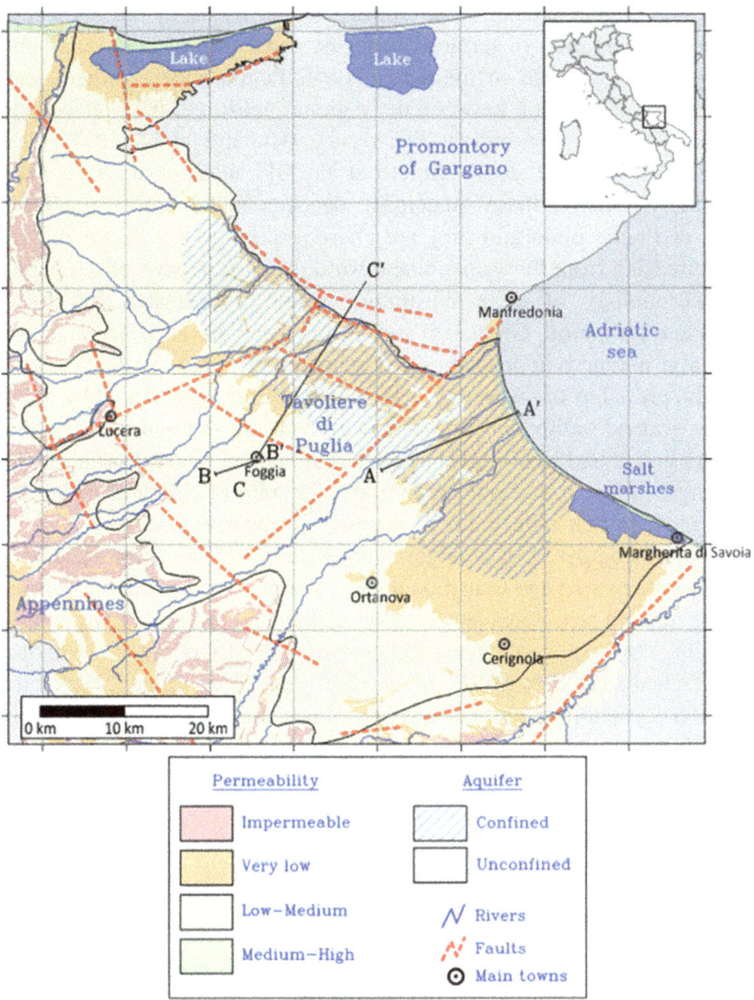

Fig. 7.3 Study area: Hydrogeology

chain and to the East by the Northern sector of the Apulian foreland represented by the Gargano Promontory.

Hundreds of meters thick Pliocene and Quaternary deposits cover the Cretaceous calcareous substratum belonging to the Apulian foreland and tectonically lowered toward the Appennine.

The older units in this foredeep sector consist of shallow-marine carbonate deposits ("Calcarenite di Gravina" formation) passing upward to a thick silty-clayey layer ("Argille subappennine" formation), which were sedimented during the Pliocene and early Pleosticene stages of subsidence (Pieri et al. 1996).

During the Quaternary period an uplift phase of the Bradanic Foredeep determined the sedimentation of terraced deposits, consisting of both continental and marine synthems grouped in the Tavoliere di Puglia super-synthem, each one resulting by the interaction between the regional uplift and the glacioeustatic sea level changes during the middle Pleistocene (Ciaranfi et al. 2011; Gallicchio et al. 2002; Moretti et al. 2010; Moretti et al. 2011; Pieri et al. 2011; Gallicchio et al. 2014). Actually, these quaternary deposits widely outcrop all over the Tavoliere di Puglia plain and they are characterized by a variable thickness, in general increasing from the Appennine toward the eastern edge of the plain, based on the morphology of their substratum mainly represented by the Argille Subappennine formation. Grain size and texture also vary all over the area mainly due to local differences in deposition environment. Generally in the Western sector of the plain, closed to the Appennine, the synthems are represented by debris flow and coarse grained sediments deposited in alluvial fun settings, whereas they pass eastward to gravel and sand-gravel deposited in braided alluvial plain. At the same time, changes in grain size and texture are also observed among synthems with different ages, so that older alluvial deposits are coarser and with poorly sorted structure when compared to younger synthems. In the easternmost sector of the study area some marine and transitional subsynthems outcrop, mainly composed by coarsening upward sequences deposited in deltaic and proximal marine settings.

According to the geological settings of the Tavoliere di Puglia, three hydrogeological structures can be detected in the subsoil (Fig. 7.4): (a) the deep karst aquifer located in the cretaceous calcareous substratum, which can be found down to 300–600 m deep; (b) the deeper porous aquifer located in some sandy portions within the Pliocene clay formation; (c) the shallow porous aquifer which lies in the quaternary alluvial and marine deposits.

No vertical hydraulic connection is detected among the overlapping aquifers and important differences are shown in the groundwater flow patterns and in the geochemical properties of groundwater (Maggiore et al. 1996). The shallow aquifer of

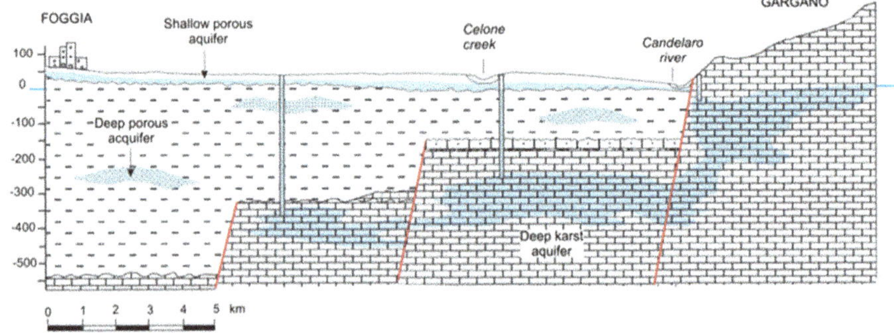

Fig. 7.4 Cross section C-C′: Hydrogeological complexes in the Tavoliere di Puglia area (Masciale et al. 2011, modified)

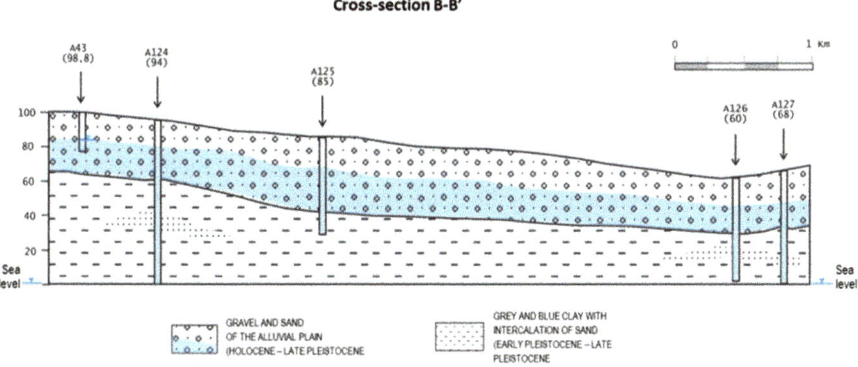

Fig. 7.5 Cross-section B-B′: Unconfined aquifer in the upstream part of the study area

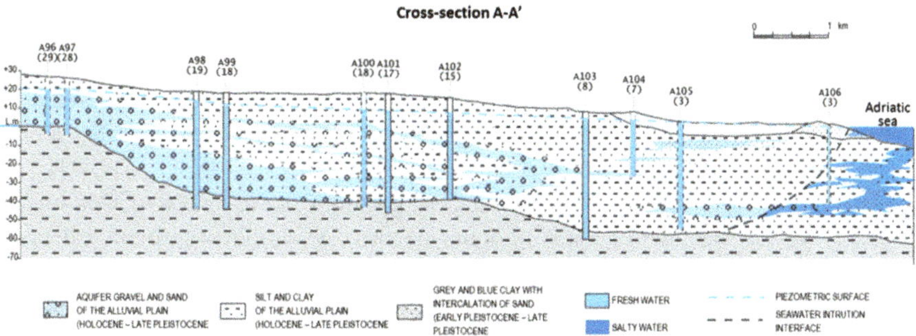

Fig. 7.6 Cross-section A-A′: Confined aquifer in the lower part of the study area

the *Tavoliere di Puglia* is mostly exploited for agricultural uses and it represents the main source of the district water supply. Due to its geological settings it consists of a complex alternation of lenticular alluvial gravel, sand and sandy-loamy sediments frequently interbedded, resulting in a multi-layered aquifer. Moving to the external sectors of the alluvial basin increasing thickness of clay and silt layers are recorded by the borehole stratigraphic data, becoming aquitards that confine the aquifer layers.

On the contrary, the coarse grained sediments prevail in upstream sector of the aquifer located in the South-western part of the study area, at the front of the Apennines chain, where the main component of the aquifer recharge by infiltrating surface water take place due to the high permeability of the outcropping lithology (Tadolini et al. 1989).

Consequently, groundwater circulates in unconfined conditions in the upstream part of the study area (Fig. 7.5) and in confined conditions in the middle-low part (Fig. 7.6). Finally, the presence of these silty-clayey sediments in the confined sector of the aquifer reduces the direct groundwater recharge, whereas an important recharge contribution comes as a baseflow from the South-western, unconfined, sector of the aquifer, probably after long traveling times.

A significant groundwater recharge component is also represented by the river water infiltration during the wet periods (Cotecchia 1956; Maggiore et al. 1996).

However, given the limited lateral continuity of the confining beds, the different aquifer layers seem to be hydraulically interconnected to each other and they are considered as a single complex groundwater flow system.

The bottom of the shallow porous aquifer is always represented by the top of the Lower Pleistocene clay unit (Argille Subappennine formation) and its morphology influences the pattern of the piezometric surface and the groundwater flow.

The groundwater flows mainly in the SW-NE direction under an average hydraulic gradient of about 0.5 %. The top of the clay formation gradually deepens eastward and closed to the coast, where it is about 60 m deep from the ground level, some aquifer layers are located below the sea level and they are affected by saline contamination due to seawater intrusion. The aquifer's size, its hydrogeologically stressed status, and hydrogeochemical characteristics have made it, in the course of time, an ideal experimental field in the framework of several national and international research projects (Barca et al. 2006a, b; Passarella et al. 2006; Lo Presti et al. 2010; Masciale et al. 2011). Because of the intense exploitation of such a resource, local water authorities have been monitoring this water body since the 1990s. The groundwater level monitoring network consists of 61 wells, almost evenly distributed over the study area (Fig. 7.7).

During the last years, monitored values of groundwater levels have been used in order to assess seasonal changes of water resource availability through a detailed estimation of the groundwater table all over the considered area with the main managerial objective of regulating the access to the resource for agricultural uses. The rainfall monitoring network provides valuable information in order to support the local farmers in planning the use of the water resource. It is made of 27 rain gauges spread over the study area and its close neighbouring.

3.2 Groundwater Level Base Statistics

Groundwater levels were measured in the 61 wells of the regional groundwater monitoring network established in 2007 within the project TIZIANO. Considering the extent of the study area, the average competence area for each monitoring site is about 46 km^2. From the starting year up to 2011, ten surveys were carried out providing two measures of water level per year, usually related to the minimum (summer) and maximum (winter) expected values. These values have been aggregated in every well, year by year, in order to produce 61 short time-series, each long 5 years. Finally, each yearly spatial dataset has been used in order to estimate the groundwater table over the whole study area by using the cockriging.

Nevertheless, a general statistical assessment of the measured values has been preventively performed in order to identify possible outliers, correlations and trends. In the following section, a short summary of the results of the statistical analysis is provided and discussed.

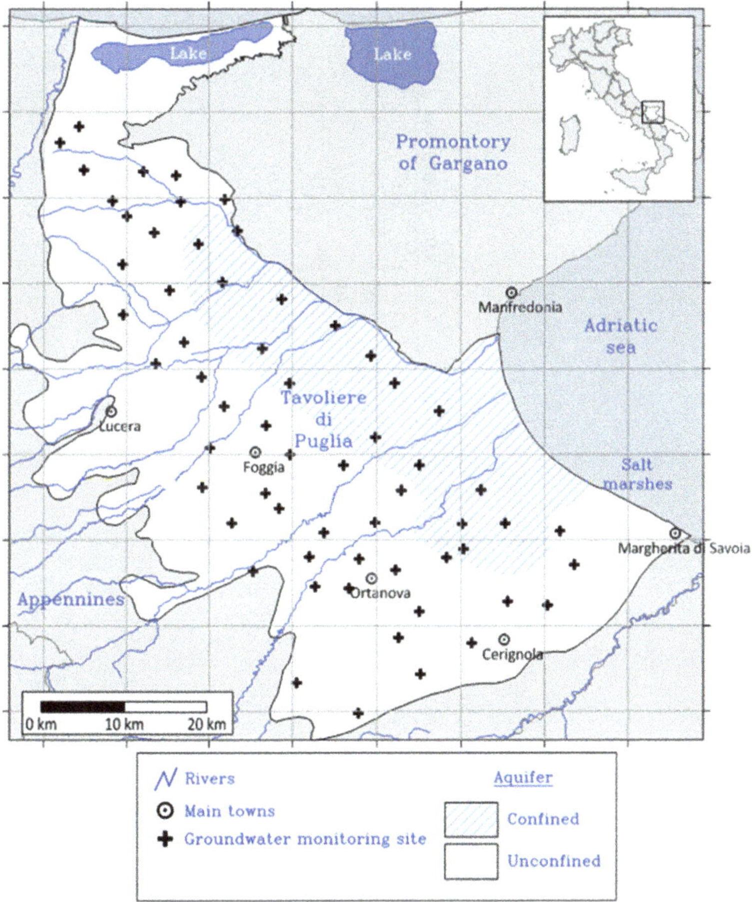

Fig. 7.7 Study area: Groundwater monitoring network

Summary statistics of measured values of groundwater level are reported in Table 7.1, for each year of the monitoring period.

Let's highlight three main points arising from values in Table 7.1.

1. Arithmetic means and standard deviations do not change evidently during the considered period, suggesting an almost constant behaviour of the groundwater system during this period.
2. Within any single year, means and medians are slightly different. Furthermore, skewness and kurtosis values significantly deviate from zero to three, respectively. This indicates that the annual distributions of piezometric levels are not symmetrical.
3. Finally, the relatively small values of the coefficient of variation do not exclude the spatial autocorrelation of each yearly dataset.

Table 7.1 Summary statistics of the measured values of groundwater levels in the aquifer of Tavoliere

Year	N	Mean (m.a.s.l.)	Median (m.a.s.l.)	Minimum (m.a.s.l.)	Maximum (m.a.s.l.)	Std. dev. (m.a.s.l.)	Coef. var.	Skewness	Kurtosis
2007	61	54.51	47.52	−32.33	283.61	56.80	1.04	1.76	5.02
2008	59	55.73	48.29	−29.62	283.77	56.88	1.02	1.77	5.09
2009	60	54.87	47.69	−38.09	284.80	57.76	1.05	1.68	4.84
2010	61	56.04	49.80	−44.12	284.57	58.34	1.04	1.55	4.25
2011	60	55.87	50.02	−36.70	284.63	57.67	1.03	1.67	4.75

Table 7.2 Moran's Index of the measured values of groundwater levels

Year	Moran's index	*z*-score	*p*-value(%)
2007	0.368	2.536	1.12
2008	0.371	2.560	1.05
2009	0.381	2.617	0.89
2010	0.754	5.059	0.00
2011	0.391	2.684	0.73

Table 7.3 Correlation matrix of groundwater levels in the aquifer of Tavoliere

Year	2007	2008	2009	2010	2011
2007	1.000	**1.000**	**1.000**	**0.999**	**0.999**
2008	**1.000**	1.000	**1.000**	**0.999**	**0.999**
2009	**1.000**	**1.000**	1.000	**0.999**	**0.999**
2010	**0.999**	**0.999**	**0.999**	1.000	**1.000**
2011	**0.999**	**0.999**	**0.999**	**1.000**	1.000

In order to test the actual spatial autocorrelation, the Moran's index I (Moran 1950; Li et al. 2007) has been computed and tested. Table 7.2 reports the Moran's indices of the measured values of groundwater levels, per year.

Given the *z*-scores reported in Table 7.2 there is a less than 5 and 1 % likelihood (*p*-values) that the related clustered patterns could be the result of random chance for the first 2 years and the remaining three, respectively. In practice, Table 7.2 points out that groundwater levels are well clustered, which means that they are well spatially autocorrelated. This last evidence allows us using geostatistics for estimating groundwater tables.

Another important information, that we can use in order to define the most reliable geostatistical tool to be applied, can be derived from the assessment of the bivariate correlation. A basic statistical tool for checking if yearly dataset are correlated each other is the correlation matrix. Table 7.3 shows the correlation coefficients related to each pair of yearly groundwater levels dataset, at a significance level of $\alpha = 0.05$. The strong time correlation of the considered variable is evident.

This further result allows us to choose cokriging as geostatistical method, which should improve the estimation accuracy. Finally, any geostatistical analysis cannot neglect to check if the observed phenomenon is characterized by a spatial trend (*drift*). This can be made by checking for the correlation between measured values and the related location (x, y). In the case at hand, a second order spatial model (Eq. 7.8) has been proved to be the most reliable to explain the spatial behaviour of the considered variables by means of appropriate statistical tests.

$$z = a + bx + cy + dx^2 + ey^2 + fxy \tag{7.8}$$

3.3 Estimating Groundwater Table by Ordinary Cokriging

Summarizing the statistical analysis of the groundwater levels, measured in the 61 wells of the aquifer of *Tavoliere*, the following information can be outlined:

- Yearly datasets are evidently spatially autocorrelated. Consequently, geostatistical methods are suitable for estimating the yearly groundwater tables, during the considered period;
- Yearly datasets are characterized by a second order drift;
- Yearly datasets are evidently correlated in time. Consequently, ordinary co-kriging can be a reliable tool for improving the estimation using the datasets of the previous and following years as secondary variables.

Actually, ordinary cokriging has been used to estimate the groundwater table of each year from 2007 to 2011. The set of secondary variables, used to support the estimation, was constantly made by the three most correlated yearly datasets.

Given that the overall dataset was made by four variables, the structural analysis has been a little bit challenging. Four direct- and six cross-variogram models have been assessed per year. The variography evidenced a strong anisotropy mainly oriented in direction SW-NE. Anisotropic, spherical models proved to be the most suitable in order to describe the spatial behaviour of the groundwater level measures. All the models positively passed through a cross-validation stage, graphically summarized in Fig. 7.8, which shows the deviation between the bisector and the regression line of the scatterplot of the predicted values versus the measured ones. Obviously, the lesser this deviation is, the better the model represents the actual behaviour of the considered variable.

Figure 7.9 shows the results of the cokriging application. The five maps of groundwater table over the whole study area, from 2007 to 2011, confirm the well-known behaviour of the groundwater flow regimes of the considered aquifer. Recharge areas are located in the South-Western part of the aquifer, at the Apennines slope and the main flow direction is SW-NE.

3.4 Calibrating the Groundwater Balances

In analogy with what defined above, let's call hydrogeological balance $B_1^{'}$ the available long time series derived from previous studies (AA.VV. 2009; Portoghese et al. 2010). $B_1^{'}$ considered all the classical hydrogeological balance terms apart from the water volumes exchanged at the boundary with other water bodies. On the other hand, the hydrogeological balance $B_2^{'}$ has been assessed for a shorter time period over the same aquifer.

In this case, a different approach has been followed consisting in computing, by subtraction, the gross volumes included between the estimated groundwater tables related to two subsequent years. Obviously, $B_2^{'}$ is overestimated since it represents

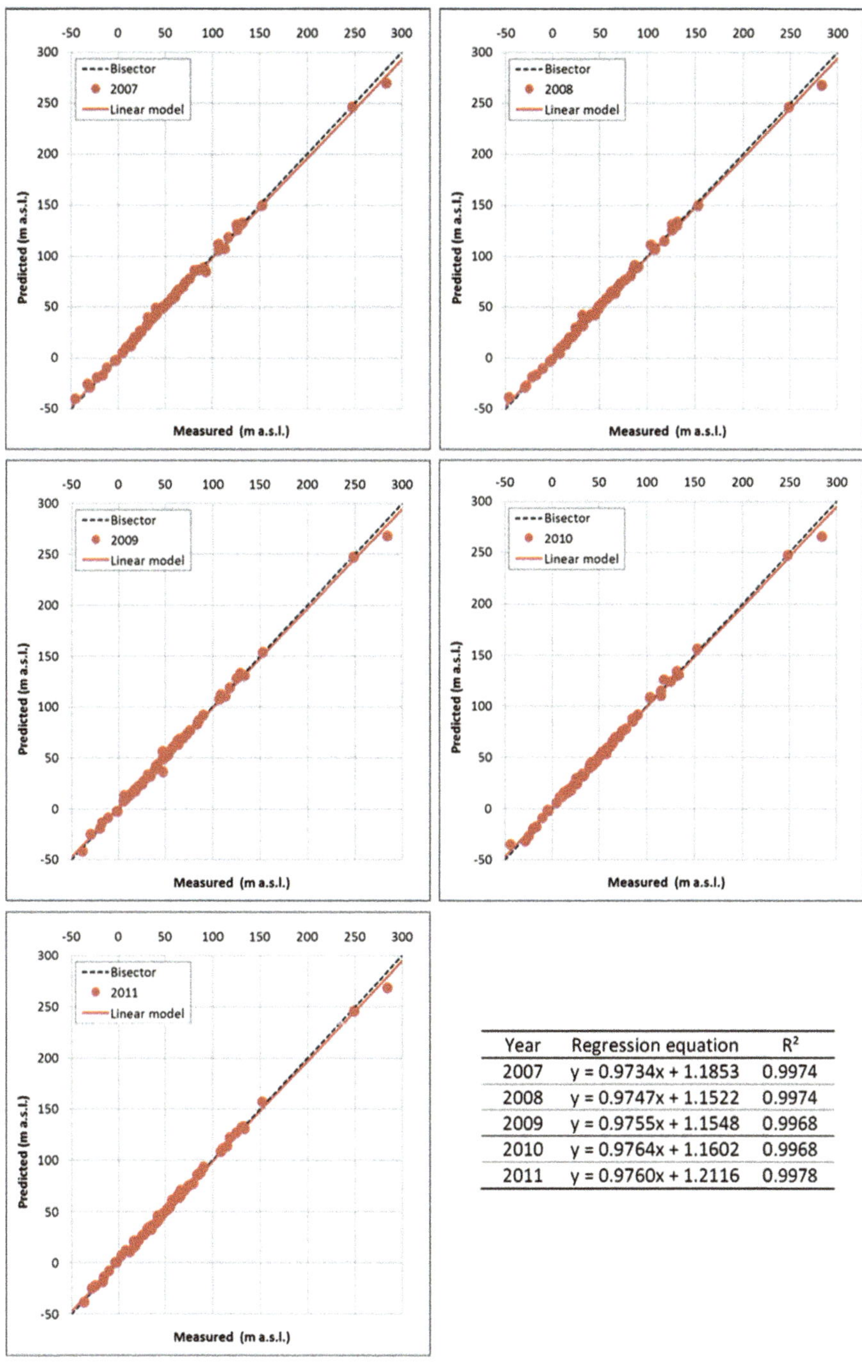

Year	Regression equation	R^2
2007	y = 0.9734x + 1.1853	0.9974
2008	y = 0.9747x + 1.1522	0.9974
2009	y = 0.9755x + 1.1548	0.9968
2010	y = 0.9764x + 1.1602	0.9968
2011	y = 0.9760x + 1.2116	0.9978

Fig. 7.8 Graphical summary of the cross-validation stage

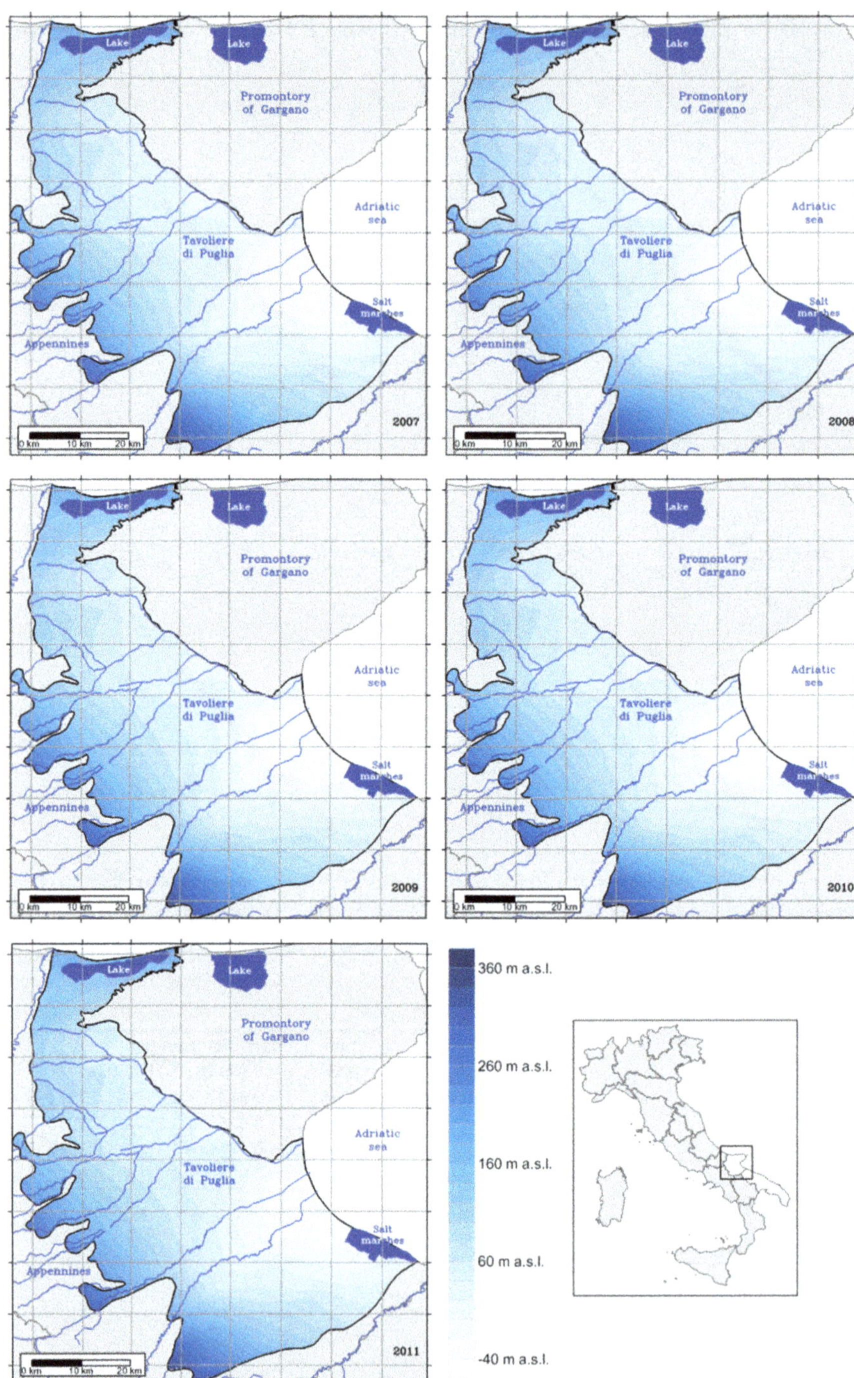

Fig. 7.9 Groundwater tables estimated by means of ordinary cokriging from 2007 to 2011 in the aquifer of the Tavoliere di Puglia

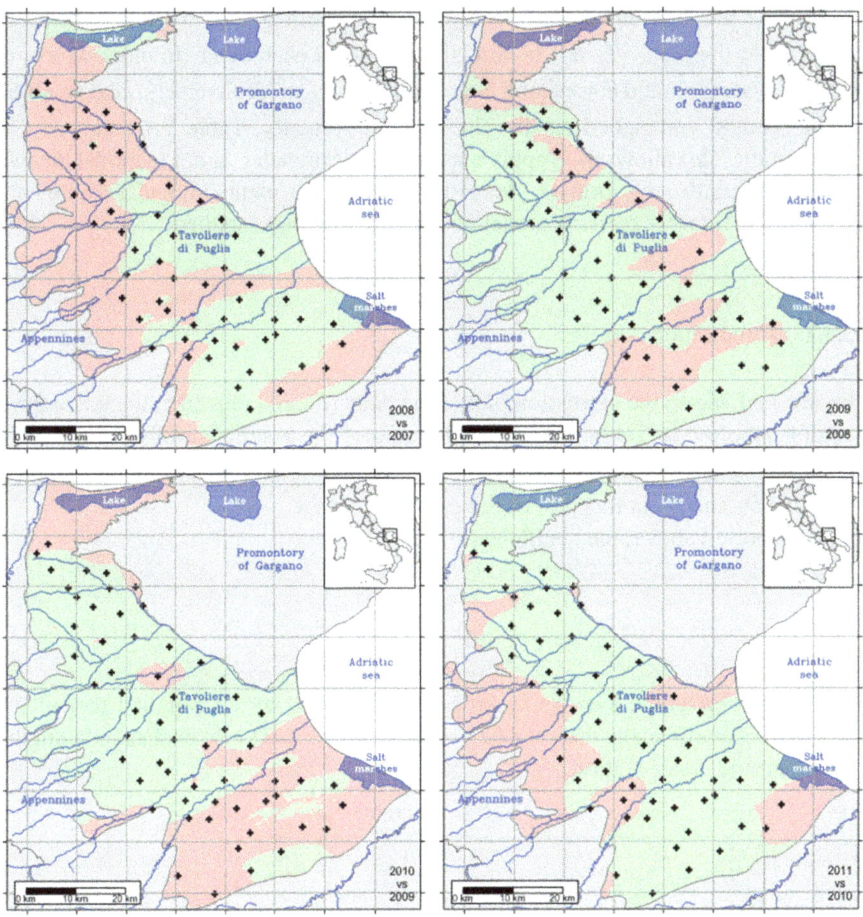

Fig. 7.10 Areas of positive (*green*) and negative (*light red*) hydrogeological balances B_2'

Table 7.4 Pairs (B_2', B_1') for the considered years (m^3)

	Unconfined		Confined	
Year	B_2'	B_1'	B_2'	B_1'
2008	−396.42	−38.05	594.54	−22.72
2009	1121.09	45.30	689.26	−17.23
2010	255.69	13.81	337.71	−32.38
2011	−308.71	−36.34	1561.98	9.59

the total volume and not the specific yield (S_y). Figure 7.10 shows the areas characterized by positive (green) and negative (ligth red) hydrogeological balances B_2', over the whole aquifer and for the considered period.

Table 7.4 reports the pairs (B_2', B_1') for the 4 years considered in this study.

Despite the "inaccuracy" in the estimation of both $B_1^{'}$ and $B_2^{'}$, it is realistic to expect that the pairs $(B_2^{'}, B_1^{'})$ be directly correlated each other. In other words, at large $B_1^{'}$ values should correspond large values of $B_2^{'}$ and, conversely, at at small $B_1^{'}$ values should correspond small values of $B_2^{'}$. Actually, Table 7.4 confirms this assumption. This allows us to apply a regression approach in order to model the two variables, jointly. As formally defined above, in the methodological framework section, the linear regression model of Eq. (7.1) has been applied.

3.4.1 Unconfined Aquifer

Figure 7.11 shows the regression plot of the pairs $(B_2^{'}, B_1^{'})$ related to the unconfined aquifer reported in Table 7.4. The high value of the coefficient of determination, $R^2 = 0.96$ and the slope of the regression line confirm a good direct correlation between $B_1^{'}$ and $B_2^{'}$ in this part of the considered area.

As reported above, the coefficients:

$$a = 0.0569$$

$$b = -13.373$$

can be viewed as the average specific yield ($S_y = a$) of the confined part of the aquifer of Tavoliere and the average volume of water ($AD = b$) exchanged with the neighbouring water bodies.

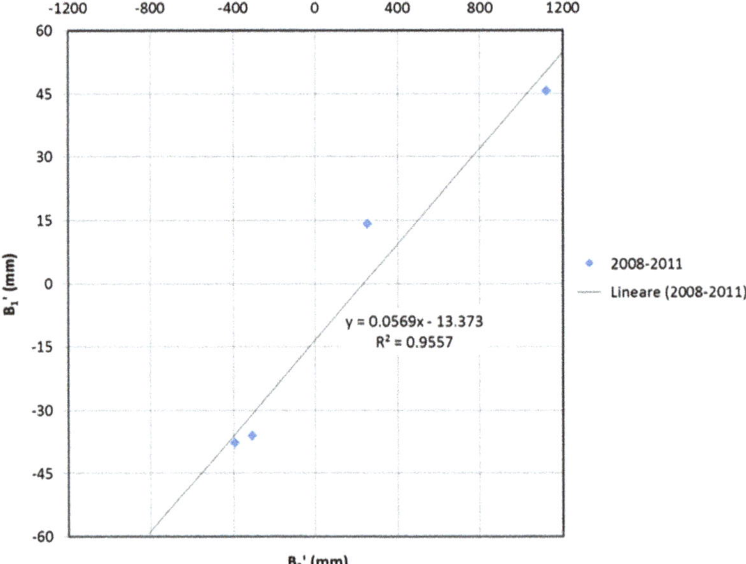

Fig. 7.11 Linear regression model of $(B_2^{'}, B_1^{'})$ for the unconfined aquifer

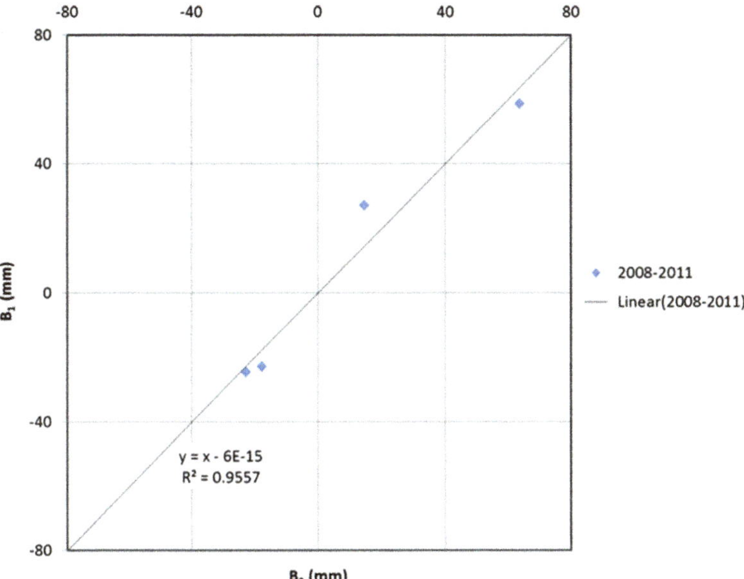

Fig. 7.12 Linear regression model of (B_2, B_1) for the unconfined aquifer

The proposed simultaneous calibration of the hydrogeological balances consists in making the transformations of Eqs. (7.3) and (7.4), that is to say purging $B_1^{'}$ of the quantity *AD* and reducing $B_2^{'}$ by S_y. The transformed balances B_1 and B_2 are plotted in Fig. 7.12. As expected, the regression line has now traslated to the origin of the axes and rotated to the bisector of the first quadrant. This means that, on average, the two balances can be now considered equivalent even though they have been assessed by means of two different and independent approaches.

3.4.2 Confined Aquifer

As for the unconfined aquifer, the pairs $(B_2^{'}, B_1^{'})$ related to the confined aquifer, have been modelled by means of a regression model. Nevertheless, given the specific geological framework described above, we used rainfall values related to the unconfined part of the Tavoliere for assessing $B_1^{'}$. In fact, the presence of clay-sandy sediments in this part of the aquifer prevents direct groundwater recharge, which comes from the unconfined part. Figure 7.13 shows the resulting regression plot of the pairs $(B_2^{'}, B_1^{'})$ related to the confined aquifer.

At first glance, Fig. 7.13, shows an evident violation of the main working assumption of direct correlation between $B_1^{'}$ and $B_2^{'}$. Again, the knowledge of the geological framework of the aquifer suggests the explanation for this result. The expected long travel times of recharging water from the upstream unconfined

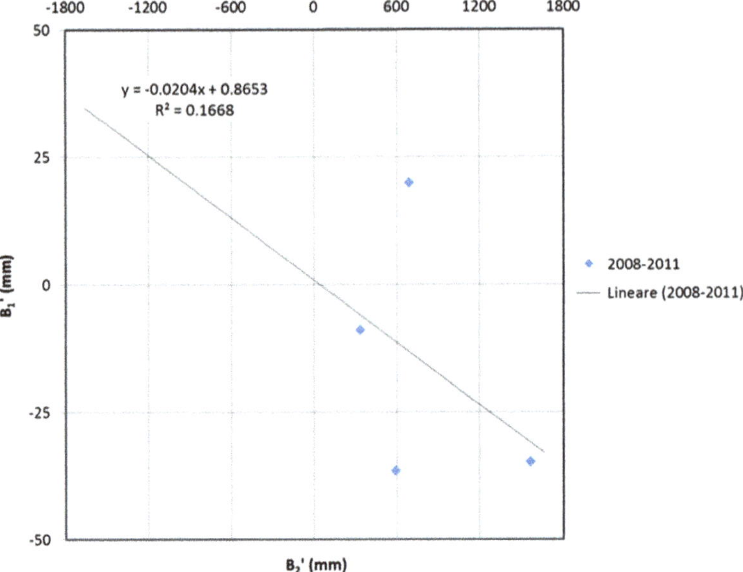

Fig. 7.13 Linear regression model of (B'_2, B'_1) for the confined aquifer. $B'_1(t)$ assessed based on $R(t)$

aquifer to the considered confined part, would suggest to introduce in the B'_1 equation a delay between the rainfall term and the balance itself.

Figure 7.14, then, shows the regression plot of the pairs $(B'_2, B'_1(t))$, where $B'_1(t)$ has been assessed based on the rainfall term related to the previous year, $R(t-1)$. Once again, it fails the working assumption of direct correlation between B'_1 and B'_2. Consequently, we introduced a 2 years delay between $B'_1(t)$ and the rainfall term, $R(t-2)$. The related regression plot is shown in Fig. 7.15, which, finally, proves the accordance between the two balances series. In fact, the high value of $R^2 = 0.99$ and the slope of the regression line confirm a good direct correlation between B'_1 and B'_2. This first result allows us to affirm that the effect of the rainfall above the unconfined part of the considered aquifer becomes evident in the confined part after about 2 years.

Then, also in this case, the coefficients:

$$a = 0.033$$
$$b = -42.497$$

can be viewed as the average specific yield ($S_y = a$) of the confined part of the aquifer of Tavoliere and the average volume of water ($AD = b$) exchanged with the neighbouring water bodies.

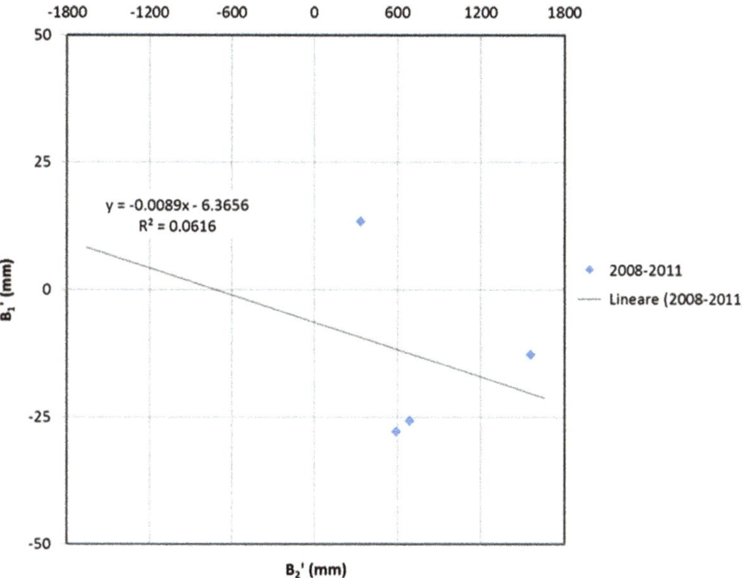

Fig. 7.14 Linear regression model of $(B_2^{'}, B_1^{'})$ for the confined aquifer. The $B_1^{'}(t)$ assessment is based on $R(t-1)$

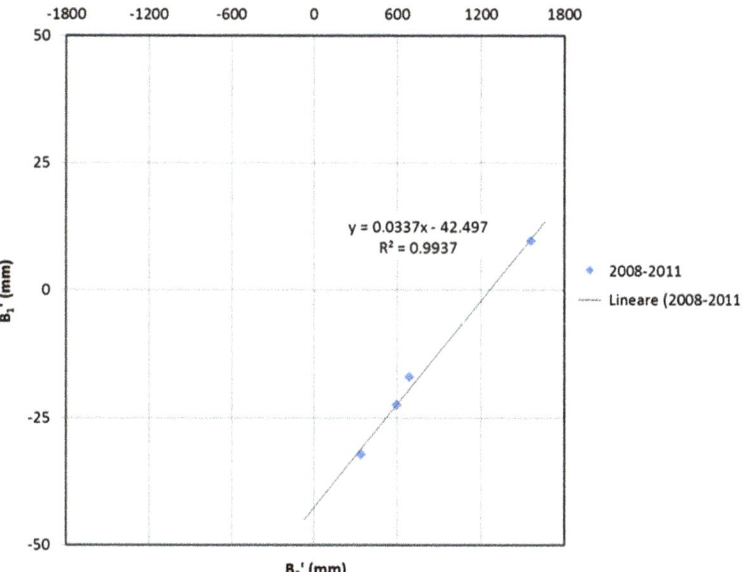

Fig. 7.15 Linear regression model of $(B_2^{'}, B_1^{'})$ for the confined aquifer. The $B_1^{'}(t)$ assessment is based on $R(t-2)$

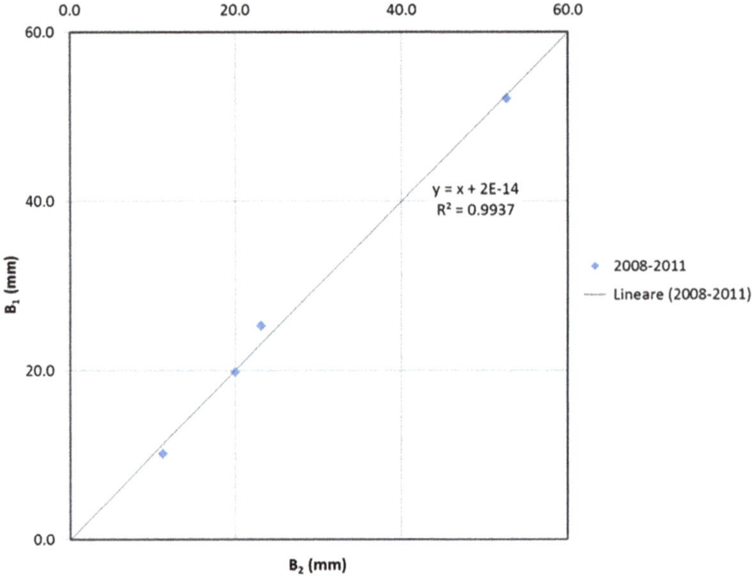

Fig. 7.16 Linear regression model of (B_2, B_1) for the confined aquifer

As for the unconfined case, the calibration of both the hydrogeological balances has been carried out. The transformed balances B_1 and B_2 are plotted in Fig. 7.16. As expected, the regression line now traslated to the origin of the axes and rotated to the bisector of the first quadrant. Once more, this means that, on average, the two balances can be now considered equivalent even though they have been assessed by means of two different and independent approaches.

3.5 Assessing a Rainfall Threshold for Predicting Stress Status of Groundwater Systems

Let's now investigate the relationship between the adjusted groundwater balance B_1 and corresponding rainfall term R. Obviously, being rainfall the main term of groundwater recharge, we expect a strong correlation between B_1 and R.

Once more, a regression model has been applied to describe such a relationship in order to assess the rainfall threshold, below which the considered aquifer can undergo to a stress status. In these cases, in fact, the recharge cannot be sufficient to restore the natural and man-induced water losses.

Figures 7.17 and 7.18 show the regression models related to both the unconfined and confined groundwater systems. In particular, both the long time series (1950–2011) of B_1 and the shorter one B_2 are plotted versus the related rainfall values. In these plots, also the 95 % confidence bands have been added in order to take into account of the regression uncertainty. These plots bear a valuable

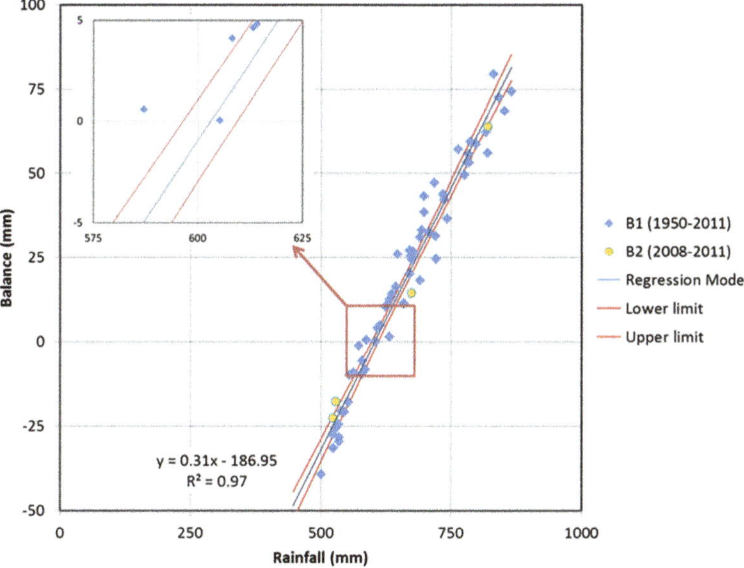

Fig. 7.17 Scatterplot and confidence bands of Rainfall and Balances for the unconfined part of the aquifer of Tavoliere di Puglia

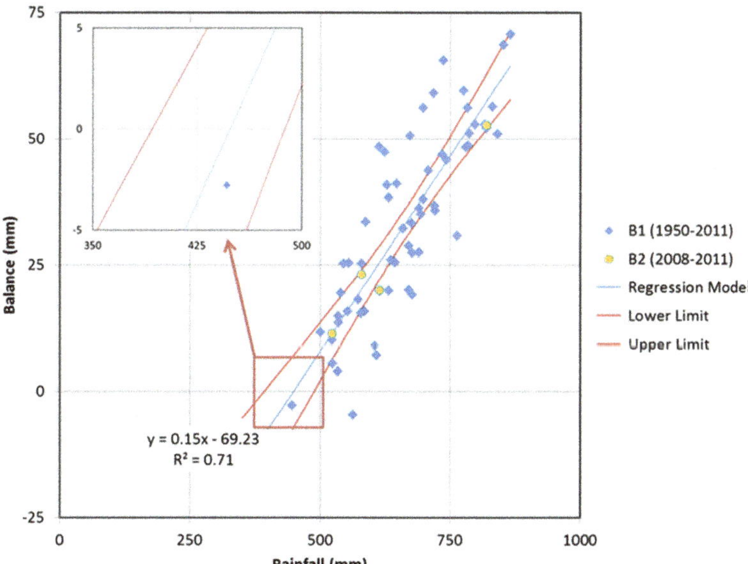

Fig. 7.18 Scatterplot and confidence bands of Rainfall and Balances for the confined part of the aquifer of Tavoliere di Puglia

information concerning the rainfall threshold for assessing possible stress status of groundwater systems. In fact, the intersection of the confidence bands with the rainfall axis, provides an interval of rainfall values below which the aquifer can undergo to a stress status with a likelihood of 95 %.

For the unconfined part of the considered aquifer, the resulting yearly rainfall interval ranges from 596 to 609 mm (Fig. 7.17); similarly it ranges from 392 to 488 mm, for the confined part (Fig. 7.18).

Let's briefly comment the difference between the coefficients of determination related to plots of Figs. 7.17 and 7.18.

These figures, in fact, show that the regression *R*-*B* of the unconfined part of the aquifer is extremely good, differently from that related to the confined part, which is slightly worse. In our opinion, the reason of the difference mainly depends on the complexity of the confined system.

As said above, the confined part of the aquifer is fundamentally recharged by rainfall in the upstream unconfined part, but the effects of this contribution becomes evident after about 2 years. Nevertheless, direct rainfall over this part of the aquifer, even though not contributing to the groundwater recharge at all, strongly impacts the irrigation. In this framework, the assessment of B_1 has been carried out considering the positive terms of the unconfined part, delayed by 2 years, and the current negative terms assessed in the confined part.

In conclusion, we guess that the physical and computational complexity of the system increases the global uncertainty of the proposed model and then lowers, in some way, the accordance between rainfal and groundwater balance, which, is still good anyway. A more detailed evaluation of the considered parameters and, in particular, a shortening of the aggregation periods (from year to season or month) could allow us tuning the assessment of the correct time delay to be considered as travel time of groundwater from the upstream part to the confined part. This should reduce the global uncertainty of the approach and provide more accurate results.

4 Results and Discussion

In this work we propose a calibration method based on the evident differences between the hydrogeological balance of the aquifer of *Tavoliere delle Puglie* assessed by two different and indipendent approaches. In the case at hand, $B_1^{'}$ has been assessed for a long time series, from 1950 to 2011, by means of the classical hydrogeological balance model, while $B_2^{'}$ was assessd as difference between the groundwater tables estimated for two subsequent years, but only from 2008 to 2011. Nevertheless, such two balance estimations were both affected by two different forms of inaccuracies:

- $B_1^{'}$ does not embody in the balance equation the term inflow/outflow from/to neighbouring water bodies (*AD*);
- $B_2^{'}$ represents the total gross volume (V_T) included between two subsequent yearly, estimated water tables. It should be reduced to the specific yield (S_y) in order to assess the actual volume of water draining from V_T.

Now, in theory, $B_1^{'}$ and $B_2^{'}$ should represent the same quantities and, consequently, they should be equal. Nevertheless, the lack of information related to *AD* and S_y, makes $B_1^{'}$ and $B_2^{'}$ different.

The proposed method compares the two short timeseries (2008–2011) by means of a regression model, which represents the above described deviation between $B_1^{'}$ and $B_2^{'}$ by a roto-traslation of the regression line. Intercept and slope of such a regression model have been interpreted as the average values of *AD* and S_y, respectively. The study aquifer was split into two connected parts, an upstream unconfined part and a downstream confined one.

Table 7.5 reports the estimated values of the intercept and slope and of the related confidence limits (95 %), for both the considered aquifer parts. The slope (specific yield) ranges from 0.06 to 0.03 for the unconfined and the confined part of the aquifer, respectively. Previous field tests carried out in the framework of a comprehensive study aimed to the mathematical modellization of the regional groundwater bodies (AA.VV. 1987), assessed values of specific yields ranging from 0.01 to 0.05 and from 0.001 to 0.005 for the unconfined and confine parts respectively. Such values seem to confirm, approximately, those found in this work even though better in the unconfined part.

Concerning the intercept (net water volumes exchanged with other water bodies), let's highlight a couple of notes. First of all, the central values must be considered as average values over the considered period, and this justifies the relatively wide confidence range, partly also due to the shortness of the considered timeseries.

Furthermore, the confined aquifer, always characterized by a range of negative values, well reflects the knowledge about the quantitative status of this water body, which is extremely stressed.

Finally, the assessed range of *ADs* have been downscaled and assigned to the yearly balances of the whole long timeseries on the basis of the related estimated recharge rates. This allowed us to put in relationship the yearly rainfall rates and the related values of B_1. Once again, the linear regression model has been used to explore such a relationship. As expected, the coefficients of determination of the regressions (Table 7.6), confirm a significant correlation between these two

Table 7.5 Estimated values of S_y (slope) and AD (intercept) and related confidence limits

Aquifer	N	R^2	Lower limit	Slope	Upper limit	Lower limit	Intercept	Upper limit
Unconfined	4	0.96	0.00	0.06	0.11	−47.10	−13.37	20.36
Confined	4	0.98	0.01	0.03	0.04	−54.67	−42.50	−30.32

Table 7.6 Hydrogeological balance B_1 versus Rainfall

Aquifer	R^2	Slope	Intercept	Threshold lower limit (mm/year)	Average threshold (mm/year)	Threshold upper limit (mm/year)
Unconfined	0.97	0.310	−187.2	596	603	609
Confined	0.71	0.128	−55.38	392	448	488

Regression parameters and threshold of groundwater deficit

quantities, particularly for the unconfined part of the considered aquifer. The confidence bands of the regression line allows also one to assess the uncertainty of the regression model (Figs. 7.17 and 7.18). These relationships, can be viewed as a quick and easy predictive tool for assessing a range of plausible water balances in a given groundwater body when the total annual rainfall is known. In particular, the range of rainfall rates corresponding to the intersection of the confidence band with the x-axis, indicates a threshold interval, below which the considered aquifer can undergo to a stress status. In these cases, in fact, the recharge cannot be sufficient to restore the natural and man-induced water losses. In Table 7.6, the rainfall threshold intervals for the considered aquifers are reported. The unconfined part of the considered aquifer results to be characterized by a higher rainfall thresholds, probably because of its own hydrogeological conditions (higher permeabilities).

5 Conclusions

In this work, several quick and easy regressive tools have been proposed capable of supporting water authorities in the assessment of yearly groundwater avilability. All the proposed tools embody a comprehensive assessment of the models uncertainty, in order to provide the manager with a reliable decision support tool. The methodologies have been applied to a complex confined-unconfined aquifer of the Apulia Region located in the South Italy. The proposed method provided results in accordance with physical knowledge of the study area, even though the physical complexity of the confined groundwater system increases the global uncertainty of the proposed model. A more detailed evaluation of the considered parameters such as a shortening of the aggregation periods should reduce the global uncertainty of the approach and provide more accurate results. Further methodological and practical improvements and refinements of the proposed methodology are currently under study.

Acknowledgements Piezometric data used in the present work have been collected within the project "TIZIANO" for the qualitative and quantitative monitoring of the regional groundwater bodies. Thanks are due to the Water Protection Department of the "Regione Puglia" who kindly provided the permission for publishing the data.

References

AA.VV. (1987) Modello Matematico dei Sistemi Acquiferi Sotterranei dei Bacini dei Fiumi Ofanto e Fortore. Ministero de Lavori Pubblici, Provveditorato alle Opere Pubbliche per la Puglia

AA.VV. (2009) Piano di Tutela delle Acque della Regione Puglia (*Water Protection Plan of the Apulia Region*)

Barca E, Passarella G, Lo Presti R, Masciale R, Vurro M (2006) HarmoniRiB River Basin data documentation. Chapter 7—Candelaro River Basin. HarmoniRiB project deliverables, vol D6.3. pp 1–47

Barca E, Passarella G, Lo Presti R, Masciale R, Vurro M (2006) Candelaro River Basin, Italy—A HarmoniRiB case study. HarmoniRiB project deliverables, vol D7.6. pp 1–54

Caputo M, Passarella G, Vurro M, Giuliano G (1998) Water and the environment: innovation issues in irrigation and drainage. In: Managing the effects of agricultural practices on groundwater quality, vol 111, ISBN: 0-419-23710-0

Castrignanò A, Giugliarini L, Risaliti R, Martinelli N (2000) Study of spatial relationships among some soil physico-chemical properties of a field in central Italy using multivariate geostatistics. Geoderma 97(1):39–60

Ciaranfi N, Loiacono F, Moretti M (2011) Note Illustrative della Carta Geologica d'Italia alla scala 1:50,000, Foglio 408 'Foggia', 69 pp. ISPRA, Roma

Cotecchia V (1956) Gli aspetti idrogeologici del Tavoliere delle Puglie. L'Acqua 11–12:168–180

Gallicchio S, Pieri P, Festa E, Moretti M, Tropeano M (2002) Caratteri geologici del Foglio 407 "San Bartolomeo in Galdo". Cartografia Geologica. Bologna III:136–139

Gallicchio S, Moretti M, Spalluto L, Angelini S (2014) Geology of the middle and upper Pleistocene marine and continental terraces of the northern Tavoliere di Puglia plain (Apulia, southern Italy). J Maps. doi:10.1080/17445647.2014.895436

Isaaks EH, Srivastava RM (1989) Applied geostatistics. Oxford University Press, New York, p 561

Kehew AE, Straw WT, Steinman WK, Bárrese PG, Passarella G, Peng WS (1996) Ground-water quality and flow in a shallow glaciofluvial aquifer impacted by agricultural contamination. Groundwater 34(3):491–500

Li H, Calder CA, Cressie N (2007) Beyond Moran's I: testing for spatial dependence based on the spatial autoregressive model. Geogr Anal 39(4):357–375

Lo Presti R, Barca E, Passarella G (2010) A methodology for treating missing data applied to daily rainfall data in the Candelaro River Basin (Italy). Environ Monit Assess 160(1–4):1–22

Maggiore M, Nuovo G, Pagliarulo P (1996) Caratteristiche idrogeologiche e principle differenze idrochimiche delle falde sotterranee del Tavoliere di Puglia. Mem Soc Geol It 51:669–684

Masciale R, Barca E, Passarella G (2011) A methodology for rapid assessment of the environmental status of the shallow aquifer of "Tavoliere di Puglia" (Southern Italy). Environ Monit Assess 177(1–4):245–261

Masciopinto C, Passarella G, Vurro M, Castellano L (1994) Numerical simulations for the evaluation of the free surface history in porous media. Comparison between two different approaches. Adv Eng Softw 21(3):149–157

Matheron G (1970) The theory of regionalised variables and its application. In: Les Cahiers du Centre de Morphol. Mathemat. de Fointainebleau. Ecole des Mines, Paris, p 211

Moran PA (1950) Notes on continuous stochastic phenomena. Biometrika 37(1–2):17–23

Moretti M, Gallicchio S, Spalluto L (2010) Evoluzione geologica del settore settentrionale del Tavoliere di Puglia (Italia meridionale) nel Pleistocene medio e superiore. Il. Quaternario 23 (2):181–198

Moretti M, Pieri P, Ricchetti G, Spalluto L (2011) Note Illustrative della Carta Geologica d'Italia alla scala 1:50.000, Foglio 396 'San Severo', 145 pp. ISPRA, Roma

Neter J, Kutner MH, Nachtsheim CJ, Wasserman W (1996) Applied linear regression models. Irwin, Chicago, p 1050

Passarella G, Vurro M, D'agostino V, Barcelona MJ (2003) Cokriging optimization of monitoring network configuration based on fuzzy and non-fuzzy variogram evaluation. Environ Monit Assess 82(1):1–21
Passarella G, Barca E, Lo Porto A (2006) Collection and elaboration of data for the pilot area of the Candelaro River (Italy). Development of completed database accompanied by appropriate software package for the retrieval, analysis and processing of flood related data. Description of the Candelaro Catchment. In: FloodMed project deliverables 2.2. p 1–57
Pieri P, Sabato l, Tropeano M (1996) Significato geodinamico dei caratteri deposizionali e strutturali della Fossa Bradanica nel Pleistocene. Mem Soc Geol It 51:501–515
Pieri P, Gallicchio S, Moretti M (2011) Note Illustrative della Carta Geologica d'Italia alla scala 1:50.000, Foglio 407 'San Bartolomeo in Galdo', 103 pp. ISPRA, Roma
Portoghese I, Matarrese R, Milella P (2010) Studio del bilancio riguardante i principali corpi Idrici sotterranei della puglia e gli acquiferi Superficiali in ambiente carbonatico Della penisola salentina. Convenzione tra l'Autorita' di Bacino della Puglia e l'IRSA-CNR per l'"Aggiornamento del Bilancio Idrogeologico dei Corpi Idrici Sotterranei della Regione Puglia"
Tadolini T, Sdao F, Ferrari G (1989) Valutazioni sul grado di protezione della falda superficiale del Tavoliere di Foggia nei confronti dei rilasci in superficie di corpi inquinanti e sulle modalità di propagazione degli stessi in seno all'acquifero. Atti delle giornate di studio su Analisi Statistica di Dati Territoriali, 1989, Bari: 461–472

Chapter 8
Evaluation of Submarine Groundwater Discharge as a Coastal Nutrient Source and Its Role in Coastal Groundwater Quality and Quantity

Henrietta Dulai, Alana Kleven, Kathleen Ruttenberg, Rebecca Briggs, and Florence Thomas

Abstract Globally, submarine groundwater discharge (SGD) is responsible for 3–4 times the water discharge delivered to the oceans by rivers. Moreover, nutrient concentrations in SGD are usually elevated in comparison to river fluxes. Here we review the major advances in the field of SGD studies and related nutrient fluxes to the coastal ocean. To demonstrate the significance of SGD as terrestrial nutrient pathway we compare stream and submarine groundwater discharge rates in a watershed on the windward side of Oahu, one of the major islands of the Hawaii archipelago. Our analysis of Kaneohe Bay, which hosts the largest coral reefs on the island revealed that SGD in the form of total (fresh+brackish) groundwater discharge was 2–4 times larger than surface inputs. Corresponding DIN and silicate fluxes were also dominated by SGD, while DIP was delivered mostly via streams. We quantified bulk nutrient uptake in coastal waters and also demonstrated that nutrients were quickly removed from the bay due to fast coastal flushing rates. This study demonstrates the need to understand SGD-derived nutrient fluxes in order to evaluate land-based coastal nutrient and pollution sources.

H. Dulai (✉) • A. Kleven
Department of Geology and Geophysics, University of Hawaii-Manoa, Honolulu, HI, USA
e-mail: hdulaiov@hawaii.edu

K. Ruttenberg
Department of Geology and Geophysics, University of Hawaii-Manoa, Honolulu, HI, USA

Department of Oceanography, University of Hawaii-Manoa, Honolulu, HI, USA

R. Briggs
Department of Oceanography, University of Hawaii-Manoa, Honolulu, HI, USA

F. Thomas
Hawaii Institute of Marne Biology, University of Hawaii-Manoa, Honolulu, HI, USA

A. Fares (ed.), *Emerging Issues in Groundwater Resources*, Advances in Water Security, DOI 10.1007/978-3-319-32008-3_8

Abbreviations

A_{Rn_cw}	Coastal water radon activity
A_{Rn_gw}	Groundwater radon activity
C_T	Terrestrial nutrient concentration
CB	Central Kaneohe Bay
CI	Coconut Island
DIN	Dissolved inorganic nitrogen
DIP	Dissolved inorganic phosphorus
DON	Dissolved organic nitrogen
DOP	Dissolved organic phosphorus
dpm	Decays per minute
GPS	Global positioning system
gw	Groundwater
HFP	Heeia Fishpond
I	Effective terrestrial end-member nutrient concentration
K_h	Horizontal eddy diffusion coefficient
L	Length
n	Number
NB	Northwest Kaneohe Bay
Q_T	Terrestrial water flux
Q_{SGD}	Submarine groundwater discharge flux
R	Nutrient removal rate
Ra	Radium
Ra_i	Nearshore water radium activity
Ra_o	Offshore water radium activity
Rn	Radon
SGD	Submarine groundwater discharge
STE	Subterranean estuary
sw	Surface water
t	Time, residence time, flushing rate
$T_{1/2}$	Radionuclide half-life
T1	Transect 1
T2	Transect 2
T3	Transect 3
TDN	Total dissolved nitrogen
TDP	Total dissolved phosphorus
Th	Thorium
U	Uranium
V	Volume
λ	Radionuclide decay constant

1 Introduction

1.1 SGD: General Description

Submarine groundwater discharge (SGD) consists of fresh meteoric water and recirculated seawater that flows through the coastal aquifer into coastal waters (Taniguchi et al. 2002). Most often it emerges as a mixture of fresh and saline water masses resulting in a full salinity range (Michael et al. 2005; Santos et al. 2012; Gonneea and Charette 2014). Generally elevated SGD is associated with certain characteristics of coastlines such as steep topography with permeable geology and high rainfall (Bokuniewicz et al. 2003). However, groundwater fluxes have been identified on all seven continents and can occur under non-typical conditions. For example, prolific meteoric water discharge has been found associated with otherwise desert-like watersheds (Johnson et al. 2008), seeping inconspicuously under estuaries (Moore 1997; Dulaiova et al. 2006; Peterson et al. 2009; Wang et al. 2014) and on coastlines with absent freshwater fluxes dominated by seawater recirculation (Kiro et al. 2013).

Based on coastal hydrological principles, SGD distribution in a cross-shore direction is expected to decrease exponentially with increasing distance offshore (Taniguchi et al. 2003). The presence of confining layers however, allows water discharge to occur kilometers from shorelines and at significant ocean depths (Moore and Wilson 2005). The majority of reported SGD studies have been performed in shallow coastal regions where its impact is the most significant in terms of pollution. Here the SGD signature is magnified by less dilution due to lesser water volumes and longer coastal residence times. As a consequence, the literature is biased towards SGD in the nearshore region (Bratton 2010).

The focus of SGD studies has diverged in multiple directions, including hydrological, geochemical, ecological, and coastal management aspects of SGD. Studies of biogeochemical processes in the subterranean estuary (STE), a subsurface zone of mixing between fresh groundwater and recirculated seawater (Moore 1999), helped to explain the composition of discharging fluids influenced by nutrient transformations (Kroeger and Charette 2008; Santos et al. 2008, Kim et al. 2012), trace metal cycling (Charette et al. 2005; Beck et al. 2009, 2013; Gonneea et al. 2008), and microbial activity (Santoro et al. 2006) in the STE. It has become clear that groundwater geochemical signatures undergo significant changes in the STE just before discharging into the coastal zone. There has also been progress in the understanding and description of terrestrial and marine driving forces of SGD (Michael et al. 2005; Robinson et al. 2006, 2007; Li and Jiao 2013; Gonneea et al. 2008). Gonneea and Charette (2014) illustrated that in addition to the terrestrial drivers such as precipitation and groundwater extraction, sea-level anomalies had a quantifiable effect on the magnitude and composition of SGD. Relevant to this marine forcing is the expected effect of the advancing sea level rise, which in addition to increased seawater intrusion is predicted to change biogeochemical interactions in the STE and may result in increased SGD solute fluxes (e.g. Roy et al. 2010). While most of the early literature focused on nutrient (e.g. Slomp and Van Cappellen 2004; Andersen et al. 2007; Bowen and Valiela 2001) and trace metal SGD fluxes (e.g. Beck et al. 2009), there is an emerging trend of a more interdisciplinary focus on SGD and its ecological consequences, including ocean acidification and its effect on coral reefs (Cyronak et al. 2014; Santos et al. 2013), linking SGD nutrient inputs to enhanced primary productivity (Waska and Kim 2011) and algal proliferation leading to coral reef degradation (Dailer et al. 2010; Smith et al. 2001).

1.2 SGD-Derived Nutrient Fluxes

Total SGD consists of meteoric fresh groundwater, which is responsible for supplying allochthonous, new terrestrial nutrients to the coastal zone, and recirculated seawater, which either carries nutrients with it from the sea or acquires them as the water flows through the STE and the seabed. Local remineralization of marine organic matter is the origin of nutrients in the latter case, which is not considered a new but an autochthonous nutrient source to the coastal waters; it is still significant, however, because it mobilizes recycled nutrients. Upon exiting the STE, the fate of SGD-derived coastal nutrients is analogous to those in river estuaries in that one may expect the same chemical continuity between groundwater and ocean water as in estuaries. Therefore it is evident that nutrients are being processed through two estuaries, once in the subsurface (the STE) and once on the surface where brackish groundwater plumes mix into the coastal water.

Nutrient fluxes are estimated by quantification of SGD (Moore 2010) and by multiplication of the water discharge by STE nutrient concentrations. This approach requires the assumption that no nutrient uptake processes or sources occur between the STE and the SGD discharge point. But as Moore (2010) points out, while SGD is relatively easy to quantify, constituent fluxes within the STE are so variable that the largest uncertainties in SGD-derived nutrient fluxes stem from the determination of the proper solute nutrient end-member. The most commonly used methods of SGD assessments are geochemical tracer techniques (Charette et al. 2008), thermal imaging (Johnson et al. 2008), geophysical techniques (Dimova et al. 2012), hydrological and watershed models (Gonneea and Charette 2014), and direct measurements using seepage meters (Lee 1977).

In the simplest scenario there is conservative mixing between groundwater nutrients and seawater resulting in a linear trend of nutrients with either salinity (Knee et al. 2010) or a groundwater tracer (Moore 2006) across the STE. This approach assumes that the recirculated seawater is nutrient-poor, and its role in the STE is mainly as a dilution agent. Typically there would also be a limited amount of organic matter remineralization within the STE, resulting in the absence of added recycled nutrients. In this case the SGD-derived nutrient flux is simply the product of fresh SGD and freshwater nutrient concentrations (Knee et al. 2010).

In more complex STE settings various nitrogen attenuation processes have been described with concurrent removal of nitrate and ammonium (Kroeger and Charette 2008; Santos et al. 2010, 2012; Glenn et al. 2013). As a result, SGD has lower nitrogen concentrations than the terrestrial end-member upstream of the STE. Similarly, reactive phosphate readily interacts with solids containing iron and aluminum oxides resulting in its removal and cycling within the STE (Spiteri et al. 2008). Gonneea and Charette (2014) reported a net removal of phosphate within the STE and a change of N:P ratios in terrestrial groundwater before and after flowing through the STE.

Nutrient additions may occur via seawater circulation through organic-rich benthic sediments. This addition is very typical for salt marshes where tidal pumping is one of the major nutrient recycling pathways (Weston et al. 2006; Wilson and Gardner 2006; Wankel et al. 2009). In several cases multiple SGD signatures have been found in the coastal zone suggesting discharges of different water masses with different nutrient compositions. Geochemical tracer balances have been used to identify and quantify these sources (Moore 2003; Charette 2007).

1.3 Nutrient Removal in the Coastal Zone

We can make an analogy between river estuaries and SGD plumes mixing into the coastal ocean. They are different in that estuaries are surficially-expressed, semi-enclosed bodies while SGD discharges along any type of coastline geometry—enclosed embayments as well as well-flushed coastal margins. They are similar, however, in that they are reaction vessels through which terrestrial solutes and solids must pass before entering the ocean (Kaul and Froelich 1984). It is therefore important to understand how SGD-derived nutrients are affected during their passage through the coastal ocean. In some instances nutrients are not stripped from SGD plumes because their transit time is too short with respect to phytoplankton cell division times (Tomasky et al. 2013). In many examples, however, there is significant coastal biological uptake resulting in deviations from conservative estuarine mixing models analogous to those described in rivers (Kaul and Froelich 1984). For example, primary production sustained by SGD-derived nutrients has been documented to result in non-conservative nitrate and silicate mixing trends and a removal of 40–90 % of nutrients within the coastal zone of small islands (Kim et al. 2011).

1.4 Case Study of Coastal Nutrient Fluxes

In this paper we present a case study that demonstrates the combined use of the most commonly applied natural radioisotopic techniques using radon and radium isotopes (Charette et al. 2008). The derived SGD is then used to estimate corresponding nutrient fluxes.

Terrestrial nutrient fluxes are investigated in two parts of Kaneohe Bay, Hawaii. The first is in a section of a watershed where the STE is presumed to play very little role in nutrient removal and a conservative nutrient behavior is expected; the second is a region of the watershed where a coastal wetland significantly alters groundwater and stream nutrient concentrations just before these discharge into the ocean. We illustrate the significance of SGD for coastal nutrient budgets in Kaneohe Bay through the following steps:

- we compare nutrient fluxes via SGD to stream inputs in different sectors of the bay
- we estimate coastal nutrient inventories and coastal residence times
- we study the estuarine behavior of SGD-derived nutrients and estimate net nutrient removal rates.

Due to the unique island watershed characteristics described below, our study site is not typical for continental margins but is a good representative of large islands, which account for the majority of SGD inputs into the Pacific Ocean. On the downstream end of the watershed a coral reef along with associated native and invasive algal communities co-exist in a delicate nutrient balance. In addition, our study site includes a fishpond, built by early Hawaiians who recognized the parts of the coastline where ample nutrient delivery by streams and groundwater discharge could sustain a vibrant aquaculture. With population growth and increased anthropogenic nutrient and sediment fluxes, the pond and the coral reef have been threatened by euthrophication, excess sediment loads, and algal overgrowth. We illustrate that among the various nutrient delivery mechanisms SGD plays a pivotal role in these systems.

2 Methods

2.1 *Study Site*

The Kaneohe watershed is located on the northeast, windward side of Oahu, the third largest island of the Hawaiian archipelago. It consists of several stream-eroded, amphitheater-headed valleys with steep headwalls and alluvial deposits. The deposits are iron and aluminum rich clays with high affinity for phosphate, radium isotopes, and ammonium. Orographic rainfall is typical for all sectors of the steep watershed. Our study focused on the northwest part, which is divided into Waikane, Waiahole and Kaaawa/Hakipuu sub-watersheds (area 36 km^2, precipitation 2.6×10^5 m^3 d^{-1}, 55 % of watershed stream runoff) and Kaneohe (area 56 km^2, precipitation 3.6×10^5 m^3 d^{-1}, 39 % of watershed stream runoff) in the central sector of the bay (Shade and Nichols 1996). The watershed has high-level, dike impounded groundwater and a basal lens which is connected to the coastal zone. Land-use is agriculture and preservation land in the northwest, low-intensity developed and preservation in the central sector, with most urban development located along the coastline and the southern part of the watershed. The coastal plain of the Heeia sector has a 0.81 km^2 wetland. Kaneohe bay is a semi-enclosed embayment with a barrier reef as a seaward boundary. In the lagoon there are patch reefs and fringing reefs, many of which have been substantially modified by the growth of fleshy algae. The fringing reef flats receive land-derived mud, sand, and rubble. The bottom sediments in the nearshore region are mostly noncalcareous clays, while sand bars and hard bottom are more typical outside of the lagoons (Smith et al. 1981).

We selected three shore-perpendicular transects (T1–T3) along fringing reefs in Kaneohe Bay (Fig. 8.1). The array of transects was selected based on specific characteristics of their location that we believed might influence the delivery of nutrients (Table 8.1). For example, Transects 2 in Waiahole and 3 in Kaaawa/ Hakipuu (T2 and T3) are located proximal to freshwater input via stream runoff. In contrast, Transect 1 in Waikane (T1), the northernmost transect, is located in a region with minimal input from surface runoff, and in the most pristine (lowest apparent anthropogenic impact) sector of the bay. Sampling locations in Central Bay (CB Fig. 8.1) only covered the northernmost tip of this subwatershed.

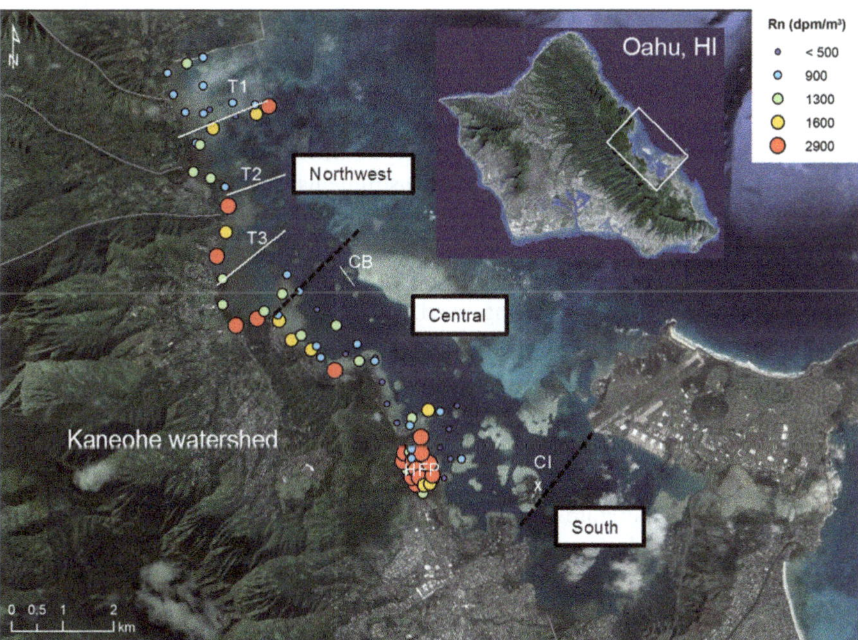

Fig. 8.1 Kaneohe Watershed is located on the windward side of Oahu, HI. The watershed consists of several sub-watersheds (divided by *grey lines*). Kaneohe Bay is composed of three sectors: northwest (NB), central (CB) and south. Radium samples were collected along transects T1–T3 in NB, at a location indicated by CB and in the Heeia Fishpond (HFP). Surface water radon activities (indicated by *colored circles*) were measured along the coastline in NB and CB as well as in HFP. Radon time-series monitoring was performed on Coconut Island (CI) located in the central sector of the bay

In our study we also included Heeia fishpond, which is a 0.39 km^2 walled estuary downstream of the Heeia wetland in the central sector of the bay. It receives water from the Heeia stream and the ocean through channels. It receives approximately 50 % of the stream flow measured at the Haiku stream gauge (USGS 16275000) (Young 2011). The pond is shallow, on average 0.5 m (Timmerman et al. 2015), and is used for aquaculture.

2.2 *Radium Isotopic Sampling and Analysis*

In order to quantify groundwater fluxes and nutrient distribution in the northwest sector of the bay, we examined three transects (T1–T3, Fig. 8.1) that extended from the coastline out to ocean salinities (2000–3000 m). In the central sector we collected only three samples at 2000 m from the shoreline (location CB, Fig. 8.1); these points were not aligned on a transect. Samples were collected in both sectors on August 17, 2010 during a dry period and in the northwest sector following the first big storm on November 4, 2010. Heeia fishpond was sampled on November 19, 2013, when we collected five samples that covered most representative salinity ranges across the pond. Surface water samples were collected into 20-L carboys for radium isotopic analysis, and for nutrient analysis (described below). Salinity was measured at the top and bottom of the water column at the time of sampling using a YSI multiparameter conductivity meter. Radium samples were weighed, filtered through MnO_2-coated acrylic fibers and analyzed on a Radium Delayed Coincidence Counter (Scientific Instruments) for short-lived ^{224}Ra, ^{223}Ra, and ^{228}Th. Excess ^{224}Ra was calculated by subtracting dissolved ^{228}Th activities before decay correction and all ^{224}Ra reported from here on refer to excess ^{224}Ra. ^{227}Ac was below detection limit of our method in all samples and all ^{223}Ra reported here is assumed to be excess ^{223}Ra. Long-lived ^{226}Ra and ^{228}Ra were measured on ashed fiber samples using a high purity germanium detector (Ortec, GEM40).

2.3 *Surface Water Profiling and Nutrient Sampling*

Water column temperature and salinity was profiled using a YSI 6200v Sonde, which was manually lowered off the side of the boat at a steady rate during constant data logging for a continuous profile of these parameters from surface to bottom water. Depth profiles were used to determine the surface mixed layer thickness by determining the depth at which salinities increased to offshore levels. Discrete water samples in the bay and fishpond were collected from surface waters and immediately transferred to shore for filtration. Nutrient samples were filtered through pre-weighed 0.2 μm polycarbonate filters and frozen until analysis. Nutrient samples from T1, T2, T3 were analyzed for dissolved PO_4^{3-}, $Si(OH)_4$, NO_3^-, NO_2^-, NH_4^+, total dissolved nitrogen (TDN), and total dissolved phosphorus (TDP) on a Technicon AutoAnalyzer II® following well-established analytical methods at the Water Center at the University of Washington. Water samples collected in 2012 from HFP were analyzed on a Seal Analytical AA3® following well-established analytical methods at the SOEST Laboratory for Analytical Biogeochemistry at the University of Hawaii. Dissolved organic phosphorus (DOP) and dissolved organic nitrogen (DON) were determined as the difference between TDP and TDN and the dissolved inorganic P and N pools.

2.4 *Radon Survey*

A surface water radon survey along the coastline, extending from the northernmost tip of the bay to the central-south sector boundary, was performed on August 17, 2010. The survey was done at high tide because the shallow parts of the reef were not accessible at low tide. We used an autonomous in situ radon detector (Rad-Aqua) into which water was pumped from about 0.2 m below water surface. The unit was housed on a dinghy moving at <5 km h^{-1} speed. A radon survey was also performed in Heeia fishpond on November 19, 2013 during which we followed the entire perimeter of the pond and a central transect in the pond covering most of the pond area. Salinity and GPS coordinates were recorded every 30 s along the surveys (Fig. 8.1). Radon data were processed using methods described in Dulaiova et al. (2010).

2.5 *Radon Time Series*

A 1-h resolution radon (Rn) time-series of Rn in surface waters off Coconut Island (Fig. 8.1) was set-up for the period of January 26, 2012 and March 22, 2012. Water from 0.2 m below surface was pumped into a Rad-Aqua instrument housed in a land-based structure. Salinity, temperature and wind speed was monitored along with the radon time-series. Radon data were processed using methods described in Burnett and Dulaiova (2003).

2.6 *Wetland and Groundwater Sampling*

We collected groundwater samples in the wetland using piezometers from 0.5 to 2.0 m depth and surface water from the stream and irrigation ditches. Groundwater samples in the upper watershed were collected from wells operated by the Board of Water Supply. Samples were analyzed for nutrients as described above and radon activities using a RAD-H2O system (Durridge Inc).

3 Results

3.1 Radium Isotope Distribution

In northwest Kaneohe Bay radium isotope enrichment was higher, in general, in the nearshore region (Fig. 8.2) although the long-lived ^{226}Ra had the opposite trend along T3, which was more pronounced in November 2010. We attribute this result to radium addition to the offshore section of the transect from nearby streams and/or removal in the nearshore water column by sorption to iron rich suspended particles (see below). Radium activities were 2–12 dpm m^{-3} ^{223}Ra, 2–45 dpm m^{-3} ^{224}Ra, and 20–160 dpm m^{-3} ^{226}Ra. In general the activities were higher in August than November 2010. Radium distribution against salinity showed that radium activities dropped significantly by the seaward end of the transects but have not reached ocean end-member levels of zero excess (Fig. 8.2). Salinities along these transects ranged from 17 to 35.7 and the water column was stratified reaching ocean salinities near the bottom.

Radium isotopes in the fishpond were 2–8 dpm m^{-3} ^{223}Ra, 13–40 dpm m^{-3} ^{224}Ra, and 43–90 dpm m^{-3} ^{226}Ra. Here salinity ranged from 11 to 33.5 and near the stream mouth we observed significant stratification of the water column with a buoyant freshwater plume with a thickness of 0.2–0.3 m. Groundwater ^{226}Ra activities across the watershed were 60–200 dpm m^{-3} (n = 8).

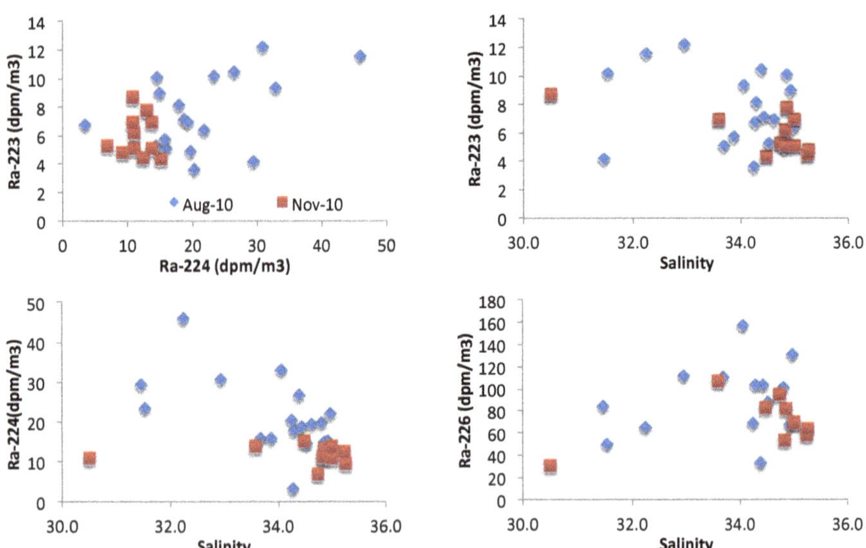

Fig. 8.2 Radium isotope distribution against salinity along transects T1–T3 in northwest Kaneohe Bay

3.2 Radon Tracer Distribution

The radon survey in northwest Kaneohe Bay in August 2010 was performed during high-and falling tides therefore it does not reflect average SGD inputs, rather a lower limit of coastal radon activities. It can be expected that highest SGD occurs along the shoreline so we focused our survey parallel to the shoreline and also made several transects in a cross-shelf direction (Fig. 8.1). Radon activities varied from 250 to 2100 dpm m^{-3} and were moderately (10x) elevated over parent-supported activities of 100–160 dpm m^{-3} estimated from dissolved ^{226}Ra measurements. In the central section of the bay activities were similar in magnitude along the coastline as well as in the Heeia fishpond.

In the fishpond the survey was performed at high tide and the measured activities were 870–2800 dpm m^{-3}. The highest activities clustered in the SW corner of the pond and near the stream mouth.

Radon time series measurement in the coastal water was performed at Coconut Island, which is in central Kaneohe Bay about 750 m from the main shoreline (Fig. 8.1). The measurements confirmed a significant tidal influence on coastal radon activities (Fig. 8.3). The observed activities ranged between 0 and 1800 dpm m^{-3}. During high tide radon levels decreased to ocean levels of ~60 dpm m^{-3} and during low tide the activities increased due to intensifying SGD and lower mixing with offshore waters. There was a period (February 1–18, 2012) of elevated baseline radon when even during periods of high tide radon activities did not decrease to supported levels of 60 dpm m^{-3}. This period showed the highest SGD fluxes during the 2-month monitoring program.

Groundwater radon concentrations were $150{,}000 \pm 190{,}000$ dpm m^{-3} in high-level groundwater ($n = 8$) and $90{,}000 \pm 80{,}000$ dpm m^{-3} in the wetland groundwater ($n = 11$).

3.3 Nutrient Distribution

Nutrient concentrations in the watershed and the coastal ocean are summarized in Table 8.1. As expected, groundwater and streams were more enriched in nutrients than coastal waters in the bay and the fishpond. Nitrate concentrations were similar in northwest Kaneohe Bay in August and November, 2010. The wetland water masses were significantly reduced in silicate and contained more reduced ammonium than nitrate. There was a large variability in DIN concentrations in the wetland. Nitrogen was mostly in the form of highly variable ammonium with an average concentration of 21 ± 140 μM; nitrate + nitrite concentrations averaged 0.3 ± 35 μM. Dissolved oxygen varied between 1 and 15 % saturation and phosphate concentrations were 0.6 ± 0.3 μM. Surface water, including the stream in the wetland, had <0.5 μM nitrate + nitrite, 11 ± 80 μM ammonium, and 0.2 ± 0.1 μM of phosphate. Dissolved oxygen concentrations were 56 ± 50 % saturation.

Table 8.1 Surface and groundwater dissolved inorganic nitrogen (DIN), dissolved inorganic phosphorus (DIP), and silicate concentrations (μM) in Kaneohe watershed in Waikane stream located in the northwest sector (Smith et al. 1981), Kaneohe Bay surface water (sw) August, Kaneohe Bay surface water November, He'eia fishpond surface water, He'eia wetland surface water, He'eia wetland groundwater (gw), and Waikane groundwater (all determined during this study)

	Sampling date	Number of samples	DIN	DIP	Silicate	DIN: DIP
Waikane sw	2/6/2013	7	12 ± 3	1.6 ± 0.5	540 ± 92	8
Kaneohe sw	8/17/2010	9	0.79 ± 0.43	0.051 ± 0.033	14.8 ± 6.0	15
Kaneohe sw	11/5/2010	11	0.68 ± 0.36	0.073 ± 0.100	6.9 ± 3.7	9
He'eia FP sw	11/19/2013	9	0.64 ± 0.53	0.11 ± 0.09	65 ± 75	6
He'eia wl sw	2–5/2013	16	11 ± 5	0.22 ± 0.14	220 ± 81	50
He'eia wl gw	2–5/2013	8	21 ± 144	0.56 ± 0.25	170 ± 53	38
Waikane gw	2/6/2013	7	12 ± 3	1.6 ± 0.5	540 ± 92	8

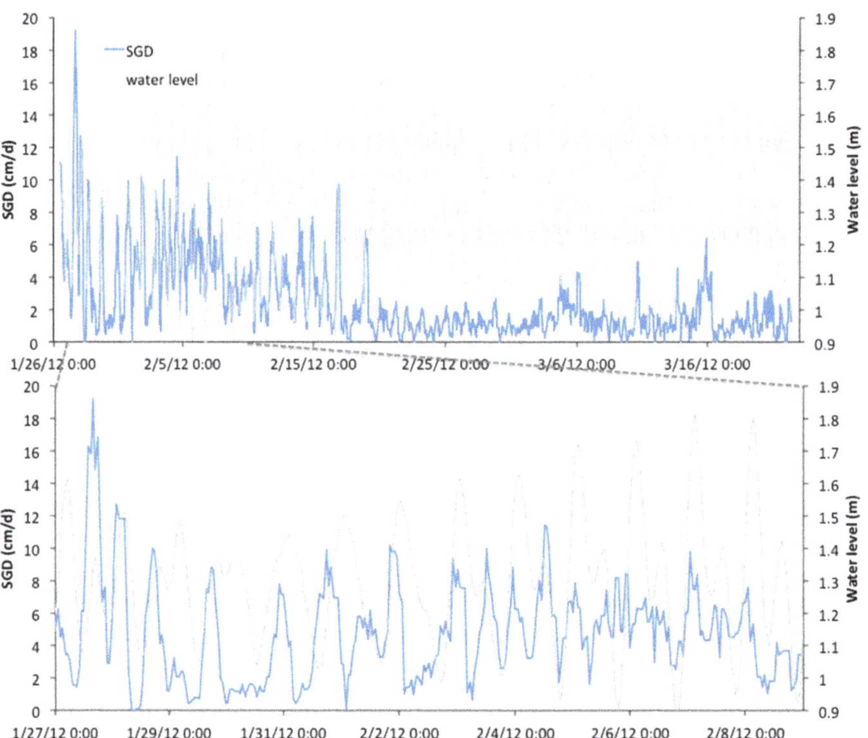

Fig. 8.3 A 1-h resolution SGD (cm d^{-1}) derived from a coastal radon record collected at Coconut Island in Kaneohe Bay, HI. The radon monitor was located on the shore of the island about 750 m from the major coastline's groundwater sources

4 Discussion

Coastal groundwater fluxes can be estimated indirectly using a mass balance of naturally occurring radioactive isotopes (Charette et al. 2008). We employ a multi-tracer approach utilizing radium isotopes to estimate SGD (Sects. 4.1–4.3) as described in Moore (1996) and Moore (2000a) then complement these estimates with radon derived groundwater fluxes (Sects. 4.4 and 4.5) utilizing methods described in Dulaiova et al. (2010) and Burnett and Dulaiova (2003). These methods are based on the continuous regeneration of radon and radium isotopes from U- and Th-bearing minerals via radioactive decay in aquifers. Groundwater becomes enriched in these isotopes because the water-to-solids surface area ratio is small and the water layer around the solids effectively captures the recoiling newly produced radionuclides. Owing to the absence of major sources, surface waters have orders of magnitude lower activities of these isotopes. A coastal mass balance can then be formulated to calculate the amount of the tracer derived by SGD after accounting for all other sources of these isotopes.

Since one of the goals of SGD studies is to demonstrate the effect of SGD on coastal water quality, some important aspects to consider are coastal nutrient inventories (Sect. 4.6) and the SGD-derived nutrient fluxes (Sect. 4.7), the rates of their dilution by mixing, and the degree to which these inputs get taken up by biological and inorganic processes. The processes listed above depend on the rate of delivery and residence time of nutrients in the coastal water. In Sect. 4.8 we use radium isotopes to estimate the age of SGD-derived conservative solutes as a measure of coastal residence time (Moore 2000b; Moore et al. 2006). This is the amount of time it takes for the nutrients to leave the investigated water body either by along-shore currents or by mixing into the offshore ocean. In Sect. 4.9 we use these residence times in combination with terrestrial fluxes to compare coastal nutrient mass balances in different sectors in Kaneohe Bay. Finally estuarine net nutrient removal rates are derived based on marine nutrient profiles and the combined SGD and stream nutrient fluxes (Sect. 4.10).

4.1 Horizontal Mixing Rates in Northwest Kaneohe Bay

The presence of excess radium isotopes in the nearshore region is an indication of coastal radium inputs from a combination of groundwater, stream, and suspended particulate sources at a rate at which elevated radium concentrations persist despite their radioactive decay (^{224}Ra half-life is 3.7 days and ^{223}Ra is 11.4 days) and mixing losses. The short- and long-lived radium distribution and mass balance can be used to determine SGD, in which the first step is to calculate horizontal mixing rates. For the same mixing ^{223}Ra returns estimates of smaller relative uncertainties on horizontal diffusion coefficient estimates than ^{224}Ra because of its longer half-

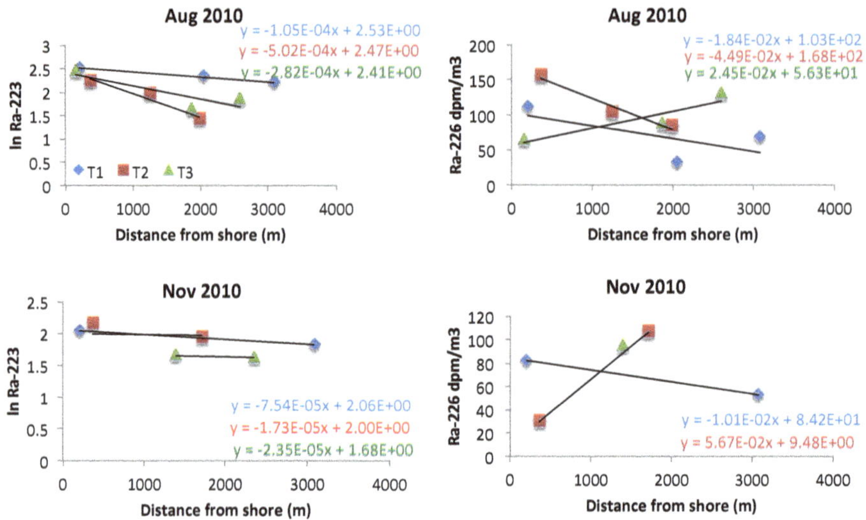

Fig. 8.4 Short and long-lived radium distribution over distance form shoreline on transects T1–T3 in the northwest sector of Kaneohe Bay in August and November, 2010

life and more gradual cross-shore gradients (Knee et al. 2011). Therefore we used ^{223}Ra as an independent mixing tracer for the nearshore region of Kaneohe Bay.

In a system controlled by eddy diffusion, we can use the distribution of ^{223}Ra to calculate a horizontal eddy diffusion coefficient (K_h, Moore 2000a) under the assumption that radium distribution depends on two processes, radioactive decay and mixing. We only had three sampling points on each transect so were not able to evaluate the influence of advection (concave or convex shape, break in slope), and we assumed that the system was controlled by eddy diffusion and that the dominant water transport was in a cross-shore direction neglecting alongshore currents. The water column was stratified along all 3 transects preventing significant radium diffusion inputs from benthic sources.

We derived gradients of the natural logarithm of ^{223}Ra from the three transects. If the above-stated assumptions are accurate then the ln^{223}Ra slope depended only on the decay constant λ_{223} and eddy diffusion coefficient K_h (Moore 2000a):

$$slope = \sqrt{\frac{\lambda_{223}}{K_h}}. \tag{8.1}$$

The slopes for the three transects (T1, T2, T3; Fig. 8.4) were calculated individually (and averaged -2.96×10^{-4} m^{-1}) resulting in an average eddy diffusion coefficient K_h of 25 m^2 s^{-1} and 57 m^2 s^{-1} in August and November 2010, respectively. The relative uncertainty on these diffusion coefficients was 17–30 % resulting from having only 3 sampling points on each transect, R^2 of slopes >0.9 and a ^{223}Ra measurement error of 17–22 %.

4.2 Coastal ^{226}Ra Fluxes in Northwest Kaneohe Bay

Next, we used the eddy diffusion coefficient and the concentration gradient of ^{226}Ra along the same transects (Fig. 8.1) to estimate the coastal flux of ^{226}Ra to the ocean. Ra-226 in this case represents a conservative tracer as due to its long half-life ($T_{1/2} = 1600$ year) it does not decay on the time scale of coastal transport processes.

The linear ^{226}Ra gradients (Fig. 8.4) were -1.0×10^{-2} dpm m^{-3} m^{-1} for T1 and -1.8×10^{-2} dpm m^{-3} m^{-1} for T2 in August 2010, while T3 had a positive slope. In November 2010, we could only used data from T1, as T2 had a positive slope and T3 was incomplete with only one sampling point. The positive slopes suggested that there was either an offshore SGD source near T3, some lateral transport of high-radium water masses from upstream coastal areas, or that radium was removed in the coastal region by sorption onto suspended particles. In this case we surmise that the positive slope can be explained by radium delivered by the Waihee River (fluxes derived from USGS station indicated in Table 8.2) located just south of T3, which may preferentially flow along a coral patch to the offshore sections of T2 and T3. Alternatively, it is possible that there was some radium sorption to particles as stream particle inputs at the beginning of T3 resulted in coastal suspended sediment load of 0.017 g L^{-1} and 0.012 g L^{-1} in August and November, respectively. We observed that the suspended particles in this watershed were enriched in iron (surmised based on color and elevated dissolved iron concentration in groundwater, Dulaiova unpublished results) most probably in form of iron (oxy)hydroxide precipitates that have shown to attract radium even at elevated salinities (Gonneea et al. 2008).

On average the surface mixed layer was 1 m thick at T1 and 1.4 m thick at T2 as determined from salinity depth profiles along the transects. Individual aquifer sector coastline lengths were used to calculate the offshore ^{226}Ra fluxes of 1.6×10^{8} dpm d^{-1}(T1) and 1.3×10^{7} dpm d^{-1} (T2) in August and 1.7×10^{8} dpm d^{-1} (T1) in November. These ^{226}Ra fluxes were supported by streams, groundwater discharge and desorption from suspended particles delivered by streams. Stream water and sediment fluxes were scaled according to the coastline length of each transect (Table 8.2). River discharge and suspended particulate flux from the streams was determined using USGS stream gauges and relationships derived by Hoover et al. (2009). The estimated sediment load was 170,000 g d^{-1} in August and 61,500,000 g d^{-1} in November. We used literature values (Krest et al. 1999) to estimate radium desorption from sediments recognizing that this input may actually be much smaller due to the iron enrichment of the particles that would not release radium as easily as the sediment types studied by Krest et al. (1999), which were suspended and bottom sediments from the Mississippi delta. We estimate a suspended particle desorption flux for ^{226}Ra of 1.1×10^{5} dpm d^{-1} and 1.29×10^{7} dpm d^{-1} in August and November 2010, respectively. Inspection of all sources revealed that the total ^{226}Ra flux was clearly dominated by groundwater inputs as river and suspended particle inputs were 1–3 orders of magnitude lower than the total radium flux. Any uncertainty in these estimates therefore was insignificant in terms of the final SGD fluxes.

Table 8.2 Characteristics of the study domains in the northwest sector of Kaneohe Bay (NB), central Kaneohe Bay (CB) not including He'eia fishpond, and the He'eia fishpond (HFP)

			Coastl. length	Stream	SGD-Rn		SGD-Ra-223/6		SGD-Ra-226		Mixing coefficient	Residence time	
												Mixing	Ra age
			m	$m^3\ d^{-1}$	$m^3\ d^{-1}$	$m^3\ m^{-1}\ d^{-1}$	$m^3\ d^{-1}$	$m^3\ m^{-1} d^{-1}$	$m^3\ d^{-1}$	$m^3\ m^{-1} d^{-1}$	$m^2\ d^{-1}$	d	d
NB	Aug-10	T1	1530	1.9E+4	5.5E+4	36	6.0E+5	393	1.7E+4	11	5.5E+6	0.9	3.4–6.6
		T2	840	3.0E+4	2.3E+4	27	3.3E+4	40	1.1E+5	133	2.4E+5	8	0–3.2
		T3	2800	1.8E+4	2.9E+4	10			4.2E+5	150	7.6E+5	4	0.1–2.8
NB total Aug			5170	6.8E+4	1.1E+5	20	1.1E+6	217	5.1E+5	98		8	6.6
NB	Nov-10	T1	1530	1.7E+5			6.3E+5	413	2.2E+5	141	1.1E+7	0.4	0.9–1.3
		T2	840	2.6E+5					–		2.0E+8	0.01	0.1–3.6
		T3	2800	1.6E+5					–		1.1E+8	0.03	0–3.2
NB total Nov			5170	6.0E+5			2.1E+6	413	7.3E+5	141		0.4	3.6
CB Aug-10			7700	1.0E+4	4.6E+4	6			2.1E+5	27			0–10
HFP Nov-13			1250	2.1E+3	2.5E+3	2							2–6

Transects T1–T3 are shown in Fig. 8.1. Coastline length and stream discharge are based on USGS stream gauges in the watershed, SGD fluxes (Sects. 4.1–4.5) and residence times are derived from the diffusion coefficients and apparent radium ages (Sect. 4.8), and are indicated for the individual parts of the sectors, including two different time-periods for NB. For comparison, the volume of water exchanged between average low and high tides is $10 \times 10^6\ m^3$ in NB and $18 \times 10^6\ m^3$ in CB (Bathen 1986)

4.3 Submarine Groundwater Discharge in Northwest and Central Kaneohe Bay Estimated via Radium Approaches

SGD-derived radium fluxes were calculated by subtracting stream and particle desorption Ra fluxes from total ^{226}Ra fluxes. The net groundwater derived flux was then divided by groundwater ^{226}Ra concentrations of 200 dpm m^{-3}. The individual SGD estimates (Table 8.2) had ~36 % relative uncertainty, which was calculated via error propagation of the individual terms: diffusion coefficients 30 %, Ra flux 35 %, groundwater Ra activity 10 %.

We normalized SGD estimates from T1 and T2 as discharge per meter shoreline and extrapolated SGD to the whole northwestern bay. SGD was 20 times higher than stream inputs in August and 3 times higher in November, 2010 (Fig. 8.5). We observed a change in the SGD:stream discharge ratio, which was expected as during

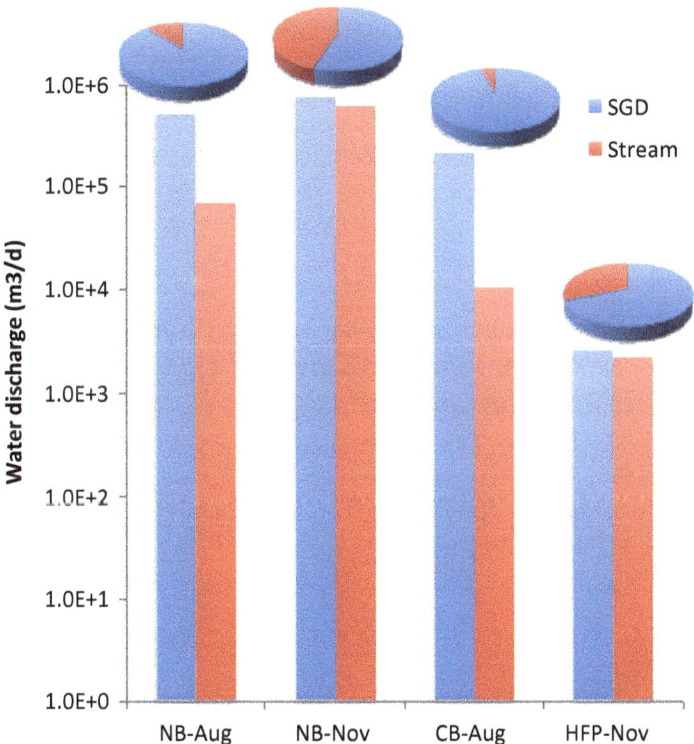

Fig. 8.5 The bars represent the magnitude of SGD (*blue*) and stream (*red*) discharge (m^3/d) for northwestern Kaneohe Bay in August (NB-Aug) and November (NB-Nov), central bay in August (CB-Aug), and Heeia fishpond in November (HFP-Nov). The *y*-axis is a logarithmic scale for better comparison between NB and HFP locations. The pie charts indicate relative DIN contribution of SGD and streams to the bay

the dry season in August groundwater flux was the major terrestrial water source to the coastline and the streams were only fed by baseflow. The November sampling was performed following the first big storm after the dry season, which resulted in increased streamflow. The groundwater aquifer responded much more slowly to the increased recharge and we did not observe an immediate increase in SGD. In fact we have shown that SGD had a 2–3 month lag in responding to increase in recharge in the Kaneohe Watershed (Leta et al. 2015).

We also used another approach to calculate SGD that involved a ^{226}Ra mass-balance and residence times (Moore 1996). Coastal activities of ^{226}Ra were multiplied by the volume of the water mass and divided by the coastal residence time derived in Sect. 4.6. In this way a replacement rate of radium was estimated which, at a steady-state, should be equal to ^{226}Ra inputs via streams and SGD. Stream and particle sources were subtracted from the total radium fluxes and finally the SGD-derived radium flux was divided by groundwater ^{226}Ra activity. The resulting SGD was 2–4 times lower than the estimates described above (Table 8.2) but still dominated over stream water discharge.

A point to keep in mind is that SGD is a mixture of brackish groundwater and recirculated seawater, and as a consequence cannot be compared in terms of freshwater fluxes to stream inputs. The salty fraction of SGD has been shown to represent 40–80 % of total fluxes in Hawaii (Street et al. 2008; Kleven 2014; Glenn et al. 2013; Mayfield 2013), seawater recirculation through the coastal aquifer therefore contributes significantly to total SGD. Yet brackish and salty SGD play an equal role as nutrient pathways and as we show below, nutrient fluxes via total SGD must be considered in nutrient coastal mass balances.

4.4 Submarine Groundwater Discharge in Northwest and Central Kaneohe Bay Estimated Using a Radon and Radium Mass Balance

A coastal radon survey was performed to determine surface water radon inventories in August 2010 (Fig. 8.1). The measured concentrations were used to calculate SGD fluxes based on the following equation:

$$Q_{SGD} = \frac{A_{Rn\_cw} * V}{\tau * A_{Rn\_gw}}, \tag{8.2}$$

where Q_{SGD} is total submarine groundwater discharge (m^3 d^{-1}), A_{Rn_cw} are coastal water radon activities corrected for non-SGD sources (diffusion from sediments, in situ radioactive production from ^{226}Ra, offshore inputs) and losses (evasion to the atmosphere, radioactive decay), A_{Rn_gw} is groundwater radon activity (dpm m^{-3}), V is the volume of the coastal water box that the measurement represents (m^3) and τ is the flushing rate of the coastal zone, which in this case was a tidal cycle (Dulaiova et al. 2010).

SGD determined by the radon survey amounted to 1.1×10^5 m^3 d^{-1} or 20 m^3 m^{-1} d^{-1} in the northwestern sector and 4.6×10^4 m^3 d^{-1} or 6 m^3 m^{-1} d^{-1} in central Kaneohe. We expect that SGD would be associated with lower salinity of bay water. The lowest salinities observed during the survey were at T2 (17–29) and the average salinity in the northwestern part of the bay was 32.1, the minimum in central Kaneohe Bay was 31.3 and the average 34.7.

We applied the ^{226}Ra mass balance (Moore 1996, 2003) in the central sector of the bay and calculated 2.1×10^5 m^3 d^{-1} or 27 m^3 m^{-1} d^{-1} of SGD. In all sectors SGD fluxes calculated via radon were 2–10 times lower than the radium mass balance derived fluxes.

At a later date in January–March 2012, post-dating the transect work, a time-series radon-monitoring station was installed southeast of T3 at Coconut Island. This station likely only captured SGD emanating locally from the 117,000 m^2 area of the small island's freshwater lens. The record revealed tidal as well as longer-term patterns of change in the radon signature reflecting changes in groundwater discharge and coastal conditions (SGD, rain, wind, mixing). The estimated groundwater advection rates, derived by using methods described by Burnett and Dulaiova (2003), ranged from 0 at high tide to a maximum of 19 cm d^{-1}, which occurred in the earlier part of the deployment in January 2012 (Fig. 8.3). These advection rates were relatively low compared to other observed rates around the island and reflect only a localized SGD from the island's small aquifer. The records showed that SGD had a tidal signature with higher fluxes at low and rising tides and lower advection rates at high tide. There are two reasons for this observed pattern: (1) at high tide the hydraulic gradient between the ocean and the coastal aquifer is smaller or even reversed relative to low tide, resulting in a smaller driving force hence less SGD; and/or (2) during flood tide the coastal SGD chemical signature is diluted because groundwater is discharged into a larger water mass. Over the time-scale of the deployment, the record showed higher SGD in January and the first half of February with a decrease of baseline fluxes in the second half of February and throughout March.

The inset on Fig. 8.3 shows a selected time interval during which neap tide switched to spring tide and the SGD dynamics mimicked the tidal progression, showing diurnal and semi-diurnal patterns. Because tides were not the only driving force behind the hydraulic gradient (precipitation, hydraulic conductivity, groundwater withdrawals and other forcing factors), we did not observe a corresponding pattern between the magnitude of SGD and tidal range.

The fact that SGD dynamics is so strongly influenced by tides has implications for the radon survey as it was performed in the tidally influenced nearshore region and only reflected a snap-shot SGD at the time of measurement. Since the northwest Kaneohe Bay survey was done at high tide, when SGD was at its lowest point, it represents a below-average estimate of SGD. The radium techniques employed here on the other hand integrated SGD over the time period of the water residence time along the sampled transects and better represented the overall groundwater fluxes. We therefore conclude that the twofold to tenfold difference in SGD estimates via radon and radium isotopes in northwest Kaneohe Bay was due to the different spatial and temporal sensitivity of these approaches. Both methods revealed large spatial heterogeneity in SGD (Fig. 8.1 and Table 8.2).

4.5 SGD into Heeia Fishpond: Central Kaneohe Bay via a Combined Radon and Radon Mass Balance

Measured radon inventories and radium isotope-derived water ages (see below) were applied in Heeia fishpond to derive SGD using Eq. (8.2). We corrected the measured radon inventories for atmospheric evasion losses and radioactive decay. The groundwater radon activities applied here were derived from the wetland located directly upstream of the fishpond. The resulting SGD flux was 2500 m^3 d^{-1}, which was about equal to the estimated contribution of stream discharge into the fishpond (2200 m^3 d^{-1} derived from a USGS stream gauge at Haiku, Fig. 8.5). A salinity mass balance calculated for the pond was also calculated which suggested that 88 % of SGD was brackish water contributed by recirculated seawater (Kleven 2014). There was elevated SGD along the seawall in the pond suggesting either a presence of a breached impermeable layer forcing groundwater to discharge offshore or that porewaters were pushed out of the sediments by tidally driven hydraulic gradient set up across the sea wall. This portion of the SGD, even though was not identified as fresh water discharge, may be a significant contributor of nutrients because porewaters in the bottom sediments in the fishpond are enriched in nitrogen and silicates (Briggs et al. 2013).

4.6 Watershed Nutrient Concentrations

Nutrient concentrations in groundwater within the individual sub-watersheds of Kaneohe Bay varied with land-cover between the watersheds. We collected samples from the high level aquifer in the upper watershed ($n = 8$) where the nutrient concentrations were uniform with little variation: 12.1 ± 2.9 μM of nitrate + nitrite, 1.6 ± 0.5 μM of inorganic phosphate and 545 ± 96 μM of silicate (Table 8.1). Oxygen concentrations were >90 % saturation and organic nutrient species were a negligible part of totals. Due to the porous, highly conductive nature of the basalt there is a direct surface water groundwater interaction and, except after significant storm events, baseflow is a major contributor to streamflow and nutrients (Izuka et al. 1994). As a consequence of this connection, stream nutrient concentrations equaled groundwater concentrations. Hoover (2002) derived discharge vs. nutrient concentration relationships and concluded that at baseflow the streams had the same silicate, nitrate and phosphate concentrations as groundwater—on average 400–500 μM silicate, 5–10 μM nitrate + nitrite, and 0.5–1.2 μM of phosphate. Surface runoff only diluted silicate concentrations but nitrate and phosphate concentrations did not change significantly after storm events (Hoover et al. 2009).

The conservative nutrient behavior in all of Kaneohe watershed assumed by Hoover et al. (2009) contrasted greatly with our observations in the Heeia watershed where the stream and groundwater flowpaths are intercepted by a coastal wetland before draining into central Kaneohe Bay. Silicate concentration in the stream and groundwater were only 170 ± 50 μM and 220 ± 80 μM, respectively. The lower silicate concentration in groundwater is likely due to its uptake by wetland grasses and macrophytes that are abundant in the wetland, and which are known to draw down silicate (Schoelynck et al. 2010). Although our study did not specifically target wetland nutrient uptake mechanisms, there was no obvious variation in wetland nutrient concentrations with stream discharge.

4.7 Nutrient Fluxes

Stream-derived nutrient fluxes into northwest Kaneohe Bay were calculated by multiplying baseflow and groundwater nutrient concentrations by stream discharge determined from USGS stream gauges and methods described by Hoover (2002). In the central bay, Heeia stream discharge was multiplied by the measured stream and wetland surface water nutrient concentrations.

Groundwater derived nutrient fluxes were calculated by multiplying SGD (derived using the radium-transect method) and well groundwater nutrient concentrations for northwest Kaneohe Bay ($m^3/d \times mol/m^3 = mol/d$), and SGD multiplied by wetland groundwater nutrient concentrations in central Kaneohe Bay (Table 8.3). This approach assumes that nutrients do not undergo any biogeochemical removal along the groundwater flowpath between the sampled location and their discharge at the coastline.

Our data indicate that brackish pore water nutrient values for the fishpond were in the same range as wetland fresh groundwater nutrient concentrations (Briggs et al. 2013; Kleven 2014). Also, Smith et al. (1981) reported nutrient concentrations in northwest and central Kaneohe Bay in porewater in the upper 0.3 m of lagoon sediments that were comparable to well groundwater concentrations, 80 ± 27 and 145 ± 60 of DIN and 16 ± 4 and 9 ± 5 of phosphate for northwest and central bay, respectively. This suggests that recirculated seawater has similar nutrient levels as fresh groundwater. This observation is in contrast to, for example, observations on the Kona coast of the Hawaii Island where linear mixing relationship was found between nutrients and salinity in coastal aquifers implying that recirculated seawater diluted groundwater nutrient concentrations (Paytan et al. 2006). As our groundwater and porewater nutrient comparison shows, dilution of groundwater nutrients by recirculated seawater does not seem to occur in Kaneohe Bay, where recirculated seawater flows through organic rich alluvial sediments rather than young basalt, and is equally enriched in nutrients. Nevertheless we acknowledge that biogeochemical transformations may remove nitrogen and phosphorus from porewaters along the groundwater recirculation path and our nutrient flux estimates may be higher than actual fluxes. Also, unlike stream discharge and fresh SGD, recirculated seawater does not contribute new nutrients to the bay, it only recycles autochthonous nutrients released by remineralization of buried organic matter.

The derived DIN fluxes were 5–10 times higher via SGD than surface runoff in all regions (Fig. 8.5), proving SGD to be a significant contributor to coastal nitrogen

Table 8.3 Nutrient fluxes by stream surface inputs and total SGD derived from SGD fluxes indicated in Table 8.2 and nutrient end-member concentrations listed in Table 8.1

		Nutrient fluxes (mol d^{-1})											
		Stream			SGD Rn-derived			SGD Ra-223/6-derived			SGD Ra-226-derived		
		DIN	DIP	Silicate	DIN	DIP	Silicate	DIN	DIP	Silicate	DIN	DIP	Silicate
NB	Aug-10	229	31	10,310	656	87	29,509	7221	963	324,961	207	28	9309
		360	48	16,202	271	36	12,211	401	54	18,064	1342	179	60,404
		221	29	9942	344	46	15,487				5026	670	226,165
NB total Aug		810	108	36,453	1271	170	57,207	13,436	1791	604,624	6080	811	273,608
NB	Nov-10	2018	269	58,858				7576	1010	340,913	2595	346	116,753
		3171	423	92,492									
		1946	259	56,756									
NB total Nov		7135	951	208,106				25,599	3413	1,151,974	8767	1169	394,520
CB Aug-10		115	2	2291	502	10	16,190				2522	336	113,494
HFP Nov-13		24	0.5	471	53	1	425						

budgets. Phosphate fluxes were lower via SGD than surface runoff which can be explained by the coupling between iron and phosphorus chemistry and phosphate occlusion in iron and aluminum (hydr)oxides in the aquifer (e.g. Spiteri et al. 2008). Silicate fluxes via SGD and surface runoff were comparable in northwest Kaneohe Bay, and were dominated by surface runoff in the central bay and in the fishpond (Table 8.2). This is expected as we see a significant silicate uptake in wetland groundwater samples, therefore SGD-derived silicate is lower than stream discharge that is fed partially by baseflow and surface runoff. Silicate fluxes in terms of SGD and surface runoff ratios were comparable to findings in the same (Hoover et al. 2009) and similar watersheds in other studies (e.g. Garrison et al. 2003; Mayfield 2013).

4.8 Coastal Residence Times

Coastal inventories of conservative solutes depend on the magnitude of their terrestrial inputs (SGD, stream, benthic fluxes) as well as on coastal mixing and dilution driven by oceanic processes. The higher the terrestrial fluxes the higher the coastal inventory and, the more effective the mixing with offshore water the less likely it is that a pollutant will have an impact on the coastal ecosystem.

The Ra-derived horizontal diffusion coefficients provide a measure of lateral mixing and can also be used to quantify residence times. In our setting, a residence time estimate is also analogous to a flushing rate (flow rate/volume). Windom et al. (2006) have shown that coastal residence time of geochemical components approximated by Ra can be related to the mixing coefficient using the following relationship:

$$t = \frac{L^2}{2K_h} \tag{8.3}$$

where t is residence time (days), K_h is the mixing coefficient (m^2d^{-1}) and L is the length of the transect over which the mixing coefficient was estimated (m). In northwest bay, application of this method resulted in residence times of 0.9–8 days in August and 0.01–0.4 days in November, 2010. Indeed, in August the observed radium inventories were higher than in November (Fig. 8.2). Radium isotope ratios have also been used as a measure of coastal residence time by determining so called apparent radium ages (Moore 2000b; Kelly and Moran 2002). In this approach a short-lived isotope is normalized to a long-lived isotope of radium and because of their chemically identical behavior only their radioactive decay drives changes in their activity ratio once the water mass is isolated from their parent nuclides. This method requires that there is a uniform activity ratio in all contributing radium sources (SGD, streams, diffusion) and this assumption does not seem to hold true in Kaneohe Bay. Measurements in Heeia stream resulted in $^{224}Ra/^{223}Ra$ activity ratio of 1.1 similar to observations in other locations on Oahu (Wailupe by Holleman 2011;

Waimanalo, Dulaiova, unpublished results). The activity ratio for these same isotopes was closer to 7 in brackish SGD and to 4 in fresh SGD (Kleven 2014). A higher activity ratio is a result of faster regeneration rate of ^{224}Ra, and hence its higher enrichment in comparison to ^{223}Ra in groundwaters of short residence time (such as recirculated seawater) in which radioactive equilibrium has not been established. Since at any time, coastal water is a mixture of stream inputs, fresh SGD, and brackish SGD, it was only possible to relate apparent radium ages of offshore water masses to the composite coastal isotope signature rather than the individual groundwater and stream activity ratios. The composite coastal activity ratio was ~4 in August when coastal radium inventories were driven by SGD inputs and ~2 in November when stream discharge also contributed significantly to coastal radium inventories. Apparent radium ages were then estimated using the following equation (Moore 2000b):

$$t = \ln \frac{^{224}Ra_i/^{223}Ra_i}{^{224}Ra_o/^{223}Ra_o} * \frac{1}{\lambda_{224} - \lambda_{223}}. \tag{8.4}$$

These ages are based on the faster decay of the short-lived ^{224}Ra (λ_{224} is its decay constant in days^{-1}) in comparison to ^{223}Ra (λ_{223} decay constant of ^{223}Ra) in an offshore water mass (Ra_o) assuming a composite nearshore end-member (Ra_i) and that both isotopes are subjected to the same dilution by mixing. The resulting average apparent radium ages were 2.7 days and 1.4 days in August and November, respectively. The uncertainty on these ages was estimated to be 50–100 % according to evaluations described by Knee et al. (2011) because the activity ratio method is not very sensitive for water ages below ~3.5 days. Both methods suggested a faster mixing rate in November than in August 2010 (Table 8.2). Lowe et al. (2009) used a coupled wave circulation numerical model and showed that wave forcing is the dominant mechanism driving currents and flushing in Kaneohe Bay. According to their result, for the conditions observed on August 16, 2010 (wave height 1–2 m and wave direction 90°; obtained for Mokapu Buoy at http://cdip.ucsd.edu) and November 4, 2010 (2–3 m and 360°) the residence times were 1.3 and 0.8 days, respectively.

Apparent water ages in the Heeia Fishpond varied from <2 days in the section that was well flushed by the incoming stream, to about 6 days in the SW corner of the pond that has a restricted circulation (Kleven 2014).

4.9 Coastal Nutrient Budgets

Terrestrial nutrient contributions to coastal inventories were calculated as the sum of stream and SGD inputs. To estimate coastal inventories we used the measured coastal nutrient concentrations and estimated water volumes of the lower salinity water layer along our radium and radon survey transects in northwest and central

Table 8.4 Coastal parameters with terrestrial and recycled nutrient inputs from streams and SGD, conservative geochemical residence time (T_{Ra}), DIN residence time with respect to coastal inventory and stream + SGD fluxes (T_N), volume of the coastal domain (V), and DIN inventory N (mol)

	Input fluxes	Coastal parameters
	NB_Aug	
SGD ($m^3\ d^{-1}$)	100,000(Rn)–1,400,000(Ra)	T_{Ra}: 6.6–8 & T_N: 0.5–4.5 d
SGD DIN ($mol\ d^{-1}$)	1330–17,400	V: 13 × 10e6 m^3
Stream ($m^3\ d^{-1}$)	68,000	N: 10,200 mol
Stream DIN ($mol\ d^{-1}$)	810	
	NB_Nov	
	2,000,000(Ra)	T_{Ra}: 0.4–3.6 & T_N: 0.3 d
	24,000	V: 13x10e6 m^3
	600,000	N: 8800 mol
	7100	
	CB_Aug	
	91,000(Rn)	T_{Ra}: 10 & T_N: 12.4 d
	1000	V: 19 × 10e6 m^3
	10,000	N: 13,800 mol
	110	
	HFP_Nov	
	2500(Rn)	T_{Ra}: 2–6 & T_N: 2.3 d
	53	V: 28 × 10e4 m^3
	2100	N: 179 mol
	24	

NB is northwest sector of Kaneohe bay, *CB* is the central sector, and *HFP* is He'eia fishpond

Kaneohe Bay, and for the fishpond the volume of the whole pond was considered. For the transects, this approach encompassed most of the northwest and central bay area affected by SGD and stream inputs. Table 8.4 illustrates terrestrial nitrogen fluxes, coastal inventories, and conservative coastal residence times for each investigated sector in Kaneohe Bay: northwest and central Kaneohe Bay, and Heeia Fishpond. The radium diffusion coefficient-derived SGD DIN fluxes were an order of magnitude higher than stream inputs, suggesting that SGD was an overwhelming source of DIN to the bay. The radon mass-balance approach resulted in DIN fluxes more comparable to stream fluxes and the ^{226}Ra mass balance-derived fluxes fall in between the two estimates.

Because of its volume being the largest, central Kaneohe Bay represented the highest inventory of nitrogen, although coastal DIN concentrations were comparable across all three sectors (Table 8.1). DIN fluxes in all sectors were dominated by groundwater inputs. Calculated SGD fluxes were highest in northwest Kaneohe Bay regardless of the method used, contributing as much as 10^5–10^6 $m^3\ d^{-1}$ of brackish and recirculated seawater discharge and adding 10^3–10^4 $mol\ d^{-1}$ of DIN. This result is consistent with reported water budgets (Shade and Nichols 1996) in which the northwestern sector receives 60 % of the total surface and groundwater runoff in Kaneohe Bay. Because we evaluated SGD in two contrasting periods (dry and after the first large storm) we can compare SGD and stream discharge responses. While stream inputs increased tenfold between August and November, SGD barely doubled for the same time period and we surmise that the reason for this difference was the delay in water recharge into the aquifer. Nevertheless, SGD stayed the dominant DIN source also in November.

DIN fluxes were significantly lower in the central sector, with SGD contributing 10^3 and streams 100 mol d^{-1}. In Heeia Fishpond DIN fluxes amounted to 80 mol d^{-1} in a 1:2 ratio of stream and SGD contribution.

We calculated coastal DIN residence times with respect to terrestrial inputs by dividing coastal inventories by the sum of stream and SGD fluxes. DIN residence times were lower than those derived by radium isotopes suggesting a nitrogen removal by biological uptake or abiotic processes (Drupp et al. 2011). The northwest section of Kaneohe Bay had DIN residence times <1 day and geochemical Ra-derived residence time of 3–8 days. In the central bay both estimates are ~10 days. In the fishpond DIN residence time is about 3 times less than Ra-derived estimates, again, suggesting DIN removal from the water column.

The northwest bay had shorter residence times in November, which could at least theoretically be related to nutrient uptake in the bay. Lucas et al. (2009) suggested that after nutrient limitation, estuarine retention or transit time is the major factor determining nutrient uptake. From our nutrient distribution, it was impossible to quantify any differences in DIN uptake between November and August when residence times were <1 day vs. 6–8 days, respectively (Table 8.2).

The dominance of SGD over terrestrial surface water inputs is typical for Hawaii coastlines. According to Zekster (2000) SGD from large islands is disproportionally greater than from most continental areas and represents at least 50 % of all SGD in the Pacific Ocean (Zektser 2000). This result is a consequence of high rainfall, steep topography, and permeable fractured rocks with large hydraulic conductivity that are typical for many islands. The global fresh SGD flux is estimated to be only <10 % of river discharge (Taniguchi et al. 2002). Early watershed budget estimates for Kaneohe bay for annual average stream and groundwater flow were 240 and 22×10^3 m^3 d^{-1} in 1978–1979, matching closely the global SGD to river discharge ratio of 1:10 (Smith et al. 1981). Follow-up studies of the Kaneohe watershed showed that recharge to groundwater aquifers is actually 1.5 times the volume of surface runoff (Shade and Nichols 1996) suggesting that fresh SGD may potentially be 1.5 times (less contribution to baseflow) the stream runoff. None of the water balance techniques are able to account for recirculated seawater and the discharge of brackish groundwater, however. The most often applied methods for total (fresh and saline) SGD measurements are geochemical approaches. These have been applied on local (Moore 2000a, b), regional (Windom et al. 2006), and ocean basin scales (Moore et al. 2008). For example, using the ^{228}Ra isotope mass balance Moore et al. (2008) estimated that total SGD represents 80–160 % of river discharge in the Atlantic Ocean. Another study showed that in the Atlantic and Indo-Pacific oceans, total SGD is 3 to 4 times greater than riverine freshwater fluxes

(Kwon et al. 2015). Our findings are in agreement with these results with SGD: stream ratios ranging between 1.2 and 4, with one outlier as high as 16. These findings demonstrate the importance of SGD for coastal geochemical budgets and nutrient contribution to the oligothrophic Pacific Ocean surrounding the Hawaiian Islands.

4.10 Coastal Nutrient Uptake Rates

The effects of SGD-derived nutrients on coastal water quality and ecosystems manifest themselves to various degrees across the islands. Several studies show that dilution and fast exchange with offshore waters results in conservative mixing trends of DIN, DIP, and silicate without any coastal biological uptake (Dollar and Atkinson 1992; Johnson et al. 2008). Other studies document nutrient utilization by nearshore phytoplankton in coastal groundwater plumes (Johnson and Wiegner 2014), nitrogen uptake by algal species identified from their $\delta^{15}N$ and C:N ratios (Dailer et al. 2010), and in some cases algal blooms dominated by invasive species were attributed to excess nutrient loads from SGD (Smith et al. 2001). In Kaneohe Bay the majority of studies focus on streams as the major sources of nutrients to the bay along with sewer outfalls, which were eliminated in 1978. The significance of stream inputs in driving nutrient and sediment concentrations and consequent phytoplankton response has been documented in several studies (e.g. Drupp et al. 2011; Ringuet and Mackenzie 2005; Hoover 2002, Young 2011). While these studies focused on stream fluxes the authors noted that groundwater fluxes were likely contributing to the observed nutrient pulses.

Our surface water dataset only covered salinities between 17 and 35.2 in the bay and 11 and 33.4 in the fishpond. Within this narrow salinity range silicate seemed to be the most conservative, plotting along a conservative mixing line in the pond with a 0 salinity intercept of 348 μM (Fig. 8.6). This value fell between upland groundwater and stream water, reflecting a mixture of water sources to the pond and confirming our findings of silicate uptake within the watershed, as high-level aquifers had 540 μM of silicate. In the bay the data were more scattered, but most of the silicate concentrations were bracketed within mixing lines fitted between stream and upland groundwater on the terrestrial side and published values of ocean concentrations (Laws 1980). The August and November data showed similar patterns except for the presence of lower salinity samples in November. DIP and DIN concentrations displayed a large degree of scatter and most samples plotted below the defined mixing lines (Fig. 8.6). If we accept our identified stream and groundwater sources as the end-members for these samples then there must be significant nutrient removal in the bay. There are at least three kinds of removal processes that could be important in this environment: (1) a biological filter—including any biological nutrient uptake; (2) a physical filter—sorption on iron-rich particulates and clays which are especially effective in capturing P species; (3) tidal exchange that significantly influence coastal nutrient concentrations via dilution

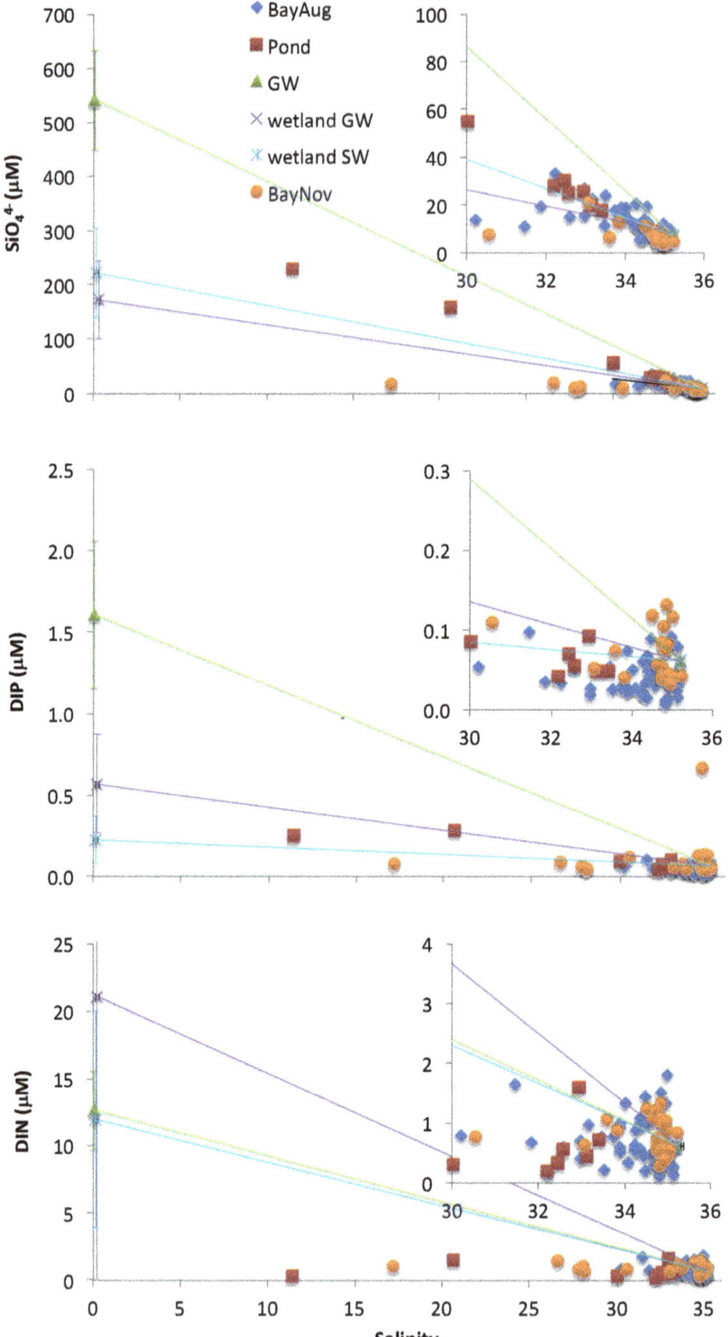

Fig. 8.6 Nutrient distribution in Kanehoe Bay in August and November 2010 and in the Heeia fishpond in November 2013. Terrestrial end-members are Waikane groundwater (GW), Heeia wetland groundwater, and Heeia wetland surface water as indicated in Table 8.1. Oceanic end-members are literature values from Laws (1980)

Table 8.5 Nutrient removal rates (R_{S+SGD}) derived from the combined inputs of stream and SGD fluxes (Q_{S+SGD}) and effective end-member intercepts (I) derived from Fig. 8.5

		I	Stream + SGD ($m^3 d^{-1}$)			C_{R+SGD}	Estuarine removal rate ($mol\ d^{-1}$)		
		μM	$Q_{S+SGDRn}$	$Q_{S+SGDRa223/6}$	$Q_{S+SGDRa226}$	μM	$R_{S+SGDRn}$	$R_{S+SGDRa223/6}$	$R_{S+SGDRa226}$
DIN	NB_Aug	2.7	174,127	1,187,863	574,869	12	−1612	−11,000	−5323
	NB_Nov	1.2		2,733,881	1,331,188	12		−29,553	−14,390
	HFP	0.6	4639			16	−71		
DIP	NB_Aug	0.09	174,127	1,187,863	574,869	1.6	−264	−1798	−870
	NB_Nov	0.01		2,733,881	1,331,188	1.6		−4341	−2114
	HFP	0.4	4639			0.2	1		
Silicate	NB_Aug	127	174,127	1,187,863	574,869	540	−71,914	−490,588	−237,421
	NB_Nov	24.9		2,733,881	1,331,188	540		−1,408,222	−685,695
	HFP	348	4639			195	710		

The actual end-member concentration (C_{R+SGD}) was derived as the average of stream and groundwater end-members. Nutrient removal rates were calculated using Eq. (8.5); negative values indicate removal and positive values are additions of nutrients. *NB* is northwest sector of Kaneohe bay, *HFP* is He'eia Fishpond and mixing. The rate of bulk nutrient removal encompassing all of these processes can be estimated using the effective terrestrial end-members based on the intercepts of the individual nutrient distributions. To make this estimate we use the following relationship (Maeda and Windom 1982):

$$R = IQ_T - C_T Q_T \tag{8.5}$$

where R is the nutrient removal rate in the coastal zone, I is the effective terrestrial end-member derived as the *y*-axis intercept of the measured bay and pond nutrient concentrations, Q_T is terrestrial water flux (stream + SGD), and C_T is the terrestrial nutrient concentration. Silicate removal rates were 73–500 mol d^{-1} and 690–1400 mol d^{-1} in August and November, respectively in the northwestern sector of the bay. There were similar significant removal rates for DIP and DIN in both time periods (Table 8.5). We did not collect enough bay nutrient data to derive nutrient removal rates for the rest of central bay.

In Heeia fishpond, DIN removal rate was 0.71 mol d^{-1} but calculated values for silicate and DIP suggest addition rather than removal (Table 8.5). In case of DIP this addition can be explained by phosphate release from suspended particles as they encounter saline waters of the pond, or benthic flux by diffusion from sediments into the shallow pond water column (Briggs et al. 2013). Silicates may be added to the water via the remineralization of the silicate rich organic matter in the wetland, contribution of high-level groundwater into SGD, or benthic flux by diffusion from sediment within the pond. Potentially, there could be remineralization and recycling of silicate within the pond water column, as well.

Our analysis shows that 78–95 % of silicate, 98 % of DIP and 83–90 % of DIN delivered to the bay and 96 % of DIN delivered to the pond is removed via biotic and/or inorganic processes. Due to the large nutrient contribution from SGD these removal rates are orders of magnitude higher than previous estimates (Smith et al. 1981).

5 Conclusions

We compared stream and submarine groundwater discharge in the northwestern and central sectors of Kaneohe bay, as well as in Heeia Fishpond. Our analysis showed that SGD in form of total (fresh+brackish) groundwater discharge was 2–4 times larger than surface inputs. We found large differences in SGD derived using different techniques for the same area. This inconsistency can be partially explained by the different time scales that the tracer techniques represent. Corresponding DIN and silicate fluxes were dominated by SGD, DIP on the other hand was delivered mostly via streams. While we observed nutrient uptake in coastal waters, nutrients were also relatively quickly removed by mixing resulting in fast coastal flushing rates.

Our study provides several major insights:

- Even our lower estimates of SGD indicate that this process must be considered in a coastal nutrient balance as groundwater delivers significant quantities of terrestrial new—as well as a recycled nutrients
- Coastal nutrient inventories are determined by the combination of SGD and coastal flushing rates. The two processes work together to influence inventory buildup and mixing offshore. Residence times vary significantly spatially and temporally in the bay.
- Seventy-eight to ninety-nine percent of nutrients delivered to the coastline are removed by biotic and abiotic processes in inner Kaneohe Bay.

Acknowledgements The authors acknowledge the help of the following individuals from the University of Hawaii: James Bishop, Kim Falinski, Christine Waters, Sam Wall. We are grateful to Kako'o'iwi and Paepae o He'eia who provided access to the study sites, logistical support, and field assistance. This paper was funded in part by a grant from the NOAA, Project R/IR-19, which is sponsored by the University of Hawaii Sea Grant College Program, SOEST, under Institutional Grant No. NA09OAR4170060, NA14OAR4170071 from NOAA Office of Sea Grant, Department of Commerce. The views expressed herein are those of the author(s) and do not necessarily reflect the views of NOAA or any of its subagencies. UNIHI-SEAGRANT-BC-12-03.

References

Andersen MS, Baron L, Gudbjerg J, Gregersen J, Chapellier D, Jakobsen R, Postma D (2007) Discharge of nitrate-containing groundwater into a coastal marine environment. J Hydrol 336:98–114

Bathen KH (1968) A descriptive study of the physical oceanography of Kaneohe Bay, Oahu, Hawaii. University of Hawaii, Hawaii Institute of Marine Biology Technical Report No. 14, p 353

Beck AJ, Cochran JK, Sañudo-Wilhelmy SA (2009) Temporal trends of dissolved trace metals in Jamaica Bay, NY: importance of wastewater input and submarine groundwater discharge. Estuar Coasts 32:535–550

Beck AJ, Charette MA, Cochran JK, Gonneea ME, Peucker-Ehrenbrink B (2013) Dissolved strontium in the subterranean estuary—implications for the marine strontium isotope budget. Geochim Cosmochim Acta 117:33–52

Bokuniewicz H, Buddemeier R, Maxwell B, Smith C (2003) The typological approach to submarine ground-water discharge (SGD). Biogeochemistry 66:145–158

Bowen JL, Valiela I (2001) The ecological effects of urbanization of coastal watersheds: historical increases in nitrogen loads and eutrophication of Waquoit Bay estuaries. Can J Fish Aquat Sci 58:1489–1500

Bratton JF (2010) The three scales of submarine groundwater flow and discharge across passive continental margins. J Geol 118(5):565–575

Briggs RA, Ruttenberg KC, Ricardo AE, Glazer BT (2013) Constraining sources of organic matter to tropical coastal sediments: consideration of non-traditional end members. Aquat Geochem 19(5–6):543–563

Burnett WC, Dulaiova H (2003) Estimating the dynamics of groundwater input into the Coastal Zone via continuous Radon-222 measurements. J Environ Radioact 69(1–2):21–35

Charette MA (2007) Hydrologic forcing of submarine groundwater discharge: insight from a seasonal study of radium isotopes in a groundwater-dominated salt marsh estuary. Limnol Oceanogr 52(1):230–239

Charette MA, Sholkovitz ER, Hansel CM (2005) Trace element cycling in a subterranean estuary, part 1: geochemistry of the permeable sediments. Geochim Cosmochim Acta 69:2095–2109

Charette MA, Moore WS, Burnett WC (2008) Uranium- and thorium-series nuclides as tracers of submarine groundwater discharge. Radioact Environ 13:155–191

Cyronak T, Schulz KG, Santos IR, Eyre BD (2014) Enhanced acidification of global coral reefs driven by regional biogeochemical feedbacks. Geophys Res Lett 41(15):5538–5546

Dailer ML, Knox RS, Smith JE, Napier M, Smith CM (2010) Using δ15N values in algal tissue to map locations and potential sources of anthropogenic nutrient inputs on the island of Maui, Hawai'i, USA. Mar Pollut Bull 60:655–671

Dimova NT, Swarzenski PW, Dulaiova H, Glenn C (2012) Utilizing multi-channel electrical resistivity methods to examine the dynamics of the freshwater–saltwater interface in two Hawaiian groundwater systems. J Geophys Res 117. doi:10.1029/2011JC007509

Dollar SJ, Atkinson MJ (1992) Effects of nutrient subsidies from groundwater to near shore marine ecosystems off the island of Hawaii. Estuar Coast Shelf Sci 35:409–424

Drupp P, DeCarlo EH, Mackenzie FT, Bienfang P, Sabine CL (2011) Nutrient inputs, phytoplankton response, and CO2 variations in a semi-enclosed subtropical embayment, Kaneohe Bay, Hawaii. Aquat Geochem 17:473–498

Dulaiova H, Burnett WC, Wattayakorn G, Sojisuporn P (2006) Are the groundwater inputs into river-dominated areas important? The Chao Phraya River-Gulf of Thailand. Limnol Oceanogr 51:2232–2247

Dulaiova H, Camilli R, Henderson PB, Charette MA (2010) Coupled radon, methane and nitrate sensors for large-scale assessment of groundwater discharge and non-point source pollution to coastal waters. J Environ Radioact 101(7):553–563. doi:10.1016/j.jenvrad.2009.12.004

Garrison GH, Glenn CR, McMurtry GM (2003) Measurement of submarine groundwater discharge in Kahana Bay, Oahu, Hawaii. Limnol Oceanogr 48:920–928

Glenn CG, Whittier RB, Dailer ML, Dulaiova H, El-Kadi AI, Fackrell J, Kelly JL, Waters CA, Sevadjian J (2013) Lahaina groundwater tracer study—Lahaina, Maui, Hawaii. Final Report prepared for the State of Hawaii Department of Health, US EPA and US Army Engineer Research and Development Center. 502 pp

Gonneea ME, Charette MA (2014) Hydrologic controls on nutrient cycling in an unconfined coastal aquifer. Environ Sci Tech 48:14178–14185

Gonneea ME, Morris P, Dulaiova H, Charette MA (2008) New perspectives on radium behavior within a subterranean estuary. Mar Chem 109:250–267

Holleman K (2011) Comparison of submarine groundwater discharge, coastal residence times, and rates of primary productivity, Manualua Bay, Oahu and Honokohau Harbor, Big Island, Hawaii, USA. M.S. Thesis, University of Hawaii

Hoover DJ (2002) Fluvial nitrogen and phosphorus inputs to Hawaiian coastal waters: storm loading, particle-solution transformations and ecosystem impacts. Unpublished Ph.D. discussion, Department of Oceanography, University of Hawaii

Hoover DJ et al (2009) Fluvial Fluxes of water, suspended particulate matter, and nutrients and potential impacts on tropical coastal water biogeochemistry: Oahu, Hawaii. Aquat Geochem 15:547–570

Izuka SK, Hill BR, Shade PJ, Tribble GW (1994) Geohydrology and possible transport routes of polychlorinated biphenyls in Haiku valley, Oahu, Hawaii. Water-Resources Investigations Report 92-4168. U.S. Geological Survey

Johnson EE, Wiegner TN (2014) Surface water metabolism potential in groundwater-fed coastal waters of Hawaii Island, USA. Estuar Coasts 37:712–723. doi:10.1007/s12237-013-9708-y

Johnson AG, Glenn CR, Burnett WC, Peterson RN, Lucey PG (2008) Aerial infrared imaging reveals large nutrient-rich groundwater inputs to the ocean. Geophys Res Lett 35(15), L15606

Kaul LW, Froelich PN (1984) Modeling estuarine nutrient geochemistry in a simple system. Geochem Cosmochim Acta 48:1417–1433

Kelly RP, Moran SB (2002) Seasonal changes in groundwater input to a well-mixed estuary estimated using radium isotopes and implications for coastal nutrient budgets. Limnol Oceanogr 47:1796–1807

Kim G, Kim J-S, Hwang D-W (2011) Submarine groundwater discharge from oceanic islands standing in oligotrophic oceans: implications for global biological production. Limnol Oceanogr 56(2):673–682

Kim TH, Waska H, Kwon EH, Suryaputra IGN, Kim G (2012) Production, degradation, and flux of dissolved organic matter in the subterranean estuary of a large tidal flat. Mar Chem 142–144:1–10

Kiro Y, Weinstein Y, Starinsky A, Yechieli Y (2013) Groundwater ages and reaction rates during seawater circulation in the Dead Sea aquifer. Geochim Cosmochim Acta 122:17–35. doi:10.1016/j.gca.2013.08.005

Kleven A (2014) Coastal Groundwater Discharge as a Source of Nutrients to He'eia Fishpond, O'ahu, HI. Undergraduate thesis, Global Environmental Science Program, University of Hawaii at Manoa. XB-12-02

Knee KL, Street JH, Grossman EE, Boehm AB, Paytan A (2010) Nutrient inputs to the coastal ocean from submarine groundwater discharge in a groundwater-dominated system: relation to land use (Kona Coast, Hawai'i, USA). Limnol Oceanogr 55:1105–1122

Knee KL, Garcia-Solosna E, Garcia-Orellana J, Boehm AB, Paytan A (2011) Using radium isotopes to characterize water ages and coastal mixing rates: a sensitivity analysis. Limnol Oceanogr Methods 9:380–395

Krest JM, Moore WS, Rama (1999) ^{226}Ra and ^{228}Ra in the mixing zones of the Mississippi and Atchafalaya Rivers: indicators of groundwater input. Mar Chem 64:129–152

Kroeger KD, Charette MA (2008) Nitrogen biogeochemistry of submarine groundwater discharge. Limnol Oceanogr 53:1025–1039

Kwon EY, Kim G, Primeau F, Moore WS, Cho H-M, DeVries T, Sarmiento JL, Charette MA, Cho Y-K (2015) Global estimate of submarine groundwater discharge based on an observationally constrained radium isotope model. Geophys Res Lett 41:8438–8444. doi:10.1002/2014GL061574

Laws EA (1980) Effects of waste-water discharges on phytoplankton communities. In: University of Hawaii, Sea Grant Cooperative Report UNIHI-SEAGRANT-CR-80-1. pp. 13–40

Lee DR (1977) A device for measuring seepage flux in lakes and estuaries. Limnol Oceanogr 22:140–147

Leta OT, El-kadi A, Dulaiova H, Ghazal K (2015) Applicability of the SWAT model for hydrological modeling of a small-scale watershed in Hawaii under scarcity of hydrological data. J Hydrol Eng (in press)

Li HL, Jiao JJ (2013) Quantifying tidal contribution to submarine groundwater discharges: a review. Chin Sci Bull 58(25):3053–3059

Lowe R, Falter JL, Monismith SG, Atkinson MJ (2009) A numerical study of circulation in a coastal reef-lagoon system. J Geophys Res 114:C06022. doi:10.1029/2008JC005081

Lucas LV, Thompson JK, Brown LR (2009) Why are diverse relationships observed between phytoplankton biomass and transport time? Limnol Oceanogr 54(1):381–390

Maeda M, Windom HL (1982) Behavior of uranium in two estuaries of the south-eastern United States. Mar Chem 11:427–436

Mayfield KK (2013) A summary of the submarine groundwater discharge (SGD) in Kahana Bay: spatial and intra-daily variability. Undergraduate thesis, Global Environmental Science Program, University of Hawaii at Manoa. 48 p

Michael HA, Mulligan AE, Harvey CF (2005) Seasonal oscillations in water exchange between aquifers and the coastal ocean. Nature 436(7054):1145–1148

Moore WS (1996) Large groundwater inputs to coastal waters revealed by 226Ra enrichments. Nature 380:612–614

Moore WS (1997) The effects of groundwater input at the mouth of the Ganges-Brahmaputra Rivers on barium and radium fluxes to the Bay of Bengal. Earth Planet Sci Lett 150:141–150
Moore WS (1999) The subterranean estuary: a reaction zone of ground water and sea water. Mar Chem 65:111–126
Moore WS (2000a) Determining coastal mixing rates using radium isotopes. Cont Shelf Res 20:1993–2007
Moore WS (2000b) Ages of continental shelf waters determined from 223Ra and 224Ra. J Geophys Res 105:22117–22122
Moore WS (2003) Sources and fluxes of submarine groundwater discharge delineated by radium isotopes. Biogeochemistry 66:75–93
Moore WS (2006) The role of submarine groundwater discharge in coastal biogeochemistry. J Geochem Explor 88:389–393
Moore WS (2010) The effect of submarine groundwater discharge on the ocean. Ann Rev Mar Sci 2:59–88
Moore WS, Wilson AM (2005) Advective flow through the upper continental shelf driven by storms, buoyancy, and submarine groundwater discharge. Earth Planet Sci Lett 235:564–576
Moore WS, Blanton JO, Joye SB (2006) Estimates of flushing times, submarine groundwater discharge, and nutrient fluxes to Okatee Estuary, South Carolina. J Geophys Res 111. doi:10. 1029/2007jc004199
Moore WS, Sarmiento JL, Key RM (2008) Submarine ground-water discharge revealed by 228Ra distribution in the upper Atlantic Ocean. Nat Geosci 1:309–311
Paytan A, Shellenbarger GG, Street JH, Gonneea ME, Davis K, Young MB, Moore WS (2006) Submarine groundwater discharge: an important source of new inorganic nitrogen to coral reef ecosystems. Limnol Oceanogr 51:343–348
Peterson PN, Burnett WC, Taniguchi M, Chen J, Santos IR, Ishitobi T (2009) Radon and radium isotope assessment of submarine groundwater discharge in the Yellow River delta, China. J Geophys Res 113:1–14
Ringuet S, Mackenzie FT (2005) Controls on nutrient and phytoplankton dynamics during normal flow and storm runoff conditions, southern Kaneohe Bay, Hawaii. Estuaries 28:327–337
Robinson C, Li L, Prommer H (2006) Tide-induced recirculation across the aquifer-ocean interface. Water Resour Res 43(7), W07428
Robinson C, Li L, Barry DA (2007) Effect of tidal forcing on a subterranean estuary. Adv Water Resour 30(4):851–865
Roy M, Martin JB, Cherrier J, Cable JE, Smith CG (2010) Influence of sea level rise on iron diagenesis in an east Florida subterranean estuary. Geochim Cosmochim Acta 74 (19):5560–5573
Santoro AE, Boehm AB, Francis CA (2006) Denitrifier community composition along a nitrate and salinity gradient in a coastal aquifer. Appl Environ Microbiol 72(3):2102–2109
Santos IR, Burnett WC, Chanton J, Suryaputra B, Dittmar T (2008) Nutrient biogeochemistry in a Gulf of Mexico subterranean estuary and groundwater-derived fluxes to the coastal ocean. Limnol Oceanogr 53(2):705–718
Santos IR, Eyre BD, Glud RN (2010) Influence of porewater advection on denitrification in carbonate sands: evidence from repacked sediment column experiments. Geochim Cosmochim Acta 96:247–258
Santos IR, Eyre BD, Huettel M (2012) The driving forces of porewater and groundwater flow in permeable coastal sediments: a review. Estuar Coast Shelf Sci 98:1–15
Santos IR, Glud RN, Maher D, Erler D, Eyre BD (2013) Diel coral reef acidification driven by porewater advection in permeable carbonate sands, Heron Island, Great Barrier Reef. Geophys Res Lett 38(3)
Schoelynck J, Bal K, Backx H, Okruszko T, Meire P, Struyf E (2010) Silica uptake in aquatic and wetland macrophytes: a strategic choice between silica, lignin and cellulose? New Phytol 186 (2):385–391

Shade PJ, Nichols WD (1996). Water budget and the effects of land-use changes on ground-water recharge, Oahu, Hawaii. USGS Professional Paper 1412-C

Slomp CP, Van Cappellen P (2004) Nutrient inputs to the coastal ocean through submarine groundwater discharge: controls and potential impact. J Hydrol 295:64–86

Smith SV, Kimmerer W, Laws E, Brock RE, Walsh T (1981) Kaneohe Bay sewage diversion experiment: perspectives on ecosystem responses to nutritional perturbation. Pac Sci 35(4)

Smith J, Smith C, Hunter C (2001) An experimental analysis of the effects of herbivory and nutrient enrichment on benthic community dynamics on a Hawaiian reef. Coral Reefs 19(4):332–342

Spiteri C, Slomp CP, Charette MA, Tuncay K, Meile C (2008) Flow and nutrient dynamics in a subterranean estuary (Waquoit Bay, MA, USA): field data and reactive transport modeling. Geochim Cosmochim Acta 72:3398–3412

Street JH, Knee KL, Grossman EE, Paytan A (2008) Submarine groundwater discharge and nutrient addition to the coastal zone and coral reefs of leeward Hawai. Mar Chem 109:355–376

Taniguchi M, Burnett WC, Cable JE, Turner JV (2002) Investigation of submarine groundwater discharge. Hydrol Process 16:2115–2129

Taniguchi M, Burnett WC, Smith CF, Paulsen RJ, O'Rourke D, Krupa SL, Christoff JL (2003) Spatial and temporal distributions of submarine groundwater discharge rates obtained from various types of seepage meters at a site in the Northeastern Gulf of Mexico. Biogeochemistry 66:35–53

Timmerman A, Young C, McManus MA, Ruttenberg KC, D'Andrea B (2015). Seasonal dynamics of freshwater input to a tropical coastal marine embayment. Estuar Coast Shelf Sci

Tomasky G, Valiela I, Charette MA (2013) Determination of water mass ages using radium isotopes as tracers: implications for phytoplankton dynamics in estuaries. Mar Chem 156:18–26

Wang G, Wang Z, Zhai W, Moore WS, Li Q, Yan X, Qi D, Jiang Y (2014) Net subterranean estuarine export fluxes of dissolved inorganic C, N, P, Si, and total alkalinity into the Jiulong River estuary, China. Geochim Cosmochim Acta 149:103–114

Wankel SD, Kendall C, Paytan A (2009) Using nitrate dual isotopic composition (dN and dO) as a tool for exploring sources and cycling of nitrate in an estuarine system: Elkhorn Slough, California. J Geophys Res 114:G01011. doi:10.1029/2008JG000729

Waska H, Kim G (2011) Submarine groundwater discharge (SGD) as a main nutrient source for benthic and water-column primary production in a large intertidal environment of the Yellow Sea. J Sea Res 65:103–113

Weston NB, Porubsky WP, Samarkin V, MacAvoy S, Erickson M, Joye SB (2006) Pore water stoichiometry of terminal metabolic products, sulfate, and dissolved organic carbon and nitrogen in intertidal creek-bank sediments. Biogeochemistry 77:375–408

Wilson AM, Gardner LR (2006) Tidally driven groundwater flow and solute exchange in a marsh: numerical simulations. Water Resour Res 42:W01405. doi:10.1029/2005WR004302

Windom HL, Moore WS, Niecheski LFH, Jahnke R (2006) Submarine groundwater discharge: a large, previously unrecognized source of dissolved iron to the South Atlantic Ocean. Mar Chem 102:252–266

Young CW (2011) Perturbation of nutrient inventories and phytoplankton community composition during storm events in a tropical coastal system: He'eia Fishpond, O'ahu, Hawaii. M.S. Thesis, The Department of Oceanography, University of Hawaii, 400 pp

Zektser IS (2000) Groundwater and the environment: applications for the global community. Lewis Publishers (CRC Press LLC)

Chapter 9
Quantifying Groundwater Export from an Urban Reservoir: A Case Study from Coastal South Carolina

L. Peterson, R. Peterson, E. Smith, and S. Libes

Abstract Climatic and anthropogenic factors can have a significant influence on groundwater resources, calling into question the future quality and quantity of the commodity. In this chapter, we discuss current and emerging issues concerning groundwater scarcity. These concepts are demonstrated using a case study from an urban reservoir that serves as a stormwater conduit to the nearshore ocean. Quantitative estimates of groundwater interaction with the reservoir were determined via direct tracer techniques which are rarely, if ever, used by urban hydrologists. Continuous time-series records of dissolved ^{222}Rn were collected to evaluate the volumetric percentage of groundwater within the reservoir from 2012 through 2013. Using high-resolution sampling, we are able to characterize groundwater and reservoir response on event and seasonal time scales, while also offering general assessments of the hydrologic conditions during the study. When rainfall was not occurring, evapotranspiration served as the primary driver of overall hydrologic characteristics, directly influencing the water table and subsequent groundwater discharged from the reservoir. However, during storm events, hydrologic factors influencing the amount of groundwater within the reservoir were found to be more complex, including event duration, magnitude, and antecedent conditions. Seasonally, rainfall patterns were largely responsible for the magnitude of groundwater present within the reservoir and quasi-related to peak export to the coastal ocean. Most notably, we observed a decline in the volumetric percentage of groundwater within the reservoir as a result of increased groundwater residence time within the aquifer—a likely function of reduced aquifer recharge that would result from more efficient stormwater management.

L. Peterson (✉) • R. Peterson • S. Libes
Coastal Carolina University, School of Coastal and Marine Systems Science,
Conway, SC 29526, USA
e-mail: lepeters@g.coastal.edu

E. Smith
University of South Carolina, North Inlet Winyah Bay National Estuarine Research Reserve,
Georgetown, SC 29442, USA

A. Fares (ed.), *Emerging Issues in Groundwater Resources*, Advances in Water Security, DOI 10.1007/978-3-319-32008-3_9

1 Water Origin Assessment

In natural settings, rainwater delivered to the land surface is either transported overland via surface pathways or infiltrated through the ground surface and conveyed through subsurface pathways. To avoid potential flooding in urban areas, rainwater fallen on impervious surfaces (roads, parking lots, buildings, and driveways) is funneled through engineered infrastructure toward larger receiving basins. Unfortunately, bypassing the natural infiltration processes eliminates any potential biological and chemical solute transformations that would otherwise occur in the subsurface prior to discharge, and also exposes surface runoff to urban contaminants. Hutchins et al. (2014) reported drastic differences between urban groundwater and surface water in terms of nutrient and dissolved carbon (DOC) concentrations, which were attributed entirely to transport pathway. In fact, urban surface runoff was found to contain over 300-fold more nitrate + nitrite ($NO_2^- + NO_3^-$) than shallow groundwater collected from a nearby aquifer (Fig. 9.1). As such, it is ideal to promote subsurface flowpaths as opposed to channelized surface flows to minimize the contamination potential and material loading to urban reservoirs. Quantifying the amount of water contributed by various sources to a reservoir can therefore be used as a diagnostic tool to interpret community shifts and water quality changes within a receiving water body.

The goal of many stormwater managers is to design infrastructure to transport rainwater away from the urban landscape toward impoundment reservoirs that act to slowly release the flood pulse into larger systems that ultimately convey the water downstream. The holding capacity of these impoundment reservoirs allows the flood pulse to be more slowly discharged downstream, thereby decoupling the

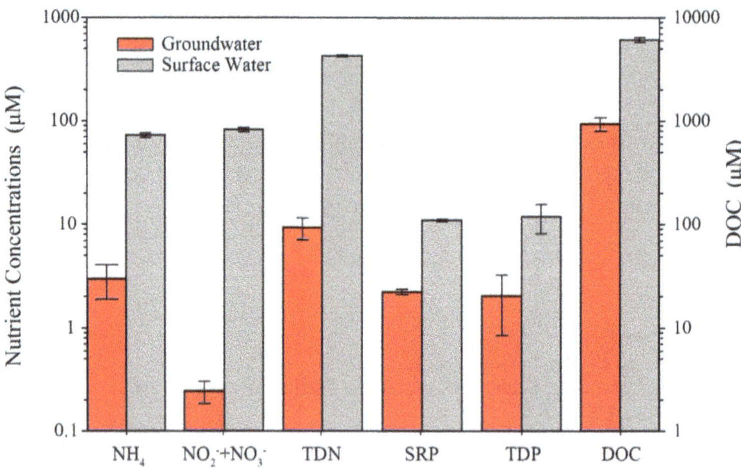

Fig. 9.1 Ammonium, nitrate + nitrite, total dissolved nitrogen (TDN), soluble reactive phosphorus (SRP), total dissolved phosphorous (TDP), and dissolved organic carbon (DOC) concentrations in urban surface water and groundwater (data from Hutchins et al. 2014). Error bars represent $1-\sigma$ standard deviation of all sample measurements

export rate of water leaving an urban reservoir from input rates. For this reason, it is important to not only consider source-specific contributions to a reservoir but also the loading potential attributed to each source being exported downstream. This concept is illustrated by Hutchins et al. (2014) whereby amendment experiments were performed using urban groundwater and surface water samples. Urban runoff was found to not only contain the highest nutrient and DOC concentrations but also to yield the highest primary and secondary production rates when amended with seawater (Fig. 9.2). These results are demonstrative of contrasting metabolic stimuli inherent to surface runoff and groundwater and the potential biological ramifications of such differences when discharged into seawater. The purpose of this chapter is to present and discuss methods used to estimate the volumetric role of groundwater within an urban reservoir and to examine the quantity and contribution of groundwater exported from such a reservoir into the nearshore ocean.

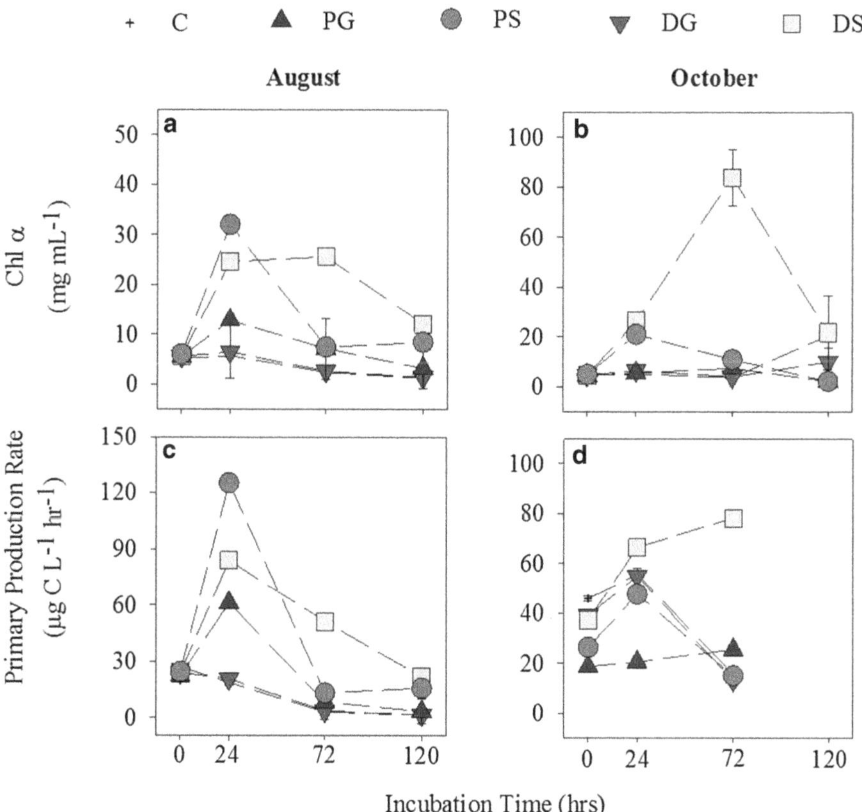

Fig. 9.2 Chlorophyll *a* concentrations (**a**, **b**) and primary production (**c**, **d**) over incubation period for control water (**c**), pristine groundwater (PG), pristine surface water (PS), developed groundwater (DG), and developed surface water (DS) treatments. Data from Hutchins et al. (2014)

Water budget delineations quantify the hydrologic components influencing a system. In this way, input/output rates may be used to characterize the degree of influence on a system by various water sources. Prior knowledge of the system allows the researcher to focus on water sources or sinks associated with a particular research question. For this case study, we selected a coastal reservoir discharging directly into the nearshore ocean to examine the export characteristics of groundwater from the system and associated driving forces. Our approach is distinguished from the aforementioned water budget construction approaches in that we are not quantifying input rates to the reservoir, but rather the amount of two distinct water sources present and available for export. The case study described throughout this chapter considers the relative volumes of groundwater and direct precipitation within a reservoir and explains how one could determine source-specific volumetric outputs. More detail regarding true water budget and water balance approaches may be found in Chap. 7.

1.1 Groundwater Measurement Techniques

Much of the literature concerning urban hydrology has focused on hydrographic studies (Meriano et al. 2011), channel geomorphology (Vietz et al. 2014), and associated ecological responses (Grimm et al. 2008). The abundance of urban reservoir data concerning physical, chemical, and biological indices of ecosystem health rarely consider source-specific hydrology as a possible influencing factor of urban water quality (Brabec et al. 2002). Researchers who have performed comprehensive studies regarding urban hydrology, including subsurface pathways have utilized indirect measurement techniques, focusing on surface water parameters directly and resolving groundwater contributions by difference (Haase 2009; Meriano et al. 2011). Additionally, hydrologic models can simulate complex fluid interactions between surface and subsurface pathways (Arnold and Allen 1996; Arnold et al. 2000), however, these methods must be validated with observational data to ensure accurate representation of both the above- and below-ground water transport pathways.

While groundwater fluxes to receiving waters may be characterized by a variety of direct measurement and tracer techniques (e.g., seepage meters, stable isotopes, and natural radiotracers), many of the researchers exercising these techniques have focused efforts on rural and suburban coastal ocean (Burnett et al. 2006), riverine (Mullinger et al. 2007), estuarine (Santos et al. 2010), and lacustrine (Dimova and Burnett 2011) environments. To address the existing gap between environment and technique, we present a case study utilizing a direct tracer measurement technique to evaluate the magnitude of groundwater influence on an urban reservoir.

1.2 Basin Description

For the case study presented here, we selected an urban reservoir situated on the northeastern coast of South Carolina within the Grand Strand region, a stretch of semi-contiguous shoreline between Winyah Bay and Little River containing the urban center of Myrtle Beach. This region includes approximately 100 km of sandy beach and supports over 14 million seasonal tourists annually. Extensive development in the area has led to significant engineering projects to channelize stormwater into coastal reservoirs and outfall pipes which then transport stormwater offshore. In total, 15 coastal reservoirs exist throughout the Grand Strand ranging from non-tidal to tidally flushed, which serve as open-air stormwater conduits designed to capture stormwater runoff from the urban landscape and ultimately convey water toward the ocean. Although the collective waters discharged from these engineered features is referred to as stormwater, these waters likely represent a mixture of groundwater, direct precipitation, and surface (overland) runoff.

1.2.1 Dogwood Reservoir

Data presented here were collected within the Dogwood Reservoir drainage basin, a ~4.7 km^2 area located within the town of Surfside Beach (Fig. 9.3).

The drainage basin is of low relief, with mostly residential, recreational, and municipal land uses (65 % of the basin area) and a significant region of vegetated uplands (23 %), as well as miscellaneous land uses including a golf course and croplands (12 %). Prior to the 1970s, this area drained into a natural reservoir, which traversed the beachface to ultimately discharge into Long Bay, the nearshore waters of the Grand Strand. Circa 1970, Dogwood Reservoir was dammed to create an aquatic amenity for the surrounding development thus forcing unidirectional freshwater flow offshore. By design, the reservoir is shallow (average depth ~1.3 m), created by drag lining from the shoreline with an upstream settling pond installed in the 2000s. Presently, Dogwood Reservoir functions as a stormwater detention pond, effectively engineered to increase water residence time by reducing water outputs to evapo(transpi)ration, infiltration, and discharge over a weir (Schueler 2000; Drescher et al. 2007). Inland reaches of the reservoir incise a vegetated landscape creating a natural bank comprised of soils and sediments rooted by grasses and shrubs. A 40 m wide broad-crested weir situated 200 m inland of the coastline directs water out of the reservoir into a concrete-lined channel that directly links Dogwood Reservoir to Long Bay. Only within the concrete-lined channel is marine water mixed with water exported from Dogwood Reservoir before discharging into Long Bay, with mixing occurring primarily during flood stages of spring tide (Fig. 9.4). Inland of the weir, the influence of the tides is effectively reduced to zero with no detectable effects noted on reservoir levels and water table heights.

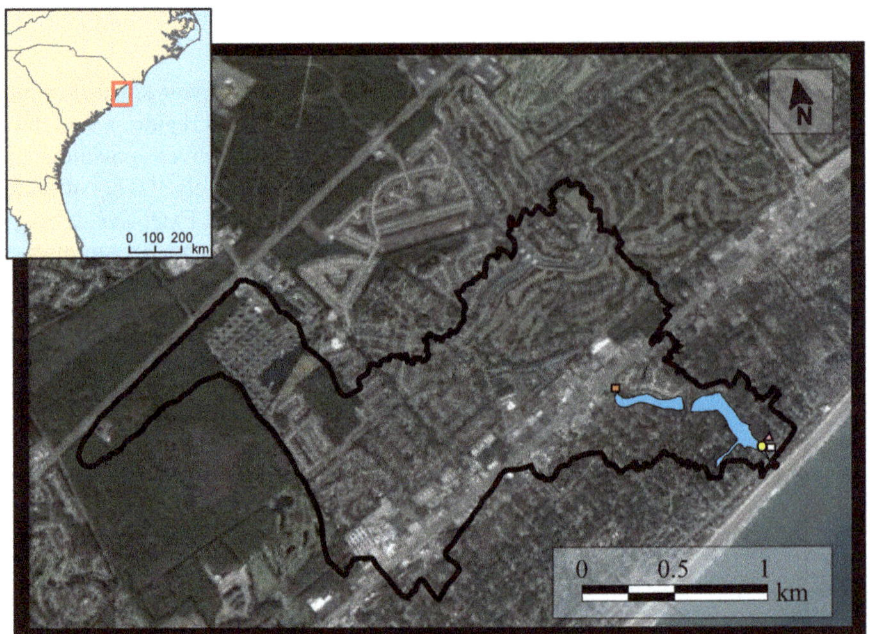

Fig. 9.3 Schematic diagram of Dogwood Reservoir and surrounding catchment region (*black polygon*). The time-series sampling station is denoted with a *circle*, and the rain gauge location with a *triangle* while end-member sampling locations are marked with *rectangles* for upstream (*dark*) and downstream (*white*) locations

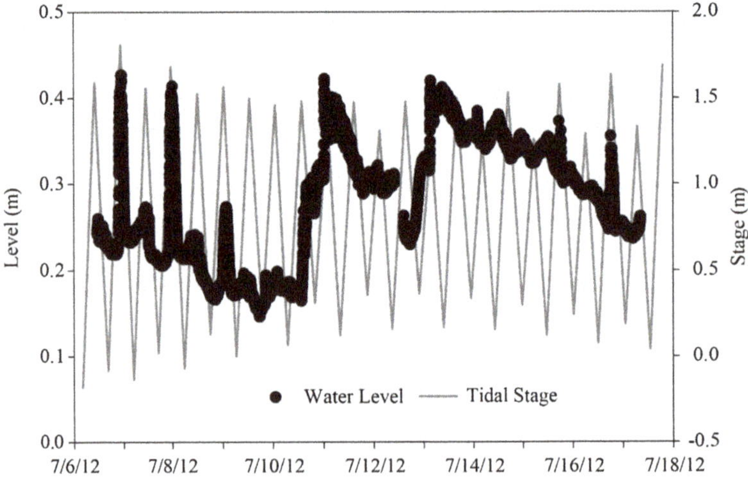

Fig. 9.4 Water level and tidal stage data collected ~5 m downstream of the Dogwood Reservoir weir and from Springmaid Pier, respectively. Note that data were collected downstream of the weir (which effectively acts as a baffle between marine and terrestrial water) and do not suggest tidal effects are propagated further upstream

1.3 Source Water Delineation

Dogwood Reservoir was chosen as the focus of this case study because of the relatively simple hydrology influencing the water body. Water sources to the reservoir (inland of the weir) include (1) groundwater, (2) direct precipitation, and (3) surface runoff which are introduced both directly to the reservoir as well as via a number of tributary creeks. Water removal mechanisms include (1) evapotranspiration, (2) infiltration to the local groundwater, (3) suspected pumping for individual irrigation purposes, and (4) discharge over the weir. Because we are interested in characterizing the export of groundwater from Dogwood Reservoir via discharge over the weir, we focus on discharge as the water export mechanism and groundwater and direct precipitation (used primarily to augment our interpretations of the groundwater data) as the inputs discussed. To evaluate export rates, we first determine the volumetric contribution of groundwater and direct precipitation to the total volume of the reservoir.

The volumetric percentages of each source term are then determined by dividing each source-specific volume within the reservoir (V_i) by the total reservoir volume V_T. Hence forth, terms will be presented as percentages given by:

$$\%V_i = \left(\frac{V_i}{V_T}\right) * 100 \quad (9.1)$$

While we recognize other water removal mechanisms may influence Dogwood Reservoir, we will discuss only the system export via discharge over the weir as our focus is to evaluate the reservoir connection to the coastal ocean.

1.4 Temporal and Spatial Scales

Time-series hydrologic data are useful to interpret various temporal scales at event and seasonal scales while providing an opportunity to evaluate general characteristics for comparison. In this case study, hydrologic time-series records of Dogwood Reservoir were constructed from October 2011 through August of 2013 and will be used to demonstrate the utility of such long-term groundwater discharge data. Discrete sampling efforts intended to capture seasonality and event-driven trends stand the chance of potentially missing the intended signal by interference from other forcing factors operating at different frequencies. For example, Dimova et al. (2013) documented no significant seasonal difference in groundwater fluxes to small Florida lakes determined via four sampling campaigns designed to capture seasonality. While those researchers may have indeed accurately characterized the system, there also exists the possibility that their field excursions serendipitously occurred during time periods that may not have captured the full variability of the systems. Alternatively, general trends may also be obscured by event scale impacts on water flow. For example, during an effort to generally describe groundwater fluxes out of the Sebastian River in Florida, Peterson et al. (2010) conducted a field excursion after a large storm passed through the area, and therefore their findings may have been biased due to system recovery following the event. The only way to avoid such potential biases when examining hydrological systems is to conduct continuous, long-term time series data collections.

The presented case study examines high-resolution time-series records to gain insights into both (1) intra-basin source-specific volumetric compositions, and (2) resulting export rates. Intra-basin volumetric source-specific compositions are reported as a percentage, representing the volumetric magnitude of source water presence relative to the total volume of the reservoir (Eq. 9.1). Conversely, basin export fluxes are reported as a volume per time for each water source applying the fraction of source water to the total rate of water exported from the reservoir.

Depending on one's research goals, it could be conceived that either intra-basin or basin export parameters would be of interest but it should be noted that determination of source-specific export requires knowledge of volumetric percentages of each source within the reservoir as per this approach. The presented case study is part of a larger effort to determine source-specific nutrient loading to the coastal ocean and as such, required determination of both intra-basin source-specific volumetric compositions and reservoir export terms.

2 Analytical Approaches

The presented case study examines an aspect of the hydrology influencing Dogwood Reservoir by determining the percent volume within the total reservoir of groundwater and direct precipitation as well as associated export rates for each source term. In this section, we describe the processes used to constrain each term required to perform the discussed computations.

2.1 Reservoir Water Origins

2.1.1 Drainage Basin and Reservoir Dimensions

We used a variety of remote sensing tools to constrain the dimensions of Dogwood Reservoir and surrounding drainage basin. First, the drainage basin boundaries were determined using information obtained from local stormwater management design plans. Aerial images of the region and GIS tools were then used to determine total drainage area and the open water area of the reservoir, which offered the opportunity to determine the dimensions of subregions characterized as impervious and vegetated cover.

To determine the total volume of Dogwood Reservoir, we conducted a bathymetric survey using a single beam sonar system outfitted with an RTK GPS antenna. In short, survey data were collected using a canoe outfitted with a mounted GPS antenna and single beam sonar transducer fixed just below the water surface to measure water depths along survey lines. All bathymetry data were measured relative to NAVD 88. To map the reservoir shallows, we traversed the region with a GPS antenna mounted pack. All data were offset corrected and discrete elevations were imported into ARC GIS to create a bathymetric surface defined by the surface area of the reservoir (Fig. 9.5).

Water volume estimates for the time of the survey were generated using ARC GIS (3D Analyst: Functional Surface) treating the elevation of the reservoir surface as a bounding surface to determine the total volume between the bathymetric surface and that of the reservoir water surface elevation. Following this procedure, we then regressed a full range of reservoir water elevations against water volume estimates to determine the coefficients of variance. From this, all recorded reservoir surface elevations throughout the project (measured using an elevation-corrected Solinst LTC Levelogger Junior CTD data logger) could be easily converted to reservoir volume, a crucial component for this analysis.

In order to determine the reservoir volume for all times throughout the study period, we made some assumptions regarding terms used in our calculations. The first assumes the water level recorded at the CTD deployment location is representative of the entire reservoir surface (i.e., the water surface lies at a constant elevation across the domain). Secondly, we assume the reservoir bank to have a slope of 90° (i.e., a flood plain is not considered at high reservoir volumes). We recorded a range in reservoir levels of only 0.22 m throughout the study, since a weir serves as a flow control structure, but we recommend higher resolution bank

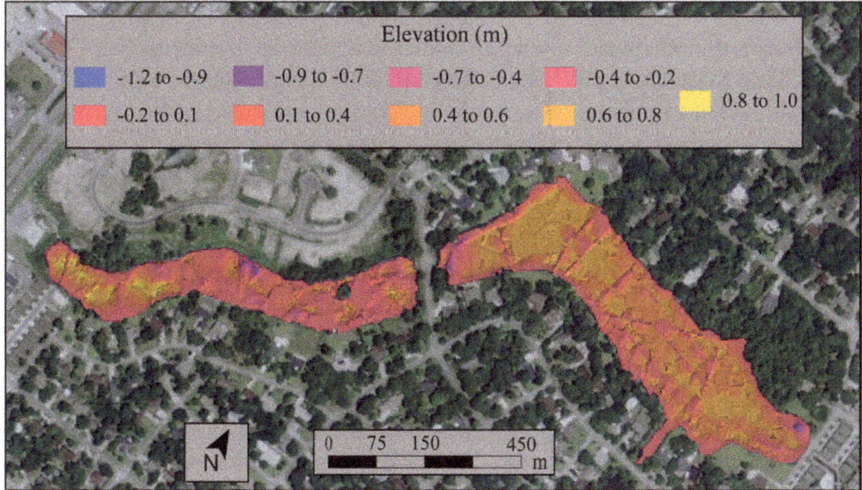

Fig. 9.5 Bathymetric map of Dogwood Reservoir. All elevation data relative to NAVD 88 datum

surveys for regions with large, frequent water level fluctuations. Lastly, and perhaps most importantly, we assume there is no net change in reservoir bathymetry during our sampling campaign (i.e., equal bounding surface elevations will always yield equal volumes). We feel it appropriate for this effort as a sediment settling pond was implemented upstream of Dogwood Reservoir, effectively trapping any new solid materials before entering the reservoir, and flows through the reservoir are likely not of sufficient force to cause bottom erosion. Nonetheless, presenting data as percentages of the total reservoir volume allow for changes in volume to occur without necessarily influencing the percent volume contributed by each source.

2.1.2 Direct Precipitation

To resolve the volumetric percentage of precipitation delivered directly to the open water area of Dogwood Reservoir, we divided the volume of precipitation falling directly on the reservoir by the total reservoir volume. Precipitation accumulations were measured locally using an automated ISCO sampler and accompanying rain gauge located ~50 m from the weir. The volume of rainfall directly delivered to the reservoir was determined by multiplying accumulated rainfall by the surface area of the reservoir. These volumetric rainfalls were averaged for each month to normalize rainfall totals to the elapsed time within each data interval. For these calculations, we assume a constant area of open water. We feel this is appropriate due to the minimal change in reservoir levels throughout the study.

2.1.3 Groundwater

Relative to surface runoff and meteoric water, groundwater is naturally enriched in ^{222}Rn (radon) through interaction with radium-rich sediments, making radon an ideal tracer of groundwater discharge studies in aquatic environments (e.g., Burnett et al. 2006; Swarzenski 2007; Charette et al. 2008). Furthermore, well established methods and instrumentation exist for converting measured radon activities to quantitative groundwater estimates after appropriate corrections are applied (Cable et al. 1996; Corbett et al. 1999; Burnett et al. 2001). Traditional radon mass balance approaches (e.g., Burnett and Dulaiova 2003) evaluate sources and sinks of ^{222}Rn affecting a water body to ultimately determine groundwater discharge rates. Often, these models assume steady-state with regard to radon inventory (dpm m^{-2} $time^{-1}$) whereby a change in measured radon activity between consecutive measurements must equal a change in radon input and/or output. Such mass balances are often applied in coastal ocean settings (e.g., Burnett et al. 2008; Santos et al. 2009) where relatively homogeneous discharge flowpaths are integrated through coastal mixing processes but little concern exists for the ultimate fate of the discharged groundwater. Several of the radon sources and sinks considered in these traditional radon mass balance models (e.g., radioactive ingrowth and decay, atmospheric degassing, diffusion from bottom sediments) are a function of the residence time of water and radon within the system. In tidal systems, corrections for these source/sink terms are often considered over tidal time scales.

For the case of inland reservoirs, however, such an approach is not ideal if the ultimate fate of the discharged groundwater is of interest. Groundwater is likely discharged along the entire upstream-downstream flowpath of such a reservoir, making an assessment of the radon residence time (a function of discharge location and reservoir flow velocities) difficult to estimate. To avoid these potential complications associated with temporal corrections, Peterson et al. (2010) proposed a model to determine groundwater inputs to a river system whereby the radon activity in surface water fluxing past a measurement site is subjected to a range of temporal corrections in order to place reasonable boundary conditions on the resulting groundwater input estimates. That model (which serves as the basis for our approach) is governed by the same principles as the traditional radon mass balance approaches, but instead of evaluating the rate of groundwater discharge into the reservoir, it resolves the volumetric amount of recently discharged groundwater present at a given time. This model is notably distinguished from traditional approaches (described above) in that we do not rely on radon mass change over time (which requires an assumption of steady-state to derive groundwater inputs), but rather treat each measurement as an individual 'sample' and therefore no assumption of steady-state is necessary. Nonetheless, both procedures account for sources of ^{222}Rn including groundwater discharge, diffusion from sediments, and production via decay of dissolved ^{226}Ra as well as sinks to atmospheric evasion and radioactive decay. Each measurement of dissolved ^{222}Rn (radon) in the reservoir is corrected for these effects, then divided by the ^{222}Rn concentration within the groundwater end-member to determine the fraction of groundwater present in the reservoir for a given interval. Section 2.1.3 describes how each ^{222}Rn source and sink term is evaluated in this case study.

To resolve the volumetric percentage of groundwater present within the reservoir, we follow the approach outlined by Peterson et al. (2010) using ^{222}Rn as an isotopic proxy for recently discharged groundwater. This approach is largely governed by two extreme assumptions regarding the residence time of ^{222}Rn in a reservoir prior to detection. As will be demonstrated, timing is crucial for work involving radioactive isotopes which must account for in situ decay and production of a tracer with a relatively short half-life. For this reason, Peterson et al. (2010) suggest using two different ^{222}Rn models in an attempt to constrain accurate results.

The first extreme condition in this model assumes all groundwater-derived ^{222}Rn was input proximal to the monitoring station (i.e., minimal residence time) and therefore requires no correction for decay or loss to the atmosphere, offering a minimum estimate of groundwater within those sampled waters. Minimum volumetric groundwater percent composition estimates are therefore given by:

$$G_{MIN} = \left[\frac{Res.\, Rn\, \left(\frac{dpm}{m^3}\right) - Bkgd.\, Ra\, \left(\frac{dpm}{m^3}\right)}{Gw.\, Rn\, \left(\frac{dpm}{m^3}\right)} \right] * 100 \tag{9.2}$$

where the minimum volumetric percentage of groundwater within the reservoir (G_{MIN}) is equal to the ^{222}Rn concentration in the reservoir (*Res. Rn*) minus the concentration of dissolved ^{226}Ra within the reservoir (*Bkgd. Ra*—that which is not derived from groundwater inputs) divided by the ^{222}Rn concentration within the groundwater (*Gw. Rn*).

Conversely, maximum volumetric percent groundwater composition estimates account for sinks of ^{222}Rn, assuming all radon entered the reservoir many days prior to detection and as such has been subject to loss from radioactive decay and diffusion to the atmosphere. Maximum percent groundwater estimates are then generated by:

$$G_{MAX} = \left[\frac{\left[\left[Res.\ Rn \left(\frac{dpm}{m^3}\right) + \left[\frac{Atm.\ Evas.\ \left(\frac{dpm}{m^2 sec.}\right) * R\ (sec.)}{Depth\ (m)} \right] - Bkgd.\ Ra \left(\frac{dpm}{m^3}\right) \right] e^{\lambda R} \right]}{Gw.\ Rn \left(\frac{dpm}{m^3}\right)} \right] *100 \quad (9.3)$$

Here, measured ^{222}Rn concentrations within the reservoir (*Res. Rn*) are assumed to be that which remain of the initial concentration after loss to atmospheric evasion (*Atm. Evas.*) and decay ($e^{\lambda R}$) since discharge prior to detection. Radon lost to decay was calculated using the decay constant of ^{222}Rn (λ) ($1.56*10^4\ s^{-1}$) and ^{222}Rn residence time in the reservoir (R). To appropriately correct our measured ^{222}Rn concentrations for decay and evasion since discharge, we use the mean life of ^{222}Rn (5.54 days), the average lifetime of a given atom within a population, to represent the residence time of radon within the reservoir. Mean life (τ) is related to the decay constant by:

$$\tau = \frac{1}{\lambda} \quad (9.4)$$

Loss of ^{222}Rn from a reservoir to the atmosphere is largely driven by a strong concentration gradient across the air/water interface and enhanced by turbulence (Burnett and Dulaiova 2003). Diffusive fluxes of ^{222}Rn to the atmosphere (F_{Rn}) were calculated by:

$$F_{Rn} = k(Rn_w - \alpha Rn_a) \quad (9.5)$$

where k is the gas transfer velocity, Rn_w and Rn_a are the ^{222}Rn concentrations in water and air, respectively, and α is Ostwald's solubility coefficient (MacIntyre et al. 1995). The gas transfer velocity is then determined by:

$$k = \frac{0.45\ u^{1.6} \left(\frac{Sc}{600}\right)^{-0.5}}{100} \quad (9.6)$$

where u is the wind speed 10 m above the water surface and Sc is the dimensionless Schmidt number representing a ratio of kinematic viscosity to a molecular diffusion coefficient. We treated Sc as a constant equal to 8.658 for a kinematic viscosity of $1.0043*10^{-6}$ m^2 s^{-1} and a molecular diffusion constant of $1.16*10^{-7}$ m^2 s^{-1}. Ostwald's solubility coefficient is then given by:

$$\alpha = 0.105 + 0.405e^{(-0.05\ T)} \tag{9.7}$$

where T is water temperature in °C. Air temperature and wind speed data (recorded every 15 min) used to determine maximum volumetric percent groundwater compositions within the reservoir were obtained from a public meteorological database containing continuous local recordings throughout the duration of the project (www.ysieconet.com). Acquisition of parameters used to generate additional correction terms will be discussed in detail throughout this section.

Long-term records of ^{222}Rn concentrations within the reservoir were measured using an automated radon system (Burnett and Dulaiova 2003) fixed to a stationary platform situated ~5 m from the nearest bank. In short, a submersible pump deployed 0.2 m above the reservoir bottom transferred water to a degassing chamber (RAD-Aqua; Durridge Co.) effectively bubbling ^{222}Rn from the water until air/water equilibrium was achieved. Radon-rich air was then pumped through desiccant to a radon-in-air monitor (RAD7; Durridge Co.) where activities were determined via alpha counting. Atmospheric ^{222}Rn activities were measured using the same instrumentation with an open air loop where ambient air was pumped through desiccant and into the counter. We then averaged these data to yield a mean atmospheric ^{222}Rn concentration of 262 dpm m^{-3}.

End-Member Characterization

The concentration of ^{222}Rn in groundwater largely varies as a function of ^{226}Ra content of the aquifer materials and groundwater residence time within the aquifer. These two factors contribute significantly to temporal and spatial variability in groundwater ^{222}Rn concentrations.

For this reason, most published studies using ^{222}Rn to estimate groundwater interaction with receiving waters discuss the importance of selecting an appropriate end-member sampling strategy. We chose to directly sample groundwater from piezometers constructed using PVC pipe with 10 cm screened sections installed 1.5, 1.25, 1.0, and 0.5 m below ground surface. To properly characterize groundwater end-member concentrations used to estimate minimum and maximum volumetric percentages of groundwater, we sampled groundwater from a nest of wells proximal to our time-series deployment (used in Eq. 9.2) and ~1 km upstream of our monitoring site (used in Eq. 9.3) (Fig. 9.3). Consequences of spatial heterogeneity on minimum and maximum groundwater discharge estimates were minimized by applying end-member data collected near the monitoring location to the minimum estimates and end-member data collected upstream to the maximum estimates. All groundwater samples were collected in 250 mL glass bottles (WAT-250 system; Durridge Co.) using a peristaltic pump and analyzed using standard RAD7 protocols. End-member data reported represent a depth-averaged concentration for each sampling effort and location in an attempt to best represent an integrated signal of groundwater entering the reservoir.

Researchers have established methods to distinguish the effects of sedimentary ^{226}Ra content and groundwater residence time on in situ end-member ^{222}Rn concentrations via experimental manipulation. Under one such method, water residence time is controlled for all treatments and differences in equilibrated ^{222}Rn activities represent geologic heterogeneity in ^{222}Rn production via ^{226}Ra decay (Corbett et al. 1998). We collected four sediment samples during piezometer installation and incubated 100 g of dry material in gas-tight 300 mL reaction flasks with radium-free tap water. Experiments were terminated after secular equilibrium (~21 days) was achieved between sedimentary ^{226}Ra and dissolved ^{222}Rn, i.e.:

$$N_{222} = \frac{\lambda_{226}}{\lambda_{222}} N_{226} \tag{9.8}$$

where the number of ^{222}Rn atoms (N_{222}) equals the number of ^{226}Ra atoms (N_{226}) multiplied by the ratio of decay constants for ^{226}Ra (λ_{226}) and ^{222}Rn (λ_{222}). After ~3 weeks, samples were measured using similar instrumentation and procedures as for groundwater field samples. Results represent maximum ^{222}Rn activities that could be produced by the aquifer materials. For this reason, experimental end-member ^{222}Rn concentrations should always be greater than or equal to those of field samples.

Additional Corrections

Sources of ^{222}Rn to the reservoir are defined as groundwater input, ^{222}Rn production via dissolved ^{226}Ra decay, and diffusion from bottom sediments. Both minimum and maximum volumetric percent groundwater composition calculations (Eqs. 9.2 and 9.3) require consideration of each of these terms. Excess ^{222}Rn activity, after accounting for production of ^{222}Rn via ^{226}Ra decay and sedimentary diffusion, represents that which was delivered to the reservoir via groundwater discharge (Burnett and Dulaiova 2003). To account for background levels of ^{222}Rn within the reservoir attributed to ^{226}Ra decay, we collected ~60 L of water from the reservoir and filtered it through MnO_2-impregnated acrylic fibers that quantitatively adsorb radium (Moore and Reid 1973). These fibers were then analyzed for total ^{226}Ra using methods outlined by Peterson et al. (2009).

While additional ^{222}Rn may be contributed to the system via diffusive fluxes from bottom sediments (e.g., Corbett et al. 1998), previous studies indicate that these contributions are minimal in advective systems (Lambert and Burnett 2003; Gleeson et al. 2013). To demonstrate the potential influence of diffusion of radon from bottom sediments, we estimated a range in radon contributed using end-member and reservoir radon concentrations, water depth and temperature, as well as porosity following recommendations by Cable et al. (1996), Corbett et al. (2000), and Burnett et al. (2003) (and references therein). As a result, mean minimum diffusive ^{222}Rn fluxes were found to be 5.91 dpm m^{-2} 30 min^{-1} (less than 1 % of the measured reservoir concentration), equivalent to 1178.25 dpm m^{-2} over the maximum defined residence time of 5.54 days (approximately 13 % of the measured reservoir concentration). Considering the range of analytical uncertainties inherent to the radon monitoring instrumentation used over the measurement period chosen, the possible radon sourced from sediment diffusion is within this limit and not considered as an independent term in the calculations. Neglecting any potential input of radon from sedimentary diffusion thus leads to our 'maximum' groundwater estimate truly serving as an upper bound on the groundwater percentage estimate. Further, any effect of the additional radon (especially concerning our maximum estimates) is reduced by averaging results from the two approaches.

Desired Resolution

The Dogwood Reservoir case study is intended to demonstrate the utility of time-series records of a direct tracer measurement method used to quantify the volumetric percentage of groundwater within an urban water body. We elected to sample ^{222}Rn continuously for approximately 2 years at 30 min intervals. While the RAD7 sampling frequency is programmed by the user, we recommend at least 30 min cycles to optimize counting statistics and measurement confidence, while providing temporal resolution sufficient to resolve short-term processes affecting the reservoir. When possible, Dulaiova et al. (2005) recommend using several RAD7s arranged in parallel to further enhance counting statistics for groundwater calculations.

Assumptions

In addition to those specifically stated throughout this section, there are several assumptions regarding our approach to evaluate the volumetric percentage of groundwater within Dogwood Reservoir. Sinks of ^{222}Rn from the reservoir were defined as atmospheric evasion and decay of ^{222}Rn. For this case study, we did not have a hydraulic scenario under which ^{222}Rn loss was occurring via groundwater recharge (due to elevated hydraulic gradients between the aquifer and the reservoir), though such a loss may exist in other settings. If aquifer recharge was indeed occurring, despite our assumption, volumetric percentages of groundwater would remain unchanged so long as the radon concentration in the reservoir is homogenous as per our approach since it would be the well mixed reservoir water (*Res. Rn* Eqs. 9.2 and 9.3) recharging the aquifer. However, quantifying a traditional water budget/balance would require knowledge of this term.

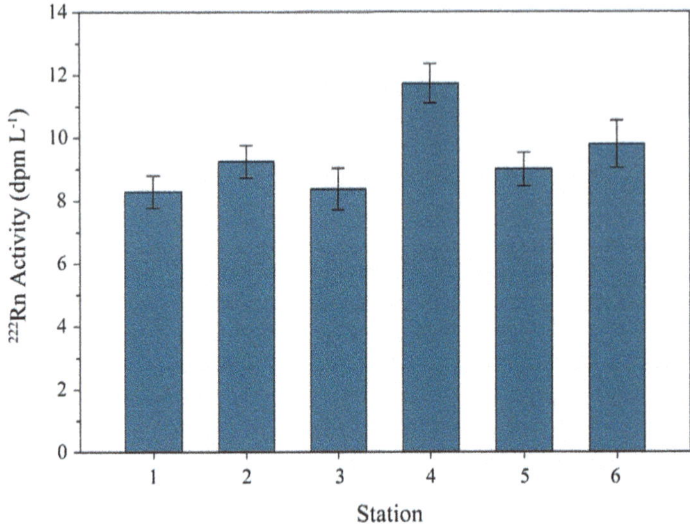

Fig. 9.6 ^{222}Rn activities of six grab samples collected around the perimeter of Dogwood Reservoir during February of 2012

Before selecting a location for our time-series deployment site, we collected several grab samples from various locations around the reservoir to examine for any lateral heterogeneity across the reservoir, and therefore potential bias any one location may have on our ^{222}Rn measurements. Because all samples exhibited similar ^{222}Rn concentrations (Fig. 9.6), we assume the data recorded at our monitoring station are representative of the entire water body when applying our extreme boundary condition approaches (Eqs. 9.2 and 9.3) to constrain volumetric percent groundwater compositions.

In determining the flux of ^{222}Rn to the atmosphere, we assume the water column is well mixed, and therefore evasional losses are distributed throughout the entire water column. Given that the average depth of the reservoir was 1.3 m, we feel this assumption is reasonable. For systems in which this assumption may be invalid, alternative options may be to characterize the depth of groundwater influence using salinity for stratified systems where fresh groundwater is discharging into marine water (Dulaiova et al. 2010). Any overestimation of ^{222}Rn lost to atmospheric degassing would lead to a further increase in calculated maximum volumetric percent compositions. Given these assumptions behind the maximum volumetric percent composition calculations, we present a conservative volumetric estimate of groundwater present within the reservoir by averaging each result generated by Eqs. 9.2 and 9.3 for a given time measurement interval.

Using ^{222}Rn as a groundwater proxy offers a robust tool to detect recently discharged groundwater within receiving water bodies. Given the half-life of ^{222}Rn (3.8 days), the mean life of any one particular atom is 5.5 days (Eq. 9.4), meaning that after ~6 days, there is a 50 % chance of detection since discharge. For this reason, ^{222}Rn can be an excellent tool for researchers interested in the role of groundwater nutrient loading over geochemically relevant time-scales.

Since it is impossible to accurately assess the residence time of each particular radon atom within the reservoir for Eq. 9.3, we elected to calculate the volumetric percentage of groundwater within the reservoir based on this mean life of 5.5 days. This time correction factor was chosen to represent the likelihood of detection for a radon atom within the reservoir, and therefore our groundwater discharge estimates represent only the 'recently discharged' groundwater component (<5.54 days since discharge). This time frame also has the added advantage that it represents the likely range of time that many solutes delivered via groundwater discharge would reside in the reservoir before biogeochemical alteration. However, within reason, one may select a correction time that suits the interest of the project or that best represents the system. For example, one could examine groundwater discharge to a receiving water body that is no older than a predicted nutrient turnover time by solving Eq. 9.3 where R equals the predicted nutrient turnover time. Figure 9.7 demonstrates the effect of varying values of R on resulting maximum volumetric percent groundwater computation estimates.

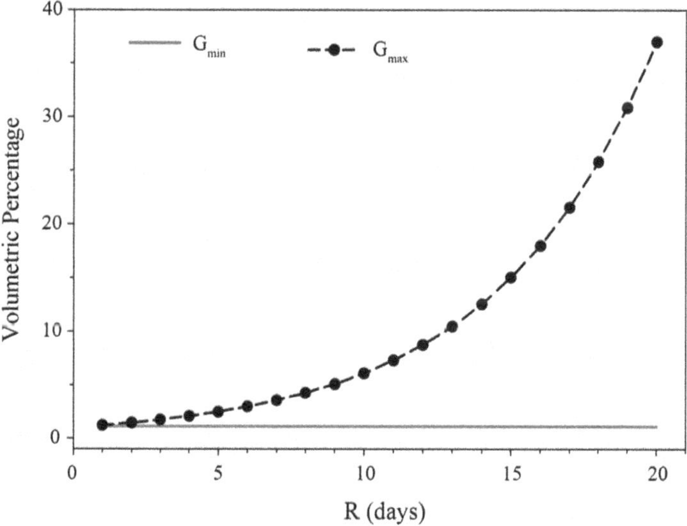

Fig. 9.7 Hypothetical volumetric percentages of groundwater within a reservoir as determined by Eqs. 9.4 (G_{MIN}) and 9.5 (G_{MAX}) to demonstrate the effect of R, the decay correction factor constant on groundwater estimates. Data plotted were generated assuming a constant concentration of ^{222}Rn within the reservoir (10 dpm m^{-3}) and end-member (1000 dpm m^{-3}), and that loss to evasion and production via ^{226}Ra decay equal zero. Note G_{MIN} remains constant as no decay correction factor is required for this term

2.2 Reservoir Export

In this section, we describe our approach to constrain discharge rates of water leaving Dogwood Reservoir. We have defined discharge over the weir as the only export term of interest for this case study, evaluating the relative contributions of direct precipitation and groundwater discharged from the reservoir and ultimately conveyed to the coastal ocean. For a broad-crested weir as exists in Dogwood Reservoir, the discharge rate of water flowing over the structure (Q) is mathematically expressed as (Hornberger et al. 1998):

$$Q = \left(\frac{8}{27}g\right)^{\frac{1}{2}} * h_{weir}^{\frac{3}{2}} * w_c \quad (9.9)$$

where g represents gravitational acceleration (9.81 m s^{-2}), h_{weir} is the difference in elevation between the water surface and weir crest, and w_c is the width of weir over which water is flowing. To evaluate the water surface elevation needed to determine h_{weir}, we deployed a submersible CTD data logger (Solinst, LTC Levelogger Junior) at a constant elevation in a PVC housing secured 20 cm above the reservoir bottom. Water levels were continuously recorded throughout the time-series effort and barometrically compensated using data from a barometer (Solinst Barologger Gold) deployed just above the water surface. We used RTK GPS instrumentation to determine the elevation of the CTD data logger so that each recorded water level could be converted to a water surface elevation.

Upon visual inspection, we noted the elevation of the weir crest was highly variable along the weir width (Fig. 9.8). For this reason, we used RTK GPS to record weir crest elevations in 10 cm increments to best represent the true shape of the structure. Values for h_{weir} were then determined for each elevation across the weir structure, and corresponding values for w_c associated with each value of h_{weir} greater than zero using the cumulative length of each equal elevation segment along the weir crest. Empirically-derived discharge estimates were independently verified using a handheld Acoustic Doppler Velocimeter (ADV) FlowTracker (Sontek/YSI Inc.) in the concrete-lined channel just downstream of the weir.

Due to the structural irregularities of the weir at Dogwood Reservoir, we assumed the elevation between measurement intervals (Fig. 9.8) was constant and equal to that of the nearest measurement. The conditions we encountered with regard to structural complexity of the weir are unique and do not likely represent the majority of engineered weir constructions, but nonetheless demonstrate a successful application of Eq. 9.9. If this approach were to be applied in a setting with a broad-crested weir with constant elevation along its width, deriving each term needed to constrain discharge should be less complicated and not involve this assumption.

We differentiate between source-specific volumetric percent compositions within the reservoir and source-specific export fluxes from the reservoir by way of the reservoir discharge rate. In this way, fluctuating discharge rates and changes

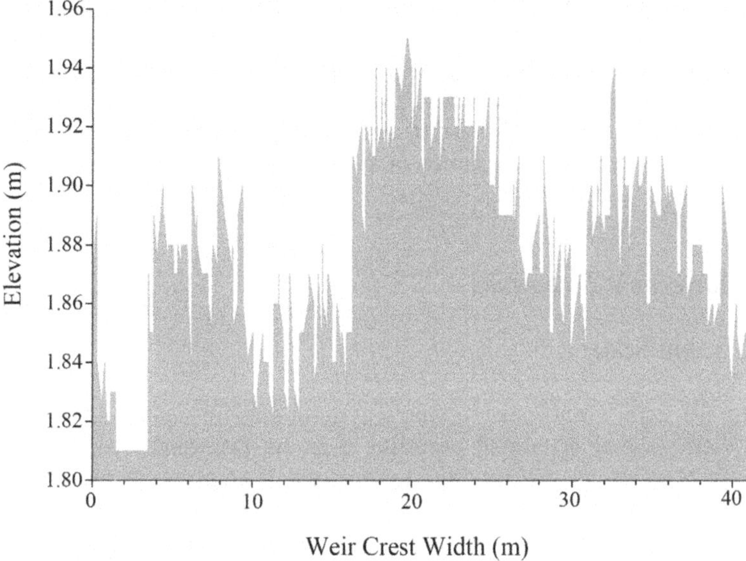

Fig. 9.8 Weir crest elevation (meters relative to NAVD 88 datum) with respect to structure width, increasing toward the south

in volumetric percentages of each source within Dogwood Reservoir can influence the source-specific export rates. This concept is especially apparent when comparing changes between two sequential time periods for both the volumetric percentage of each source within the reservoir and for source-specific export fluxes. To demonstrate, consider two subsequent measurement cycles in which the volumetric percentage of groundwater in Dogwood Reservoir increases from 10 to 10.5 % and corresponding water discharge rates decrease from 1.5 to 1.0 $m^3\ s^{-1}$. In this case, inferring export behaviors based entirely on an increase in volumetric groundwater composition alone would be misleading, since the actual export flux of groundwater significantly decreases (from 0.150 to 0.105 $m^3\ s^{-1}$). Rather than assuming that changes within the reservoir accurately represent changes in export from the reservoir, a more comprehensive assessment would consider the total volumetric rate of water exported in addition to the volumetric percent compositions of each source.

3 Data Analysis and Interpretation

In this section, we present example analyses intended to examine change in Dogwood Reservoir with regard to different temporal scales, as a demonstration of insights that can be gleaned from a long-term continuous record of groundwater interaction with an urban reservoir. Interpretations based on this data set highlight a potential emerging threat to groundwater in the region, as diminishing groundwater discharged from the reservoir reflects a reduced volumetric percentage of groundwater within the catchment, suggesting reduced aquifer recharge rates. Similar to the organizational style of Sect. 2, we first present data and interpretations concerning the volumetric percent of direct precipitation and groundwater within the basin and follow with discussion of source-specific export fluxes.

3.1 Reservoir Dynamics

3.1.1 Event Scale

We use the term ‘event scale’ to represent quasi-random weather-driven phenomena which resulted in rainfall occurring over the Dogwood Reservoir drainage basin. In this section, we examine the hydrologic impacts of one of the many recorded rain events during our 2 year sampling effort compared to a relatively dry period to highlight reservoir responses to precipitation perturbations.

Dry Period

The period between September 28 and October 3, 2013 represents a 6-day dry period preceded by relatively dry antecedent conditions. During this time, we observed clear rhythms in reservoir water levels and water table heights (Fig. 9.9). On average, water levels within the reservoir changed by 1.5 cm day^{-1} (ranging from 0.69 to 0.71 m), while the average change in water table height was observed to be nearly a factor of two larger (2.7 cm day^{-1}), ranging from 1.05 to 1.08 m. Although the magnitude of change between reservoir levels and water table heights was different, both were found to oscillate at a similar frequency, reaching daily maximum levels in early morning (between 02:00 and 07:00) and minima in the evening (between 15:00 and 18:00). Water table changes slightly preceded changes in reservoir levels suggesting reservoir hydrology was likely controlled by groundwater flows. We suspect the periodicity of water level and water table change was driven by evapotranspiration following daily photosynthetic rhythms (Lautz 2007; Winter 1999). Higher evapotranspiration rates during the day lower both the reservoir elevation as well as that of the water table. During nighttime hours, evapotranspiration is greatly reduced, allowing the local water table to rebound as it is recharged from upgradient aquifer pressures. This enhanced hydraulic gradient likely drives more groundwater discharge into the reservoir, thereby leading to an increase in reservoir levels. Although we did not observe significant trends in the volumetric percentages of groundwater present within the reservoir over this period, the covariance of the two water level variables (reservoir and water table) suggest evapotranspirative forcings dominate the hydrology of the reservoir and water table during rain-free conditions. It is also conceivable that

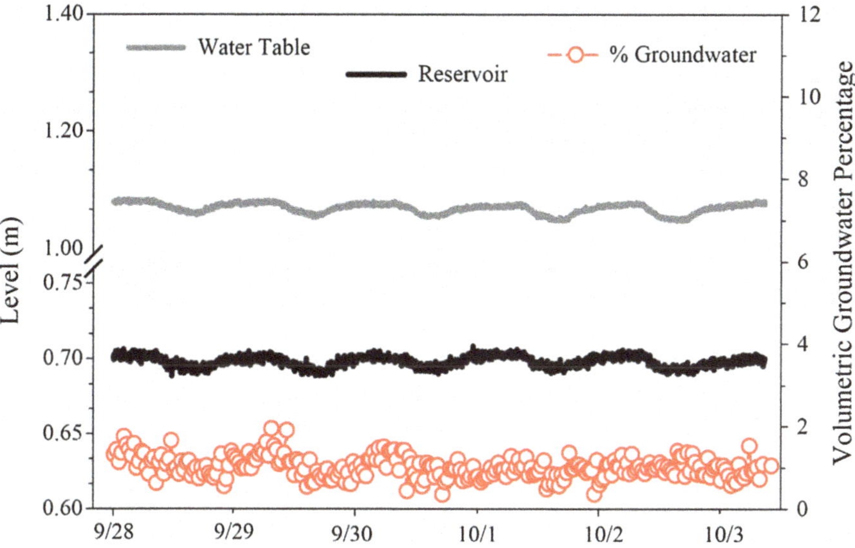

Fig. 9.9 Water table and reservoir levels with volumetric estimates of groundwater percentages during a rain-free period in 2013. *Note*: *Y*-axis ranges have been scaled appropriately for comparison with Fig. 9.10

incorporating an additional radiotracer with a shorter half life than that of ^{222}Rn (e.g., ^{220}Rn; $T_{1/2}$: 55 s) would have offered additional insights to the system dynamics over such time scales.

Rain Event Period

From June 2 through June 9, 2013, a total of 90.7 mm of rainwater fell on the Dogwood drainage basin. Rainfall primarily occurred in three isolated events on June 3rd, 6th, and 7th following dry antecedent conditions. Out of hundreds of observed rainfall events throughout our study period, these dates were selected to avoid the effects of storm piggybacking since the rain events highlighted here were preceded by 10 rain-free days. Storm piggybacking refers to residual effects from one storm event that influence the subsequent event, and would therefore impact the hydrologic response of the reservoir to the storm (Hancock et al. 2010).

Rather than examining this period as one large event, we felt it more appropriate to consider each time period associated with a single storm to evaluate the hydrologic response to event magnitude and antecedent conditions. The June 3rd event delivered 25.7 mm of rainwater, the June 6th event produced 10.9 mm of rainfall, and the June 7th event delivered more than twice as much rain as the event on June 3rd (54.1 mm), providing a range of event magnitudes for analysis. In general, event-driven change to water table height was a factor of 3 greater than water table fluctuations associated with evapotranspiration during normal conditions (Figs. 9.9 and 9.10). Within 1 h after rainfall, reservoir levels and water table heights responded to the addition of rainwater, however, the water table required substantially more time to equilibrate after the event concluded due to the slow nature of flow through a porous medium (Fig. 9.10).

The percentage of groundwater within the reservoir increased following each event as rainwater recharged the surficial aquifer, likely increasing discharge into the reservoir (Fig. 9.11). However, post-event groundwater percentage maxima were observed to occur at variable lag times behind rainfall events, a likely result of continual but variable draining of the aquifer driven by an increased hydraulic gradient between the water table and reservoir level. Additional evidence for a delay in groundwater inputs was found by comparing the equilibration time required for reservoir and water table heights to stabilize following rainfall perturbations. Because reservoir levels stabilized within approximately 1.5 h after rainfall, and water table levels were observed to continually decline for many hours (and in some cases days) after an event concluded, we surmise the effects of these events on reservoir hydrology would persist for many days after a single event, largely resulting from continued groundwater input.

While the response time of the reservoir and water table levels remained fairly constant following each of the three selected events, the magnitude and timing of changes in the volumetric percentage of groundwater present were unique to each event. The volumetric percentage of groundwater began to increase (above a pre-event average of 2.0 %) 26 h after the event began and continued to increase to a maximum of 9.2 % approximately 56 h post-event. Despite higher rainfall totals for the first event relative to the second, we observed the lowest mean event-driven

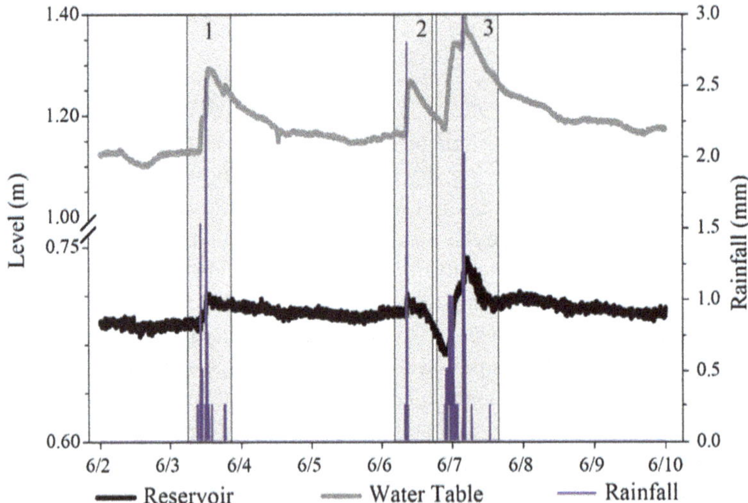

Fig. 9.10 Water table and reservoir level data with rainfall totals (cumulative over 2 min intervals) for three isolated rain events occurring in June of 2013

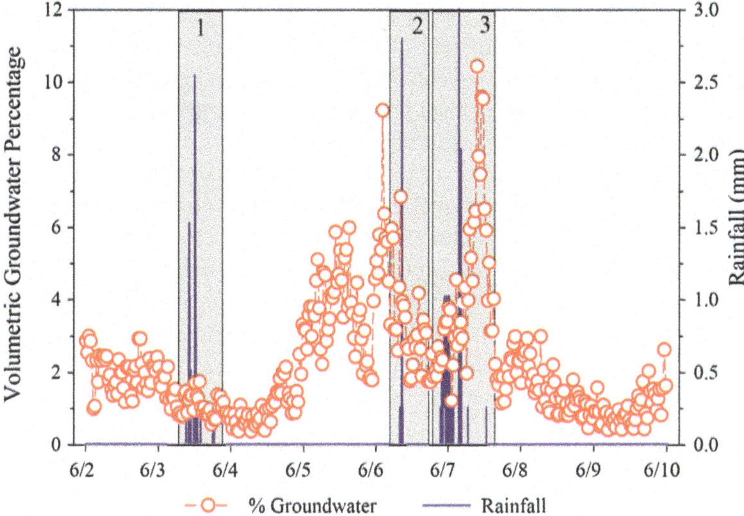

Fig. 9.11 Volumetric groundwater percentages and rainfall totals (cumulative over 2 min intervals) for three isolated rain events occurring in June of 2013

groundwater percentage within the reservoir following the first event. This is likely attributed to the dry antecedent conditions prior to rainfall resulting in the lowest observed water table heights prior to the onset of the first event. Reduced water table heights would require greater rainfall volumes to reach a threshold for discharge to occur causing lower volumetric percentages of groundwater within the reservoir.

The maximum volumetric percentage of groundwater (6.8 %) following the second event was less than that associated with the first event, albeit greater than the average percentage prior to rainfall. We attribute this result to storm piggybacking whereby elevated groundwater discharge rates into the reservoir resulting from the first event continued into the onset of the second event, greatly increasing the mean event-driven volumetric percentage of groundwater present. We suspect continued groundwater discharge (from the first event) is driven by the elevated water table height as compared to that observed prior to the first event. While the maximum volumetric groundwater percentage occurrence associated with the second event was observed almost immediately following the onset of rainfall, we suspect this is attributed to another instance of storm piggybacking as the second event appears to have resulted in a reduction to the percentage of groundwater present several hours after the event.

Higher rainfall totals for the third event resulted in a maximum observed water table height recorded to be 10 cm and 14 cm higher than maxima associated with the first and second events, respectively (Fig. 9.10). The increased water table height subsequently increased the amount of time necessary for water table levels to equilibrate, which likely increased the rate of groundwater discharging into the reservoir as shown by a sharp increase in the percentage of groundwater present (maximum 10.4 %). Approximately 40 h after the third event, we observed the co-occurrence of the minimum event groundwater percentage and the equilibration of the water table (albeit higher than pre-event levels) suggesting a reduced rate of groundwater discharge into the reservoir less than 2 days post event.

Evaluating a period of rainy days as well as sequential events in isolation demonstrates a complexity in the Dogwood Reservoir groundwater/reservoir hydrology. While general trends regarding the percentage of groundwater within the reservoir were apparent, the occurrence, duration, magnitude, and preceding conditions of isolated events were found to significantly influence the timing and relative magnitude of appearance. The observed complexity demonstrates the potential for misinterpretations regarding groundwater/reservoir interaction if only discrete samplings are used to characterize the relationship.

3.1.2 Seasonal Scales

It should be acknowledged that a 2 year data set is not necessarily ideal for examining seasonal dynamics. However, we feel using continuous time-series data in this way is much more appropriate than the use of single point measurements used to represent similar scales, so these data are intended to demonstrate the potential insights that such data sets can offer. As shown in Sect. 3.1.2, groundwater interaction with a surface reservoir can exhibit a high degree of temporal variability, and as such, discrete data regarding various hydrological flowpaths should be interpreted with caution.

To examine seasonal effects on the volumetric percentage of groundwater within Dogwood Reservoir, we grouped many data into quarter year categories representing each of the four seasons experienced by South Carolina. We define winter as December through February, spring as March through May, summer as June through August, and fall as September through November. In total, we sampled through seven seasons including two winters, springs, and summers while capturing only fall of 2012.

In general, seasonal changes were evident in the volumetric percentages of groundwater within Dogwood Reservoir. On average, the volumetric percentage of groundwater within the reservoir was highest during spring (7.3 %) and lowest in fall (3.8 %) (Fig. 9.12). The volumetric percentage of direct precipitation was highest during summer ($4.5 * 10^{-3}$ %) when mean rainfall rates were highest (0.142 mm h^{-1}) and lowest in fall ($2.0 * 10^{-3}$ %) when rainfall rates were lowest (0.064 mm h^{-1}). With the exception of summer, the volumetric percentage of groundwater corresponded with that of direct precipitation within the reservoir.

Considering infiltration of rainwater to be the primary recharge mechanism for a surficial aquifer, it seems reasonable that higher seasonal rainfall totals enhanced groundwater discharge into the reservoir, increasing the volumetric percentage present. However, in summer, average rainfall rates were 30 % higher than spring, yet the percentage of groundwater present within the reservoir reached a seasonal

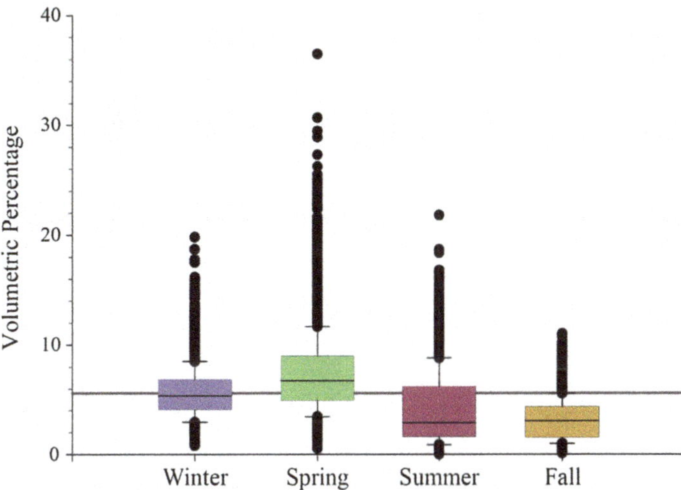

Fig. 9.12 Volumetric groundwater percentages for winter, spring, summer, and fall seasons. Solid *black line* represents the mean long-term volumetric percentage of groundwater within Dogwood Reservoir

minimum (Fig. 9.12). Data from long-term water table monitoring efforts in Briarcliffe Acres (approximately 28 km NE of Dogwood Reservoir) showed prominent seasonal trends in water table height, having reached maximum levels in summer (Fig. 9.13). We suspect increased rainfall rates and elevated water table heights reduced soil infiltration rates and subsequently groundwater discharge rates over the course of the season. This counter-intuitive association has been shown as a strong negative relation between infiltration rates and soil moisture content (Xue and Gavin 2008; Cerda 1997). As a result, greater amounts of rainwater were conveyed overland, reducing the volumetric percentage of groundwater present within Dogwood Reservoir in summer by 25 % of the long-term average.

Exploring potential seasonal trends in the 2 year time-series data record also provides the opportunity to examine changes between years on seasonal scales. In general, large amounts of rainfall occur during spring and summer seasons in coastal South Carolina, however, maximum monthly mean rainfall rates occurred earlier in 2013 than in 2012 (Fig. 9.14). In 2012, maximum mean spring and summer rainfall rates occurred in May (0.202 mm h^{-1}) and August (0.194 mm h^{-1}), respectively, whereas 2013 mean seasonal maxima appear in April (0.244 mm h^{-1}) and June (0.202 mm h^{-1}).

To then compare monthly volumetric percentages of groundwater, we averaged data collected from all months defined within a particular season. We required at least 70 % of the month be represented by valid data for each source water term to adequately characterize monthly conditions (Table 9.1). Data from Briarcliffe Acres provides regional-scale evidence as maximum water table heights shifted from August 31 in 2012 to July 31 in 2013 (Libes and Peterson 2012—present), suggesting forces contributing to the observed seasonal water budget changes in

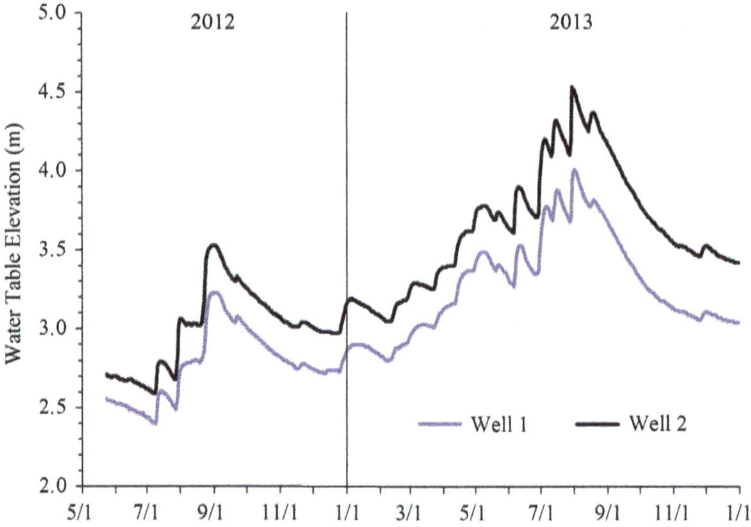

Fig. 9.13 Water table elevations (relative to NAVD 88) measured from two inland wells in Briarcliffe Acres from 2012 through 2013. Data were obtained from an online public database (http://bccmws.coastal.edu/bagw/)

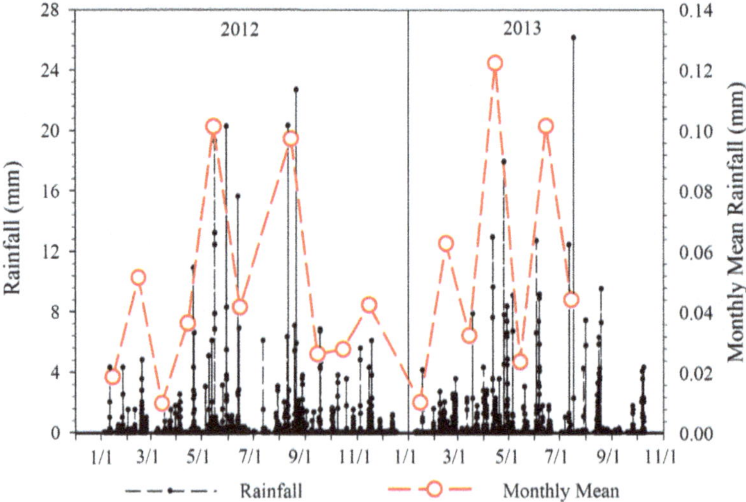

Fig. 9.14 Local long-term rainfall records (2 min sampling interval) and monthly mean rainfall totals from 2012 through 2013

Dogwood Reservoir impacted an area larger than that of the studied system. As a result, we observed a subsequent shift in maximum spring volumetric percentages of groundwater within Dogwood Reservoir from May 2012 at 8.4 % to April 2013 at 7.5 %.

Table 9.1 Monthly mean volumetric percentages of groundwater and direct precipitation as well as monthly mean total, groundwater, and direct precipitation discharge rates

		Mean groundwater (%)	Mean direct precipitation (%)	Mean total discharge ($m^3\ s^{-1}$)	Mean groundwater discharge ($m^3\ s^{-1}$)	Mean direct precipitation discharge ($m^3\ s^{-1}$)
2012	Jan.	6.46	$1.16 * 10^{-3}$	$2.15 * 10^{-2}$	$1.34 * 10^{-3}$	$2.50 * 10^{-7}$
	Feb.	6.98	$3.26 * 10^{-3}$	$2.02 * 10^{-2}$	$1.53 * 10^{-3}$	$6.57 * 10^{-7}$
	Mar.	7.67	$6.02 * 10^{-4}$	$3.55 * 10^{-2}$	$2.96 * 10^{-3}$	$2.14 * 10^{-7}$
	Apr.	8.17	$2.29 * 10^{-3}$	$3.06 * 10^{-2}$	$2.54 * 10^{-3}$	$6.99 * 10^{-7}$
	May	8.41	$6.42 * 10^{-3}$	$3.79 * 10^{-2}$	$3.37 * 10^{-3}$	$2.43 * 10^{-6}$
	Jun.	8.28	$2.62 * 10^{-3}$	$2.80 * 10^{-2}$	$2.34 * 10^{-3}$	$7.34 * 10^{-7}$
	Aug.	3.85	$6.23 * 10^{-3}$	$2.60 * 10^{-2}$	$1.50 * 10^{-3}$	$1.77 * 10^{-6}$
	Sept.	2.94	$1.65 * 10^{-3}$	$2.22 * 10^{-2}$	$6.53 * 10^{-4}$	$3.66 * 10^{-7}$
	Oct.	4.27	$1.75 * 10^{-3}$	$1.55 * 10^{-2}$	$6.46 * 10^{-4}$	$2.71 * 10^{-7}$
	Nov.	4.04	$2.67 * 10^{-3}$	$2.57 * 10^{-2}$	$1.05 * 10^{-3}$	$6.87 * 10^{-7}$
2013	Jan.	4.86	$6.29 * 10^{-4}$	$3.53 * 10^{-2}$	$1.71 * 10^{-3}$	$2.22 * 10^{-7}$
	Feb.	5.18	$3.94 * 10^{-3}$	$4.51 * 10^{-2}$	$2.45 * 10^{-3}$	$1.78 * 10^{-6}$
	Mar.	5.80	$2.02 * 10^{-3}$	$5.44 * 10^{-2}$	$3.21 * 10^{-3}$	$1.10 * 10^{-6}$
	Apr.	7.51	$7.70 * 10^{-3}$	$4.79 * 10^{-2}$	$3.70 * 10^{-3}$	$3.69 * 10^{-6}$
	May	6.40	$1.48 * 10^{-3}$	$2.41 * 10^{-2}$	$1.84 * 10^{-3}$	$3.57 * 10^{-7}$
	Jun.	2.56	$6.44 * 10^{-3}$	$2.24 * 10^{-2}$	$5.58 * 10^{-4}$	$1.44 * 10^{-6}$
	Jul.	2.24	$2.78 * 10^{-3}$	$2.55 * 10^{-2}$	$5.91 * 10^{-4}$	$7.10 * 10^{-7}$

3.1.3 General Characteristics

While examination of various temporal scales offers an opportunity to explore storm event and seasonal forcings, evaluation of the entire time-series record provides an integrated assessment of the general hydrology affecting Dogwood Reservoir. In this section, we describe general characteristics of the reservoir and compare observations from 2012 to 2013 to illustrate a potentially significant threat to groundwater resources of the study region.

We observed considerable variance in ^{222}Rn activities within the reservoir ranging from 0.13 to 22.75 dpm L^{-1} (mean 6.19 dpm $L^{-1} \pm 0.65$ dpm L^{-1}) (Fig. 9.15).

Mean minimum volumetric groundwater percentages of the reservoir (derived using Eq. 9.2) were 4.22 % ±2.66 % while mean maximum estimates (derived using Eq. 9.3) were 6.38 % ±4.6 % (Fig. 9.16).

On average, model results for minimum and maximum volumetric percent groundwater estimates varied by ±2.16 %, and have been averaged to offer a conservative estimate for each time interval. In so doing, we determined the volumetric percentage of groundwater present ranged from 0.03 % to 36.48 % (mean 5.6 %) while direct precipitation percentages ranged from 0.0 % to $7.7 * 10^{-3}$ % (mean $3.2 * 10^{-3}$ %) (Fig. 9.17 and Table 9.1).

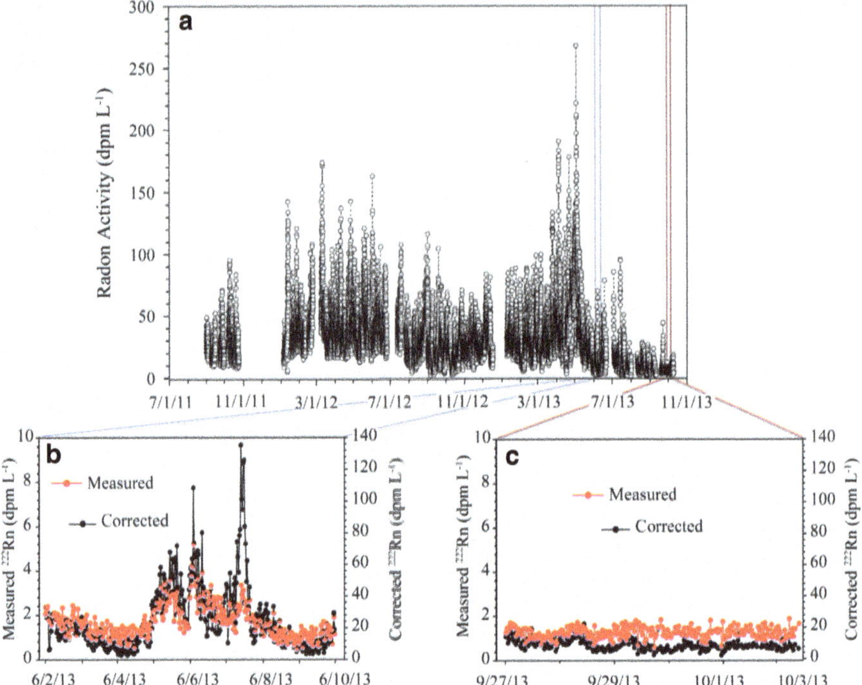

Fig. 9.15 Mean reservoir ^{222}Rn (radon) activities recorded at 30 min intervals representing the average measured (Eq. 9.2) and corrected (Eq. 9.3) radon activities from October, 2011 through October, 2013 (**a**). Measured and corrected radon activities for the time periods discussed in Sects. 3.2.1 (**b**) and 3.1.1 (**c**). Corrected activities account for radon loss over 5.54 days to decay and atmospheric evasion

The relatively low volumetric percentages of groundwater is inherently related to the reservoir turnover time (with respect to discharge over the weir) and our definition of groundwater. We found the mean turnover time of Dogwood Reservoir to be 78.2 days $\pm$ 69.4 days after analyzing approximately 300,000 measurements from 2013 (Fig. 9.18). The exceptionally high range in turnover times is a result of an infinite value when reservoir levels are too low to discharge over the weir. Nonetheless, by defining our groundwater estimates to consider only groundwater (and ^{222}Rn) that is no older than ~6 days since discharge, most of the remaining reservoir volume, especially when stormwater runoff is not actively discharging into the reservoir, is remnant water from previous events and overland flow. However, utilization of additional tracers and/or increasing the value of R (Eq. 9.3) would allow assessment of 'older' groundwater.

To examine annual volumetric groundwater percent characteristics, we averaged all monthly data from 2012 to 2013 with temporally overlapping records. As a result, we only consider data collected from January through June in our yearly interpretations. From this, we observed a decrease in the average volumetric percentage of groundwater within Dogwood Reservoir by 2.7 % from 7.7 % in

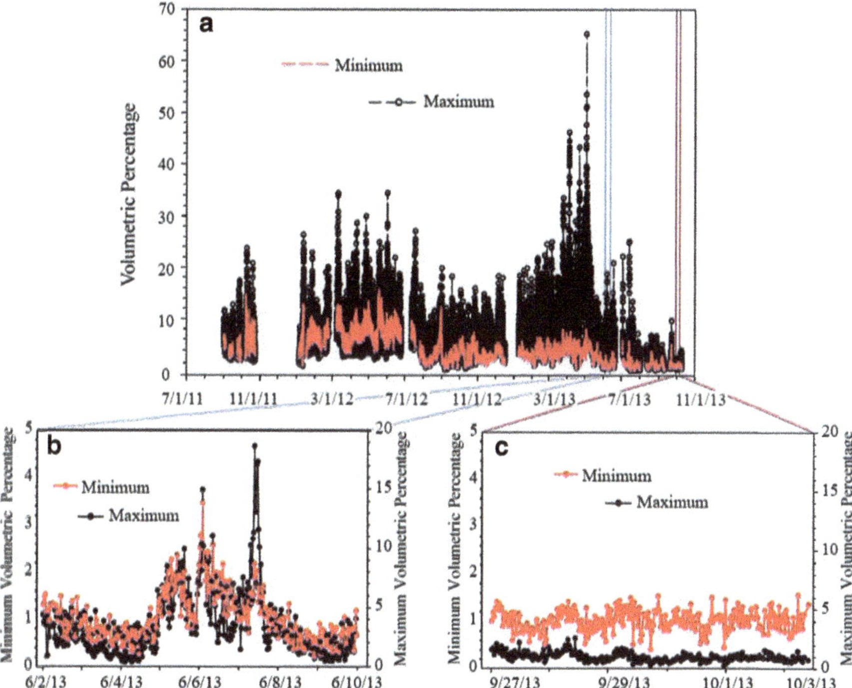

Fig. 9.16 Minimum and maximum volumetric percentages of groundwater from 2011 through 2013 determined using Eqs. 9.2 and 9.3, respectively (**a**). Minimum and maximum groundwater percentages for time periods discussed in Sects. 3.2.1 (**b**) and 3.1.1 (**c**)

2012 to 4.9 % in 2013 (Table 9.2). However, mean rainfall rates increased from 0.086 mm h^{-1} in 2012 to 0.116 mm h^{-1} in 2013 as did the volumetric percentage of direct precipitation within the reservoir (from 2.7×10^{-3} % in 2012 to 3.6×10^{-3} % in 2013). Although our approach to constrain direct precipitation percentages is inherently dependent upon total reservoir volume, the total volume of water within the reservoir remained nearly constant, suggesting increased percentages of direct precipitation within the reservoir largely reflected increased rainfall rather than a change in reservoir volume.

To further understand the volumetric percent decline of groundwater within Dogwood Reservoir from 2012 to 2013 despite more rainfall occurring in 2013, we compared our mean laboratory incubation ^{222}Rn downstream groundwater end-member (414.2 dpm L^{-1}) to ^{222}Rn concentrations measured in 64 groundwater field samples collected throughout the study period (minimum 73.7 dpm L^{-1}; maximum 253.5 dpm L^{-1}) to evaluate groundwater residence times (end-member data presented in Table 9.3). Groundwater residence time (T) is then given by:

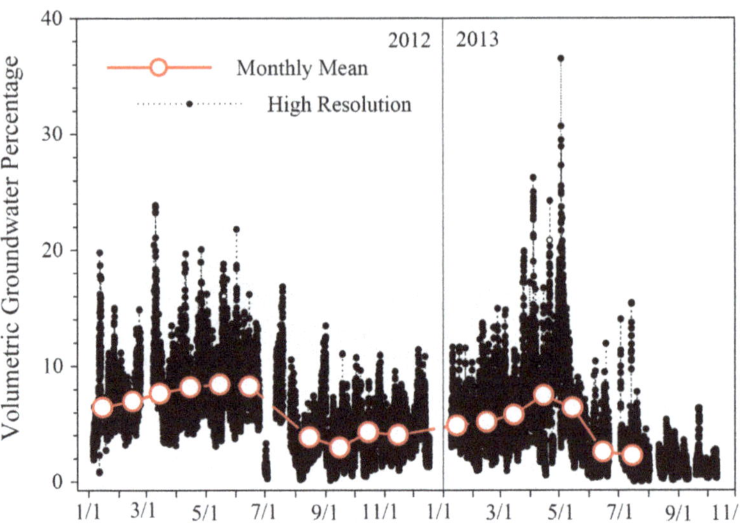

Fig. 9.17 Monthly mean and high resolution (30 min sampling resolution) volumetric groundwater percentage estimates from 2012 through 2013

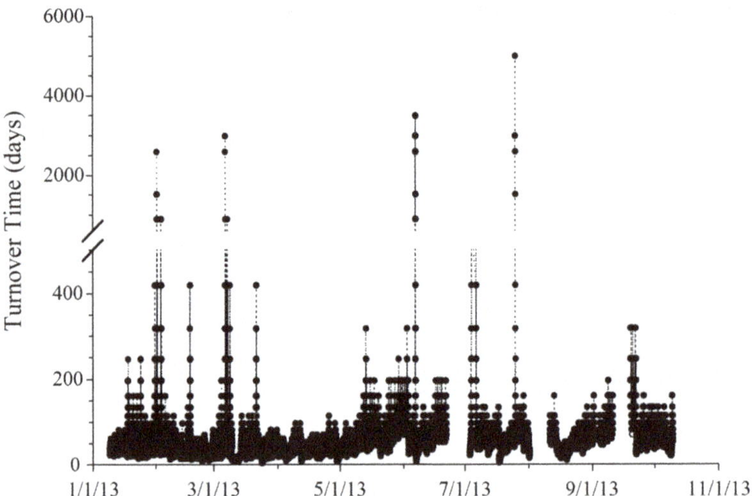

Fig. 9.18 Reservoir turnover times for all 2013 water level and total discharge data. Turnover time was determined by dividing total reservoir volume by corresponding total discharge rates. Note turnover time effectively reaches infinity when discharge equals zero $m^3\ s^{-1}$

$$N_D = \frac{\lambda_P}{\lambda_D - \lambda_P} {N^0}_P \left(e^{-\lambda_P T} - e^{-\lambda_D T}\right) \quad (9.10)$$

Table 9.2 Mean yearly parameters for volumetric percentages of groundwater and direct precipitation as well as total and corresponding source-specific discharge values

	Groundwater (%)	Direct precipitation (%)	Total discharge ($m^3\ s^{-1}$)	Groundwater discharge ($m^3\ s^{-1}$)	Direct precipitation discharge ($m^3\ s^{-1}$)
2012	7.66	$2.72 * 10^{-3}$	$2.89 * 10^{-2}$	$2.35 * 10^{-3}$	$8.31 * 10^{-7}$
2013	4.93	$3.57 * 10^{-3}$	$3.64 * 10^{-2}$	$2.01 * 10^{-3}$	$1.33 * 10^{-6}$
% change	−35.60	31.07	25.81	−14.32	59.99
Difference	−2.73	$8.47 * 10^{-4}$	$7.47 * 10^{-3}$	$3.36 * 10^{-4}$	$4.98 * 10^{-7}$

Percent change and difference calculations were performed chronologically where positive values indicate an increase and negatives indicate a decrease from 2012 to 2013. Yearly values represent data collected from January through June of each respective year

Table 9.3 Depth averaged groundwater ^{222}Rn activities (end-member concentrations) reported for downstream (applied in Eq. 9.2) and upstream (applied in Eq. 9.3) sampling locations and $1-\sigma$ standard deviations of samples collected 1.0, 1.25, and 1.5 m below ground

Collection date	Downstream ^{222}Rn activity (dpm L^{-1})	Upstream ^{222}Rn activity (dpm L^{-1})
9/14/2011	73 ± 30	436 ± 183
11/17/2011	113 ± 66	666 ± 128
12/12/2011	115 ± 33	615 ± 102
2/1/2012	139 ± 34	740 ± 167
2/19/2012	151 ± 55	696 ± 234
5/2/2012	99 ± 29	767 ± 253
5/10/2012	141 ± 69	603 ± 120
7/17/2012	107 ± 19	614 ± 199
8/8/2012	181 ± 78	552 ± 214
10/18/2012	124 ± 30	627 ± 103
11/16/2012	229 ± 46	647 ± 127
3/12/2013	177 ± 63	446 ± 163
4/10/2013	195 ± 64	568 ± 159
6/4/2013	253 ± 20	975 ± 58
7/10/2013	138 ± 14	441 ± 69
9/18/2013	109 ± 28	594 ± 110

Average measurement uncertainties for all samples were 12.7 % of the reported activities

where N_D represents the number of ^{222}Rn atoms within a discreet groundwater sample, λ_P and λ_D are the decay constants of ^{226}Ra and ^{222}Rn, respectively, and $N^0{}_P$ is number of ^{226}Ra atoms present at equilibrium. Because the activity of two isotopes are equal after attaining secular equilibrium, we know the activity of -surface-bound ^{226}Ra capable of contributing ^{222}Rn to porewaters was 414.2 dpm L $^{-1}$ (based on our equilibrium incubations) and can then determine the number of ^{226}Ra atoms present by:

$$A = \lambda N \tag{9.11}$$

where A is the isotope activity, λ is the decay constant of that isotope and N is the number of atoms present. Using a depth averaged ^{222}Rn end-member concentration representative of each downstream groundwater sampling to determine a corresponding residence time, we solved Eq. 9.10 for residence time (T) and found a significant increase in groundwater residence time within the aquifer from 2012 to 2013, from 1.8 days to 3.2 days (Fig. 9.19). As a result, groundwater velocities were decreased, reducing the rate of groundwater discharge into the reservoir, which resulted in decreased volumetric percentages of groundwater from 2012 to 2013 (Fig. 9.17 and Table 9.2).

3.2 System Export

In this section, we examine total and source-specific export rates from Dogwood Reservoir using total and source-specific discharge rates over various time periods. Similar to the layout of Sect. 3.1, we describe event and seasonal temporal scales and conclude with a general assessment of Dogwood Reservoir throughout our study period.

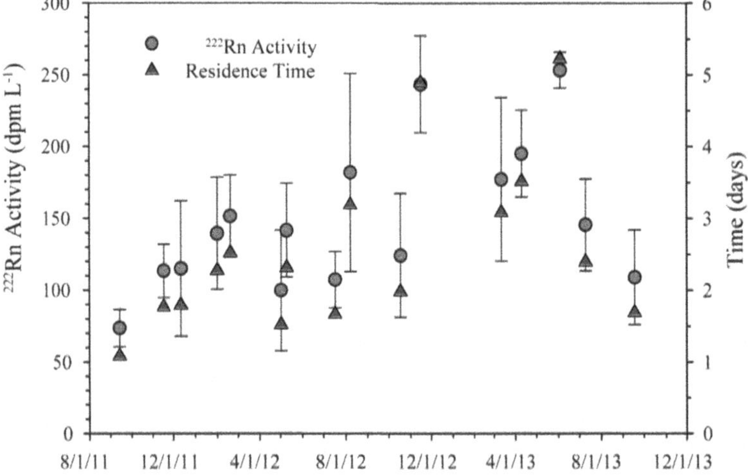

Fig. 9.19 Downstream depth averaged groundwater end-member ^{222}Rn activities and corresponding groundwater residence times within the surficial aquifer. Averages represent data collected 1.0, 1.25, and 1.5 m below ground. Error bars represent $1 - \sigma$ standard deviations

3.2.1 Event Scale

During relatively dry periods, evapotranspiration-driven changes in reservoir levels and water table heights contribute significantly to the pattern observed in total reservoir export rates. As described earlier, these forcings correspond with relatively unchanged volumetric percentages of groundwater within the reservoir. Although the percentage of groundwater within the reservoir remained constant during normal conditions, total reservoir export rates reach maximum values from 02:00 through 07:00, resulting in the greatest rates of groundwater exported from the system during these times (Fig. 9.20).

However, during a storm period including three isolated rain events (Fig. 9.10), total system export is largely driven by the addition of new water to the reservoir from various sources associated with the event (Fig. 9.21). It was found that during these times, wind direction may also be a significant driver of total export rates. Soon after the second event, total discharge rates sharply declined as the winds shifted from SW to SE (Libes 2006—present). Because the long-axis of Dogwood Swash is oriented NW to SE, a SE wind forces water away from the weir, effectively reducing total system export rates (to 0 $m^3\ s^{-1}$ for the presented case). At the onset of the third event, the prevailing SW wind returned, dramatically increasing discharge rates from the reservoir by several orders of magnitude (from $3.7 * 10^{-4}$ to $1.9 * 10^{-1}$ $m^3\ s^{-1}$) and subsequently increasing the groundwater export rate.

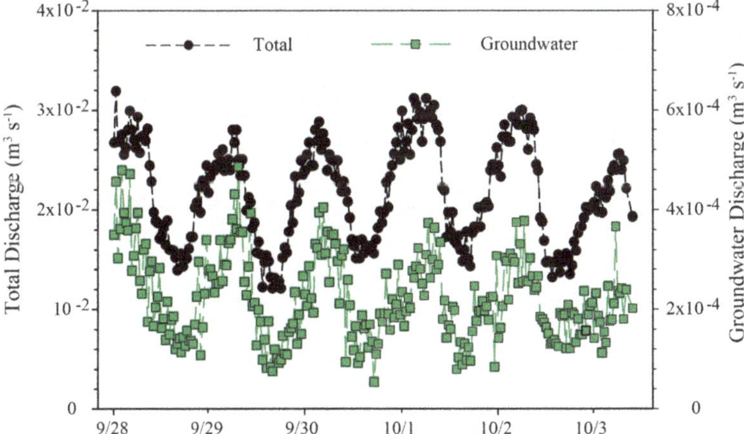

Fig. 9.20 Total and groundwater discharge rates out of the reservoir via weir discharge during a rain-free period in 2013

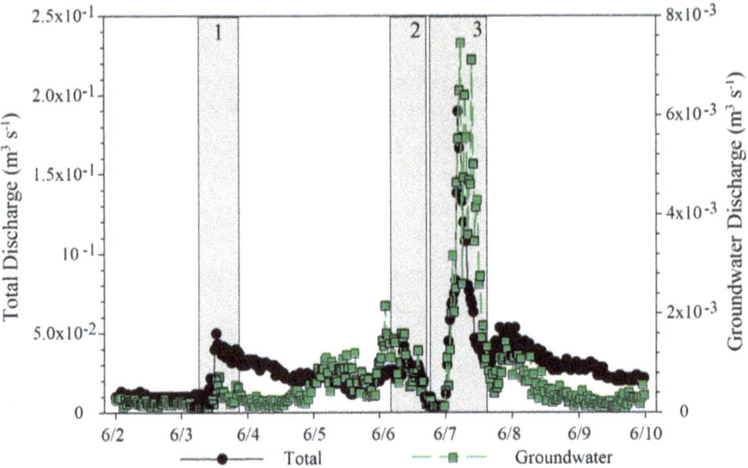

Fig. 9.21 Total and groundwater discharge rates out of the reservoir via weir discharge for three isolated rain events occurring in June of 2013

3.2.2 Seasonal Scales

Much like the seasonal scale changes observed in the volumetric percentage of groundwater within Dogwood Reservoir, we observed changes in total and groundwater export rates (Fig. 9.22). Mean total discharge rates were highest in spring ($3.84 * 10^{-2}$ m^3 s^{-1}) with the second lowest seasonal value occurring during summer (Fig. 9.22a). We suspect this result is a consequence of high evaporative losses during this time resulting in decreased discharge rates despite high rainfall rates. Work by Grimmond and Oke (1986) and Hanrahan et al. (2010) emphasize the importance of seasonality in evaporative losses representing a significant removal mechanism from surface water bodies during summer months.

Governed by increased total discharge, highest groundwater export rates were observed during spring when the volumetric percentage of groundwater and the total export from the system were highest (Fig. 9.22b). We also observed earlier peak discharge rates in 2013 relative to 2012 suggesting a shift in the timing of maximum water and material delivery to the nearshore (Fig. 9.23).

3.2.3 General Export Characteristics

In general, total discharge rates from the reservoir were quite variable from 2012 through 2013 (Fig. 9.24). The mean over this period was found to be $3.0 * 10^{-2}$ m^3 s^{-1}, ranging from 0.00 m^3 s^{-1} to $5.5 * 10^{-1}$ m^3 s^{-1}. For comparison, Little River, a major estuarine inlet located ~46 km NE of Dogwood Reservoir, discharged an average rate of 16.12 m^3 s^{-1} into Long Bay over the same interval (US DOI, USGS 2015). While the average discharge rate from Dogwood Reservoir represents a small fraction of that

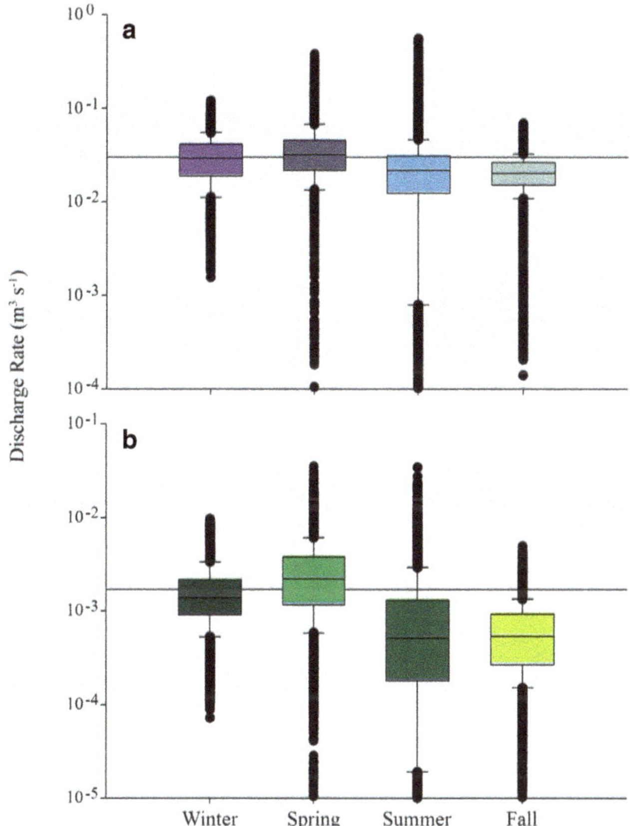

Fig. 9.22 Total (**a**) and groundwater (**b**) discharge rates for winter, spring, summer, and fall seasons. Solid *black line* represents the mean long-term rates for total and groundwater discharge from Dogwood Reservoir, respectively

from Little River, extrapolating this rate to each of the similar 15 coastal stormwater catchments draining into Long Bay throughout the Grand Strand estimates that 35.5 % of the discharge from Little River during the study period was exported offshore from the urban reservoirs of the Grand Strand.

From 2012 to 2013, the total discharge rate increased by 26 % from $2.9 * 10^{-2}$ to $3.6 * 10^{-2}$ m^3 s^{-1} (Fig. 9.23 and Table 9.2). Although total discharge rates were lower in 2012, more intense discharge pulses were observed during this year, reaching a maximum of $5.6 * 10^{-1}$ m^3 s^{-1} as compared to the 2013 maximum of $3.2 * 10^{-1}$ m^3 s^{-1}. While direct precipitation to the reservoir surface often contributes negligible volumetric percentages, (always less than $8.0 * 10^{-3}$ %), increased total discharge rates from Dogwood Reservoir in 2013 were likely driven by increased rainfall totals and subsequently increasing total groundwater and stormwater input as the drainage basin conveyed water into the reservoir. Increased rainfall in 2013, however, did not correspond to increased volumetric percentages

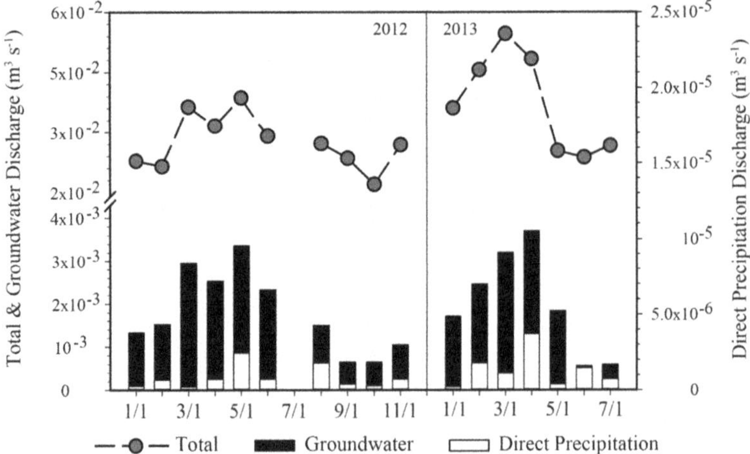

Fig. 9.23 Monthly mean total, groundwater, and direct precipitation discharge rates from January 2012 through July 2013

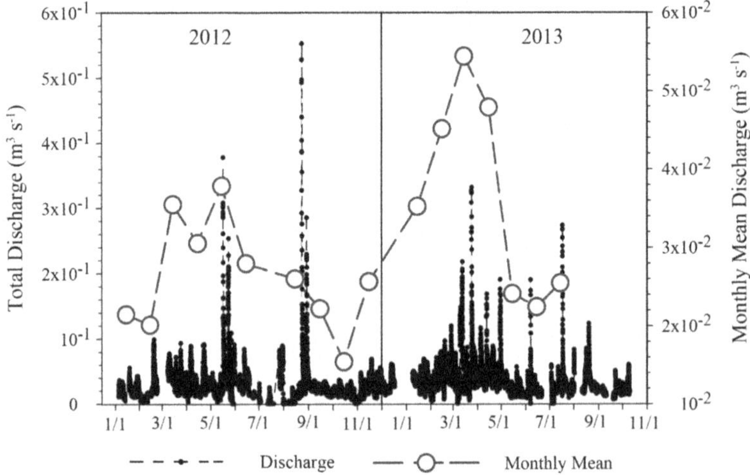

Fig. 9.24 High resolution and monthly mean total discharge rates from Dogwood Reservoir from 2012 through 2013

of groundwater within the reservoir nor to an increase in the groundwater-derived component of total discharge. In fact, groundwater fluxes out of the reservoir decreased by 14 % from $2.4 * 10^{-3}$ to $2.0 * 10^{-3}$ $m^3 s^{-1}$ suggesting an increase in stormwater drainage efficiency by reducing groundwater infiltration rates under increased precipitation rates (Table 9.2).

4 Summary

The Dogwood Reservoir case study demonstrates the utility of time-series data to perform various temporal analyses on comprehensive data records. Using a natural tracer of groundwater interactions with an urban reservoir provided an integrated view of the hydrologic response to storm and seasonal scale forcings, while offering a continuous, long-term record with which results of various temporal analyses may be compared. As urbanization continues to dominate the nation and especially coastal regions, we can expect changes to aquatic systems and the water sources that discharge into them. While we do not have definitive proof that the threats to groundwater resources demonstrated by way of this case study are related to anthropogenic manipulation of the natural plumbing that supplies water to Dogwood Reservoir, we do have evidence that a decline in groundwater resources may be a significant issue in this region. Perhaps observed changes in the volumetric percentage of groundwater within Dogwood Reservoir were only part of a larger cycle, oscillating at a frequency less than that of our sampling period, and not evidence of a serious threat to groundwater resource. Nonetheless, hydrologic alterations associated with urbanization and anthropogenic landscape modifications may not result in negative impacts but certainly will consequent change.

Key Points

- ^{222}Rn is an effective tool to directly assess the amount of recently discharged groundwater within an urban reservoir.
  - Consider correcting measured concentrations for the addition of ^{222}Rn via decay of dissolved ^{226}Ra and diffusion from bottom sediments as well as accounting for any suspected loss via diffusion across the air/water interface and/or radioactive decay.
- Time-series data are extremely valuable to capture complex phenomena that appear highly variable in time.
  - Evapotranspiration-driven change in water table height was found to govern the reservoir Dogwood Reservoir hydrology, ultimately responsible for increased export rates of groundwater during normal conditions. However, during rain events of various magnitudes and storm durations, a number of environmental parameters (including wind speed and direction, time since last rainfall, and water table height prior to the event) contribute to the percentage of groundwater present within the reservoir and therefore the amount exported.
  - Seasonal drivers associated with rainfall and evaporation increased the volumetric percentage of groundwater within the reservoir during spring and increased total discharge rates resulting in maximum seasonal export of groundwater and direct precipitation despite lower rainfall totals (as compared to summer).

- In general, volumetric percentages of groundwater were found to be significant but reduced from 2012 to 2013.
  - Groundwater was found to comprise a modest but significant amount of the Dogwood Reservoir budget with a mean volumetric percentage of 5.6 % but with significant variance (up to 36.45 %).
  - From 2012 to 2013, the volumetric percentage of groundwater in Dogwood Reservoir decreased by 2.7 % due to increased aquifer residence times, in spite of greater precipitation rates. As a result, groundwater export to the coastal ocean was reduced

References

Arnold JG, Allen PM (1996) Estimating hydrologic budgets for three Illinois watersheds. J Hydrol 176(1–4):57–77

Arnold JG, Muttiah RS, Srinivasan R et al (2000) Regional estimation of base flow and groundwater recharge in the Upper Mississippi river basin. J Hydrol 227(1–4):21–40

Brabec E, Schulte S, Richards PL (2002) Impervious surfaces and water quality: a review of current literature and its implications for watershed planning. J Plan Lit 16(4):499–514

Burnett WC, Dulaiova H (2003) Estimating the dynamics of groundwater input into the coastal zone via continuous radon-222 measurements. J Environ Radioact 69(1–2):21–35

Burnett WC, Kim G, Lane-Smith D (2001) A continuous monitor for assessment of ^{222}Rn in the coastal ocean. J Radioanal Nucl Chem 249(1):167–172

Burnett WC, Cable JE, Corbett DR (2003) Radon tracing of submarine groundwater discharge in coastal environments. In: Taniguchi M, Wang K, Gamo T (eds) Land and marine hydrogeology. Elsevier, Amsterdam, pp 25–43

Burnett WC, Aggarwal PK, Aureli A et al (2006) Quantifying submarine groundwater discharge in the coastal zone via multiple methods. Sci Total Environ 367(2–3):498–543

Burnett WC, Peterson R, Moore WS et al (2008) Radon and radium isotopes as tracers of submarine groundwater discharge—results from the Ubatuba, Brazil SGD assessment intercomparison. Estuar Coast Shelf Sci 76(3):501–511

Cable JE, Burnett WC, Chanton JP et al (1996) Estimating groundwater discharge into the northeastern Gulf of Mexico using radon-222. Earth Planet Sci Lett 144(3–4):591–604

Cerda A (1997) Seasonal changes of the infiltration rates in a Mediterranean scrubland on limestone. J Hydrol 198(1–4):209–225

Charette MA, Moore WS, Burnett WC (2008) Uranium- and thorium-series nuclides as tracers of submarine groundwater discharge. In: U-Th series nuclides in aquatic systems. Elsevier, Amsterdam, pp 155–191

Corbett DR, Burnett WC, Cable PH et al (1998) A multiple approach to the determination of radon fluxes from sediments. J Radioanal Nucl Chem 236(1):247–253

Corbett DR, Chanton J, Burnett WC et al (1999) Patterns of groundwater discharge into Florida Bay. Limnol Oceanogr 44(4):1045–1055

Corbett DR, Dillon K, Burnett W et al (2000) Estimating the groundwater contribution into Florida Bay via natural tracers ^{222}Rn and CH_4. Limnol Oceanogr 45(7):1546–1557

Dimova NT, Burnett WC (2011) Evaluation of groundwater discharge into small lakes based on the temporal distribution of radon-222. Limnol Oceanogr 56(2):486–494

Dimova NT, Burnett WC, Chanton JP et al (2013) Application of radon-222 to investigate groundwater discharge into small shallow lakes. J Hydrol 486:112–122

Drescher SR, Messersmith MJ, Davis BD et al (2007) State of knowledge report: stormwater ponds in the coastal zone. South Carolina Department of Health and Environmental Control—Ocean and Coastal Resource Management

Dulaiova H, Peterson R, Burnett WC et al (2005) A multi-detector continuous monitor for assessment of ^{222}Rn in the coastal ocean. J Radioanal Nucl Chem 263(2):361–365

Dulaiova H, Camilli R, Henderson PB et al (2010) Coupled radon, methane, and nitrate sensors for large-scale assessment of groundwater discharge and non-point pollution to coastal waters. J Environ Radioact 101(7):553–563

Gleeson J, Santos IR, Maher DT et al (2013) Groundwater-surface water exchange in a mangrove tidal creek: evidence from natural geochemical tracers and implications for nutrient budgets. Mar Chem 156:27–37

Grimm NB, Faeth SH, Golubiewski NE et al (2008) Global change and the ecology of cities. Science 319(5864):756–760

Grimmond CSB, Oke TR (1986) Urban Water-balance 2. Results from a suburb of Vancouver, British Columbia. Water Resour Res 22(10):1404–1412

Haase D (2009) Effects of urbanization on the water balance—a long-term trajectory. Environ Impact Assess Rev 29(4):211–219

Hancock GS, Holley JW, Chambers RM (2010) A field-based evaluation of wet retention ponds: how effective are ponds at water quality control? J Am Water Resour Assoc 46(6):1145–1158

Hanrahan JL, Kravtsov SV, Roebber PJ (2010) Connecting past and present climate variability to the water levels of Lakes Michigan and Huron. Geophys Res Lett 37(1):L01701. doi:10.1029/2009GL041707

Hornberger GM, Raffensperger JP, Wiberg PL et al (1998) Open chanel hydraulics. In: Elements of physical hydrology. The Johns Hopkins University Press, Baltimore, pp 73–97

Hutchins P, Smith E, Koepfler E et al (2014) Metabolic responses of estuarine microbial communities to discharge of surface runoff and groundwater from contrasting landscapes. Estuar Coasts 37(3):736–750

Lambert MJ, Burnett WC (2003) Submarine groundwater discharge estimates at a Florida coastal site based on continuous radon measurements. Biogeochemistry 66(1):55–73

Lautz LK (2007) Estimating groundwater evapotranspiration rates using diurnal water-table fluctuations in a semi-arid riparian zone. Hydrogeol J 16(3):483–497

Libes SM (2006 – present) Long Bay Hypoxia Monitoring Consortium time series. http://bccmws.coastal.edu/lbos/. Accessed 12 Dec 2015

Libes SM, Peterson RN (2012 – present) Briarcliffe Acres groundwater and lake level monitoring program. http://bccmws.coastal.edu/bagw/. Accessed 17 Aug 2013

MacIntyre S, Wanninkhof R, Chanton JP (1995) Trace gas exchange across the air–sea interface in freshwater and coastal marine environments. In: Matson PA, Harris RC (eds) Biogenic trace gases: measuring emissions from soil and water. Blackwell Science, Boston, pp 52–97

Meriano M, Howard KWF, Eyles N (2011) The role of midsummer urban aquifer recharge in stormflow generation using isotopic and chemical hydrograph separation techniques. J Hydrol 396(1–2):82–93

Moore WS, Reid DF (1973) Extraction of radium from natural waters using manganese impregnated acrylic fibers. J Geophys Res 78(36):8880–8886

Mullinger NJ, Binley AM, Pates JM et al (2007) Radon in Chalk streams: spatial and temporal variation of groundwater sources in the Pang and Lambourn catchments, UK. J Hydrol 339 (3–4):172–182

Peterson RN, Burnett WC, Dimova N et al (2009) A comparison of measurement methods for radium-226 on manganese-fiber. Limnol Oceanogr Methods 7(2):196–205

Peterson RN, Santos IR, Burnett WC (2010) Evaluating groundwater discharge to tidal rivers based on a Rn-222 time-series approach. Estuar Coast Shelf Sci 86(2):165–178

Santos IR, Dimova N, Peterson RN et al (2009) Extended time series measurements of submarine groundwater discharge tracers (^{222}Rn and CH_4) at a coastal site in Florida. Mar Chem 113 (1–2):137–147

Santos IR, Peterson RN, Eyre BD et al (2010) Significant lateral inputs of fresh groundwater into a stratified tropical estuary: evidence from radon and radium isotopes. Mar Chem 121 (1-4):37–48

Schueler T (2000) Influence of groundwater on performance of stormwater ponds in Florida: the practice of watershed protection. Watershed Prot Tech 2(4):525–528

Swarzenski PW (2007) U/Th series radionuclides as coastal groundwater tracers. Chem Rev 107 (2):663–674

US Department of the Interior, US Geological Survey (2015) USGS current conditions for South Carolina. http://waterdata.usgs.gov/sc/nwis/uv? Accessed 18 July 2013

Vietz GJ, Sammonds MJ, Walsh CJ et al (2014) Ecologically relevant geomorphic attributes of streams are impaired by even low levels of watershed effective imperviousness. Geomorphology 206:67–78

Winter TC (1999) Relation of streams, lakes, and wetlands to groundwater flow systems. Hydrogeol J 7(1):28–45

Xue J, Gavin K (2008) Effect of rainfall intensity on infiltration into partly saturated slopes. Geotech and Geol Eng 26(2):199–209

Chapter 10
Evaluating Hydrogeological and Topographic Controls on Groundwater Arsenic Contamination in the Middle-Ganga Plain in India: Towards Developing Sustainable Arsenic Mitigation Models

Sushant K. Singh, Stefanie A. Brachfeld, and Robert W. Taylor

Abstract We investigated the spatial distribution and severity of groundwater arsenic contamination in three previously un-studied villages located near the confluence of the Rivers Ganges and Sone, within the Maner block of Patna district in the Bihar State, India. We also gathered information on the demographic, socioeconomic and health issues of local residents in order to identify at-risk populations due to the exposure to elevated concentrations of arsenic. Arsenic concentrations were measured in 157 drinking water sources, which were tested using field-tests kits. Spatial patterns in arsenic distribution were compared with local physiographic and hydrogeologic parameters. Arsenic levels exceeding the WHO and the BIS standards (10 μg/L and 50 μg/L respectively) were found in all three villages, with a maximum of 300 μg/L. The shallow aquifers ($\leq$50 m below ground surface) and older hand pumps were found to be arsenic contaminated. The deeper aquifers (>50 m) exhibited arsenic levels within permissible limits. Elevated arsenic levels are observed close to the River Ganges. However, a moderate ($r = 0.240$, $p = 0.031$) positive correlation with the surface water flow direction indicates that arsenic migrates from south to north and from west to east in the study area. This suggests that River Sone alluvium is a potential source of arsenic contamination in Bihar.

Highlights

- Arsenic concentrations exceeding WHO and BIS standards are documented in three villages in the Maner Block of Patna district, Bihar, India.
- Arsenic contaminated hand pumps in Suarmarwa, Rampur Diara, and Bhawani Tola are 10–15 years older than the arsenic-free hand pumps.

S.K. Singh (✉) • S.A. Brachfeld • R.W. Taylor
Department of Earth and Environmental Studies, Montclair State University,
1 Normal Avenue, Montclair, NJ 07043, USA
e-mail: sushantorama@gmail.com

A. Fares (ed.), *Emerging Issues in Groundwater Resources*, Advances in Water Security, DOI 10.1007/978-3-319-32008-3_10

- Arsenic concentration correlates with the depth of tube wells, flow direction of surface water, distance to the rivers, and distance to the drainage point.
- The River Sone is a potential source of arsenic contamination in the Middle-Ganga Plain in India.
- The National Chemical Laboratory-Arsenic Field Test Kits (NCL-FTK) is a reliable and quick field method to analyze arsenic levels in water.

1 Introduction

Geogenic groundwater arsenic contamination is referred to as "the biggest global mass poisoning in human history," with the maximum incidence in South Asia (Nordstrom 2002; Rahman et al. 2014; Singh and Vedwan 2015). Arsenic is a group 'A' carcinogen (IARC-WHO 1999, 2001). It occurs in five valence states, among which arsenic *III* (arsenite, the most toxic form) and arsenic *V* (arsenate) are the most abundant species in groundwater (Ravenscroft et al. 2009). Exposure to arsenic through drinking and food induces carcinogenic and non-carcinogenic health effects (Phan et al. 2010; Singh 2011; Singh et al. 2014). Children are particularly susceptible to arsenic-induced health problems such as reduction in IQ levels, among other ailments (Nahar et al. 2014; Wasserman et al. 2014). The impacts of arsenic hazards on human health are exacerbated by poor socioeconomic conditions, which make impoverished people especially vulnerable to elevated arsenic level in drinking water (Curry et al. 2000; Singh and Vedwan 2015).

Alleviating arsenic contamination presents a major challenge to researchers, communities, and policy makers. India continues to follow the World Health Organization's (WHO) old standard, set in 1963, of 50 μg/L of arsenic as the maximum concentration for "safe" drinking water (Yamamura 2001; Singh and Vedwan 2015), even though the Bureau of Indian Standards (BIS) later set a lower acceptable limit of 10 μg/L (BIS 2012). Approximately three decades ago, millions of hand pumps were installed in Bangladesh and India to extract groundwater for domestic use (Opar et al. 2007). At that time, groundwater was viewed as a safer alternative to unhygienic surface water sources and considered a means to protect millions of people from a variety of water-borne diseases (Opar et al. 2007; Singh and Vedwan 2015). However, in the Middle-Ganga Plain (*MGP*), approximately 87 % of the tested groundwater sources were found to be contaminated with arsenite (Mukherjee et al. 2012). Previous studies have addressed health impacts related to consumption of arsenic-contaminated groundwater and food (Chakraborti et al. 2003; Singh 2011; Singh and Ghosh 2011, 2012; Singh et al. 2014). However, the total exposed population to groundwater arsenic and the consequences of exposure are still unknown in the Bihar State in the *MGP*.

Groundwater and surface water arsenic contamination have been linked to both natural geochemical processes (geogenic) and anthropogenic activities. The latter includes mining and application of arsenic-bearing fertilizers and pesticides (Mukherjee et al. 2006; Ravenscroft et al. 2009). Natural geogenic arsenic is common in India, Bangladesh, Nepal, and Pakistan (Smedley and Kinniburgh 2002; Singh and Vedwan 2015). Recent studies suggest that microbially-driven or chemically-driven reductive dissolution of arsenic-bearing Fe oxyhydroxides in organic-rich sediment is the main source of arsenic in shallow aquifers (Nickson et al. 1998, 2000; Dowling et al. 2002; Mahanta et al. 2015; Stuckey et al. 2015).

The use of physiographic, hydrogeologic and soil properties have been explored as predictors of heavy metal contamination in shallow aquifers using probability models (Twarakavi and Kaluarachchi 2005; Winkel et al. 2008; Yang et al. 2014). The heterogeneous natures of these properties, both at the surface and at depth, make aquifer-specific investigations vital to understanding arsenic mobility at the local scale. Winkel et al. (2008) found a positive correlation between organic-rich Holocene fluvial and deltaic sediments and groundwater arsenic levels above 10 μg/L. Similarly, Yang et al. (2014), found that deltaic, organic rich sediments were the strongest predictors of elevated arsenic. In both the studies the data sets were restricted to areas of flat topography, and neither the well depth nor the surface elevation was included in the model.

Additional parameters need to be incorporated into the next generation of predictive models. These include flow direction, flow accumulation, and distance to drainage point. These parameters are derived from Digital Elevation Models (DEM). "Flow direction" gives the cardinal direction of the surface water flow. "Flow accumulation" is a dimensionless parameter that represents the cumulative flow into each cell from all upstream cells in the DEM. "Drainage points" are cells with the highest flow accumulation. These parameters and methods utilize Geographic Information Science (GIS) and remote sensing (RS) techniques, which are well suited to identifying spatial and temporal patterns. These new parameters can be used to investigate whether arsenic transport is a contributing factor to elevated arsenic concentrations in groundwater.

Moreover, easily interpretable presentation of results for decision-makers in the form of statistical models or GIS maps is lacking in Bihar (Singh and Vedwan 2015). A recent study presented a composite vulnerability index and GIS-based vulnerability map for the arsenic contaminated areas and the communities living in those areas (Singh and Vedwan 2015). This map, combined with predictive models, would be helpful for decision-makers to prioritize areas for arsenic-mitigation programs and predict the probability of success of an arsenic-mitigation policy in a given area.

The objectives of this study are to: (a) evaluate the arsenic contamination levels in drinking water sources in three (two previously unstudied) villages in the Bihar; (b) assess the population size at risk due to exposure to arsenic contamination; and (c) assess the hydrogeologic and topographic relationships with arsenic concentrations in groundwater.

2 Study Area

The Bihar State is located in the *MGP* of India (Fig. 10.1). Approximately 12 million inhabitants have been reported as at-risk in the state (Singh et al. 2014; Singh 2015a, b; Singh and Vedwan 2015). Groundwater arsenic contamination in Bihar was first detected a decade ago in the Bhojpur district (Chakraborti et al. 2003). As of 2014, groundwater arsenic contamination has been documented in 22 of 38 districts (Singh et al. 2014). The remaining 16 districts have either not been investigated or results not yet reported in mainstream scientific literature.

Multiple research groups and agencies have investigated the origin and distribution pattern of arsenic in the *MGP* (Ghosh et al. 2009; Saha et al. 2011; Mukherjee et al. 2012; Saha and Shukla 2013). There exists a wide spatial distribution of groundwater arsenic levels in the region. The levels frequently exceed 1000 µg/L (100 times higher than the WHO standard) in several districts located within 10 km of the River Ganges including Buxar, Bhojpur, Patna, Samastipur, and Bhagalpur (Ghosh et al. 2009; Saha 2009; Saha et al. 2009; Singh and Vedwan 2015).

This study investigated three villages covering 70 acres; Suarmarwa (30 acres), Rampur Diara (25 acres), and Bhawani Tola (15 acres) in the Maner block of Patna

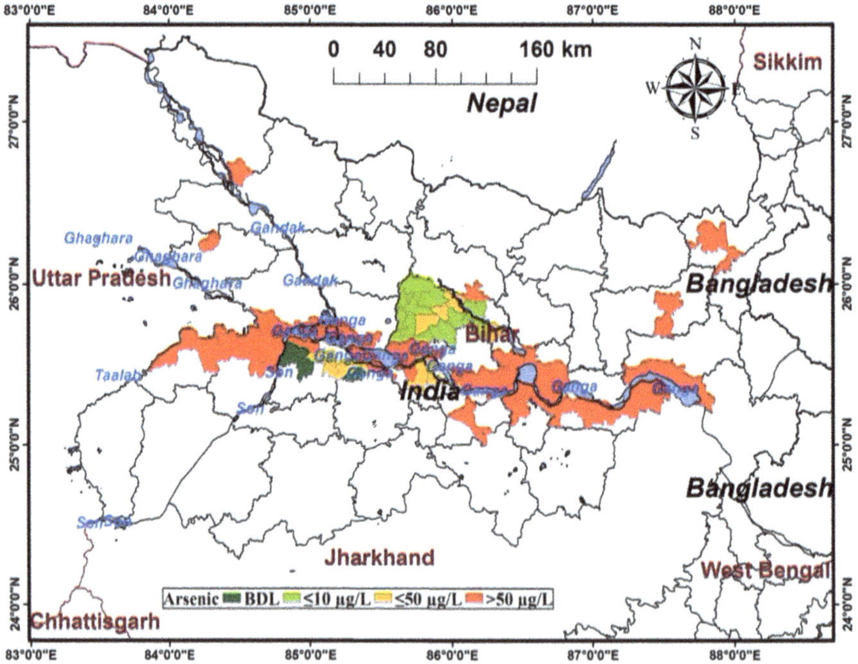

Fig. 10.1 Arsenic contaminated community blocks of Bihar and neighboring regions (Singh 2015a)

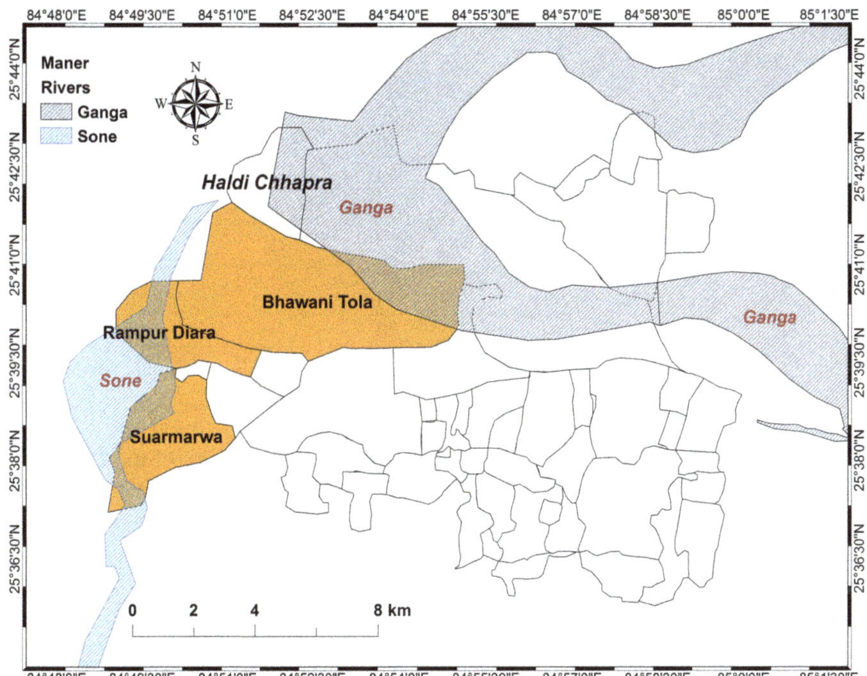

Fig. 10.2 The study area showing the three villages in Maner block of Patna district, Bihar. Bhawani Tola is bordered by the River Ganges to the north. Rampur Diara and Suarmarwa are situated on the bank of the River Sone. Haldi Chhapra is a highly arsenic contaminated village located immediately north of the study area. Neither Suarmarwa or Bhawani Tola has been investigated for arsenic contamination

district of Bihar State (Fig. 10.2). The block is one of the worst arsenic affected blocks in the state (SOES 2004; Singh 2011; Singh and Ghosh 2012). The groundwater based drinking sources in most of the villages of this block, which are within 10 km of the River Ganges, have elevated levels of arsenic (SOES 2004; Ghosh et al. 2009; Singh 2011; Singh and Ghosh 2012).

3 Materials and Methods

3.1 Arsenic Testing

To sample over a large area and investigate the maximum number of drinking sources possible, we used reliable and pre-tested field test-kits (FTK) for arsenic analysis. FTKs have been widely used for initial screening of arsenic contamination in drinking water (Nickson et al. 2007; Ghosh et al. 2009; George et al. 2014; van Geen et al. 2014). We used an arsenic FTK developed by the National Chemical

Table 10.1 Comparison of NCL-FTK and AAS arsenic concentrations

Sample	NCL-FTK (μg/L)	AAS (μg/L)
1	BDL (<10 μg/L)	BDL (<5 μg/L)
2	60	70
3	50	70
4	40	40
5	50	60
6	BDL (<10 μg/L)	BDL (<5 μg/L)
7	40	40
8	50	50
9	70	80
10	40	10

BDL below detection limit; for FTK: <10 μg/L and for AAS <5 μg/L

Laboratory (NCL), Pune, India, which is a constituent laboratory of the Council of Scientific and Industrial Research (CSIR), Government of India. The NCL-FTK protocols for field-based arsenic measurements have been approved by the Government of Bihar, the Government of Uttar Pradesh, and the United Nations Children's Fund (UNICEF) (Nickson et al. 2007; Ghosh et al. 2009). The NCL-FTK detects arsenic levels between 10 and 110 μg/L, with an approximate 15 min reaction period and a cost of $0.83 per sample (based on 2013 USD rate) (NCL 2002). Arsenic concentrations exceeding 110 μg/L must be analyzed by Atomic Absorption Spectrophotometry (AAS) or Inductively Coupled Plasma Mass Spectrometry (ICP-MS). We selected 22 samples at random for AAS analysis in order to assess the range of arsenic concentrations in the study area (Tables 10.1 and 10.2). Samples were handled according to the protocols of Nickson et al. 2007. AAS analyses were conducted by the Sriram Institute for Industrial Research (An ISO-9001:2008 certified Institute in Delhi, India).

The FTK uses a cotton plug saturated with lead (Pb)-acetate to filter impurities. The color of the plug was monitored after every analysis. High arsenic levels (>50 μg/L) caused the plug to turn black, in which case the plug was immediately replaced. A fresh plug was used after every ten analyses, and an arsenic standard of 50 μg/L was measured after every 20 analyses.

The accuracy of the NCL-FTK was previously reported at 67 %, i.e., an R^2 of 0.67 was determined for a set of ten samples analyzed using both the NCL-FTK and AAS (Ghosh et al. 2009). In this study, we obtained an R^2 value of 0.82 for a set of ten samples analyzed by NCL-FTK and AAS (Tables 10.1 and 10.2; Fig. 10.3). Although some samples are off the trend line, the NCL-FTK can be utilized to identify low contamination (arsenic levels <10 μg/L), moderately contaminated water (10–50 μg/L), and highly contaminated (>50 μg/L) water.

A total of 157 water samples (Suarmarwa = 57, Rampur Diara = 50, Bhawani Tola = 50) were collected from hand pumps and tested immediately using the NCL-FTK. One hundred and fifty-four samples were taken from shallow aquifers (<50 m) and three from deep tube wells (>50 m). We also tested open wells/dug wells (Fig. 10.4), which were abandoned by the communities. Though open well results

Table 10.2 Arsenic concentration in groundwater samples from Atomic Absorption Spectrophotometry

Sample no.	Village name	[Arsenic] μg/L
1	Rampur Diara	5<
2	Rampur Diara	70
3	Rampur Diara	80
4	Rampur Diara	80
5	Rampur Diara	70
6	Rampur Diara	40
7	Rampur Diara	70
8	Rampur Diara	20
9	Rampur Diara	60
10	Rampur Diara	<5
11	Rampur Diara	300
12	Rampur Diara	40
13	Rampur Diara	40
14	Rampur Diara	<5
15	Rampur Diara	5<
16	Bhawani Tola	80
17	Rampur Diara	40
18	Rampur Diara	40
19	Rampur Diara	50
20	Bhawani Tola	250
21	Rampur Diara	10
22	Rampur Diara	30
Min		<5
Max		300
Mean		62
Std. dev.		75

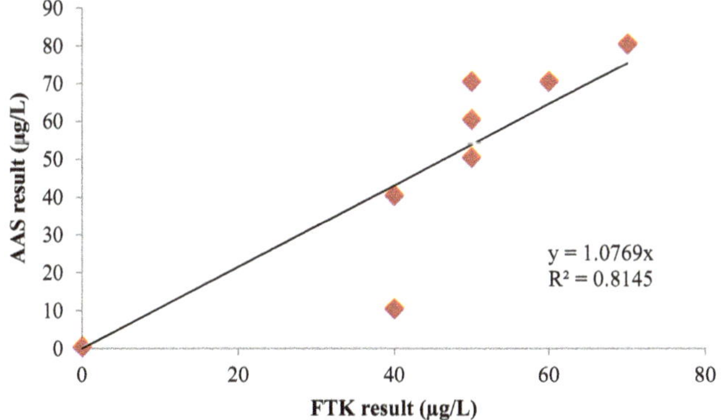

Fig. 10.3 Comparison of NCL-FTK and AAS arsenic concentrations in water samples. Values that were below detection were plotted as 0 μg/L

are not presented here, they exhibited arsenic levels well within the WHO limit (BDL). Open wells have been reported as arsenic-free sources of drinking water and could be a cost-effective arsenic-mitigation option (Saha 2009; Singh 2015a, b).

Fig. 10.4 An open dug-well in Rampur Diara potentially targeted for renovation (Photograph by Sunil Kumar Singh 2014)

3.2 Population At-Risk and Population Exposed to Arsenic

We define an "at risk" population as one whose members consume arsenic contaminated drinking water. The "population exposed" describes individuals who have access to drinking water sources with arsenic levels exceeding safe levels. This can be calculated with either the WHO (10 μg/L) or the BIS (50 μg/L) standard (Nickson et al. 2007). At-risk population size is calculated as:

$$\textit{Population at risk} = \textit{Total population covered}_{As} \times \%\ \textit{of sources contaminated with arsenic}$$

where *total population covered*$_{As}$ is the total population that uses the tested water sources (Singh 2015a). Total population covered is calculated as the number of households surveyed multiplied by 6, which is the average household size in the three villages (Singh 2015a). Population exposed was calculated as follows:

$$\textit{Population exposed} = \textit{Total population of the village} \times \%\ \textit{of sources contaminated with arsenic}$$

3.3 GIS Mapping and Data Analysis

Arc Geographic Information System (ArcGIS) Desktop version 10.2 (ESRI 2012) was used to create all the maps. Block level shape files were obtained from the Central Ground Water Board, Mid-Eastern Region, Patna, Bihar, India. River shape files for the Bihar State were downloaded from the diva-GIS web-portal (http://www.diva-gis.org/datadown). A hardcopy of the village map of the Maner block was obtained from the Bihar Census Bureau, India. The village map was georeferenced to generate village layers. A RICOH GPS Enabled Camera was used to take photographs of water sources and enter their geographical coordinates into shape files using the GPS Photo-Link program. Only 88 of 157 water sources were processed in this manner due to weak satellite signals or failure of the GPS Photo-Link program to process the coordinates. Other information such as arsenic concentration and elevation of the water source was entered into the attribute table.

Advanced Space borne Thermal Emission and Reflection Radiometer (ASTER) Global Digital Elevation Model (GDEM) Version 2 (2011), a product of the Ministry of Economy, Trade, and Industry (METI) of Japan and the U.S. National Aeronautics and Space Administration (NASA), was used for hydrological and topographic analysis (http://earthexplorer.usgs.gov/). The surface elevation, slope, flow direction, and flow accumulation were extracted at the coordinates of each water source using the Spatial Analyst tool. ASTER GDEM and Arc Hydro Tools 10.2 were used to analyze elevation, slope, flow direction, flow accumulation, stream network geometry, drainage catchment area, and drainage points. All possible sinks in the GDEM were filled by applying the fill-sink option available within the Arc Hydro Tool. The flow direction tool was applied to determine the direction of steepest downward slope, where each of the eight possible flow directions out from the center of a 3×3 grid of cells is represented by an integer that increases in a clockwise manner in increments of 16 from 1 (East) to 128 (Northeast). The flow accumulation tool was used to determine the cumulative flow into each cell from all upstream cells. Drainage points were defined as the cells of highest flow accumulation. The probable effect of the any nearby drainage (such as the Ganges and the Sone) on the contamination level in the tested sources was analyzed. In addition, distance between tested water sources and drainage points, and the distance between arsenic contaminated and arsenic-free drinking water sources were calculated.

3.4 *Statistical Analysis*

The water quality analysis data and the DEM-derived spatial data were transferred into SPSS for statistical analysis (IBM 2012). A crosstabs analysis was performed between arsenic concentration and the villages surveyed to assess the independence. Bivariate analysis was performed to examine associations between water quality and hydrological and topographical data.

4 Results

4.1 *Groundwater Arsenic Contamination in the Surveyed Villages*

The AAS analysis results of 22 groundwater samples are given in Table 10.2. Rampur Diara had a maximum arsenic concentration of 300 μg/L. This is substantially higher than the maximum of 103 μg/L observed in a previous study in this village (Singh 2011; Singh and Ghosh 2012). The next highest level of 250 μg/L was measured in Bhawani Tola (Table 10.2).

Of the 157 groundwater sources tested by the NCL-FTK, 39 of 57 sources in Suarmarwa (68 %), 31 of 50 sources in Rampur Diara (62 %), and 18 of 50 sources in Bhawani Tola (36 %) were below the detection limit (BDL) of 10 μg/L (Table 10.3). Two sources (4 %) in Rampur Diara and two sources in Bhawani Tola (4 %) had arsenic concentrations near the WHO standard of 10 μg/L. Thirteen samples (22.8 %) in Suarmarwa, 11 (22 %) in Rampur Diara, and 10 (20 %) in Bhawani Tola, had arsenic levels between the WHO standard of 10 μg/L and the BIS standard of 50 μg/L (Table 10.3). In Bhawani Tola, 20 samples (40 %) have arsenic concentrations exceeding 50 μg/L (Table 10.3). The average arsenic concentration was 14 μg/L in Suarmarwa, 17 μg/L in Rampur Diara, and 38 μg/L in Bhawani Tola. The arsenic levels in all the three villages were significantly different from each other ($p = 0.001$) (Table 10.3).

The depths of the drinking water sources were similar ($p = 0.322$) in the three villages (Table 10.3, Fig. 10.5). Only 5 % of the water sources in Suarmarwa and 6 % each in Rampur Diara and Bhawani Tola drew from depths shallower than 20 m below the ground surface (Table 10.3). The majority of the water samples (Suarmarwa = 95 %, Rampur Diara = 90 %, and Bhawani Tola = 94 %) were drawn from depths between 20 and 50 m below ground surface (Table 10.3). Only two drinking water wells in Rampur Diara drew water from depths below 50 m below ground surface. Both the sources (one boring and one hand pump) were arsenic-free. The crosstabs analysis between the three arsenic contamination groups (Table 10.4) and the three depth groups indicates that of 90 % of the water sources (79 of 88) in the BDL group were located between 20 and 50 m.

Four water sources with 10 μg/L of arsenic were observed (Table 10.3). Two of these came from 0 to 20 m depth and two from 20 to 50 m depth. One hundred percent (100 %) of the sources ($n = 34$) with arsenic levels between 11 and 50 μg/L were collected from wells in the depth range of 20–50 m below ground. Approximately 97 % ($n = 30$) of the sources with arsenic levels >51 μg/L were collected in the same depth range of 20–50 m below ground surface. Only one sample with an

Table 10.3 Distribution of arsenic concentration and well depths (p = significance level)

Village	Arsenic (p = 0.001)				Depth (p = 0.322)		
	BDL	10 μg/L	11–50 μg/L	>51 μg/L	0–20 m	20–50 m	>50 m
Suarmarwa	39 (68 %)	0 (0 %)	13 (23 %)	5 (9 %)	3 (5 %)	54 (95 %)	0 (0 %)
Rampur Diara	31 (62 %)	2 (4 %)	11 (22 %)	6 (12 %)	3 (6 %)	45 (90 %)	2 (4 %)
Bhawani Tola	18 (36 %)	2 (4 %)	10 (20 %)	20 (40 %)	3 (6 %)	47 (94 %)	0 (0 %)
N	88	4	34	31	9	146	2

BDL below detection limit

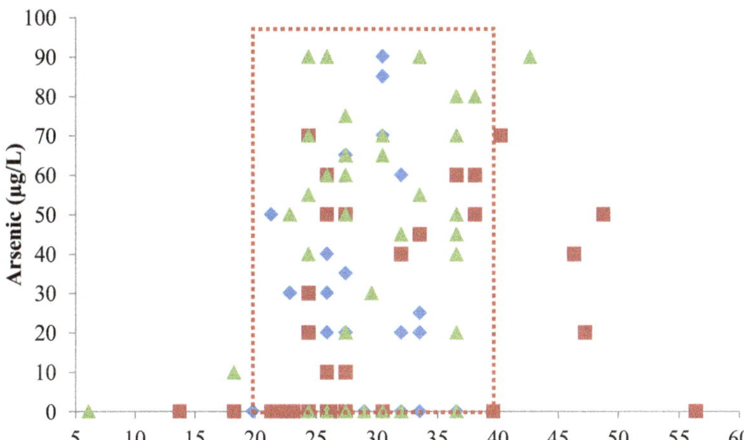

Fig. 10.5 Arsenic concentrations vs. depths of hand pumps

Table 10.4 Arsenic concentrations by depth of hand pumps

		Depths of hand pumps (m) (p = 0.007)		
		0–20 (%)	20–50 (%)	>50 (%)
Arsenic concentration (μg/L)	BDL (n = 88 of 157)	8	90	2
	10 (n = 4 of 157)	50	50	0
	11 to 50 (n = 34 of 157	0	100	0
	>51 (n = 31 of 157)	3	97	0

BDL below detection limit

arsenic level of 85 μg/L came from a shallow well in the depth range of 0–20 m below ground. Based on the Chi-square test, the distribution of arsenic was significantly different at different depths (p = 0.007), which showed a wide spatial distribution.

4.2 *Populations At-Risk and Exposed*

Approximately 10 % (n = 204) of individuals were found to be at-risk due to exposure to >10 μg/L of arsenic in drinking water. The highest population at-risk is in Suarmarwa 12 % (n = 78), followed by Rampur Diara (9 %, n = 66), and Bhawani Tola (8 %, n = 60) (Table 10.5). The total population at-risk due to the exposure to >50 μg/L of arsenic in drinking water is 9 % (n = 186), with the highest population at-risk in Bhawani Tola 17 % (n = 120), followed by Rampur Diara

Table 10.5 Populations at-risk and exposed to drinking arsenic contaminated water

Village	TPC	TP	PR >10 μg/L	PR >50 μg/L	PE >10 μg/L	PE >50 μg/L
Suarmarwa	667	4696	78	30	1071	413
Rampur Diara	769	3437	66	36	756	412
Bhawani Tola	712	3000	60	120	600	1200
Total	2148	11,133	204	186	2427	2026

TPC Total population covered, *TP* Total population, *PR* Population at risk, *PE* Population exposed

(5 %, n = 36), and Suarmarwa (4 %, n = 30) (Table 10.5). Population at risk and exposed population at >10 and >50 μg/L are completely separate populations.

A total of 22 % (n = 2427) and 18 % (n = 2026) of the inhabitants were found to be exposed to >10 μg/L and >50 μg/L of arsenic in drinking water, respectively (Table 10.5). The highest exposed population was in Suarmarwa at >10 μg/L (WHO standard) and in Bhawani Tola at >50 μg/L (BIS standard) (Table 10.5).

5 Discussion

5.1 *Spatial and Temporal Patterns of Arsenic Contamination*

We compared our arsenic concentration data with previously reported results from Maner block (Table 10.6). High concentrations of arsenic have previously been observed in Rampur Diara (Singh 2011; Singh and Ghosh 2012). However, this is the first time that we have observed arsenic concentrations above the WHO and the BIS standards in Suarmarwa and Bhawani Tola.

All of the arsenic contaminated sources draw water from the depth range of 20–50 m below the ground surface, with the exception of one 10 μg/L sample in Bhawani Tola. The findings are consistent with previous studies. It suggests that shallow aquifers are contaminated to a depth of 50 m, the maximum depth of the hand pump sources tested in this study (Saha 2009; Saha et al. 2009). We found a positive correlation between the hand pump depths (up to 50 m) and arsenic concentration ($R^2 = 0.217$, $p = 0.007$). In previous studies, no significant correlation between arsenic concentration and hand pump depths was reported, possibly due to limited sample size and smaller geographical area of study (Singh and Vedwan 2015). However, defining a clear relationship between arsenic contamination and depth requires a large number of samples at each depth horizon.

The age of the majority of the hand pumps tested was between 4 and 13 years. Twenty one percent (21 %) of the arsenic-free water sources were installed between 1950 and 2000. A slight majority (51 %) of the arsenic-free hand pumps were installed between 2001 and 2010, followed by 28 % installed in 2011 or later. A majority (59 %) of water sources with arsenic levels between 11 μg/L and 50 μg/L were installed between 2001 and 2010, followed by 35 % in 2011 or later, and only 6 % before the year 2000. Likewise, a majority (57 %) of drinking water sources with arsenic levels greater than 50 μg/L was installed between 2001 and 2010,

Table 10.6 Arsenic concentrations reported in villages of Maner Block of Patna District

Village	# samples analyzed	10–50 µg/L	>51 µg/L (%)	References
Puranka Tola	54	2 %	98	(SOES 2004)
Nayka Tola	41	0	100	(SOES 2004)
Badal Tola	74	9 %	91	(SOES 2004)
DihalRike Tola	16	0	100	(SOES 2004)
Haldichhapra	10	0	100	(Singh and Ghosh 2012)
Rampur Diara	10	40 %	60	(Singh and Ghosh 2012)
Suarmarwa	57	23 %	9	This study
Rampur Diara	50	26 %	12	This study
Bhawani Tola	50	24 %	40	This study

Table 10.7 Distance between arsenic contaminated and arsenic safe[a] sources

	Suarmarwa	Rampur Diara	Bhawani Tola
Mean (m)	49.7	50.3	203.1
Std. deviation (m)	30.9	28.9	75.9
Minimum (m)	9.0	20.0	24.0
Maximum (m)	98.0	112.0	303.0

[a]Safe represents sources with arsenic concentration of 10 µg/L or below the detection limit (BDL)

followed by 37 % sources in 2011 or later, and 6 % before the year 2000. We did not find any significant correlation between the arsenic contamination groups and the age groups (year of installation) of the hand pumps ($p = 0.254$).

A distance analysis was performed between arsenic contaminated drinking water sources (>10 µg/L) and arsenic safe drinking water sources (≤10 µg/L). The purpose was to investigate the distance that villagers would be required to travel to obtain arsenic-free water. The mean distance between the contaminated and safe sources is approximately 50 m in Suarmarwa and Rampur Diara (Table 10.7). The mean distance for Bhawani Tola is 203 m (Table 10.7). However, there may be other sources in between, which were not tested in this study.

5.2 *Hydrogeological and Topographic Assessment*

The three villages are located at elevations between 46 and 64 m above mean sea level (mamsl) (Fig. 10.6). The tested water sources have surface elevations between 52 and 62 mamsl. The slope (mean = 2.76°) of the ground surface is towards the north and east (Table 10.8). Suarmarwa (50–64 mamsl) and Rampur Diara (49–64 mamsl) have similar elevation profiles, with a mean elevation of approximately 56 mamsl and a standard deviation of 3 m. There is no relationship between elevation and arsenic concentration in Suarmarwa and Rampur Diara. Bhawani

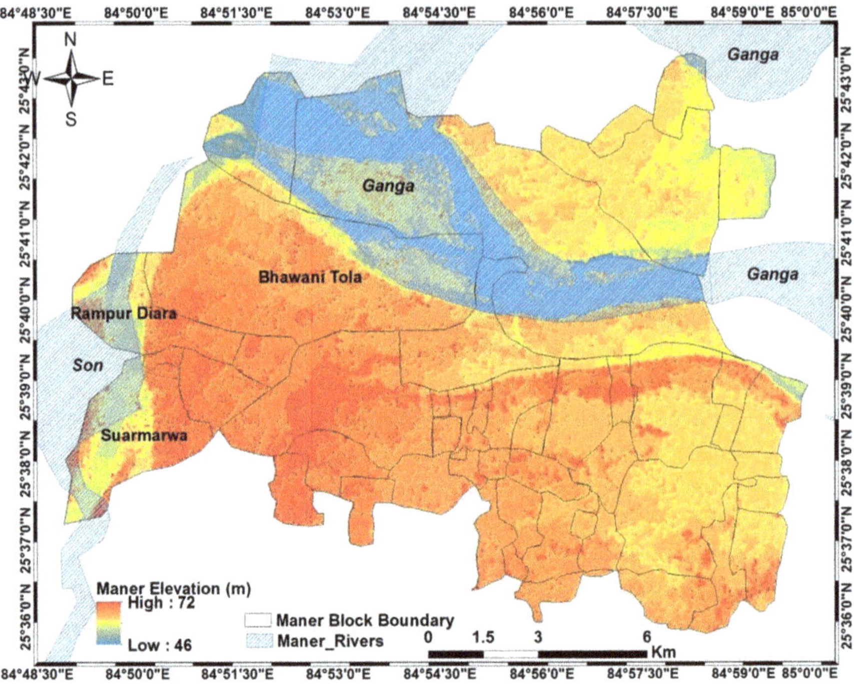

Fig. 10.6 Digital Elevation Model (DEM) of Maner Block of Patna District, Bihar, India. Data from ASTER Global Digital Elevation Model (GDEM) Version 2 (2011)

Tola is located at elevations of 46–64 mamsl with a mean value of 54.2 mamsl and a standard deviation of 3.9 m. In Bhawani Tola, higher arsenic concentrations were found at higher elevations.

The average distance between the tested sources and the River Sone was 1172 m, with a maximum distance of 1854 m (Table 10.8). The distance between the tested sources and the River Sone was less than half of the distance to the River Ganges (5209 m; Table 10.8). The average distance between the tested sources and the nearest drainage point was 636 m and average samples fall on the geographical coordinates between 25.65864° N and 84.84340° E (Table 10.8).

We observe no correlation between arsenic concentrations and elevation of the hand pumps, topographic slope, or flow accumulation (Fig. 10.7a, d). This may be due to the relatively constant topography in our study area, with only 20 m of vertical change over approximately 20 km. We note that Twarakavi and Kalauarachchi, 2005, found elevation, along with land use and soil hydrologic group, to be a statistically significant variable for arsenic prediction in the Sumas-Blaine Aquifer in Washington State. Elevation may play a more significant role in areas with a greater elevation range and steeper slopes, which influences soil texture.

Table 10.8 Descriptive statistics of hydrogeological and topographic attributes of tested drinking water sources

	Minimum	Maximum	Mean		Std. deviation
	Statistic	Statistic	Statistic	Std. error	Statistic
Surface elevation (m) extracted from DEM	52	62	55.07	0.29	2.62
Slope (degrees)	0.00	7.95	2.76	0.22	1.94
Flow direction	1	128	37.99	4.93	44.41
Flow accumulation	0	6971	290.79	147.33	1325.96
Distance to the River Sone (m)	530.3	1854.8	1171.63	42.05	378.45
Distance to the River Ganges (m)	3955.4	6980.8	5209.33	114.07	1026.62
Distance to the drainage point (m)	67.4	981.8	636.48	26.89	242.10

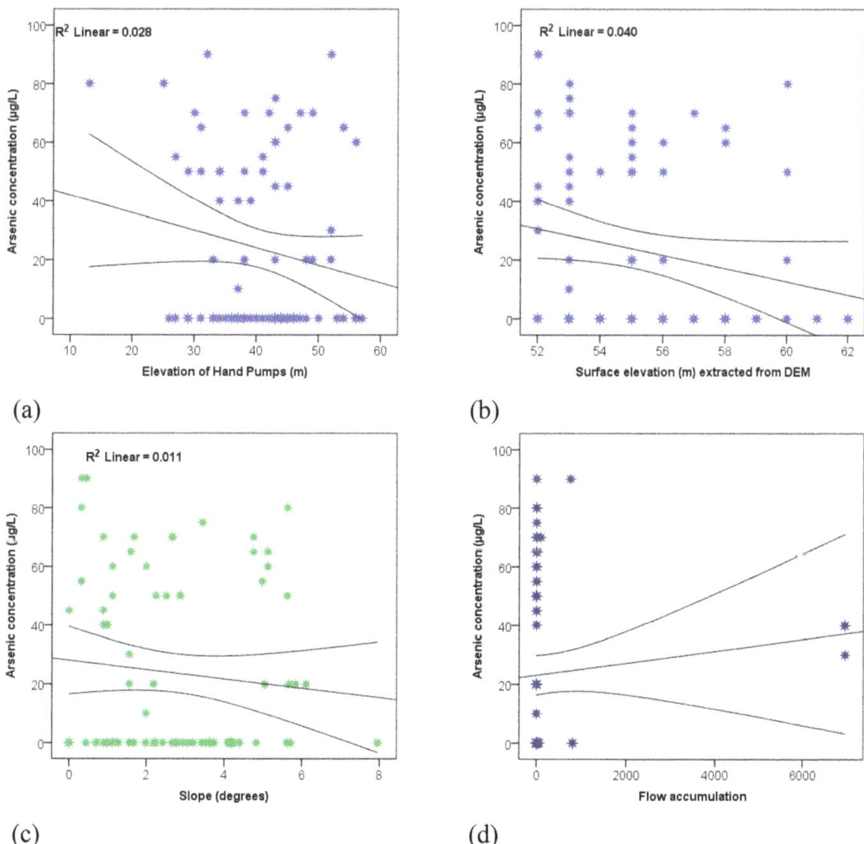

Fig. 10.7 Arsenic concentrations vs. (**a**) elevations of the tested hand pumps (**b**) surface elevations extracted from DEM (**c**) slope, and (**d**) flow accumulation, with regression line and 95 % confidence lines

A moderate positive correlation ($r = 0.240$, $p = 0.031$) was observed between arsenic concentration and flow direction (Table 10.9, Fig. 10.8a–c), i.e., arsenic concentration tends to increase in the direction of surface water flow.

Near Analysis results shows a weak but significant negative correlation ($R^2 = 0.090$, $p = 0.006$) between the arsenic concentration and the distance to the River Ganges (Fig. 10.9 and 10.10), i.e., arsenic concentrations are highest closer to the Ganges. This supports the hypothesis that the River Ganges and/or its levees and alluvium are sources of arsenic to groundwater.

A moderate positive correlation ($R^2 = 0.176$, $p < 0.001$) was observed between arsenic concentration and the distance to the River Sone (Fig. 10.9b), i.e., arsenic concentrations are lowest close to the River Sone, and increase with distance away from the River Sone. Explaining this observation requires geologic and hydrogeochemical study of River Sone water and alluvium.

A negative correlation ($r = -0.456$, $p < 0.001$) was observed between arsenic concentration and distance to the drainage points, i.e., high concentrations of arsenic occur close to the drainage points in the study area (Fig. 10.10, Table 10.9). This suggests that arsenic is concentrated along drainage lines and at drainage points (Fig. 10.11), and these are potential transport paths of arsenic to drinking water sources in their vicinity.

This finding also suggests that areas far from the drainage points are potential arsenic free sources of drinking water.

Two drainage points associated with high concentrations of arsenic are located adjacent to the River Ganges, and situated at the confluence of multiple drainage line networks that pass through the three villages and through the River Sone (Fig. 10.11). This suggests that arsenic carried by the River Sone, from alluvium and/or surface run-off, is transported towards the River Ganges. This would amplify the Ganges floodwaters as a source of arsenic contamination in the surrounding floodplain.

Arsenic concentrations are weakly but positively associated with the latitude ($r = 0.264$, $p = 0.017$) (Fig. 10.12a, Table 10.9) and moderately positively associated with the longitude ($r = 0.449$, $p < 0.001$) (Fig. 10.12b, Table 10.9). This is consistent with the flow direction results (Fig. 10.8a–c), suggesting that groundwater flows towards the northeast, from 25.64 N to 25.67 N (South to North) and 84.84 E to 84.85 E (West to East). In addition, high arsenic concentrations are prevalent in the villages situated in the North and Northeast of the study area, such as Haldi Chhapra (Fig. 10.2). There are more than 40 villages and habitations in Maner block that are located further to the North and East of our study area. If the observations reported here extend over a larger spatial scale, these villages to the northeast are predicted to have elevated levels of arsenic in their drinking water, and their water sources should be tested.

Table 10.9 Correlation matrix of arsenic concentrations and hydrogeological and topographic attributes

	Arsenic concentration (μg/L)	Distance to the River Sone (m)	Depth of hand pumps (m)	Elevation of hand pumps (m)	Surface elevation (m) extracted from DEM	Slope (degrees)	Flow direction	Flow accumulation	Distance to the River Ganges (m)	Distance to the drainage point (m)	Latitude (degree decimal)	Longitude (degree decimal)
Arsenic concentration (μg/L)	1	**0.420**[a]	**0.375**[a]	−0.168	−0.199	−0.104	**0.240**[b]	0.090	**−0.301**[a]	**−0.456**[a]	**0.264**[b]	**0.449**[a]
Distance to the River Sone (m)		1	0.009	−0.182	−0.127	−0.057	**0.274**[b]	−0.206	**−0.923**[a]	**−0.819**[a]	**0.883**[a]	**0.991**[a]
Depth of hand pumps (m)			1	0.085	0.051	0.109	0.139	−0.028	0.017	−0.017	−0.023	0.015
Elevation of hand pumps (m)				1	0.146	0.134	0.144	0.015	0.115	**0.301**[a]	−0.100	−0.182
Surface elevation (m) extracted from DEM					1	**0.640**[a]	0.010	**−0.260**[b]	0.088	0.172	−0.077	−0.130
Slope (degrees)						1	0.084	−0.193	0.052	0.052	−0.049	−0.059
Flow direction							1	−0.094	−0.195	**−0.359**[a]	0.173	**0.280**[b]
Flow accumulation								1	**0.278**[b]	0.126	**−0.290**[a]	−0.172
Distance to the River Ganges (m)									1	**0.580**[a]	**−0.996**[a]	**−0.869**[a]

(continued)

Table 10.9 (continued)

	Arsenic concentration (μg/L)	Distance to the River Sone (m)	Depth of hand pumps (m)	Elevation of hand pumps (m)	Surface elevation (m) extracted from DEM	Slope (degrees)	Flow direction	Flow accumulation	Distance to the River Ganges (m)	Distance to the drainage point (m)	Latitude (degree decimal)	Longitude (degree decimal)
Distance to the drainage point (m)										1	**−0.511**[a]	**−0.855**[a]
Latitude (degree decimal)											1	**0.819**[a]
Longitude (degree decimal)												1

[a]Correlation is significant at the 0.01 level (2-tailed)
[b]Correlation is significant at the 0.05 level (2-tailed)
Pearson Product Moment Correlation Coefficients (r) between arsenic concentrations and the hydrogeological and topographic parameters. The significance level was set at $\alpha = 0.05$. Bolded values are significant at $\alpha = 0.05$ (marked[b]) and at $\alpha = 0.01$ (marked[a])

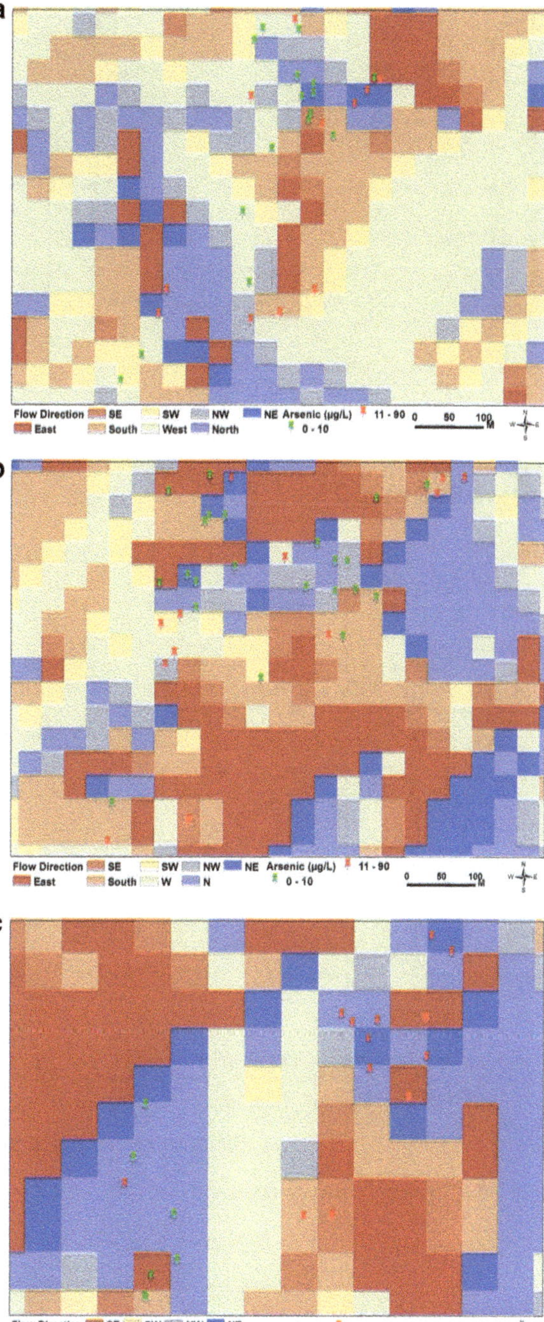

Fig. 10.8 Arsenic concentration (*red* and *green* symbols) superimposed on the flow direction map in (**a**) Suarmarwa, (**b**) Rampur Diara, and (**c**) Bhawani Tola. Each grid cell is color coded to indicate the flow direction in that cell

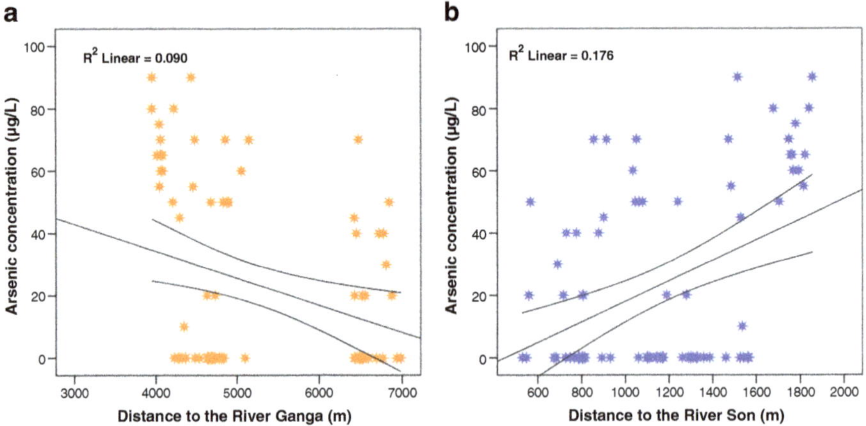

Fig. 10.9 Arsenic concentrations vs. distance to the (**a**) River Ganges and (**b**) River Sone with 95 % confidence lines

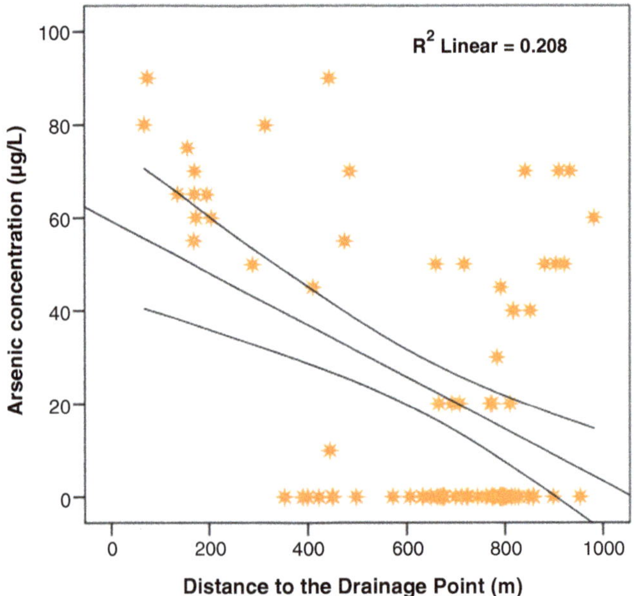

Fig. 10.10 Arsenic concentrations vs. distance from the drainage point with 95 % confidence lines

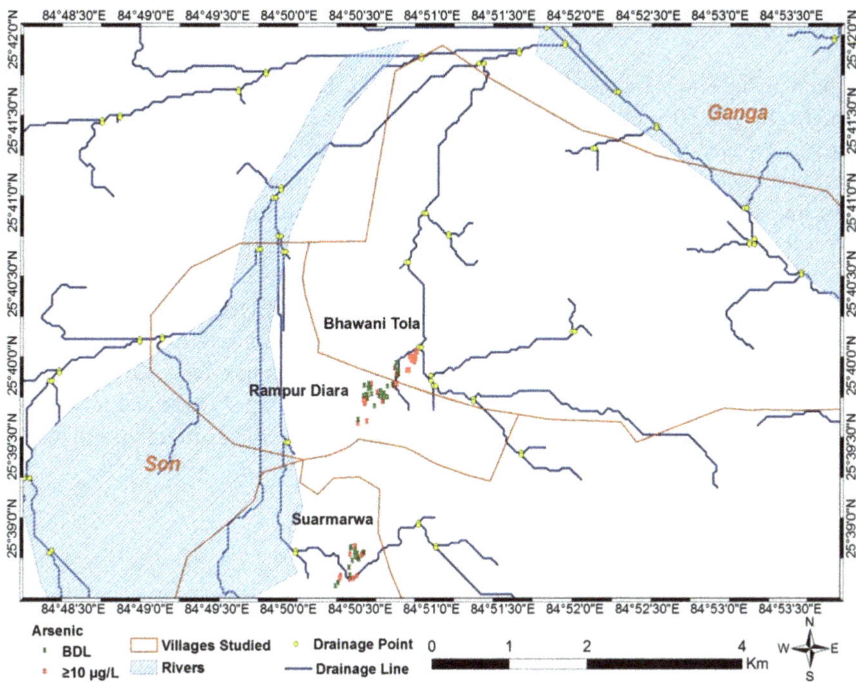

Fig. 10.11 Distribution of arsenic along drainage point and drainage line networks. *Note*: The *orange circle* at the north of the study area encloses two drainage points associated with high arsenic levels, and which sit at the confluence of drainage lines passing through all three villages and through the River Sone

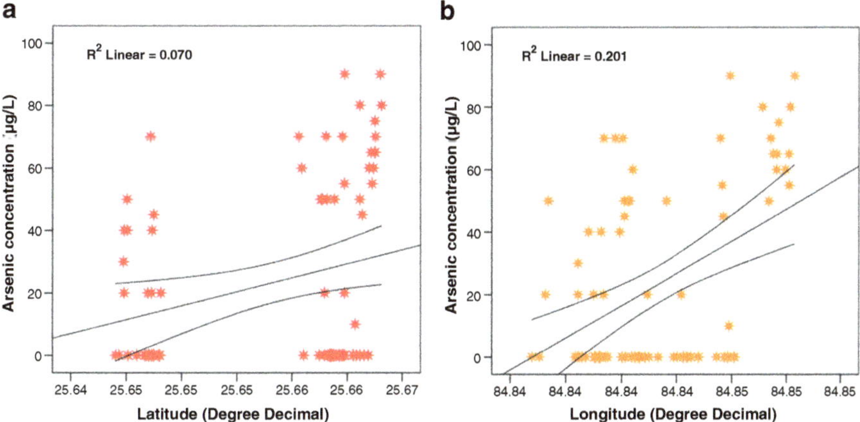

Fig. 10.12 Arsenic concentrations vs. (**a**) latitude and (**b**) longitude with 95 % confidence lines

6 Conclusions

The NCL-FTK was found to be a rapid and reliable method to identify low (<10 μg/L), moderate (10–50 μg/L), and high (>50 μg/L) levels of arsenic contamination in drinking water. Our results document arsenic contamination in aquifers that supply drinking water to Suarmarwa and Bhawani Tola, two previously unstudied villages. Arsenic levels of 250–300 μg/L were observed in Bhawani Tola and Rampur Diara. We observed correlations between arsenic levels and depth of hand pumps, land surface elevation, surface water flow direction, distance to the River Ganges, and distance to drainage points. While the River Ganges has previously been identified as a source of arsenic, our results suggest that the River Sone and/or its alluvial deposits are also potential sources of arsenic. The highest population at-risk due to drinking arsenic contained water found in Bhawani Tola, followed by Rampur Diara and Suarmarwa. Low cost arsenic mitigation strategies are likely to be the most attractive and feasible options for these rural villages.

7 Initiatives Taken

Two open dug-wells in Rampur Diara (Fig. 10.3) were identified for further renovation and installation of hand pumps. Homeowners living adjacent to water wells were approached and verbal consent obtained for operation and maintenance of the wells after renovation. Homeowners were willing to renovate the existing open wells to access arsenic-free water, and agreed to share the water with their community. The lack of funds to renovate wells is an obstacle, but can be surmounted with community contributions along with external funding and logistical and technical support from agencies such as UNICEF, WHO, the "Group for the Protection, Study, and Monitoring of the Environment (GPSME)," a US-based non-profit organization, and the Indian Institute of Sustainable Development (IISD), of New Delhi, India.

Acknowledgements The authors are grateful to Dr. Amy V. Ferdinand, Director, Environmental Health and Safety, Montclair State University, New Jersey, USA for providing a RICOH Digital GPS Camera. Thanks also extend to Dr. Dipankar Saha, Regional Director, Central Groundwater Board, Mid-Easter Region, Patna, Bihar, India for providing block-level shapefile of Bihar. Additionally, the authors are also grateful to Mr. Basant Singh, Sunil Kumar Singh, Kundan Kumar Singh, and Arvind Kumar Singh of Rampur Diara village; Ramlagan Kumar, Guddu Kumar, Rakesh Kumar, Babloo Kumar, Mohan Ray, and Jaswant Kumar of Suarmarwa village for logistic assistance in survey administration, and to the three village heads Mr. Randhir Kumar, Mr. Shailesh Singh, and Mrs. Suman Singh for their cooperation during the survey.

References

BIS (2012) Indian Standard Drinking Water—Specification (Second Revision): IS 10500:2012. New Delhi, India: Bureau of Indian Standards (BIS), Manak Bhavan, 9 Bahadur Shah Zafar Marg, New Delhi 110002, India

Chakraborti D, Mukherjee SC, Pati S, Sengupta MK, Rahman MM, Chowdhury UK et al (2003) Arsenic groundwater contamination in Middle Ganga Plain, Bihar, India: a future danger? Environ Health Perspect 111(9):1194

Curry A, et al. (2000) Towards an assessment of the socioeconomic impact of arsenic poisoning in Bangladesh. World Health Organization

Dowling CB, Poreda RJ, Basu AR, Peters SL, Aggarwal PK (2002) Geochemical study of arsenic release mechanisms in the Bengal Basin groundwater. Water Resour Res 38(9):12-11-12-18

ESRI (2012) ArcGIS for Desktop 10.1. from Environmental Systems Research Institute

George CM, Sima L, Arias MHJ, Mihalic J, Cabrera LZ, Danz D et al (2014) Arsenic exposure in drinking water: a major unrecognized health threat in Peru. Bull World Health Organ 92:565–572

Ghosh AK, Singh SK, Bose N, Roy NP, Singh SK, Upadhyay AK et al (2009) Arsenic hot spots detected in the State of Bihar (India): a serious health hazard for estimated human population of 5.5 Lakhs. In: Ramanathan AL, Bhattacharya P, Keshari PK, Bundschuh J, Chandrashekharam D, Singh SK (eds) Assessment of ground water resources and management. I.K. International Publishing House, New Delhi, pp 62–70

IARC-WHO (1999) IARC monograph on the evaluation of the carcinogenic risk of chemicals to man. International Agency for Research on Cancer (IARC), World Health Organization (WHO)

IARC-WHO (2001) IARC monographs on the evaluation of carcinogenic risks to humans. International Agency for Research on Cancer (IARC), World Health Organization (WHO)

IBM (2012) IBM SPSS statistics for Windows, version 21.0. IBM Corp, Armonk

Mahanta C, Enmark G, Nordborg D, Sracek O, Nath B, Nickson RT et al (2015) Hydrogeochemical controls on mobilization of arsenic in groundwater of a part of Brahmaputra river floodplain, India. J Hydrol Reg Stud 4:154–171

Mukherjee A, Sengupta MK, Hossain MA, Ahamed S, Das B, Nayak B et al (2006) Arsenic contamination in groundwater: a global perspective with emphasis on the Asian scenario. J Health Popul Nutr 24(2):142–163

Mukherjee A, Scanlon BR, Fryar AE, Saha D, Ghosh A, Chowdhuri S, Mishra R (2012) Solute chemistry and arsenic fate in aquifers between the Himalayan foothills and Indian craton (including central Gangetic plain): influence of geology and geomorphology. Geochim Cosmochim Acta 90:283–302

Nahar MN, Inaoka T, Fujimura M (2014) A consecutive study on arsenic exposure and intelligence quotient (IQ) of children in Bangladesh. Environ Health Prev Med 19(3):194–199

NCL (2002) Specifications for arsenic field test kit for drinking water. National Chemical Laboratries (NCL), Pune

Nickson R, McArthur J, Burgess W, Ahmed KM, Ravenscroft P, Rahmann M (1998) Arsenic poisoning of Bangladesh groundwater. Nature 395(6700):338

Nickson R, McArthur J, Ravenscroft P, Burgess W, Ahmed K (2000) Mechanism of arsenic release to groundwater, Bangladesh and West Bengal. Appl Geochem 15(4):403–413

Nickson R, Sengupta C, Mitra P, Dave S, Banerjee A, Bhattacharya A et al (2007) Current knowledge on the distribution of arsenic in groundwater in five states of India. J Environ Sci Health A 42(12):1707–1718

Nordstrom DK (2002) Worldwide occurrences of arsenic in ground water. Science(Washington) 296(5576):2143–2145
Opar A, Pfaff A, Seddique A, Ahmed K, Graziano J, Van Geen A (2007) Responses of 6500 households to arsenic mitigation in Araihazar, Bangladesh. Health Place 13(1):164–172
Phan K, Sthiannopkao S, Kim K-W, Wong MH, Sao V, Hashim JH et al (2010) Health risk assessment of inorganic arsenic intake of Cambodia residents through groundwater drinking pathway. Water Res 44(19):5777–5788
Rahman MM, Mondal D, Das B, Sengupta MK, Ahamed S, Hossain MA et al (2014) Status of groundwater arsenic contamination in all 17 blocks of Nadia district in the state of West Bengal, India: a 23-year study report. J Hydrol 518:363–372
Ravenscroft P, Brammer H, Richards K (2009) Arsenic pollution: a global synthesis, 28. Wiley, Oxford
Saha D (2009) Arsenic groundwater contamination in parts of middle Ganga plain, Bihar. Curr Sci 97(6):753–755
Saha D, Shukla R (2013) Genesis of arsenic-rich groundwater and the search for alternative safe aquifers in the Gangetic Plain, India. Water Environ Res 85(12):2254–2264
Saha D, Dwivedi S, Sahu S (2009) Arsenic in ground water in parts of middle Ganga plain in Bihar—an appraisal. Editorial Board 24(2&3)
Saha D, Sahu S, Chandra P (2011) Arsenic-safe alternate aquifers and their hydraulic characteristics in contaminated areas of Middle Ganga Plain, Eastern India. Environ Monit Assess 175 (1–4):331–348
Singh S (2011) Arsenic contamination in water, soil, and food materials in Bihar. Lambert Academic, Germany
Singh SK (2015a) Assessing and mapping vulnerability and risk perceptions to groundwater arsenic contamination: towards developing sustainable arsenic mitigation models. Ph.D. Dissertation, Montclair State University, Montclair
Singh SK (2015b) Groundwater arsenic contamination in the Middle-Gangetic Plain, Bihar (India): the danger arrived. Int Res J Environ Sci 4(2):70–76
Singh SK, Ghosh AK (2011) Entry of arsenic into food material–a case study. World Appl Sci J 13 (2):385–390
Singh SK, Ghosh AK (2012) Health risk assessment due to groundwater arsenic contamination: children are at high risk. Hum Ecol Risk Assess Int J 18(4):751–766
Singh SK, Vedwan N (2015) Mapping composite vulnerability to groundwater arsenic contamination: an analytical framework and a case study in India. Nat Hazards 75(2):1883–1908. doi:10.1007/s11069-014-1402-2
Singh SK, Ghosh A, Kumar A, Kislay K, Kumar C, Tiwari R et al (2014) Groundwater arsenic contamination and associated health risks in Bihar, India. Int J Environ Res 8(1):49–60
Smedley P, Kinniburgh D (2002) A review of the source, behaviour and distribution of arsenic in natural waters. Appl Geochem 17(5):517–568
SOES (2004) 4th report on Bihar: groundwater arsenic contamination and health effects in Maner block of Patna District, Bihar-India. School of Environmental Studies, Jadavpur University (SOES), Kolkata
Stuckey JW, Schaefer MV, Kocar BD, Benner SG, Fendorf S (2015) Arsenic release metabolically limited to permanently water-saturated soil in Mekong Delta. Nat Geosci 9:70–76
Twarakavi NK, Kaluarachchi JJ (2005) Aquifer vulnerability assessment to heavy metals using ordinal logistic regression. Groundwater 43(2):200–214
van Geen A, Win KH, Zaw T, Naing W, Mey JL, Mailloux B (2014) Confirmation of elevated arsenic levels in groundwater of Myanmar. Sci Total Environ 478:21–24
Wasserman GA, Liu X, LoIacono NJ, Kline J, Factor-Litvak P, van Geen A et al (2014) A cross-sectional study of well water arsenic and child IQ in Maine schoolchildren. Environ Health 13 (1):23

Winkel L, Berg M, Amini M, Hug SJ, Johnson CA (2008) Predicting groundwater arsenic contamination in Southeast Asia from surface parameters. Nat Geosci 1(8):536–542
Yamamura S (2001) Drinking water guidelines and standards. In: United Nations Synthesis Report on Arsenic in Drinking Water (draft report). World Health Organization, Geneva
Yang N, Winkel LH, Johannesson KH (2014) Predicting geogenic arsenic contamination in shallow groundwater of South Louisiana, United States. Environ Sci Technol 48 (10):5660–5666

Chapter 11
Groundwater and Surface Water Interactions in Relation to Natural and Anthropogenic Environmental Changes

Mohammad Safeeq and Ali Fares

Abstract Groundwater and surface water interaction is an essential component of the hydrological cycle. The hydraulic connectivity and exchange of water between surface water (e.g. rivers, lakes, wetlands) and underlying aquifers provide many ecosystem services that sustain human and ecological well-being. Climate change, increased population, and industrial growth have resulted in substantial environmental (e.g. land use and land cover, climate, groundwater) changes across the globe. As a result, decline in groundwater levels, drying of streams, shrinking lakes, wetlands, and estuaries have been observed across the world. This generates concerns about the effects of such environmental changes on groundwater and surface water interactions, and on the quality and quantity of water resources. This chapter presents an overview of groundwater and surface water interactions, pressing environmental change issues centered on natural and anthropogenic environmental changes, and available management tools that quantify the integrated groundwater and surface water flow processes. This chapter also briefly discusses exciting research opportunities enabled by satellite remote sensing. We close in with a discussion of future management challenges and strategies for sustainable use of groundwater and surface water resources. One outcome of this chapter is to provide resource managers, researchers, consultant groups, and government agencies basic understanding of the types, mechanism, and effects of natural and anthropogenic landuse changes on groundwater and surface water interactions, and available management tools for studying groundwater and surface water interactions.

M. Safeeq (✉)
Sierra Nevada Research Institute, University of California at Merced, Merced, CA 95343, USA

USDA Forest Service, PSW Research Station, Fresno, CA 93710, USA
e-mail: msafeeq@ucmerced.edu

A. Fares
College of Agriculture and Human Sciences Prairie View, A&M University,
Prairie View, TX, USA

A. Fares (ed.), *Emerging Issues in Groundwater Resources*, Advances in Water Security, DOI 10.1007/978-3-319-32008-3_11

1 Introduction

Groundwater is an important natural resource that accounts for about 30 % of the total freshwater (Table 11.1). The extent and magnitude of groundwater occurrence largely depends on prevailing climate, land use and land cover, soil, as well as the underlying geology (Fig. 11.1). The excess precipitation (i.e. precipitation minus evapotranspiration) as well as natural recharge from surface water bodies (e.g. rivers and lakes) and artificial recharge during irrigation replenishes the underlying aquifers. The water stored in the groundwater is continuously moving and exchanging naturally with other surface water bodies (e.g. rivers, lakes, and wetlands). In a natural environment, total groundwater recharge is balanced by the release from the aquifers by evapotranspiration and exfiltration to rivers, springs, lakes, and marshes. However, like other components of hydrological cycle, the movement and exchange of groundwater can be altered and often accelerated by climate change and other anthropogenic means, such as groundwater pumping, changing land use and land cover, and artificial recharge.

Groundwater is a critical source of fresh water that sustains life and ecosystems (Stromberg et al. 1996; Kløve et al. 2011a; Gleeson et al. 2012). In many environments, natural discharge and pumping from groundwater sustains aquatic habitat, wildlife, and agriculture during dry periods and prolonged droughts. About two billion people worldwide depend on groundwater supplies, which include 273 transboundary aquifer systems (Puri and Aureli 2009). However, this vital freshwater resource is under stress from rapidly growing water demands and diminishing supply of clean water (Kummu et al. 2010; Wada et al. 2010, 2011a, b; Gleeson and Wada 2013). This stress will intensify with rapidly increasing global human population and altered hydrologic cycle as a result of global climate change (Vörösmarty et al. 2000; Oki and Kanae 2006; Haddeland et al. 2014). Overexploitation of groundwater resources may lead to various issues, such as reduced river flow and baseflow, drying of fresh water bodies (i.e. lakes and wetlands) and wells, saltwater intrusion, sea level rise, and land subsidence. Most importantly, this natural water reservoir will become increasingly unavailable during periods of low or no rainfall when we need it the most.

The issues surrounding groundwater have long been recognized but most of the studies are limited in scope and location (Taylor et al. 2013). This was in part due to a lack of reconcilable global data regarding hydrogeology and aquifer conditions. In 2008, the Intergovernmental Panel on Climate Change (IPCC) report emphasized the need to improve our understanding of groundwater under climate change (Bates et al. 2008). Since then, there has been a growing interest in the scientific community in investigating the issues related to sustainable groundwater management and related ecosystem services (Dragoni and Sukhija 2008; Aeschbach-Hertig and Gleeson 2012; Shi et al. 2012; Kløve et al. 2011b; Foster and MacDonald 2014; Khan et al. 2014; Gorelick and Zheng 2015). Over the past decade, advanced global earth system modeling, data assimilation, and satellite remote sensing techniques have considerably enhanced our understanding of groundwater resources and the

Table 11.1 Estimate of global water distribution (*source*: Shiklomanov 1993)

Water source	Water volume (10^3 km^3)	Percent of freshwater	Percent of total water
Oceans, seas, and bays	1,338,000	–	96.5
Ice caps, glaciers, and permanent snow	24,064	68.7	1.74
Ground water	23,400	–	1.7
Fresh groundwater	10,530	30.1	–
Saline/brackish groundwater	12,870	–	–
Soil moisture	16.5	0.05	0.001
Ground ice and permafrost	300	0.86	0.022
Lakes	176.4	–	0.013
Fresh	91.0	0.26	–
Saline	85.4	–	–
Atmosphere	12.9	0.04	0.001
Swamp water	11.47	0.03	0.0008
Rivers	2.12	0.006	0.0002
Biological water	1.12	0.003	0.0001
Total water	**1,385,984**	**–**	**100**
Total freshwater	**35,029**	**100**	**2.53**

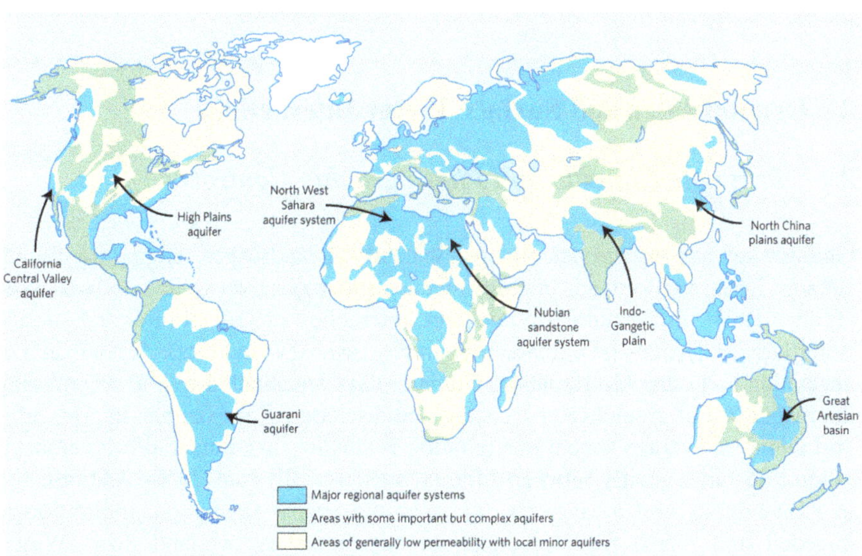

Fig. 11.1 Global aquifer systems (adopted from Taylor et al. 2012)

interactions between groundwater and surface water under climate and other environmental changes and human activities. In other words, our scientific knowledge of how groundwater responds to climate and other environmental changes is expanding rapidly (Earman and Dettinger 2011; Green et al. 2011; Gleeson et al. 2012; Holman et al. 2012; Wada et al. 2012; Fan et al. 2013; Taylor et al. 2013; Voss et al. 2013; Döll et al. 2014; Haddeland et al. 2014; Fan 2015).

As the stress on groundwater resources increases, there is a growing concern related to its impact on interrelated ecosystem services (Kløve et al. 2011b; Kløve et al. 2014; Costanza et al. 2014; Griebler and Avramov 2015). In particular, the degree to which interactions between groundwater and surface water that supports many ecosystem services is impacted by climate change (e.g. drought, flood, wildfire etc.) and other anthropogenic (e.g. land use change, deforestation, groundwater pumping etc.) activities (Tweed et al. 2009; Falke et al. 2011; Halford and Plume 2011; King et al. 2014; Adams et al. 2014; Caschetto et al. 2014; Kløve et al. 2014; Kurylyk et al. 2014; Wagner et al. 2014; Foster and Allen 2015; Tian et al. 2015). In this chapter, we aim to provide an overview of interaction between groundwater and surface water, potential role of natural and anthropogenic environmental changes on the interaction between groundwater and surface water, and discuss modeling and advancement in data acquisition techniques. Although there is a range of other activities (e.g. building levees, reservoirs, diversions, canals, hydropower, and dams) that may potentially affect the interaction between groundwater and surface water, we emphasize here the changes in land use and land cover and development of groundwater resources and aquifers, which are the most vulnerable to climate change.

2 Groundwater and Surface Water Interactions

2.1 *Principle Climatic and Physiographic Controls*

Groundwater and surface water interact in nearly all landscapes, ranging from small streams, lakes, and wetlands in headwater areas to major river valleys and seacoasts (Winter 1999). In a natural environment, interaction of groundwater and surface water largely depends on climate, topography, geology, and biotic factors such as vegetation and other bio-turbation. Climate plays an essential role in determining the magnitude of groundwater recharge and discharge. For example, in arid/semi-arid climates, average annual precipitation is smaller than potential evapotranspiration, and varies greatly between different seasons (Jolly et al. 2008). Additionally, in mountainous arid areas with dry soils and sparse vegetation, infiltration is impeded and runoff from precipitation can occur as overland flow (Winter et al. 1998). Under these conditions, groundwater recharge from surface infiltration is limited and seepage of water from rivers and exfiltration from groundwater into rivers dominate the groundwater and surface water interactions. In contrast, under humid/semi-humid climates, average annual precipitation exceeds average annual potential evapotranspiration, and leads to a net surplus of water for groundwater recharge, streamflow, and storage in other surface water bodies. In volumetric terms, the contribution of groundwater discharge to surface water could be much higher in humid/semi-humid climates than that in arid/semi-arid climates. However, in relative terms, the groundwater contribution to surface water in humid/semi-humid climates is typically less critical and lower than that in arid/semi-arid climates.

Water flow mechanisms and pathways in a landscape are governed by the topography. The mountainous landscapes are often referred to as the natural "water towers" (Viviroli and Weingartner 2008) that feed streams and serve as natural recharge areas to groundwater. The high gradient and coarse textured streambed in mountainous landscapes increase the "hyporheic exchange" across the landscape (Winter et al. 1998). The "hyporheic exchange" is defined as the mixing of surface water and shallow groundwater through, beneath, and alongside a riverbed (Tonina and Buffington 2009). In mountainous environments recharge is often dominated by the snowmelt that moves quickly through the shallow subsurface due to steep hydraulic gradient. The steep valley side and flat-floored alluvial valley often separate the groundwater recharge and discharge areas. As a result, during the rainy and snowmelt seasons water tables in lower parts of the mountainous landscapes rise quickly, and create fens and feed streams.

Underlying geology and rock type controls permeability (i.e. the ease with which a fluid moves through a material), which in turn controls the water flow pathways and discharge patterns to streams (Grant 2007; Tague and Grant 2009; Jefferson et al. 2008). Permeability varies over 13 orders of magnitude across rock and sediment types because of differences in pore size, geometry, and connectedness (Freeze and Cherry 1979; Gleeson et al. 2011). Generally, landscapes with very high permeability porous media produce groundwater fed streams (Fig. 11.2a) with muted hydrographs and very high baseflow (Tague et al. 2007; Tague and Grant 2009; Jefferson et al. 2008). Groundwater and surface water interaction is prominent in these types of landscapes. In contrast, streams in landscapes with low permeability porous media can be described as flashy with low baseflow (Fig. 11.2b). These flashy streams and other surface water bodies (e.g. fens, bogs etc.) have very limited groundwater and surface water interaction and often go dry during the dry season. In addition to the magnitude of groundwater contribution to total discharge, geology dictates the stream water chemistry through the geochemical interactions of groundwater with minerals in unconsolidated sediments, soils, and fractured bedrock. Groundwater fed streams stay colder as compared to those streams that are primarily generated from surface runoff (Tague et al. 2007) and provide habitat for many cold water species (Eaton and Scheller 1996; Jackson et al. 2001).

Land use and land cover also dictates the magnitude and direction of the groundwater and surface water interaction by modulating the hydrologic cycle. In fact, land use and land cover types not only affect the partitioning of groundwater and surface water by altering the rate of infiltration and thus the balance between runoff and recharge, but also the total amount, timing, and quality of water available in the stream and other water bodies. For example, in the southwestern US, groundwater recharge related to land use and land cover ranges from <0.1 mm per year in natural rangeland to 130–640 mm per year in irrigated

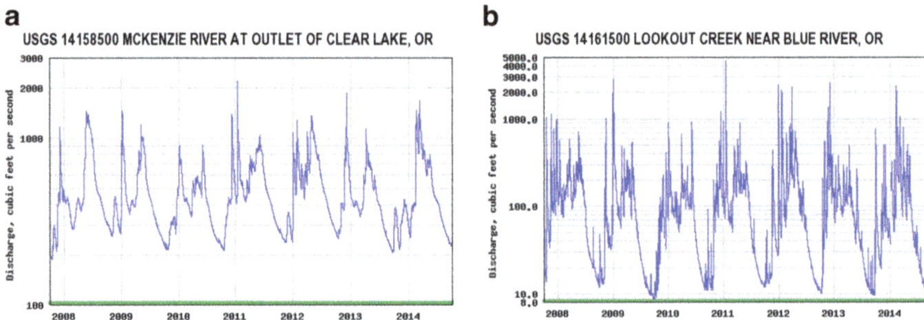

Fig. 11.2 An illustration of a (**a**) groundwater fed stream with muted hydrograph and high baseflow, and (**b**) surface water fed stream with flashy hydrograph and lower baseflow

agriculture ecosystems (Scanlon et al. 2005). On a global scale, Scanlon et al. (2007) showed higher groundwater recharge (two orders of magnitude) and streamflow (one order of magnitude) through crops than through original native forests, but of lower water quality due to mobilization of salts and salinization caused by shallow water tables. Hardison et al. (2009) showed an inverse relationship between catchment total impervious area and riparian groundwater level. They also showed an increase in channel incision with increasing catchment total impervious area and storm water runoff. Similarly, land use and land cover around riparian zone affects the hyporheic exchange and stream ecosystem metabolism (Hancock 2002). Land use and land cover also influence the magnitude of peak discharge or floods by modifying how rainfall and snowmelt are stored on, and run off, the land surface into streams (Konrad 2003). In fact, at smaller spatial scales, the impact of land use on floods supersedes the impact of climate variability (Blöschl et al. 2007).

2.2 *Mechanisms and Dominant Types*

Groundwater and surface water interactions occur by surface lateral flow through the unsaturated zone and by infiltration in to, or exfiltration from, the saturated zone (Sophocleous 2002). Following a recharge event either as rain or snowmelt, water can take different pathways before it reaches to the groundwater and stream channel. Water can move to streams quickly through infiltration-excess and saturation-excess overland flow. Subsurface flow can also enter streams quickly through interflow (near-surface flow of water within the soil profile) and return flow (subsurface water that returns to overland flow). Water can also enter streams slowly through baseflow from delayed subsurface flow and groundwater. As highlighted by Sophocleous (2002), translatory flow or piston flow (Hewlett and Hibbert 1967) and groundwater ridging (Sklash and Farvolden 1979) are a few other mechanisms by which a stream can respond to a recharge signal rather quickly. Translatory or piston flow is described as displacement of pre-event water from the soil pore space due to the increasing pressure of infiltrating new water (Lischeid et al. 2002), whereas groundwater ridging describes the large and rapid increases in hydraulic head in groundwater during storm periods (Sophocleous 2002). Translatory flow occurs where infiltration capacities far exceed the maximum rates of rainfall or snowmelt so that all water enters the soil (Harr 1976). In contrast, groundwater ridging occurs on hill slopes where the depth to groundwater table is closer to surface near the stream than it is further away from the stream.

Identification of the dominant water flow processes leading to the groundwater and surface water interactions is rather complex. Groundwater and surface water interactions may be dominated by a single process or by a combination of processes depending on the magnitude of recharge, the antecedent soil moisture conditions, the soil hydraulic properties, and the subsurface geology and aquifer characteristics. For example, sandy soil catchments show a higher proportion of baseflow than the corresponding loamy soil catchments (Andersen et al. 2001). Tague and Grant (2004) show a strong correspondence between the magnitude of baseflow and underlying catchment geology. Similarly, catchments with presence of macropores or fractures may be dominated by translatory and pipe flows (Harr 1976; Rossi et al. 2012). In the same catchment, one dominant runoff process can be replaced by another depending on the magnitude of recharge (Uchida et al. 2002).

Groundwater flow through the streambed into the stream (i.e. effluent or gaining stream) and stream water infiltrating through the streambed into the groundwater (i.e. influent or losing stream) are the two dominant types of groundwater and surface water interactions (Kalbus et al. 2006). However, groundwater and surface water interaction could also occur by flow-through and parallel-flow (Winter et al. 1998; Woessner 1998, 2000). In gaining streams, the groundwater table is higher than the water level in the stream (Fig. 11.3a). Conversely, in losing streams the groundwater is continuously connected to stream water by a saturated zone (Fig. 11.3b), and at other times groundwater and stream water are completely disconnected by an unsaturated zone (Fig. 11.3c). However, in both cases, the groundwater table is always lower than the water level in the stream (Gordon et al. 1992). A stream can gain and loose water simultaneously by flow-through when the groundwater table is higher than the water level in the stream on one bank and is lower than the water level at the opposite bank (Woessner 2000). In streams where the groundwater table and water level in the stream are equal, parallel-flow can occur where the stream is neither losing nor gaining.

In many environments, these types of groundwater and surface water connectivity can often occur in succession, and a stream may switch back and forth between losing or gaining (e.g. Safeeq and Fares 2012) depending on the hydraulic head, orientation of stream and groundwater flow field, and precipitation patterns. Rapid rise in stream level due to heavy precipitation, snowmelt, and reservoir release may create temporary bank storage (Fig. 11.3d). Similarly, temporary (e.g. beaver dams and log jams) or permanent (e.g. hydroelectric and flood control dams) water control structures could also elevate the stream water level and create bank storage or even overbank flooding. Stream bank storage is often transient and usually returns to the stream within few days or weeks. However, in the case of overbank flooding, recharge water throughout the floodplain can take months and

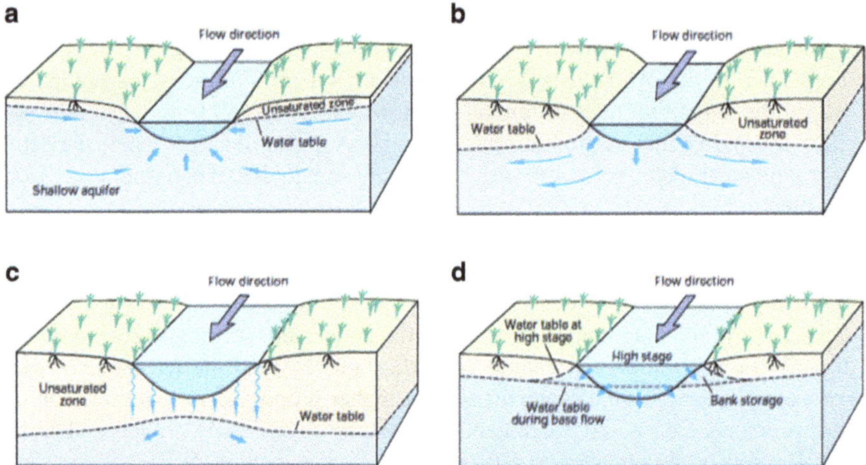

Fig. 11.3 Interactions of groundwater and surface water in a (**a**) gaining stream, (**b**) losing stream, (**c**) disconnected stream, and (**d**) stream with bank storage (adopted from Winter et al. 1998)

often years depending on the groundwater flow paths in the floodplain (Winter et al. 1998).

Similar to streams and rivers, other surface water bodies like lakes and wetlands can receive and recharge groundwater (Fig. 11.4). Although, the groundwater and surface water interaction in lakes and wetlands are similar to those in streams, the dominant processes are quite different. For example, water table in lakes and wetlands are relatively constant as opposed to streams. Additionally, in lakes and wetlands direct precipitation and evapotranspiration have a much bigger impact on groundwater and surface water interaction than those in streams. Some lakes and wetlands either receive (Fig. 11.4a) or recharge (Fig. 11.4b) groundwater throughout their entire bed. However, most lakes and wetlands simultaneously receive groundwater inflow from one portion of the bed on one side of the shoreline and recharge groundwater through another portion of the bed on the opposite side of the shoreline (Fig. 11.4c), similar to flow-through in streams (Alley et al. 1999). However, as pointed out by Winter et al. (1998), the major differences between lakes and wetlands in terms of groundwater and surface water interaction are in the amplitude and frequency of water-level fluctuations, as well as the ease with which water moves through their beds due to the differences in the presence of rooted vegetation in wetlands (Fig. 11.5).

The degree to which surface water bodies can receive or recharge groundwater depends on a number of factors including geomorphological position, hydraulic head, underlying bed geology and hydraulic conductivity, climate, and connectivity to other surface water bodies (e.g. local aquifers, streams, and rivers). A wetland can be entirely located on a nearly flat land surface instead of occupying low points and depression as in the case of streams and lakes (Fig. 11.5a). However, a wetland can also occupy slopes (Fig. 11.5b), depressions along the stream as fens (Fig. 11.5c) or upland depressions as bogs (Fig. 11.5d). A wetland located in

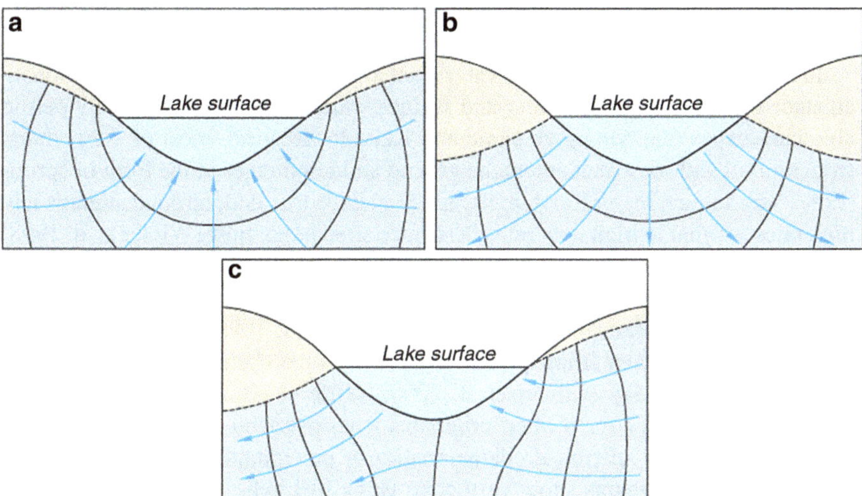

Fig. 11.4 Groundwater and surface water interactions in lakes where they can (**a**) receive groundwater inflow, (**b**) lose water as seepage to groundwater, or (**c**) receive and lose groundwater at the same time (adopted from Winter et al. 1998)

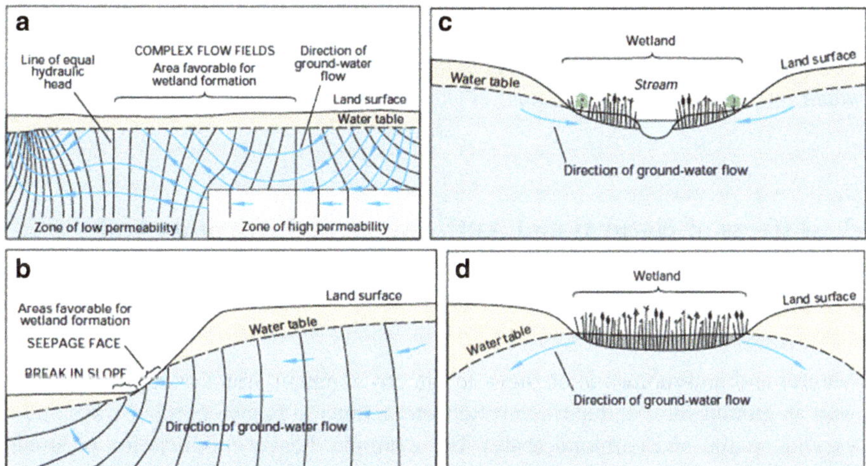

Fig. 11.5 Groundwater and surface water interaction in wetland ecosystems where (**a**) underlain groundwater can feed wetland due to complex groundwater flow field, (**b**) groundwater discharge at seepage faces and at breaks in slope of the water table can feed wetlands, (**c**) the riverine wetland is adjacent to a gaining stream, or (**d**) the wetland is directly sourced from precipitation with no stream connectivity and the groundwater gradient is away from wetland (adopted from Winter et al. 1998)

riverine environment (Fig. 11.5c) will experience frequent water level changes that will affect its hydraulic/hydrologic characteristics and wetland-groundwater interactions. Similarly, groundwater and surface water interaction will be limited if a lake is underlain by low permeability till and receives groundwater only from a local flow system or if most of the groundwater passes beneath the lake.

In many environments, groundwater discharge through springs can significantly enhance the level of groundwater and surface water interactions. In highly permeable landscapes (e.g. young volcanic and Karst landscapes), most of the recharge from rainfall and snowmelt enters the ground and re-emerges in the form of springs (Fig. 11.6). In fact, many Karst landscapes can have true underground streams with high rates (similar to high flow rates in surface streams) of flow (Winter et al. 1998). Depending on the flow paths (shallow or deep), water from these springs alter the flow timing (Jefferson et al. 2006, 2007), stream water temperature (Tague et al. 2007), and stream biogeochemistry (Burns et al. 1998; Cohen et al. 2013). The slow moving water from springs often reflects the chemical characteristics of the underlying geology (Olivier et al. 2008; Carrie et al. 2015). In contrast, fast moving water in the stream often originates from overland or shallow subsurface runoff, reflecting the chemical characteristics of precipitation and soil. Similarly, spring-dominated streams show very contrasting hydraulic and geomorphic characteristics as compared to runoff-dominated streams (Manga 1996; Sear et al. 1999). For example, spring-dominated streams have a poorly defined floodplain, near bankfull summer baseflow, and experience stable woody debris accumulation (Tague et al. 2007). In contrast, under a very similar environment, runoff-dominated streams often have well-developed floodplains, lower summer baseflow, exposed cobble and boulder bars, and unstable woody debris accumulation (Grant 2007). These differences indicate the degree of groundwater and surface water interaction in these streams.

3 Effects of Natural and Anthropogenic Environmental Changes on Groundwater and Surface Water Interactions

Natural and anthropogenic changes in the environment affect water resources, as well as groundwater and surface water interactions in many complex ways and at varying spatial and temporal scales. For example, beaver modification of stream channel or log jams following flooding can have very localized effects on groundwater and surface water interaction (Westbrook et al. 2006). However, the effects of a hydroelectric dam on groundwater and surface water interaction can often expand across basins and countries. Similarly, the effects of climate change and deforestation on the hydrologic cycle can impact the groundwater and surface water interaction at local and regional scales with varying degrees of complexities. Hence, our goal here is to provide an overview of some of the most pressing environmental change issues (natural as well as anthropogenic) in the context of groundwater and surface water interaction.

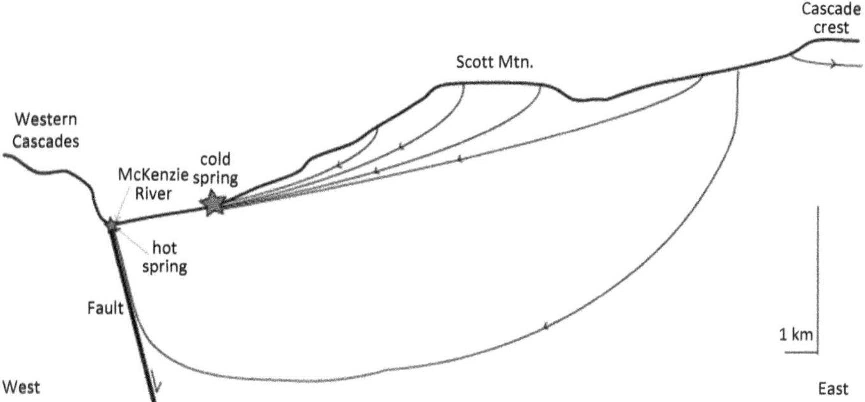

Fig. 11.6 An illustration of springs flow paths in young volcanic landscape of Oregon cascades (adopted from Jefferson et al. 2006)

3.1 Natural Environmental Changes

The nature of climate (i.e. precipitation and temperature variability) and underlying geology dictates the landuse patterns as well as characteristics (e.g. size, persistence, functioning) of surface waters. In the US, a majority of the wetlands are located in the glacial terrain of the north-central US, coastal terrain along the Atlantic and Gulf coasts, and riverine terrain in the lower Mississippi River Valley (Winter et al. 1998). With some exceptions, many of these surface water bodies are sourced from direct precipitation or subsequent runoff, making them vulnerable to inter-annual and inter-decadal precipitation variability. Many streams and wetlands often go dry during the low precipitation season. Episodes of drought alters the upstream land cover through forest fire, insect and drought related tree mortality, physically knocking over trees, growth inhibition, and other chronic problems due to poor soil aeration and other environmental changes (e.g. invasive species invasion, erosion and landslides etc.). Species distribution patterns are not only shaped by the drought (Engelbrecht et al. 2007), but interaction of fire and drought can also cause shifts in tree species composition (Moser et al. 2010; Phillips et al. 2010). These changes in landuse deteriorate the surface water quality and water balance, leading to reduced storage and increased overland flow in the subsequent years post fire (Wondzell and King 2003; Onda et al. 2008). Post fire clogging of preferential flow paths by ash and subsequent reduction in subsurface water storage has also been reported (Onda et al. 2008). Alencar et al. (2006) reported that the area burned by forest fire in the Brazilian Amazon forest during the severe drought year was 13 times greater than the area burned during the average rainfall year, and twice the area of annual deforestation. Wild forest fires not only destroy the riparian vegetation (Neary et al. 2005), but also enhance sediment transport (Rulli and Rosso 2005), debris flows (Wells 1987), and other geomorphic processes (Swanson 1981). The post fire increase in ash and sediment can limit the groundwater and surface water interactions by altering the hydrologic and geomorphic properties of riverine and wetland environments (Neary et al. 2005; Buffington and Tonina 2009). Like wild fires, droughts can severely impact natural aquatic ecosystems, including riparian and wetland vegetation and biogeochemical processes. Williamson et al. (2001) reported slightly higher (1.91 % during the drought compared to 1.12 % prior to drought) tree mortality rate during the 1997 drought in the Amazon Basin. As drought builds, baseflow declines (Fendeková and Fendek 2012), streams start desiccating (Jaeger et al. 2014), and the volume of water in wetlands and lakes drops due to reduce inflow and accelerated evaporation (Gaeta et al. 2014). This may cause a contraction in the wetted area of streambed and wetlands, and isolate marginal habitats (Bond et al. 2008).

3.2 Anthropogenic Environmental Changes

Vegetation controls large fractions of the water balance from the land surface, including both evapotranspiration and the magnitude and distribution of total precipitation reaching the ground either in the form of rain or snow. In extreme arid climates, evapotranspiration alone could account for >80 % of the total precipitation (Lu et al. 2003; Sanford and Selnick 2013). Similarly, foliage in temperate forests is widely reported to intercept and evaporate 10–40 % of annual rainfall (Maidment 1993). In tropical forests, canopy interception accounts for as much as 45 % of the rainfall, and under dry conditions the initial 2.5 mm of rainfall never hits the ground surface (Safeeq and Fares 2014). The evaporated water from forest landscapes affects the hydrologic cycle by increasing the atmospheric moisture vapor and raising the likelihood of precipitation events and increasing water yield (Ellison et al. 2012). In cold climates, forest canopies also modulate the magnitude and timing of snow accumulation and melt (Lundquist et al. 2013). Forests are also known to enhance soil moisture storage and infiltration, and reduce quick flow. Through these processes, forests recharge groundwater, maintain summer baseflow, and regulate peak flows during heavy rainfall.

3.2.1 Forest Disturbances

But even as the vegetation is modulating hydrologic response, it too is changing in hydrologically relevant ways (Cramer et al. 2001; Donohue et al. 2013). The world's forest vegetation structure and composition is changing due to climate change, deforestation, fire, and forest management (Westerling et al. 2006; Hansen et al. 2013). With a warming climate, increased fire and drought related forest mortalities have been observed and are only expected to further intensify as climates continue to change throughout the next century (Westerling et al. 2006; Williams et al. 2010, 2013). Likelihood of forests and rangelands invasion by nonnative plant species under climate change, which have already threatened the

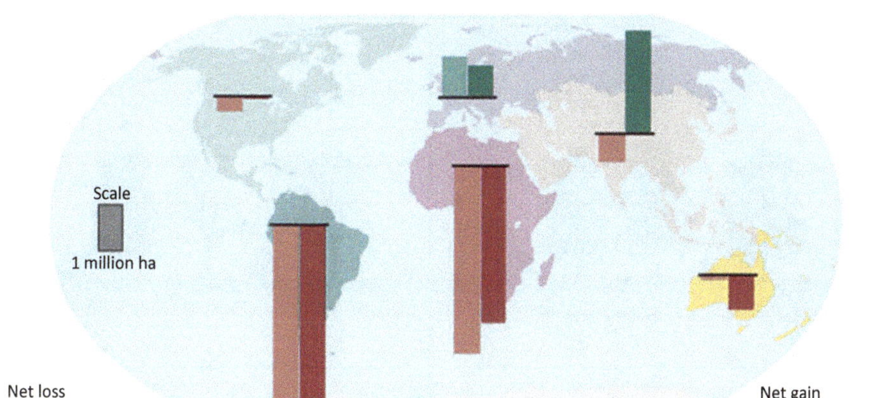

Fig. 11.7 Annual change in forest area between 1990–2000 and 2000–2010 (adopted from FAO 2010)

hydrology and ecology of many landscapes (Gordon 1998; Levine et al. 2003; Safeeq and Fares 2014), may be magnified. Anthropogenic land use change due to deforestation has been observed across the globe (Fig. 11.7). Although, in some parts of the world the rate of deforestation has declined over time, some 13 million hectares of forests were still converted annually to other uses (FAO 2010). These drivers of vegetation change are shifting the frequency, size, and type of forest disturbances, and will have a profound impact on energy, carbon, and water cycles (Fig. 11.8).

A change in climate would be expected to shift plant distribution as species expand in newly favorable areas and decline in increasingly hostile locations at a rate between 11 m/decade in lower latitudes to as much as 17/decade in higher latitudes (Kelly and Goulden 2008; Chen et al. 2011a). As the temperature continues to warm, vegetation will shift from water deficit lower elevation to currently energy limited higher elevation. In addition to the elevational range shift, vegetation compositions are also expected to change as climate continues to warm (Lenihan et al. 2003; Rustad et al. 2012; Jiang et al. 2013). In the state of California, a shift in dominance (from needle-leaved to broad-leaved life-forms) and increases in vegetation productivity, especially in the relatively cool and mesic regions of the state are predicted (Lenihan et al. 2003). Lenihan et al. (2003) also predicted greater coverage of mixed evergreen forest, especially along the western slope of the Sierra Nevada where the baseline maps shows evergreen conifer forest. These changes in vegetation are not simply an expansion and contraction of species' ranges, but rather a change in complex linkages that affect the soil–plant–atmosphere

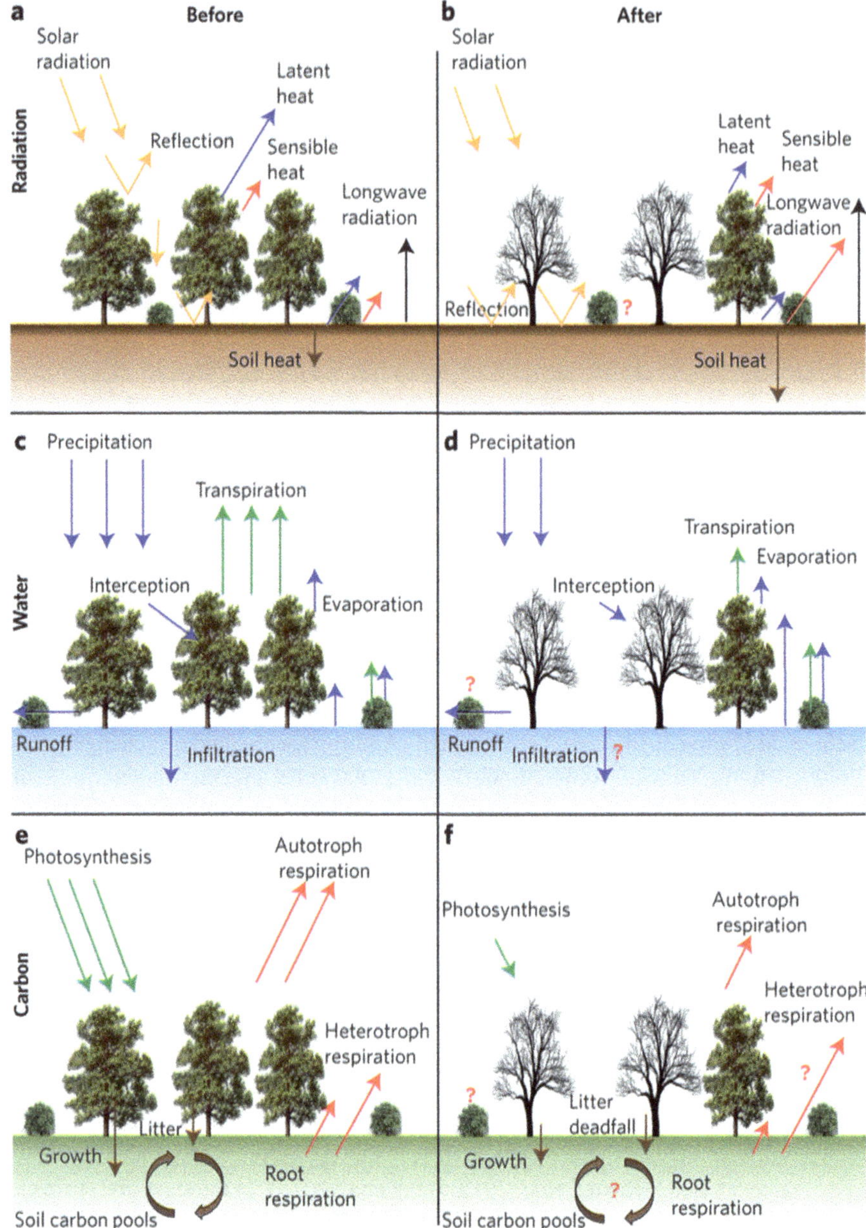

Fig. 11.8 Fluxes of radiation, water and carbon before and after widespread forest die-off (adopted from Anderegg et al. 2013)

continuum and dependent biophysical processes (Scott et al. 2014). For example, vegetation expansion to higher elevations may results in higher evapotranspiration and lower discharge (Goulden and Bales 2014) but contraction of woody plants may result in higher water yield. Similarly, changes in vegetation growth patterns in response to enhanced CO_2 in the atmosphere may result in higher plant water use efficiency (Donohue et al. 2013; Friend et al. 2014).

The changes in water balance and water quality associated with forest disturbances can influence groundwater and surface water interactions in several ways by altering the hydrologic cycle (Fig. 11.8). An increase or decrease in groundwater and/or surface water quantity and quality will not only affect the hyporheic exchange, but also the wetlands and lakes that depend on it. Following forest harvest, changes in the water table and stream baseflow are expected to follow, but the magnitude and duration of this increase vary according to the climate, forest type, geology, and topography of the landscape (Smerdon et al. 2009; Sørensen et al. 2009; Adams et al. 2012). A similar change in water table and groundwater contributions to streams are also expected under forest die-offs (Adams et al. 2012; Anderegg et al. 2013; Bearup et al. 2014). At event scale, forest harvesting also modifies the runoff pathways by which water flows to the stream channel (Moore and Wondzell 2005). Declines in canopy interception could lead to higher water tables due to an increase in the amount of water infiltrating the soil during storms (Dhakal and Sidle 2004). Increases in water tables, along with higher antecedent soil moisture conditions due to reduced evapotranspiration, following forest harvest or forest die-offs could result in higher quick flow. Changes in channel morphology (e.g. woody debris, fine sediment deposition, channel incision etc.) could substantially influence stream-aquifer interactions (Moore and Wondzell 2005). Intrusion of fines into streambed sediment and clogging of bed materials has been shown to decrease the magnitude of hyporheic exchange (Schalchli 1992; Packman and MacKay 2003). Changes in vegetation and invasive species introduction might further alter stream–forest and subsequently stream-groundwater interactions, especially in the riparian environment, due to higher water use by the invasive species as compared to native species (Cavaleri and Sack 2010).

3.2.2 Conversion of Wetlands to Agriculture

Agriculture can affect wetlands and riparian areas in many ways including nonpoint source loading from agricultural runoff and/or clearing of wetlands and riparian areas for agricultural production. On average, the world has lost 54–57 % of its wetlands since 1700 AD and 64–71 % of wetlands since 1900 AD (Davidson 2014). In the US, after rapid (53 %) decline in wetlands between 1780 and 1980, mainly due to the conversion of wetlands to agriculture, wetland losses finally outdistanced wetland gains between 2004 and 2009 (Dahl 2011). The rate of wetland loss in Europe has also declined but the wetland loss rate has remained high in Asia, where large-scale and rapid conversion of coastal and inland natural wetlands to agriculture and urban use is continuing (Davidson 2014). On global scale, the area of land occupied by surface water bodies (i.e. wetlands, lakes, swamps, marshes etc.) has declined by 6 % during the 1993–2007 period, with the largest decline of surface water coinciding with the largest increase in population (Prigent et al. 2012). Although in the past, losses have been larger and faster for inland wetlands than coastal natural wetlands (Davidson 2014). During recent times, the loss of coastal wetlands has speeded up (Dahl 2011), and the sea level rise in response to climate change may potentially reduce coastal wetlands further by 5–20 % (Nicholls 2004).

As discussed previously, processes behind exchange of groundwater and surface water in wetlands are highly variable—both in space and time. As a result, understanding of process based impacts of wetland to agriculture land conversion on groundwater and surface water exchange can be complex and may vary significantly by wetland types (Bullock and Acreman 2003). In general, conversion of wetlands to agriculture and other uses are known to affect the flood attenuation, groundwater recharge and discharge, and water quality (Reddy et al. 1999). The organic matter rich hydric soil acts as a sponge and help wetlands in retaining water during the wet season and releasing it slowly into neighboring rivers during the dry season as baseflow. However, not all wetland types support the generalized model of reduced floods, recharge groundwater, and augment low flows (Bullock and Acreman 2003). Hence, the precise nature of wetland to agriculture conversion impacts on the interaction between groundwater and surface water will depend on the wetland type and nature of water transfer mechanisms.

3.2.3 Groundwater Development

Increasing water demand and climate change has already threatened the global groundwater resources. Human groundwater withdrawal for drinking water supply, irrigation, and industrial uses has exceeded the natural recharge, especially in sub-humid to arid areas (Fig. 11.8a–c). Between 1960 and 2000 there has been a twofold increase in global groundwater abstraction (Wada et al. 2011a), causing a range of issues including water table decline (Rodell et al. 2009; Voss et al. 2013), land subsidence (Teatini et al. 2006; Calderhead et al. 2011), streamflow depletion (Safeeq and Fares 2012; Barlow and Leake 2012), salinization and saltwater intrusion (Todd and Mays 2005; Barlow and Reichard 2010), and contributing to sea level rise (Konikov and Likhodedova 2011; Wada et al. 2012).

Groundwater depletion may also conflict with lakes and wetlands (Llamas 1988; Van der Kamp and Hayashi 1998; Patten et al. 2008), and threatens many riparian ecosystems in arid and semi-arid regions of the world (Stromberg et al. 1996; Scott et al. 1999). Human alteration of hydraulic head by groundwater pumping or groundwater abstraction by tunnels may induce a "losing" condition, especially when the water table drawdown starts intersecting with water level in the stream (Safeeq and Fares 2012; Barlow and Leake 2012). Similar to the streams, changes in groundwater flow patterns due to groundwater pumping may alter groundwater and surface water interactions in lakes and wetlands. Since the quantity of water and chemical balances determine the principal characteristics and functions of

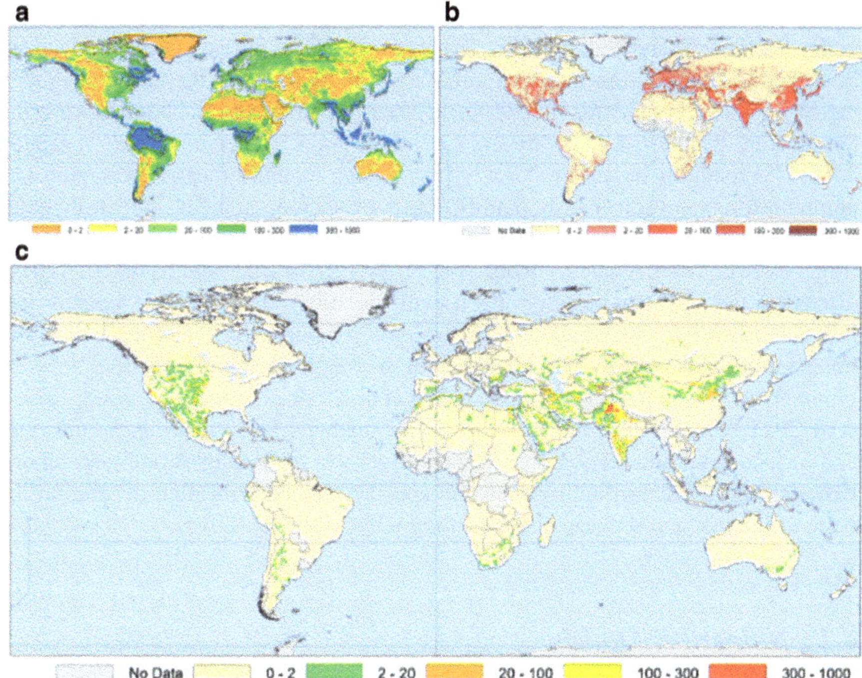

Fig. 11.9 (**a**) Simulated average groundwater recharge, (**b**) total groundwater abstraction for the year 2000 and (**c**) groundwater depletion for the year 2000 (all in mm per year) (adopted from Wada et al. 2010)

wetlands, alteration to the water quantity or quality balance negatively affects wetland functioning and characteristics, which in turn affects groundwater and surface water interaction. The most striking groundwater withdrawal trends were observed in Asia (Fig. 11.9), where wetland loss rate has remained high, which could further alter the characteristics and functioning of the region's wetlands and other surface water bodies.

4 Overview of Management Tools

4.1 Measuring Groundwater and Surface Water Interactions

Techniques of measuring interaction between groundwater and surface water are quite variable within the water resources research community. Harvey and Wagner (2000) and Kalbus et al. (2006) provided overviews of the methods that are currently used for measuring interactions between groundwater and surface water at various scales (Fig. 11.10). Kalbus et al. (2006) grouped these methods into

(1) direct measurements of water flux using a seepage meter (Lee 1977), heat pulse meter (Taniguchi and Fukuo 1993; Krupa et al. 1998), ultrasonic meter (Paulsen et al. 2001), dye-dilution meter (Sholkovitz et al. 2003), and the electromagnetic meter (Rosenberry and Morin 2004); (2) heat tracer methods by measuring the temperature profile (Stonestrom and Constantz 2003; Anderson 2005; Constantz 2008); (3) methods based on Darcy's Law, which includes piezometers, slug and pumping tests, and tracer tests; and (4) mass balance approaches, which includes hydrograph separation, environmental and heat tracers, monitoring wells, etc. (Fig. 11.10). More recently, Robinson et al. (2008) and Singha et al. (2014) discussed the opportunities for the use of electrical and magnetic geophysical instrumentation in watershed scale hydrology. Among these, electrical resistivity imaging has been particularly successful in studying the interaction between groundwater and surface water (Henderson et al. 2010; Nyquist et al. 2008; Coscia et al. 2011, 2012). All of the aforementioned methods for measuring groundwater and surface water interaction provide powerful tools for managers and researchers. However, the choice of appropriate and specific method will often depend on the scale of interest and process understanding (Bertrand et al. 2014).

4.2 Numerical Models

The development of computer based numerical models has served as a management and scientific tool for investigating the groundwater and surface water interactions. Beckers et al. (2009) conducted a comprehensive hydrologic model review with the aim to identify trade-offs between model complexity and model functionality, and guide model selection. Although, the review was conducted with the focus on climate change and forest management, many models described by Beckers et al. (2009) are capable of simulating groundwater and surface water interaction. More recently, AquaResource Inc. (2011) and Maxwell et al. (2014) conducted an integrated groundwater and surface water model review and inter-comparision, respectively. Here, we briefly summarize strengths and weaknesses of selected models from Beckers et al. (2009), AquaResource Inc. (2011) and Maxwell et al. (2014), along with a few other models that are frequently used by researchers, resource managers, and other practitioners (Table 11.2). These models are described as follows:

4.2.1 Modular Groundwater Flow (MODFLOW) Model

MODFLOW (Harbaugh 2005) is the United States Geological Survey (USGS) three-dimensional finite-difference groundwater model with capabilities to simulate solute transport, variable-density flow (including saltwater), aquifer-system compaction and land subsidence, parameter estimation, and groundwater management. MODFLOW has been extensively used for simulating hyporheic exchange in

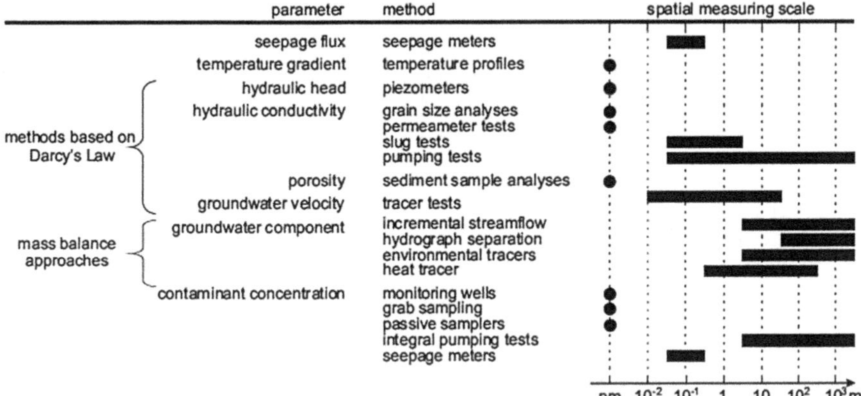

Fig. 11.10 Spatial measuring scales of the different methods to measure the interaction between groundwater and surface water. The spatial scale is given as radius or distance of influence. *Dots* represent point measurements (pm) (adopted from Kalbus et al. 2006)

streams (Cardenas 2015). However, MODFLOW often requires coupling of streamflow packages for simulating groundwater and surface water interactions (Brunner et al. 2010; Han and Endreny 2014). MODFLOW utilizes the Streamflow-Routing (SFR2) package (Niswonger and Prudic 2005) to simulate the interaction between groundwater and surface water. Streamflow is routed based on the continuity equation, assuming steady and uniform flow such that the volumetric inflow is equal to the outflow minus all sources and sinks to the channel (Niswonger and Prudic 2005). In addition to SFR2, which is an improvement over SFR1 (Prudic et al. 2004) and the STReamflow (STR1) routing package (Prudic 1989), MODFLOW has been coupled with the Branch-Network Dynamic Flow (BRANCH; Schaffranek et al. 1981) model (MODBRNCH; Swain and Wexler 1996) for studying groundwater-surface water interaction. Similar to SFR2, BRANCH uses hydraulic conductivity of streambed and differences in aquifer and stream stage to describe leakage between streams and aquifers. However, unlike SFR2, which runs within MODFLOW, simulation of seepage can be performed independently using BRANCH and imported into MODFLOW. This flexibility allows MODFLOW and BRANCH to operate in different time steps, which could be essential for accurately capturing the time lag between groundwater flow and streamflow. Scibek et al. (2007) successfully applied the coupled BRANCH and MODFLOW to study the groundwater and surface water interaction under scenarios of climate change. Similarly, Krause and Bronstert (2005, 2007) and Kraus et al. (2007) used the integrated water balance and nutrient dynamics model IWAN by coupling MODFLOW and a spatially distributed deterministic hydrological model (WASIM-ETH-I, Schulla 1997) to simulate the groundwater and surface water interactions in a lowland floodplain. Jobson and Harbaugh (1999) coupled MODFLOW with the Diffusion Analogy Surface-Water Flow (DAFLOW) Model that has been used to simulate the groundwater and surface water interaction

Table 11.2 Summary of coupled groundwater and surface water models (modified from Spanoudaki et al. 2009)

Model	Surface water flow	Groundwater flow	Coupling method	Major drawback
Streamflow Routing Module (SFR1/ SFR2), Prudic et al. (2004), Niswonger and Prudic (2005)	1-D steady and uniform stream flow, streamflow routing based on continuity equation	3-D saturated (Modflow)	Iterative coupling	Only suited for stream-aquifer interaction
MODBRANCH, Swain and Wexler (1996)	1-D streamflow, Saint Venant equations (Branch model)	3-D saturated (Modflow)	Iterative coupling	Only suited for stream-aquifer interaction
IWAN, Krause and Bronstert (2005, 2007)	2-D, TOPMODEL based, spatially distributed hydro-logical model, streamflow routing with kine-matic wave approach (WASIM-ETH-I)	3-D saturated (Modflow)	Iterative coupling	Streamflow is not simulated, only vertical fluxes (uptake and recharge) are coupled
MODFLOW/ DAFLOW, Jobson and Harbaugh (1999)	1-D streamflow, diffusion wave approximation to the Saint Venant equations (DAFLOW)	3-D saturated (Modflow)	Iterative coupling	Only suited for stream-aquifer interaction
GSFLOW, Markstrom et al. (2008)	2-D Determinis-tic, distributed-parameter water-shed model, streamflow routing is based on: (1) 1-D steady and uniform streamflow, streamflow routing based on continuity equa-tion, (2) -kinematic-wave equation that approximates the Saint-Venant equations (PRMS)	3-D saturated (Modflow)	Iterative coupling	GSFLOW operates at daily time step and does not support adaptive time stepping
InHM, VanderKwaak (1999)	1-D streamflow and 2-D overland flow, diffusion	3-D variably saturated,	Simultaneous simulation of groundwater	Lack of SVAT processes

(continued)

Table 11.2 (continued)

Model	Surface water flow	Groundwater flow	Coupling method	Major drawback
	wave approximation to the Saint Venant equations	Richards equation	and surface water flow	
HydroGeoSphere, Therrien et al. (2010)	1-D streamflow and 2-D overland flow, diffusion wave approximation to the Saint Venant equations	3-D variably saturated, modified Richards equation	Simultaneous simulation of groundwater and surface water flow	Source-code is proprietary; lack of flexibility in representing various hydrological processes at different levels of complexity
MODHMS, Panday and Huyakorn (2004)	1-D streamflow and 2-D overland flow, diffusion wave approximation to the Saint Venant equations	3-D variably saturated, Richards equation	Simultaneous simulation of groundwater and surface water flow	No explicit representation of vegetation and snowpack evolution; source-code is proprietary; lack of flexibility in representing various hydrological processes at different levels of complexity
MIKE-SHE, Graham and Refsgaard (2001)	1-D streamflow and 2-D overland flow, diffusion and kinematic wave approximation to the Saint Venant equations, respectively	3-D saturated and 1-D unsaturated using Richards equation	Non-iterative coupling	Source-code is proprietary; assumes no horizontal flow in the unsaturated zone
ParFlow.CLM, Ashby and Falgout (1996), Maxwell and Miller (2005)	2D streamflow using kinematic-wave approximation of the St. Venant equations	3-D variably saturated, Richards equation (ParFlow)	Simultaneous simulation of groundwater and surface water flow	Primarily a research code with limited user support

in ephemeral streams (Heidbüchel 2007). The coupled MODFLOW/DAFLOW simulates water exchange for each sub-reach on the basis of the stream-aquifer head difference between stream and aquifer, stream geometry (thickness and width), and hydraulic conductivity of the streambed.

4.2.2 Groundwater and Surface Water Flow (GSFLOW) Model

GSFLOW (Markstrom et al. 2008) is a coupled Groundwater and Surface-water FLOW model based on the integration of the USGS Precipitation-Runoff Modeling System (PRMS; Leavesley et al. 1983, 2005) and MODFLOW. PRMS is a hydrologic response units (HRU) based deterministic and distributed parameters watershed-modeling system for simulating streamflow, snowpack evolution, evapotranspiration, soil moisture storage, and groundwater recharge and discharge. The PRMS model serves as the land surface component of the GSFLOW. The GSFLOW model operates as "fully integrated" where equations of governing surface and subsurface flows are solved simultaneously, or as "coupled regions" where surface and subsurface systems are discretized into separate regions and the governing equations that describe flow in each region are integrated using iterative solution methods. GSFLOW operates at daily time step and can be used to simulate groundwater and surface water for small (few square kilometers, Markstrom et al. 2008) to large (several thousand square kilometers, Surfleet and Tullos 2013) watersheds. The specific applications of GSFLOW range from studying groundwater and surface water interactions in rivers (Huntington and Niswonger 2012; Waibel et al. 2013; Hassan et al. 2014) in lakes and meadows (Hunt et al. 2013; Huntington et al. 2013; Essaid and Hill 2014), under intensive agriculture (Tian et al. 2015), and under climate and land use change scenarios (Surfleet and Tullos 2013; Mateus et al. 2015).

4.2.3 Integrated Hydrology Model (InHM)

InHM (VanderKwaak 1999), is a spatially distributed, fully integrated, coupled surface-subsurface flow and transport model. InHM is predominantly a groundwater focused model and lacks explicit representation of many land surface processes (Beckers et al. 2009). The model is best suited for rain-dominated systems. Implementing the model in snow-dominated systems will require off-line calculations of snow accumulation after which snowmelt can be simulated as a specified flux rate. Additionally, the models have no explicit representation of vegetation and hence do not consider Soil-Vegetation-Atmosphere Transfer (SVAT) processes. Evapotranspiration can be calculated analytically using the Kristensen and Jensen (1975) method and is required as a meteorological input along with precipitation for continuous simulations. However, roads, lakes, and wetlands can be incorporated as part of the simulation. In terms of model application for simulating groundwater and surface water interaction, with a few exceptions (Jones et al. 2006; Mirus et al. 2007), the majority of InHM modeling has been limited to experimental plots (Li et al. 2008) and model testing (Sebben et al. 2013).

4.2.4 HydroGeoSphere

Similar to InHM, the HydroGeoSphere (Therrien et al. 2010) is a three-dimensional, physically based, spatially distributed, fully integrated surface water and groundwater model. However, the most important feature of HydroGeoSphere is its ability to not only simulate "classical" processes of the hydrological cycle (i.e. overland and streamflow, evapotranspiration, groundwater recharge, or sub-surface discharge into surface water bodies such as rivers or lakes) but also to simulate variable–density flow and transport, straight, or branching chain first-order decay reactions, reactive chemical species transport, heat transport, unsaturated flow and flow through fractures in a physically based way (Brunner and Simmons 2012). HydroGeoSphere is also equipped with functionality to incorporate roads and culverts, lakes, and wetlands that are very relevant for studying groundwater and surface interactions. As a groundwater model, HydroGeoSphere is capable of simulating subsurface flow processes in a physically based manner, including porous media, fractures, subsurface conduits, macropores, and perched water tables. Applications of HydroGeoSphere include groundwater-lake interactions (Ala-aho et al. 2015), groundwater component of streamflow (Partington et al. 2011), streamflow-storage interactions (Bhaskar and Welty 2015), pumping-induced regional land subsidence (Calderhead et al. 2011), and climate change impact assessments (Chen et al. 2011b).

4.2.5 Modular Hydrologic Modeling System (MODHMS)

MODHMS (developed by HydroGeologic Inc, Herndon, VA) model is a three-dimensional, physically based, spatially distributed, and integrated surface water and groundwater modeling framework. MODHMS model has been developed after the USGS MODFLOW model and the Hydrologic Modeling System (HEC-HMS), and is designed to be fully compatible with that MODFLOW's input and output as a separate module. Similar to InHM, MODHMS model has no explicit representation of vegetation and snowpack evolution. MODHMS model includes all of the standard MODFLOW and HEC-HMS functionality and is capable of simulating subsurface processes, lakes, wetlands, dams, weirs, culverts, and open channel flow (Beckers et al. 2009). Examples of MODHMS model applications include water table evaporation (Young et al. 2007), stream-aquifer interaction (Werner et al. 2006), seawater intrusion to aquifers (Werner and Gallagher 2006), and interactions between land use and groundwater (Barron et al. 2013).

4.2.6 MIKE-SHE

MIKE-SHE (DHI Water & Environment, Denmark) is a three-dimensional, physically based, spatially distributed, and grid based model. MIKE-SHE has been designed and developed to fully integrate groundwater and surface water flow, including SVAT processes and sophisticated water management utilities (e.g., reservoir operation). Unlike InHM and MODHMS, MIKE-SHE simulates snowpack evolution using a degree-day calculation method. A distinguishing feature of MIKE-SHE is that it is able to simulate both groundwater and surface water with a precision equal to that of predominantly groundwater or surface water focused models (Yan and Zhang 2001). The MIKE-SHE modeling system simulates most major hydrological processes of water movement, including canopy and land surface interception, snowmelt, evapotranspiration, overland flow, channel flow, and saturated/unsaturated subsurface/groundwater flow. The model is capable of simulating a range of processes (Graham and Butts 2005) including ground water and surface water interactions (Loinaz et al. 2013; Foster and Allen 2015), wetland management and restoration (Zacharias et al. 2005; Thompson 2004), and floodplain management (Wen et al. 2013).

4.2.7 Parallel Watershed Flow (ParFlow.CLM) Hydrologic Model

ParFlow (Ashby and Falgout 1996; Maxwell and Miller 2005) is a parallel, three-dimensional, and variably saturated groundwater flow model. Land-surface processes including the land-energy budget, biogeochemistry, and snowpack evolution are simulated via the coupled Community Land Model (CLM). Flow is exchanged between the root zone, deeper vadose zone, and saturated groundwater zone on the basis of interdependent equations that calculate flow and storage of water throughout the simulated hydrologic system. The model has shown great promise in exploring the feedbacks across multiple land surface and subsurface processes including groundwater (Ferguson and Maxwell 2010, 2012; Maxwell et al. 2015; Srivastava et al. 2015) and wetland-groundwater flow interactions (Nalesso 2009; Ali et al. 2015).

4.3 Remote Sensing Techniques

Introduction and advancement of remote sensing technologies over the years have significantly improved our understanding of biophysical and hydrological land surface processes in a way that was nearly impossible with ground-based observations. Satellite based Tropical Rainfall Measuring Mission (TRMM) and Global Precipitation Measurement (GPM) mission have provided much needed rainfall and snow monitoring, which are often not available and/or accessible in many parts of the world. Satellite imagery from Landsat 1–8, Moderate Resolution Imaging Spectroradiometer (MODIS), and Advanced Very High Resolution Radiometer (AVHRR) has facilitated the global scale characterization and monitoring of snow, vegetation, lakes, and wetlands. In addition, monitoring of regional scale groundwater patterns was a real challenge due to lack of long-term observation wells and inconsistency in water-level monitoring activities in both space and time.

However, since 2002 NASA's Gravity Recovery and Climate Experiment (GRACE) satellites have revolutionized the investigations about Earth's water reservoirs, aquifers, ice, and oceans by making detailed measurements of Earth's gravity field anomalies. In addition, newly launched Soil Moisture Active Passive (SMAP) satellite is expected to provide soil moisture measurements and distinguish between ground surface that is frozen or thawed. The proposed (expected to be launched by 2020) Surface Water & Ocean Topography (SWOT) satellite mission will provide valuable observations of the temporal and spatial variations in water volumes stored in rivers, lakes, and wetlands. Similarly, fine scale mapping of vegetation, snow, lakes and wetlands using Light Detection and Ranging (LiDAR) has been on the rise. All of these datasets will provide great potential for environmental monitoring, resource evaluation, data assimilation, and hydrological forecasts. For environmental monitoring, data from these satellites can be analyzed individually as well as combined across multiple sensors operating at different spatial and temporal resolutions. These satellite data can be combined with hydrologic models for better understanding of the interactions between different components of hydrologic cycle at a range of spatial and temporal scales (e.g. Ozesmi and Bauer 2002; Tweed et al. 2009; Voss et al. 2013; Proulx et al. 2013).

5 Future Management Challenges and Strategies

Increasing population, economic development, and continuing expansion of irrigated agriculture will continue to drive demand for water worldwide and affect its availability. Regional variation in surface water availability and the mismatch between the timing of natural water supply and demand create water stress in several parts of the world. Although, river flow and water stored in lakes and reservoirs can help supplement the irrigation, environmental, and urban water use during long dry seasons. Water withdrawals above a threshold can have unintended consequences, including drop in water level that could impact the aquatic ecosystem. Additionally, water stored in lakes and reservoirs are often not enough due to climate variability, and farmers and water managers have to pump groundwater in order to meet the water demand. In regions where surface water availability is limited and frequent droughts are common, overexploitation of groundwater can occur. In the natural environment, groundwater recharge from excess rainfall or snowmelt is balanced by the groundwater discharge in the form of evapotranspiration and/or exfiltration into streams, springs, wetlands, lakes, and oceans. Therefore, groundwater abstraction and subsequent lowering of groundwater water tables can have negative effects on these groundwater-fed water bodies and related ecosystems. Also, many coastal environments have become vulnerable to salt water intrusion and groundwater inundation.

Hydraulic connectivity and feedback between groundwater and surface water has long been recognized. However, water planning and management decisions often treat groundwater and surface water as separate resources. As pointed out by

Taylor et al. (2012), a comprehensive water management approach that integrates groundwater and surface water is essential in sustaining ecosystems and helping reduce human and ecosystem vulnerability to climate change. Integrated numerical models that consider the interactions between groundwater and surface water under climate change and human activities across watershed and topographic boundaries will be needed to develop water resources planning and management. Ellison et al. (2012) have drawn a nice example from the forest cover-water yield debate and argued why we need to consider forest ecosystems as "*global public goods*". As we discussed earlier, the lack of consistent observations of available water resources has been a challenge in developing, implementing, and evaluating integrated groundwater and surface water management strategies. Satellite remote sensing of terrestrial water and biophysical resources will help overcome some of the challenges in developing sustainable water resource management and adaptation strategies. However, some of the available datasets from these satellites (e.g. GRACE) are too coarse to resolve the local scale aquifer management issues. Hence, ground-based strategic water resources monitoring will not only be vital for informing the local groundwater and surface water response to climate and landuse changes but also help improve satellite observations.

On the conservation side, water managers and farmers will need to adopt conservation, reuse of gray and reclaimed water, and improved water use efficiency measures. Long-term planning for conjunctive use of combined groundwater and surface water supplies will be needed to reduce the water stress during dry periods. Reducing reliance on groundwater during the wet periods, when surface water is in abundance, would be essential. In addition, artificial recharge of groundwater systems, which has been undertaken in various parts of the world, presents both challenges and opportunities. Groundwater recharge has declined considerably due to reduced precipitation, increased atmospheric demand due to rising global temperature, increasing use of farm machinery, efficient irrigation technologies that limit recharge, and rapid growth in urban areas. Managing aquifer recharge areas and exploiting opportunities for storing surplus surface water during wet years through enhanced groundwater recharge may prove vital.

6 Conclusions

Hydraulic connectivity between groundwater and surface water is essential in providing cool and clean water for human consumption and supporting other ecosystem services. Climate change, increasing population growth and industrial development continue to threaten many groundwater aquifers and surface water bodies across the world. Deforestation, loss of inland and coastal wetlands, drying streams, and overexploitation of aquifers have been observed across the world. As a result, depletion and degradation of water-related ecosystems are rising. In a recent study, Costanza et al. (2014) estimated that the loss of global ecosystem services is \$4.3–20.2 trillion/year due to landuse change alone. This is a significant number when compared with the current global gross domestic product of $75.6 trillion in 2013 (The World Bank DataBank 2014). Currently a number of coupled numerical models are available to investigate and quantify groundwater and surface water interactions under climate and other environmental changes. Integrated numerical models aided by the ground-based monitoring and satellite remote sensing may help guide the future water resources evaluation, impact assessment, and management plan. Efforts should be directed to not only on the use of these integrated models as a required tool for water resources planning and impact evaluation but also continue to revise and improve models as further data becomes available.

Integrated water resources management planning that is based on the conjunctive groundwater and surface water storage and use will provide a vital role in the face of climate change and increasing population. Additionally, protection, management, and restoration of ecosystems, such as mountain forests and wetlands that provide critical water resources will be critical. There is a critical knowledge gap in our understanding of the magnitude and feedbacks of the interactions between groundwater and surface water under climate change and human activities. Therefore, it is necessary to understand and forecast present and future water and water-related environmental problems through monitoring and focused research. There is also a need to not only manage the water and water-related ecosystem services locally, but also view them as "*global public goods*". This will require trans-boundary research and management cooperation, local capacity building, along with advanced regulatory and economic instruments (Mejia et al. 2012) for water resources management.

References

Adams HD, Luce CH, Breshears DD, Allen CD, Weiler M, Hale VC, Huxman TE (2012) Ecohydrological consequences of drought-and infestation-triggered tree die-off: insights and hypotheses. Ecohydrology 5(2):145–159

Adams M, Smith PL, Yang X (2014) Assessing the effects of groundwater extraction on coastal groundwater-dependent ecosystems using satellite imagery. Mar Freshw Res 66(3):226–232, http://dx.doi.org/10.1071/MF14010

Aeschbach-Hertig W, Gleeson T (2012) Regional strategies for the accelerating global problem of groundwater depletion. Nat Geosci 5(12):853–861

Ala-aho P, Rossi PM, Isokangas E, Kløve B (2015) Fully integrated surface–subsurface flow modelling of groundwater–lake interaction in an esker aquifer: model verification with stable isotopes and airborne thermal imaging. J Hydrol 522:391–406

Alencar A, Nepstad D, Diaz MCV (2006) Forest understory fire in the Brazilian Amazon in ENSO and non-ENSO years: area burned and committed carbon emissions. Earth Interact 10(6):1–17

Ali M, Nussbaumer R, Ireson A, Keim D (2015) Modelling of seasonal dynamics of Wetland-Groundwater flow interaction in the Canadian Prairies Geophysical Research Abstracts. vol 17, EGU2015-3072-1

Alley WM, Reilly TE, Franke OL (1999) Effects of ground-water development on ground-water flow to and from surface-water bodies. US Geological Survey Circular, 1186

Anderegg WR, Kane JM, Anderegg LD (2013) Consequences of widespread tree mortality triggered by drought and temperature stress. Nat Clim Chang 3(1):30–36

Andersen H, Pedersen M, Jorgensen O, Kronvang B (2001) Analysis of the hydrology and flow of nitrogen in 17 Danish catchments. Water Sci Technol 44(7):63–68
Anderson MP (2005) Heat as a ground water tracer. Groundwater 43(6):951–968
AquaResource Inc (2011) Integrated surface and groundwater model review and technical guide for The Ontario Ministry of Natural Resources. http://www.waterbudget.ca/system/files/publications/2011_AquaResource_IntegratedModellingGuide.pdf
Ashby SF, Falgout RD (1996) A parallel multigrid preconditioned conjugate gradient algorithm for groundwater flow simulations. Nucl Sci Eng 124(1):145–159
Barlow PM, Leake SA (2012) Streamflow depletion by wells: understanding and managing the effects of groundwater pumping on streamflow. US Geological Survey, Reston, p 95
Barlow PM, Reichard EG (2010) Saltwater intrusion in coastal regions of North America. Hydrogeol J 18(1):247–260
Barron OV, Donn MJ, Barr AD (2013) Urbanisation and shallow groundwater: predicting changes in catchment hydrological responses. Water Resour Manag 27(1):95–115
Bates BC, Kundzewicz ZW, Wu S, Palutikof JP (eds) (2008) Climate change and water. Technical Paper of the Intergovernmental Panel on Climate Change, IPCC Secretariat, Geneva, 210 p
Bearup LA, Maxwell RM, Clow DW, McCray JE (2014) Hydrological effects of forest transpiration loss in bark beetle-impacted watersheds. Nat Clim Chang 4(6):481–486
Beckers J, Smerdon B, Wilson M (2009) Review of hydrologic models for forest management and climate change applications in British Columbia and Alberta. Forrex Forum for Research and Extension in Natural Resources, Kamloops, BC Forrex Series 25. www.forrex.org/publications/forrexseries/fs25.pdf
Bertrand G, Siergieiev D, Ala-Aho P, Rossi PM (2014) Environmental tracers and indicators bringing together groundwater, surface water and groundwater-dependent ecosystems: importance of scale in choosing relevant tools. Environ Earth Sci 72(3):813–827
Bhaskar AS, Welty C (2015) Analysis of subsurface storage and streamflow generation in urban watersheds. Water Resour Res 51(3):1493–1513
Blöschl G, Ardoin-Bardin S, Bonell M, Dorninger M, Goodrich D, Gutknecht D, Szolgay J (2007) At what scales do climate variability and land cover change impact on flooding and low flows? Hydrol Process 21(9):1241–1247
Bond NR, Lake PS, Arthington AH (2008) The impacts of drought on freshwater ecosystems: an Australian perspective. Hydrobiologia 600(1):3–16
Brunner P, Simmons CT (2012) HydroGeoSphere: a fully integrated, physically based hydrological model. Groundwater 50(2):170–176
Brunner P, Simmons CT, Cook PG, Therrien R (2010) Modeling surface water-groundwater Interaction with MODFLOW: some considerations. Groundwater 48(2):174–180
Buffington JM, Tonina D (2009) Hyporheic exchange in mountain rivers II: effects of channel morphology on mechanics, scales, and rates of exchange. Geography Compass 3 (3):1038–1062
Bullock A, Acreman M (2003) The role of wetlands in the hydrological cycle. Hydrol Earth Syst Sci Discuss 7(3):358–389
Burns DA, Murdoch PS, Lawrence GB, Michel RL (1998) Effect of groundwater springs on NO_3^- concentrations during summer in Catskill Mountain streams. Water Resour Res 34 (8):1987–1996
Calderhead AI, Therrien R, Rivera A, Martel R, Garfias J (2011) Simulating pumping-induced regional land subsidence with the use of InSAR and field data in the Toluca Valley, Mexico. Adv Water Resour 34(1):83–97
Cardenas MB (2015) Hyporheic zone hydrologic science: a historical account of its emergence and a prospectus. Water Resour Res 51(5):3601–3616
Carrie R, Dobson M, Barlow J (2015) The influence of geology and season on macroinvertebrates in Belizean streams: implications for tropical bioassessment. Freshwater Sci 34(2):648–662
Caschetto M, Barbieri M, Galassi DM, Mastrorillo L, Rusi S, Stoch F et al (2014) Human alteration of groundwater–surface water interactions (Sagittario River, Central Italy):

implication for flow regime, contaminant fate and invertebrate response. Environ Earth Sci 71 (4):1791–1807

Cavaleri MA, Sack L (2010) Comparative water use of native and invasive plants at multiple scales: a global meta-analysis. Ecology 91(9):2705–2715

Chen IC, Hill JK, Ohlemüller R, Roy DB, Thomas CD (2011a) Rapid range shifts of species associated with high levels of climate warming. Science 333(6045):1024–1026

Chen J, Sudicky EA, Gula J, Peltier WR, Park Y, Ross M (2011b). Impact of climate change on Canadian surface water and groundwater resources: a continental-scale hydrologic modelling study using multiple high-resolution RCM projections. In AGU Fall Meeting Abstracts (vol 1, p 0853

Cohen MJ, Kurz MJ, Heffernan JB, Martin JB, Douglass RL, Foster CR, Thomas RG (2013) Diel phosphorus variation and the stoichiometry of ecosystem metabolism in a large spring-fed river. Ecol Monogr 83:155–176

Constantz J (2008) Heat as a tracer to determine streambed water exchanges. Water Resour Res 44 (4):W00D10

Coscia I, Greenhalgh SA, Linde N, Doetsch J, Marescot L, Günther T, Vogt T, Green AG (2011) 3D crosshole ERT for aquifer characterization and monitoring of infiltrating river water. Geophysics 76(2):G49–G59

Coscia I, Linde N, Greenhalgh S, Vogt T, Green A (2012) Estimating travel times and groundwater flow patterns using 3D time-lapse crosshole ERT imaging of electrical resistivity fluctuations induced by infiltrating river water. Geophysics 77(4):E239–E250

Costanza R, de Groot R, Sutton P, van der Ploeg S, Anderson SJ, Kubiszewski I et al (2014) Changes in the global value of ecosystem services. Glob Environ Chang 26:152–158

Cramer W, Bondeau A, Woodward FI, Prentice IC, Betts RA, Brovkin V et al (2001) Global response of terrestrial ecosystem structure and function to CO2 and climate change: results from six dynamic global vegetation models. Glob Chang Biol 7(4):357–373

Dahl TE (2011). Status and trends of wetlands in the conterminous United States 2004 to 2009. U.S. Department of the Interior, Fish and Wildlife Service, Washington, DC 108 pp. http://www.fws.gov/wetlands/Status-And-Trends-2009/index.html

Davidson NC (2014) How much wetland has the world lost? Long-term and recent trends in global wetland area. Mar Freshw Res 65(10):934–941

Dhakal AS, Sidle RC (2004) Pore water pressure assessment in a forest watershed: Simulations and distributed field measurements related to forest practices. Water Resour Res 40, W02405, doi:10.1029/2003WR002017

Döll P, Müller Schmied H, Schuh C, Portmann FT, Eicker A (2014) Global-scale assessment of groundwater depletion and related groundwater abstractions: combining hydrological modeling with information from well observations and GRACE satellites. Water Resour Res 50 (7):5698–5720

Donohue RJ, Roderick ML, McVicar TR, Farquhar GD (2013) Impact of CO2 fertilization on maximum foliage cover across the globe's warm, arid environments. Geophys Res Lett 40 (12):3031–3035

Dragoni W, Sukhija BS (2008) Climate change and groundwater: a short review. Geol Soc Lond Spec Publ 288(1):1–12

Earman S, Dettinger M (2011) Potential impacts of climate change on groundwater resources-a global review. J Water Clim Change 2(4):213–229

Eaton JG, Scheller RM (1996) Effects of climate warming on fish thermal habitat in streams of the United States. Limnol Oceanogr 41(5):1109–1115

Ellison D, Futter MN, Bishop K (2012) On the forest cover–water yield debate: from demand- to supply-side thinking. Glob Chang Biol 18(3):806–820

Engelbrecht BM, Comita LS, Condit R, Kursar TA, Tyree MT, Turner BL, Hubbell SP (2007) Drought sensitivity shapes species distribution patterns in tropical forests. Nature 447 (7140):80–82

Essaid HI, Hill BR (2014) Watershed-scale modeling of streamflow change in incised montane meadows. Water Resour Res 50:2657–2678. doi:10.1002/2013WR014420

Falke JA, Fausch KD, Magelky R, Aldred A, Durnford DS, Riley LK, Oad R (2011) The role of groundwater pumping and drought in shaping ecological futures for stream fishes in a dryland river basin of the western Great Plains, USA. Ecohydrology 4(5):682–697

Fan Y (2015) Groundwater in the earth's critical zone—relevance to large-scale patterns and processes. Water Resour Res. 50 yr Anniversary Special Issue. doi:10.1002/2015WR017037

Fan Y, Li H, Miguez-Macho G (2013) Global patterns of groundwater table depth. Science 339 (6122):940–943. doi:10.1126/science.1229881

FAO (2010) Forest Resources Assessment 2010. http://www.fao.org/forestry/fra/fra2010/en/. Accessed 04 Mar 2015

Fendeková M, Fendek M (2012) Groundwater drought in the Nitra river basin-identification and classification. J Hydrology Hydromech 60(3):185–193

Ferguson IM, Maxwell RM (2010) Role of groundwater in watershed response and land surface feedbacks under climate change. Water Resour Res 46(10):W00F02

Ferguson IM, Maxwell RM (2012) Human impacts on terrestrial hydrology: climate change versus pumping and irrigation. Environ Res Lett 7(4):044022

Foster SB, Allen D M (2015) Groundwater—surface water interactions in a mountain-to-coast watershed: effects of climate change and human stressors. Advances in Meteorology. 1–22. ArticleID 861805

Foster S, MacDonald A (2014) The 'water security' dialogue: why it needs to be better informed about groundwater. Hydrogeol J 22(7):1489–1492

Freeze RA, Cherry JA (1979) Groundwater. Prentice-Hall, Englewood Cliffs, p 604

Friend AD, Lucht W, Rademacher TT, Keribin R, Betts R, Cadule P, Woodward FI (2014) Carbon residence time dominates uncertainty in terrestrial vegetation responses to future climate and atmospheric CO2. Proc Natl Acad Sci 111(9):3280–3285

Gaeta JW, Sass GG, Carpenter SR (2014) Drought-driven lake level decline: effects on coarse woody habitat and fishes. Can J Fish Aquat Sci 71(2):315–325

Gleeson T, Wada Y (2013) Assessing regional groundwater stress for nations using multiple data sources with the groundwater footprint. Environ Res Lett 8(4):044010

Gleeson T, Smith L, Moosdorf N, Hartmann J, Dürr HH, Manning AH et al (2011) Mapping permeability over the surface of the Earth. Geophys Res Lett 38(2):L02401

Gleeson T, Wada Y, Bierkens MF, van Beek LP (2012) Water balance of global aquifers revealed by groundwater footprint. Nature 488(7410):197–200

Gordon NB, McMahon TA, Finlayson BL (1992) Stream hydrology: an introduction for ecologists. Wiley, Chichester

Gordon DR (1998) Effects of invasive, non-indigenous plant species on ecosystem processes: lessons from Florida. Ecol Appl 8(4):975–989

Gorelick SM, Zheng C (2015) Global change and the groundwater management challenge. Water Resour Res 51:3031–3051. doi:10.1002/2014WR016825

Goulden M, Bales R (2014) Mountain runoff vulnerability to increased evapotranspiration with vegetation expansion. Proc Natl Acad Sci U S A 111(39):14071–14075

Graham N, Refsgaard A (2001) MIKE SHE: a distributed, physically based modeling system for surface water/groundwater interactions. MODFLOW (pp 321–327)

Graham DN, Butts MB (2005) Flexible, integrated watershed modelling with MIKE SHE. In: Singh V, Frevert D (eds) Watershed models. CRC Press, Boca Raton, pp 245–272

Grant G (2007) Running dry: where will the west get its water? U.S. Forest Service Science Findings No. 97. http://www.fs.fed.us/pnw/sciencef/scifi97.pdf. Accessed 04 Mar 2015

Green TR, Taniguchi M, Kooi H, Gurdak JJ, Allen DM, Hiscock KM et al (2011) Beneath the surface of global change: impacts of climate change on groundwater. J Hydrol 405(3):532–560

Griebler C, Avramov M (2015) Groundwater ecosystem services: a review. Freshwater Sci 34 (1):355–367

Haddeland I, Heinke J, Biemans H, Eisner S, Flörke M, Hanasaki N et al (2014) Global water resources affected by human interventions and climate change. Proc Natl Acad Sci 111 (9):3251–3256

Halford KJ, Plume RW (2011) Potential effects of groundwater pumping on water levels, phreatophytes, and spring discharges in Spring and Snake Valleys, White Pine County, Nevada, and adjacent areas in Nevada and Utah. U. S. Geological Survey

Han B, Endreny TA (2014) Comparing MODFLOW simulation options for predicting intra-meander flux. Hydrol Process 28(11):3824–3832

Hancock PJ (2002) Human impacts on the stream–groundwater exchange zone. Environ Manage 29(6):763–781

Hansen MC, Potapov PV, Moore R, Hancher M, Turubanova SA, Tyukavina A, Townshend JRG (2013) High-resolution global maps of 21st-century forest cover change. Science 342 (6160):850–853

Harbaugh AW (2005) MODFLOW-2005, the U.S. Geological Survey modular ground-water model-the Ground-Water Flow Process: U.S. Geological Survey Techniques and Methods 6-A16

Hardison EC, O'Driscoll MA, DeLoatch JP, Howard RJ, Brinson MM (2009) Urban land use, channel incision, and water table decline along coastal plain streams, North Carolina. J Am Water Resour Assoc 45(4):1032–1046

Harr RD (1976) Hydrology of small forest streams in western Oregon. USDA Forest Service General Technical Report PNW-55

Harvey JW, Wagner BJ (2000) Quantifying hydrologic interactions between streams and their subsurface hyporheic zones. In: Jones JA, Mulholland PJ (eds) Streams and groundwaters. Academic, New York, pp 3–44

Hassan ST, Lubczynski MW, Niswonger RG, Su Z (2014) Surface–groundwater interactions in hard rocks in Sardon Catchment of western Spain: an integrated modeling approach. J Hydrol 517:390–410

Heidbüchel I (2007) Recharge processes in ephemeral streams derived from a coupled stream flow routing/groundwater model—application to the Lower Kuiseb (Namibia). Doctoral dissertation, Thesis, Freiburg University

Henderson R, Day-Lewis F, Abarca E, Harvey C, Karam H, Liu L, Lane J (2010) Marine electrical resistivity imaging of submarine groundwater discharge: Sensitivity analysis and application in Waquoit Bay, Massachusetts, USA. Hydrogeol J 18:173–185. doi:10.1007/s10040-009-0498-z

Hewlett JD, Hibbert AR (1967) Factors affecting the response of small watersheds to precipitation in humid areas. In: Sopper WE, Lull HW (eds) Forest hydrology. Pergamon Press, New York, pp 275–290

Holman IP, Allen DM, Cuthbert MO, Goderniaux P (2012) Towards best practice for assessing the impacts of climate change on groundwater. Hydrogeol J 20(1):1–4

Hunt RJ, Walker JF, Selbig WR, Westenbroek SM, Regan RS (2013) Simulation of climatechange effects on streamflow, lake water budgets, and stream temperature using GSFLOW and SNTEMP, Trout Lake Watershed, Wisconsin. US Geological Survey Scientific Investigations Report, 5159, 118

Huntington JL, Niswonger RG (2012) Role of surface-water and groundwater interactions on projected summertime streamflow in snow dominated regions: an integrated modeling approach. Water Resour Res 48:W11524. doi:10.1029/2012WR012319

Huntington JL, Niswonger RG, Rajagopal S, Zhang Y, Gardner M, Morton CG, Reeves DM, McGraw D, Pohll GM (2013) Integrated hydrologic modeling of Lake Tahoe and Martis Valley mountain block and alluvial systems, Nevada and California, Golden, CO, MODFLOW and More 2013 Conference Proceedings, 5 pp

Jackson DA, Peres-Neto PR, Olden JD (2001) What controls who is where in freshwater fish communities the roles of biotic, abiotic, and spatial factors. Can J Fish Aquat Sci 58 (1):157–170

Jaeger KL, Olden JD, Pelland NA (2014) Climate change poised to threaten hydrologic connectivity and endemic fishes in dryland streams. Proc Natl Acad Sci 111(38):13894–13899

Jefferson A, Grant G, Rose T (2006) Influence of volcanic history on groundwater patterns on the west slope of the Oregon High Cascades. Water Resour Res 42:W12411. doi:10.1029/2005WR004812

Jefferson A, Grant GE, Lewis SL (2007) A river runs underneath it: geological control of spring and channel systems and management implications. Cascade Range, Oregon. US Forest Service Pacific Northwest Research Station General Technical Report PNW-GTR (689, Part 2), pp 391–400

Jefferson A, Nolin A, Lewis S, Tague C (2008) Hydrogeologic controls on streamflow sensitivity to climate variation. Hydrol Process 22(22):4371–4385

Jiang X, Rauscher SA, Ringler TD, Lawrence DM, Williams AP, Allen CD et al (2013) Projected future changes in vegetation in western North America in the twenty-first century. J Climate 26 (11):3671–3687

Jobson HE, Harbaugh AW (1999) Modifications to the diffusion analogy surface-water flow model (DAFLOW) for coupling to the modular finite-difference ground-water flow model (MODFLOW) (No. 99-217). US Dept. of the Interior, US Geological Survey; Federal Center

Jolly ID, McEwan KL, Holland KL (2008) A review of groundwater-surface water interactions in arid/semi-arid wetlands and the consequences of salinity for wetland ecology. Ecohydrology 1:43–58

Jones JP, Sudicky EA, Brookfield AE, Park YJ (2006) An assessment of the tracer-based approach to quantifying groundwater contributions to streamflow. Water Resour Res 42(2):W02407

Kalbus E, Reinstorf F, Schirmer M (2006) Measuring methods for groundwater—surface water interactions: a review. Hydrol Earth Syst Sci 10:873–887

Kelly A, Goulden M (2008) Rapid shifts in plant distribution with recent climate change. Proc Natl Acad Sci U S A 105(33):11823–11826

Khan MR, Voss CI, Yu W, Michael HA (2014) Water resources management in the Ganges basin: a comparison of three strategies for conjunctive use of groundwater and surface water. Water Resour Manag 28(5):1235–1250

King AC, Raiber M, Cox ME (2014) Multivariate statistical analysis of hydrochemical data to assess alluvial aquifer–stream connectivity during drought and flood: Cressbrook Creek, southeast Queensland, Australia. Hydrogeol J 22(2):481–500

Kløve B, Ala-aho P, Bertrand G, Boukalova Z, Ertürk A, Goldscheider N, Widerlund A (2011a) Groundwater dependent ecosystems. Part I: hydroecological status and trends. Environ Sci Pol 14(7):770–781

Kløve B, Allan A, Bertrand G, Druzynska E, Ertürk A, Goldscheider N et al (2011b) Groundwater dependent ecosystems. Part II. Ecosystem services and management in Europe under risk of climate change and land use intensification. Environ Sci Pol 14(7):782–793

Kløve B, Ala-Aho P, Bertrand G, Gurdak JJ, Kupfersberger H, Kværner J et al (2014) Climate change impacts on groundwater and dependent ecosystems. J Hydrol 518:250–266

Konrad CP (2003) Effects of urban development on floods. U.S. Geological Survey Fact Sheet 076-03

Konikov E, Likhodedova O (2011) Global climate change and sea-level fluctuations in the Black and Caspian Seas over the past 200 years. Geol Soc Am Special Papers 473:59–69

Krause S, Bronstert A (2005) An advanced approach for catchment delineation and water balance modelling within wetlands and floodplains. Adv Geosci 5(5):1–5

Krause S, Bronstert A (2007) Water balance simulations and groundwater–surface water-interactions in a mesoscale low-land river catchment. Hydrol Process. doi:10.1002/hyp.6182

Krause S, Jacobs J, Bronstert A (2007) Modelling the impacts of land-use and drainage density on the water balance of a lowland–floodplain landscape in northeast Germany. Ecol Model 200(3-4):475–492. doi:10.1016/j.ecolmodel.2006.08.015

Kristensen KJ, Jensen SE (1975) A model for estimating actual evapotranspiration from potential transpiration. Nord Hydrol 6:170–188

Krupa SL, Belanger TV, Heck HH, Brock JT, Jones BJ. Krupaseep-the next generation seepage meter, International Coastal Symposium (ICS 98), Special Issue 26, 210–213,1998
Kummu M, Ward PJ, de Moel H, Varis O (2010) Is physical water scarcity a new phenomenon? Global assessment of water shortage over the last two millennia. Environ Res Lett 5(3):034006
Kurylyk BL, MacQuarrie KTB, Caissie D, McKenzie JM (2014) Shallow groundwater thermal sensitivity to climate change and land cover disturbances: derivation of analytical expressions and implications for stream temperature projections. Hydrol Earth Syst Sci Discuss 11 (11):12573–12626
Leavesley GH, Lichty RW, Troutman BM, Saindon LG (1983) Precipitation-runoff modeling system—user's manual: U.S. Geological Survey Water-Resources Investigations Report 83-4238, 207 p
Leavesley GH, Markstrom SL, Viger RJ, Hay LE (2005) USGS modular modeling system (MMS)—precipitation-runoff modeling system (PRMS) MMS-PRMS. In: Singh V, Frevert D (eds) Watershed models. CRC Press, Boca Raton, pp 159–177
Lee DR (1977) Device for measuring seepage flux in lakes and estuaries. Limnol Oceanogr 22 (1):140–147
Lenihan JM, Drapek R, Bachelet D, Neilson RP (2003) Climate change effects on vegetation distribution, carbon, and fire in California. Ecol Appl 13(6):1667–1681
Levine JM, Vila M, Antonio CM, Dukes JS, Grigulis K, Lavorel S (2003) Mechanisms underlying the impacts of exotic plant invasions. Proc R Soc Lond B Biol Sci 270(1517):775–781
Li Q, Unger AJA, Sudicky EA, Kassenaar D, Wexler EJ, Shikaze S (2008) Simulating the multi-seasonal response of a large-scale watershed with a 3D physically-based hydrologic model. J Hydrol 357(3):317–336
Lischeid G, Kolb A, Alewell C (2002) Apparent translatory flow in groundwater recharge and runoff generation. J Hydrol 265(1):195–211
Llamas MR (1988) Conflicts between wetland conservation and groundwater exploitation: two case histories in Spain. Environ Geol Water Sci 11(3):241–251
Loinaz MC, Davidsen HK, Butts M, Bauer-Gottwein P (2013) Integrated flow and temperature modeling at the catchment scale. J Hydrol 495:238–251
Lu J, Sun G, McNulty SG, Amatya DM (2003) Modeling actual evapotranspiration from forested watershed across the Southeastern United States. J Am Water Resour Assoc 39(40):887–896
Lundquist JD, Dickerson-Lange SE, Lutz JA, Cristea NC (2013) Lower forest density enhances snow retention in regions with warmer winters: a global framework developed from plot-scale observations and modeling. Water Resour Res 49(10):6356–6370
Maidment DR (1993). Hydrology. In: Maidment DR (ed) Chap. 1 in Handbook in hydrology, McGraw-Hill, pp 1–1 to 1–15, and Appendix A, pp. A-1 to A-11
Manga M (1996) Hydrology of spring-dominated streams in the Oregon Cascades. Water Resour Res 32(8):2435–2439
Markstrom SL, Niswonger RG, Regan RS, Prudic DE, Barlow PM (2008) GSFLOW-coupled ground-water and surface-water flow model based on the integration of the Precipitation-Runoff Modeling System (PRMS) and the Modular Ground-Water Flow Model (MODFLOW-2005): U.S. Geological Survey Techniques and Methods 6-D1, 240 p
Mateus C, Tullos DD, Surfleet CG (2015) Hydrologic sensitivity to climate and land use changes in the Santiam River Basin, Oregon. J Am Water Resour Assoc 51(2):400–420
Maxwell RM, Miller NL (2005) Development of a coupled land surface and groundwater model. J Hydrometeorol 6(3):233–247
Maxwell RM, Putti M, Meyerhoff S, Delfs JO, Ferguson IM, Ivanov V, Sulis M (2014) Surface-subsurface model intercomparison: a first set of benchmark results to diagnose integrated hydrology and feedbacks. Water Resour Res 50(2):1531–1549
Maxwell RM, Condon LE, Kollet SJ (2015) A high-resolution simulation of groundwater and surface water over most of the continental US with the integrated hydrologic model ParFlow v3. Geosci Model Dev 8:923–937

Mejia A, Hubner MN, Sánchez ER, Doria M (2012) The United Nations World water development report–N° 4–water and sustainability (a review of targets, tools and regional cases) (vol 3). UNESCO

Mirus BB, Ebel BA, Loague K, Wemple BC (2007) Simulated effect of a forest road on near-surface hydrologic response: Redux. Earth Surf Process Landf 32(1):126–142

Moore R, Wondzell SM (2005) Physical hydrology and the effects of forest harvesting in the Pacific Northwest: a review. J Am Water Resour Assoc 41(4):763–784

Moser B, Temperli C, Schneiter G, Wohlgemuth T (2010) Potential shift in tree species composition after interaction of fire and drought in the Central Alps. Eur J For Res 129(4):625–633

Nalesso M (2009) Integrated surface-ground water modeling in wetlands with improved methods to simulate vegetative resistance to flow. FIU Electronic Theses and Dissertations. Paper 122. http://digitalcommons.fiu.edu/etd/122.

Neary DG, Ryan KC, DeBano LF (2005) Wildland fire in ecosystems: effects of fire on soils and water. Gen Tech Rep RMRS-GTR-42-vol 4, 250

Nicholls RJ (2004) Coastal flooding and wetland loss in the 21st century: changes under the SRES climate and socio-economic scenarios. Glob Environ Chang 14(1):69–86

Niswonger RG, Prudic DE (2005) Documentation of the streamflow-routing (SFR2) package to include unsaturated flow beneath streams—a modification to SFR1: U.S. Geological Survey Techniques and Methods 6-A13, 50 p

Nyquist JE, Freyer PA, Toran L (2008) Stream bottom resistivity tomography to map ground water discharge. Ground Water 46(4):561–569. doi:10.1111/j.1745-6584.2008.00432.x

Oki T, Kanae S (2006) Global hydrological cycles and world water resources. Science 313 (5790):1068–1072

Olivier J, Van Niekerk HJ, Van der Walt IJ (2008) Physical and chemical characteristics of thermal springs in the Waterberg area in Limpopo Province, South Africa. Water SA 34(2):163–174

Onda Y, Dietrich WE, Booker F (2008) Evolution of overland flow after a severe forest fire, Point Reyes, California. Catena 72(1):13–20

Ozesmi SL, Bauer ME (2002) Satellite remote sensing of wetlands. Wetl Ecol Manag 10 (5):381–402

Packman AI, MacKay JS (2003) Interplay of stream-subsurface exchange, clay particle deposition, and streambed evolution. Water Resour Res 39(4): ESG 4–1 to 4–9. doi:10.1029/2002WR001432

Panday S, Huyakorn PS (2004) A fully coupled physically-based spatially-distributed model for evaluating surface/subsurface flow. Adv Water Resour 27(4):361–382

Partington D, Brunner P, Simmons CT, Therrien R, Werner AD, Dandy GC, Maier HR (2011) A hydraulic mixing-cell method to quantify the groundwater component of streamflow within spatially distributed fully integrated surface water–groundwater flow models. Environ Model Software 26(7):886–898

Patten DT, Rouse L, Stromberg JC (2008) Isolated spring wetlands in the Great Basin and Mojave Deserts, USA: potential response of vegetation to groundwater withdrawal. Environ Manage 41(3):398–413

Paulsen RJ, Smith CF, O'Rourke D, Wong TF (2001) Development and evaluation of an ultrasonic ground water seepage meter. Ground Water 39(6):904–911

Phillips OL, Van der Heijden G, Lewis SL, López-González G, Aragão LE, Lloyd J et al (2010) Drought–mortality relationships for tropical forests. New Phytol 187(3):631–646

Prigent C, Papa F, Aires F, Jimenez C, Rossow WB, Matthews E (2012) Changes in land surface water dynamics since the 1990s and relation to population pressure. Geophys Res Lett 39, L08403. doi:10.1029/2012GL051276

Proulx RA, Knudson MD, Kirilenko A, VanLooy JA, Zhang X (2013) Significance of surface water in the terrestrial water budget: a case study in the Prairie Coteau using GRACE, GLDAS, Landsat, and groundwater well data. Water Resour Res 49(9):5756–5764

Prudic DE (1989) Documentation of a computer program to simulate stream-aquifer relations using a modular, finite-difference, ground-water flow model: U.S. Geological Survey Open-File Report 88-729, 113 p

Prudic DE, Konikow LF, Banta ER (2004) A new streamflow-routing (SFR1) package to simulate stream-aquifer interaction with MODFLOW-2000: U.S. Geological Survey Open-File Report 2004-1042, p 95

Puri S, Aureli A (eds) (2009) Atlas of transboundary aquifers—global maps, regional cooperation and local inventories. UNESCO-IHP ISARM Programme. UNESCO, Paris. [CD only] http://www.isarm.net/publications/324

Reddy KR, D'Angelo EM, Harris WG (1999) Biogeochemistry of wetlands. In: Sumner M (ed) Handbook of soil science. CRC Press, Boca Raton, pp G89–G119

Robinson DA, Binley A, Crook N, -Lewis FD, Ferré TPA, Grauch VJS, Slater L (2008) Advancing process-based watershed hydrological research using near-surface geophysics: a vision for, and review of, electrical and magnetic geophysical methods. Hydrol Process 22(18):3604–3635

Rodell M, Velicogna I, Famiglietti JS (2009) Satellite-based estimates of groundwater depletion in India. Nature 460(7258):999–1002

Rosenberry DO, Morin RH (2004) Use of an electromagnetic seepage meter to investigate temporal variability in lake seepage. Ground Water 42(1):68–77

Rossi PM, Ala-aho P, Ronkanen AK, Kløve B (2012) Groundwater–surface water interaction between an esker aquifer and a drained fen. J Hydrol 432:52–60

Rulli MC, Rosso R (2005) Modeling catchment erosion after wildfires in the San Gabriel Mountains of southern California. Geophys Res Lett 32(19):L19401

Rustad L, Cambell J, Dukes JS, Huntington T, Lambert KF, Mohan J, Rodenhouse N (2012) Changing climate, changing forests: the impacts of climate change on forests of the northeastern United States and eastern Canada. USDA Forest Service. General Technical Report NRS-99

Safeeq M, Fares A (2012) Hydrologic effect of groundwater development in a small mountainous tropical watershed. J Hydrol 428:51–67

Safeeq M, Fares A (2014) Interception losses in three non-native Hawaiian forest stands. Hydrol Process 28(2):237–254

Sanford WE, Selnick DL (2013) Estimation of evapotranspiration across the conterminous United States using a regression with climate and land-cover Data1. J Am Water Resour Assoc 49 (1):217–230

Scanlon BR, Reedy RC, Stonestrom DA, Prudic DE, Dennehy KF (2005) Impact of land use and land cover change on groundwater recharge and quality in the southwestern US. Glob Chang Biol 11(10):1577–1593

Scanlon BR, Jolly I, Sophocleous M, Zhang L (2007) Global impacts of conversions from natural to agricultural ecosystems on water resources: quantity versus quality. Water Resour Res 43 (3):W03437

Schaffranek RW, Baltzer RA, Goldberg DE (1981) A model for simulation of flow in singular and interconnected channels: U.S. Geological Survey Techniques of Water-Resources Investigations, book 7, chap. 3, 110 p

Schalchli U (1992) The clogging of coarse gravel river beds by fine sediment. Hydrobiologia 235 (236):189–197

Schulla J (1997) Hydrologische Modellierung von Flussgebieten zur Abschätzung von Folgen der Klimaänderung. Züricher Geographische Schriften, Heft 69

Scibek J, Allen DM, Cannon AJ, Whitfield PH (2007) Groundwater–surface water interaction under scenarios of climate change using a high-resolution transient groundwater model. J Hydrol 333(2):165–181

Scott ML, Shafroth PB, Auble GT (1999) Responses of riparian cottonwoods to alluvial water table declines. Environ Manage 23(3):347–358

Scott RL, Huxman TE, Barron-Gafford GA, Darrel Jenerette G, Young JM, Hamerlynck EP (2014) When vegetation change alters ecosystem water availability. Glob Chang Biol 20 (7):2198–2210

Sear DA, Armitage PD, Dawson FH (1999) Groundwater dominated rivers. Hydrol Process 13 (3):255–276

Sebben ML, Werner AD, Liggett JE, Partington D, Simmons CT (2013) On the testing of fully integrated surface–subsurface hydrological models. Hydrol Process 27(8):1276–1285

Shi F, Zhao C, Sun D, Peng D, Han M (2012) Conjunctive use of surface and groundwater in central Asia area: a case study of the Tailan River Basin. Stoch Env Res Risk A 26(7):961–970

Shiklomanov I (1993) World fresh water resources. In: Peter HG (ed) Water in crisis: a guide to the world's fresh water resources. Oxford University Press, New York

Sholkovitz E, Herbold C, Charette M (2003) An automated dye-dilution based seepage meter for the time-series measurement of submarine groundwater discharge. Limnol Oceanogr Methods 1(1):16–28

Singha K, Day-Lewis FD, Johnson T, Slater LD (2014) Advances in interpretation of subsurface processes with time-lapse electrical imaging. Hydrol Process 29:1549–1576. doi:10.1002/hyp.10280

Sklash MG, Farvolden RN (1979) The role of groundwater in storm runoff. Dev Water Sci 12:45–65

Smerdon BD, Redding T, Beckers J (2009) An overview of the effects of forest management on groundwater hydrology. J Ecosyst Manage 10(1):22–44

Sørensen R, Ring E, Meili M, Högbom L, Seibert J, Grabs T et al (2009) Forest harvest increases runoff most during low flows in two boreal streams. J Human Environ 38(7):357–363

Sophocleous M (2002) Interactions between groundwater and surface water: the state of the science. Hydrogeol J 10(1):52–67

Spanoudaki K, Stamou AI, Nanou-Giannarou A (2009) Development and verification of a 3-D integrated surface water–groundwater model. J Hydrol 375(3):410–427

Srivastava V, Graham W, Muñoz-Carpena R, Maxwell RM (2015) Insights on geologic and vegetative controls over hydrologic behavior of a large complex basin–global sensitivity analysis of an integrated parallel hydrologic model. J Hydrol 519:2238–2257

Stonestrom DA, Constantz J (eds) (2003) Heat as a tool for studying the movement of ground water near streams (No. 1260). US Dept. of the Interior, US Geological Survey

Stromberg JC, Tiller R, Richter B (1996) Effects of groundwater decline on riparian vegetation of semiarid regions: the San Pedro, Arizona. Ecol Appl 6(1):113–131

Surfleet CG, Tullos D (2013) Uncertainty in hydrologic modelling for estimating hydrologic response due to climate change (Santiam River, Oregon). Hydrol Process 27(25):3560–3576

Swain ED, Wexler EJ (1996) A coupled surface-water and ground-water flow model (MODBRNCH) for simulation of stream- aquifer interaction: U.S. Geological Survey Techniques of Water-Resources Investigations, book 6, chap. A6, 125 p

Swanson FJ (1981) Fire and geomorphic processes. In Mooney HA, Bonnicksen TM, Christensen NL, Lotan JE, Reiners WA (eds) Fire regimes and ecosystem properties. USDA Forest Service General Technical Report WO-26. USDA Forest Service. pp 401–444

Tague C, Grant GE (2004) A geological framework for interpreting the low-flow regimes of Cascade streams, Willamette River Basin, Oregon. Water Resour Res 40(4). doi:10.1029/2003WR002629

Tague C, Grant GE (2009) Groundwater dynamics mediate low-flow response to global warming in snowdominated alpine regions. Water Resour Res 45(7)

Tague C, Farrell M, Grant G, Lewis S, Rey S (2007) Hydrogeologic controls on summer stream temperatures in the McKenzie River basin, Oregon. Hydrol Process 21(24):3288–3300

Taniguchi M, Fukuo Y (1993) Continuous measurements of groundwater seepage using an automatic seepage meter. Ground Water 31(4):675–679

Taylor RG, Scanlon B, Döll P, Rodell M, Van Beek R, Wada Y et al (2013) Ground water and climate change. Nat Clim Chang 3(4):322–329

Teatini P, Ferronato M, Gambolati G, Gonella M (2006) Groundwater pumping and land subsidence in the Emilia-Romagna coastland, Italy: modeling the past occurrence and the future trend. Water Resour Res 42(1):W01406

The World Bank DataBank (2014) http://databank.worldbank.org/data/home.aspx (Retrieved 21 June 2015)

Therrien R, McLaren RG, Sudicky EA, Panday SM (2010) HydroGeoSphere: a three-dimensional numerical model describing fully-integrated subsurface and surface flow and solute transport. Groundwater Simulations Group, University of Waterloo, Waterloo

Thompson JR (2004) Simulation of wetland water-level manipulation using coupled hydrological/hydraulic modeling. Phys Geogr 25(1):39–67

Tian Y, Zheng Y, Wu B, Wu X, Liu J, Zheng C (2015) Modeling surface water-groundwater interaction in arid and semi-arid regions with intensive agriculture. Environ Model Software 63:170–184

Todd DK, Mays LW (2005) Groundwater hydrology, 3rd edn. Wiley, Hoboken

Tonina D, Buffington JM (2009) Hyporheic exchange in mountain rivers I: mechanics and environmental effects. Geogr Compass 3(3):1063–1086

Tweed S, Leblanc M, Cartwright I (2009) Groundwater–surface water interaction and the impact of a multi-year drought on lakes conditions in South-East Australia. J Hydrol 379(1):41–53

Uchida T, Kosugi KI, Mizuyama T (2002) Effects of pipe flow and bedrock groundwater on runoff generation in a steep headwater catchment in Ashiu, central Japan. Water Resour Res 38(7):24-1–24-14

Van der Kamp G, Hayashi M (1998) The groundwater recharge function of small wetlands in the semi-arid northern prairies. Great Plains Res 8:39–56

VanderKwaak JE (1999). Numerical Simulation of Flow and Chemical Transport in Integrated Surface-Subsurface Hydrologic Systems, PhD dissertation, University of Waterloo, Waterloo

Viviroli D, Weingartner R (2008) Water towers—a global view of the hydrological importance of mountains. In: Wiegandt E (ed) Mountains: sources of water, sources of knowledge, vol 31, Advances in global change research. Springer, New York, pp 15–20

Vörösmarty CJ, Green P, Salisbury J, Lammers RB (2000) Global water resources: vulnerability from climate change and population growth. Science 289(5477):284–288

Voss KA, Famiglietti JS, Lo M, Linage C, Rodell M, Swenson SC (2013) Groundwater depletion in the Middle East from GRACE with implications for transboundary water management in the Tigris-Euphrates-Western Iran region. Water Resour Res 49(2):904–914

Wada Y, van Beek LP, van Kempen CM, Reckman JW, Vasak S, Bierkens MF (2010) Global depletion of groundwater resources. Geophys Res Lett, 37(20). doi:10.1029/2010GL044571

Wada Y, Van Beek LPH, Bierkens MF (2011a) Modelling global water stress of the recent past: on the relative importance of trends in water demand and climate variability. Hydrol Earth Syst Sci 15(12):3785–3808

Wada Y, Van Beek LPH, Viviroli D, Dürr HH, Weingartner R, Bierkens MF (2011b) Global monthly water stress: 2. Water demand and severity of water stress. Water Resour Res 47(7): WO7518

Wada Y, Beek LP, Sperna Weiland FC, Chao BF, Wu YH, Bierkens MF (2012) Past and future contribution of global groundwater depletion to sea-level rise. Geophys Res Lett 39(9):L09402

Wagner MJ, Bladon KD, Silins U, Williams CH, Martens AM, Boon S et al (2014) Catchment-scale stream temperature response to land disturbance by wildfire governed by surface–subsurface energy exchange and atmospheric controls. J Hydrol 517:328–338

Waibel MS, Gannett MW, Chang H, Hulbe CL (2013) Spatial variability of the response to climate change in regional groundwater systems–examples from simulations in the Deschutes Basin, Oregon. J Hydrol 486:187–201

Wells WG (1987) The effects of fire on the generation of debris flows in southern California. Rev Eng Geol 7:105–114

Wen L, Macdonald R, Morrison T, Hameed T, Saintilan N, Ling J (2013) From hydrodynamic to hydrological modelling: Investigating long-term hydrological regimes of key wetlands in the Macquarie Marshes, a semi-arid lowland floodplain in Australia. J Hydrol 500:45–61
Werner AD, Gallagher MR (2006) Characterisation of sea-water intrusion in the Pioneer Valley, Australia using hydrochemistry and three-dimensional numerical modelling. Hydrogeol J 14 (8):1452–1469
Werner AD, Gallagher MR, Weeks SW (2006) Regional-scale, fully coupled modelling of stream–aquifer interaction in a tropical catchment. J Hydrol 328(3):497–510
Westbrook CJ, Cooper DJ, Baker BW (2006) Beaver dams and overbank floods influence groundwater–surface water interactions of a Rocky Mountain riparian area. Water Resour Res. 42(6). doi:10.1029/2005WR004560
Westerling AL, Hidalgo HG, Cayan DR, Swetnam TW (2006) Warming and earlier spring increase western US forest wildfire activity. Science 313(5789):940–943
Williams AP, Allen CD, Millar CI, Swetnam TW, Michaelsen J, Still CJ, Leavitt SW (2010) Forest responses to increasing aridity and warmth in the southwestern United States. Proc Natl Acad Sci 107(50):21289–21294
Williams AP, Allen CD, Macalady AK, Griffin D, Woodhouse CA, Meko DM et al (2013) Temperature as a potent driver of regional forest drought stress and tree mortality. Nat Clim Chang 3(3):292–297
Williamson GB, Laurance WF, Oliveira AA, Delamônica P, Gascon C, Lovejoy TE, Pohl L (2001) Amazonian tree mortality during the 1997 El Nino drought. Conserv Biol 14(5):1538–1542
Winter TC (1999) Relation of streams, lakes, and wetlands to groundwater flow systems. Hydrogeol J 7(1):28–45
Winter TC, Harvey JW, Franke OL, Alley WM (1998) Ground water and surface water: a single resource. U.S. Geological Survey Circular 1139:79
Woessner WW (1998) Changing views of stream-groundwater interaction. In: Proceedings of the Join Meeting of the XXVIII Congress of the International Association of the Hydrogeologists and the Annual meeting of the American Institute of Hydrology. American Institute of Hydrology, St. Paul, pp 1–6
Woessner WW (2000) Stream and fluvial plain ground-water interactions: re-scaling hydrogeologic thought. Groundwater 38(3):423–429
Wondzell SM, King JG (2003) Postfire erosional processes in the Pacific Northwest and Rocky Mountain regions. For Ecol Manage 178(1):75–87
Yan J, Zhang J (2001) Evaluation of the MIKE SHE modeling system. Southern Cooperative Series Bulleting, 398
Young C, Wallender W, Schoups G, Fogg G, Hanson B, Harter T et al (2007) Modeling shallow water table evaporation in irrigated regions. Irrig Drain Syst 21(2):119–132
Zacharias I, Dimitriou E, Koussouris T (2005) Integrated water management scenarios for wetland protection: application in Trichonis Lake. Environ Model Software 20(2):177–185

Chapter 12
Contemporary Methods for Quantifying Submarine Groundwater Discharge to Coastal Areas

Ram L. Ray and Ahmet Dogan

Abstract Submarine Groundwater Discharge (SGD), which represents subsurface exchange of water between land and ocean, is a major component of the hydrological cycle. Until the mid-1990s, it was generally believed that SGD rates were not large enough to influence ocean water budgets. This thought might be due to the difficulty in quantifying rates of SGD, because most SGD occurs as diffusive flow, rather than discrete spring flow. However, there is a growing recognition that the submarine discharge of fresh groundwater into coastal oceans is just as important as river discharge in some areas of the coastal ocean. Due to growing ecological concerns about SGD, there is considerable progress on research about SGD with particular emphasis on how to quantify and trace the SGD, and to develop some forecasting or predictive capability of SGD rates based on climatic and seasonal effects. This chapter presents a comprehensive overview of the methods used to quantify SGD to coastal areas and summarizes the previous studies on SGD. In addition, this chapter also discusses driving forces of groundwater flow through coastal aquifers, mechanism of groundwater seawater interaction and some other important issues that are necessary to understand the methods for quantifying SGD in coastal areas. The main goal of this chapter is to provide an overview of the applied methodologies to quantify SGD in coastal areas, which in turn will allow researchers, coastal zone managers, and others to choose appropriate methods that meet their specific project requirements.

R.L. Ray (✉)
Cooperative Agricultural Research Center, College of Agriculture and Human Sciences, Prairie View A&M University, Prairie View, TX 77446, USA
e-mail: raray@pvamu.edu

A. Dogan
Civil Engineering Department, Hydraulics Lab, Yıldız Technical University, Istanbul 34220, Turkey

A. Fares (ed.), *Emerging Issues in Groundwater Resources*, Advances in Water Security, DOI 10.1007/978-3-319-32008-3_12

Acronyms

CFC	Chlorofluorocarbons
GIS	Geographical information system
IAEA	International Atomic Energy Agency
IAPSO	International Association of Physical Sciences of the Oceans
IHP	International Hydrological Program
IOC	Intergovernmental Oceanographic Commission
LOICZ	Land-Ocean Interactions in the Coastal Zone
OBC	Optical backscattering
PMC	Particulate matter concentration
Ra	Radium
Rn	Radon
RSGD	Recirculated saline groundwater discharge
SCOR	Scientific Committee on Oceanic Research
SE	South East
SF_6	Sulphur hexafluoride
SFGD	Submarine fresh groundwater discharge
SFGD	Submarine fresh groundwater discharge
SGR	Submarine groundwater recharge
SPE	Submarine porewater exchange
TABI	Thermal airborne broadband imager
TIR	Thermal infrared

1 Introduction

Submarine Groundwater Discharge (SGD) has been known for many centuries. The Roman geographer, Strabo (63 B.C.–21 A.D.) mentioned a submarine freshwater spring 4 km from Latakia, Syria (Mediterranean) near the island of Aradus. Water from this spring was collected from a boat, utilizing a lead funnel and leather tube, and transported to the city as a source of freshwater (Kohout 1966). Sonrel (1868) also reported discharge of freshwater from submarine springs, and stated risks of using it to sailors because of water quality issues (Schluter et al. 2004; Kumar et al. 2010; Xia et al. 2012).

SGD, which can be defined as subsurface exchange of water between land and ocean, is a major component of the hydrological cycle and it is comprised of terrestrial water mixed with seawater that has infiltrated into coastal aquifers (Zektser and Loaiciga 1993; Moore 2010; Xin et al. 2014). Burnett et al. (2003) defined SGD as "any and all flow of water on continental margins from the seabed to the coastal ocean, regardless of fluid composition or driving force". In other words, SGD is the complementary portion of saltwater intrusion process, which is defined as the invasion of seawater at the bottom of the aquifer due to density difference between fresh groundwater and seawater, causing contamination of groundwater by salt (Johnson 2007).

Groundwater exchange between land and ocean takes place in two ways: as groundwater discharge from land to ocean floor and/or seawater intrusion into coastal aquifers (Zektser and Loaiciga 1993). The seawater intrusion into fresh groundwater is very common at most of the coastal regions in the world that causes salinization to the coastal groundwater (Ravindran and Ramanujam 2014). The salinization of the existing groundwater can lead to a severe deterioration of the quality of fresh groundwater (Oude Essink 2001). SGD has also been recognized as an important pathway for trace elements and nutrient cycling into the ocean, which further impacts the quality of ocean water (Santos 2008). In many areas, groundwater has already been contaminated with a variety of pollutants such as heavy metals, radio nuclides, and organic compounds (Burnett et al. 2006; Wilson and Rocha 2013) which can lead to environmental degradation due to SGD inputs to the coastal environments from those contaminated groundwater sources. According to Wilson and Rocha (2013), despite the acknowledgement of its potential impact on coastal ecosystem functioning, SGD remains a poorly understood and often overlooked process when implementing coastal monitoring and management programs. In the past, SGD has been neglected significantly because of difficulties in locating and quantifying it (UNESCO 2004). Recently, direct discharge of groundwater into coastal zone has received increased attention; it is now recognized as the process, which may represent a potentially important pathway for material transport (Taniguchi et al. 2002).

Due to significant impact of SGD in coastal areas, estimation of SGD becomes very important. However, since SGD is typically characterized by low specific flow rates, detection and quantification of SGD is difficult. On the other hand, since SGD occurs over very large areas, a significant quantities of total flux reaches to the ocean, which can be detected or quantified precisely (Burnett et al. 2003).

In coastal environments, total water discharge into the ocean consists of surface water discharge and SGD. In this chapter, the term SGD is used to represent all direct discharge of subsurface fluids across the land–ocean interface (Fig. 12.1). Submarine Porewater Exchange (SPE) occurs across the seabed by SGD and Submarine Groundwater Recharge (SGR). The SGD rate was defined by Li et al. (1999) as:

$$SGD = SGD_n + SGD_w + SGD_t \tag{12.1a}$$

where SGD_n is net groundwater discharge, which essentially comes from aquifer recharge, SGD_w is outflow due to wave-setup-induced groundwater circulation and SGD_t is tidally driven oscillating flow. Later, Taniguchi et al. (2002) presented a modified linear conceptual model to estimate SGD as:

$$SGD = SFGD + RSGD \tag{12.1b}$$

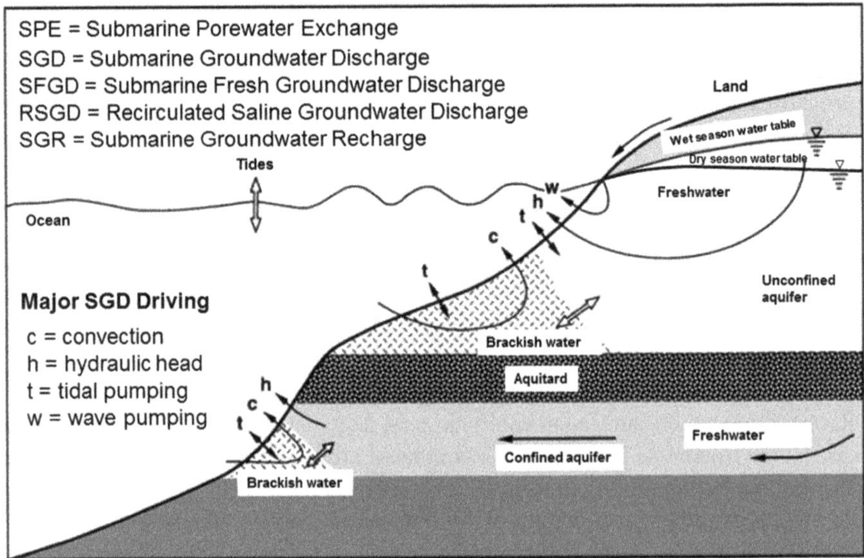

Fig. 12.1 SGD processes (not at scale). Arrows indicate fluid movement (Taniguchi et al. 2002)

where SFGD is submarine fresh groundwater discharge; RSGD is recirculated saline groundwater discharge, which comprises a number of components that may be defined as:

$$\mathrm{RSGD} = \mathrm{RSGD_w} + \mathrm{RSGD_t} + \mathrm{RSGD_c} \tag{12.2}$$

where RSGDw is recirculated water due to wave set-up, RSGDt is recirculated water due to tidally driven oscillation, and RSGDc is recirculated water due to convection (either density or thermal). Thus, total net SGD includes both net fresh groundwater and a recirculated seawater discharge component (Taniguchi et al. 2002). Recent investigations demonstrated that the interplay of different coexisting forcing factors in controlling seawater circulation is not clear (Xin et al. 2014). There are strong coupling of tide-induced and density gradient driven seawater circulations and interactions between tide and wave induced circulations, which are highly nonlinear (King et al. 2010; Xin et al. 2010; Kuan et al. 2012). Therefore, these results challenge some widely used linear conceptual SGD models (Xin et al. 2014), because the linear model does not consider the nonlinear interactions among oceanic forcing factors at varying spatial and/or temporal scales (Xin et al. 2015). Moreover, the transient behavior of the discharge under typically nonstationary oceanic circulations is unknown and unaccounted for by the conceptual linear SGD model (Xin et al. 2014).

There are some other SGD driving forces (e.g., flow and topography induced pressure, fluid shear, gas bubble upwelling, sediment compaction etc.), which are not included in Fig. 12.1, but discussed in detail in Santos et al. (2012).

This chapter provides an overview of the methods used to quantify SGD to coastal areas and also briefly discusses driving forces of groundwater flow through coastal aquifers, mechanism of groundwater seawater interaction and some other important issues that are necessary to understand the methods for quantifying SGD in coastal areas. The main goal of this chapter is to provide an overview of the applied methodologies to quantify SGD in coastal areas and allow researchers, coastal zone managers and others to choose the appropriate methods in order to meet their specific project requirements.

2 Significance of SGD to Coastal Processes and Ecosystems

Before 1965, researchers were mainly focused on the relationship between fresh groundwater inflow to the sea and seawater intrusion into fresh groundwater in coastal areas (Reilly and Goodman 1985). Johannes (1980) stated that "collectively SGD deserves more attention than it has received from marine ecologists". In agreement with this, it is now recognized that SGD is an important pathway to transport materials that includes pollutant if any present in SGD in coastal areas. Miller and Ullman (2004) conducted an aerial thermal infrared imaging survey along the southwestern margin of Delaware Bay and found evidence of abundant discharge at Cape Henlopen. SGD has potential to create estuarine conditions near the SGD locations, which can alter local benthic habitats and ecology. The movement of water across the sediment/water interfaces is very important to the ecology of aquatic habitats (Simmons 1992) because the groundwater might deliver contaminants to coastal waters. According to Barlow (2003), coastal groundwater systems have been contaminated by numerous chemical pollutants, however, the major concern is excessive nutrient loads, particularly due to nitrogen transport to the coastal environments, which occurs as a consequence of activities such as wastewater disposal from septic tanks and agricultural and urban use of pesticides and fertilizers. Nutrients loading through SGD to coastal water is the most significant factor that alters the structure and function of coastal aquatic ecosystems. Valiela et al. (1992) investigated several watersheds of Waquoit Bay in the state of Massachusetts that differ in degree of urbanization and nutrient loading rates and found groundwater flow was the major mechanism that transports nutrients to coastal waters. One of the most common effects of large nutrients inputs to coastal waters is the acceleration of eutrophication (Barlow 2003). Excess nutrients increase algal biomass, loss or change of important local habitats, change in biodiversity and reduce dissolved oxygen (Barlow 2003). One possible cause of increment of harmful algal blooms is increased nutrient supply through SGD (LaRoche et al. 1997; Hwang et al. 2005; Burnett et al. 2006). Increased macro algal biomass dominates aquatic ecosystems, which cause significant changes to coastal environments. Although SGD is considered to be lower than the total freshwater discharge to the ocean, there is increasing evidence that nutrients and other contaminants associated with SGD might play a significant role to coastal ecology than it could impact quantitatively (Valiela et al. 1992, 2002; Burnett et al. 2003; Barlow 2003). This is particularly true in coastal areas where there is high inputs of nutrients and other contaminants, from domestic, industrial, municipal, and agricultural practices to the groundwater, and topographically driven groundwater flow to shallow upland streams is low (Hays and Ullman 2007).

3 Groundwater Flow at Coastal Zones

Global hydrological cycle along with the nature of hydraulics requires groundwater flow from highlands towards the coastal zones. Freshwater originating from precipitation infiltrates through the soil profile to build up terrestrial hydraulic gradient that forces groundwater flow towards the coastal ocean, which is a common coastal process driven by a composite of climatologic, hydrogeologic, and oceanographic factors. Land slope and groundwater hydraulic gradients towards the coast that occur naturally as a result of both short and long term climatic conditions always transport both surface and groundwater toward coastal zones. In coastal waters, physical oceanographic processes such as wave set-up, tidal activities, and density-driven circulation impact these hydraulic gradients intermittently and thus affect rates of SGD. Although only fresh groundwater discharge has traditionally been accounted for in numerical simulations of coastal water budgets, discharge of recirculated saline groundwater may be equally important in terms of nutrient transport across coastal zones.

Until the mid-1990s, it was generally thought that SGD rates were not large enough to influence ocean water budgets. This thought might be due to the difficulty in quantifying rates of SGD, because most SGD occurs as diffusive flow, rather than discrete spring flow. However, there is a growing recognition that the submarine discharge of fresh groundwater into coastal oceans is just as important as river discharge in some areas of the coastal ocean. Due to growing ecological concerns about SGD, there is considerable progress on SGD researches with particular emphasis on how to quantify and trace the SGD, and to develop some forecasting or predictive capability of SGD rates based on climatic and seasonal effects.

Increasing population density and intensifying agricultural activities in coastal areas have caused transportation of nutrients and other contaminants into the coastal environment from fertilizer use, industrial practices, and wastewater discharge. SGD related scientific researches focus on how fluid flux across ocean floor and recirculation through sediments affects elemental cycling at all scales. Experiments that address more applied aspects of nutrient delivery can also yield information that is valuable for developing a more general understanding of land-ocean aquifer interactions. While previous focus of scientific researches was to examine geologic control on coastal aquifers and groundwater discharge, recently the focus has shifted towards investigation of land–sea exchange, ecosystems health, and climate-change-related processes, as well as natural geohazards (Swarzenski et al. 2004).

3.1 Driving Equations of Groundwater Flow in Coastal Zones

Driving forces of groundwater flow through coastal aquifers create a complex flow regime with significant variability in space and time. The major driving forces are terrestrial—hydraulic gradient, which are governed by both short and long term climatic conditions and marine originated forces i.e., tidal pumping, wave set-up, current-induced topographic flow, and convection. Figure 12.1 shows some typical submarine groundwater discharge phenomena with fresh and salt—groundwater recirculation mechanism at coastal zones.

A general three-dimensional groundwater flow continuity equation can be derived by invoking Darcy's law as:

$$\left[\frac{\partial\left(K_x \frac{\partial h}{\partial x}\right)}{\partial x}+\frac{\partial\left(K_y \frac{\partial h}{\partial y}\right)}{\partial y}+\frac{\partial\left(K_z \frac{\partial h}{\partial z}\right)}{\partial z}\right] \pm Q_{ext}=S_s \frac{\partial h}{\partial t} \tag{12.3}$$

where h is hydraulic head, K_x, K_y, and K_z are saturated hydraulic conductivities [LT^{-1}] in the x, y, and z directions, respectively, Q_{ext} is a volumetric source or sink term [$L^3L^{-3}T^{-1}$], and S_s is the specific storage [L^{-1}], defined as the volume of water a unit volume of saturated aquifer releases from storage for unit decline in hydraulic head (Todd 1980; Mays 2011). Equation 12.3 is the general three-dimensional groundwater flow continuity equation. This equation is only valid for constant density groundwater flow, however, in coastal boundary there is mixing zone that includes both fresh and saltwater, which has variable density based on salt concentration. Therefore, a new set of variable density groundwater flow equation should be derived by introducing a state equation, which defines density of water based on salt concentration and transforms saltwater heads into equivalent freshwater heads.

3.1.1 Hydraulic Approaches to Saltwater Intrusion Process and SGD

Saltwater-freshwater systems in coastal aquifers can be conceptualized using two basic hydraulic approaches depending on the relative thickness of the transition zone between freshwater and saltwater. The first approach is the freshwater-saltwater sharp interface approach (Henry 1959; Essaid 1990) and the second one is the variable density approach with solute transport (or hydrodynamic dispersion). The former approach ignores the mixing zone and considers the transition zone negligible compared to aquifer thickness. In this approach, saltwater intrusion phenomenon is modeled as a two-region fluid flow separated by a sharp interface. However, hydrodynamic dispersion often causes a mixing zone across the interface between saltwater and freshwater. If thickness of the mixing zone expands to a considerable extent then the sharp interface approach may not be valid. In this case, the latter approach needs to be used to model the saltwater intrusion problem using the density-dependent miscible flow and transport approaches. The latter approach assumes a mixing or transition zone and a solute transport mechanism to model the hydrodynamic dispersion process (Henry 1964; Pinder and Cooper 1970; Huyakorn et al. 1987). It represents the physical system more realistically compared to the sharp-interface approach. The appropriateness of either of these approaches and their method of analysis depends on the characteristics of the aquifer system under investigation and the problems of which solutions are being sought. The most important issue in modeling saltwater intrusion problems is the proper conceptualization of the hydrogeologic system and definition of boundary conditions. Volker and Rushton (1982) compared steady state solutions using disperse and sharp interface approaches and showed that as the coefficient of dispersion decreases the solutions of those two approaches converge.

Sharp-Interface Approach: this approach takes into account the saltwater intrusion system caused by the dynamics of saltwater and freshwater in coastal areas. It assumes saltwater and freshwater as immiscible fluids. The transition zone of two liquids acts as a common boundary, where the saltwater and freshwater flow fields are coupled through the interfacial boundary condition of continuity of flux and pressure. The sharp interface approach does not provide information about the nature of the interface zone, instead it simulates the regional flow dynamics of the groundwater system and the response of the interface to applied stresses (Dogan and Fares 2008).

Ghyben-Herzberg (Ghyben 1888; Herzberg 1901) relationship between saltwater and freshwater was given by Hubbert (1940) in which the steady state groundwater flow conditions in both salt and freshwater zones are valid:

$$z = \frac{\rho_f}{\rho_s - \rho_f} h_f - \frac{\rho_f}{\rho_s - \rho_f} h_s \tag{12.4}$$

where z is the elevation of a point on the interface at which head is measured [L], ρ_f and ρ_s are fresh and saltwater densities [ML^{-3}], respectively, h_f and h_s are the freshwater and saltwater heads [L], respectively. Hubbert's relation must hold true at the interface to ensure the continuity of pressure field at the interface.

Groundwater flow systems for the freshwater region (bounded with the interface and phreatic surface) with recharge and for the seawater region (bounded with the interface and the impervious bottom) can be defined with the following equations (Reilly 1993):

$$S_f \frac{\partial h_f}{\partial t} + \nabla \cdot \vec{q}_f - Q_f = 0 \tag{12.5a}$$

$$S_s \frac{\partial h_s}{\partial t} + \nabla \cdot \vec{q}_s - Q_s = 0 \tag{12.5b}$$

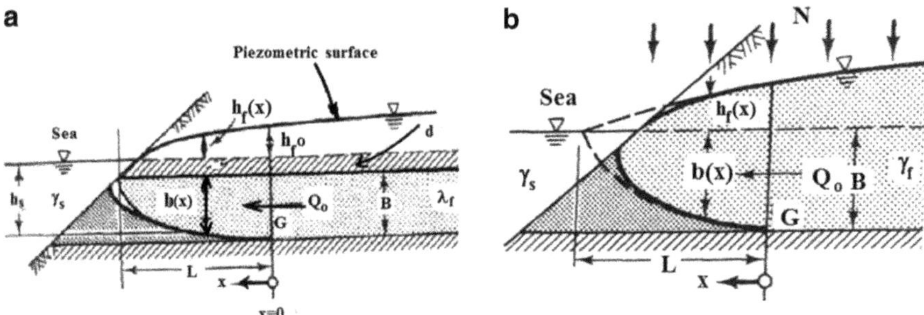

Fig. 12.2 The shape of a stationary interface by the Dupuit- Ghyben- Herzberg approximation for (**a**) confined aquifers and (**b**) unconfined aquifers (Bear 1979)

where h is the head (L), t is time (T), $\vec{q}$ is the specific discharge [LT^{-1}] determined by Darcy's law for constant density fluid as $\vec{q} = -K\nabla h$, K is hydraulic conductivity [LT^{-1}], S is the specific storage [L^{-1}], Q is the source sink term [T^{-1}], ∇h is the hydraulic gradient where $\nabla = \frac{\partial}{\partial x}i + \frac{\partial}{\partial y}j + \frac{\partial}{\partial z}k$ (where i, j, and k are unit vectors in the x, y, and z directions), and subscripts f and s refer to freshwater and saltwater, respectively. Equations 12.5a and 12.5b must be solved simultaneously for the freshwater (h_f) and saltwater heads (h_s). Then the interface elevation at any x-y location in the aquifer can be calculated using Eq. 12.4. From the definition of h_f, freshwater head along the bottom of the sea at the outflow face has to be greater than zero to be able to discharge through sea bottom. Bear (1979) showed the boundary conditions for the freshwater and seawater flow domains in terms of $h_{f\ \text{and}}$ h_s as seen in Fig. 12.2.

A stationary interface may be assumed for a steady flow condition to demonstrate the shape of the interface and the relationship between freshwater flow to the sea and the extent of seawater intrusion as seen in the Fig. 12.2 for both confined and unconfined aquifers. Assuming horizontal bottom and constant thickness B for a confined aquifer (Fig. 12.2a), let the origin $x = 0$ be located at the interface toe (point G), the seaward freshwater flow at this point is denoted as Q_o, which is the difference between total aquifer replenishment and withdrawal. In the coastal strip to the right of point G, Q_o can be formulated as:

$$Q_o = -K_f b(x) \frac{\partial h(x)}{\partial x} \tag{12.6}$$

By integrating Eq. 12.6 from $x = 0$ to L, the equation of the parabolic shape of the interface is obtained. If freshwater flow region thickness set $b(x) = 0$ at $x = L$ and head of freshwater $h_f = (B + d)/\delta$ with $\delta = \gamma_f/(\gamma_s - \gamma_f)$ substituted into the integration results then the following expression is obtained (Bear 1979):

$$Q_oL = \frac{Kh_{fo}}{2}(\delta h_{fo} - 2d) + \frac{Kd^2}{2\delta} = \frac{K}{2\delta}B^2 \quad (12.7)$$

This equation expresses the relationship among the length of the seawater intrusion (L), freshwater component of SGD (Q_o), and piezometric head h_{fo} above the toe. As freshwater component of SGD (Q_o) increases, L decreases. This means that the extent of seawater intrusion, L, is a decision variable in the management of the coastal aquifer and it can be controlled by controlling SGD (Q_o), or alternatively by controlling the recharge and/or pumping in the coastal aquifer strip.

For a phreatic aquifer (Fig. 12.2b) with uniform recharge N and assuming the steady essentially horizontal flow in the aquifer with $(x) = \delta h_f(x)$, the continuity relationship leads to:

$$Q_o + Nx = -K(b + h_f)\frac{\partial h_f}{\partial x} = -K(1+\delta)h_f\frac{\partial h_f}{\partial x} \quad (12.8)$$

After integrating Eq. 12.8 at $x = L$, $h_f = 0$ the following expression is obtained;

$$h_{fo}^2 = \frac{2Q_oL + NL^2}{K(1+\delta)}, \text{ or } Q_o = \frac{KB^2}{2L}\frac{(1+\delta)}{\delta^2} - \frac{NL}{2}, \quad h_{fo} = \frac{B}{\delta} \quad (12.9)$$

Here again, the interface has a parabolic shape and there is a relationship between L and SGD (Q_0). By controlling freshwater head, h_{fo} (i.e., by means of artificial recharge), the water table may be lowered both landward and seaward of the toe, without causing any additional seawater intrusion. Landward of the toe, water levels may fluctuate as a result of some optimal management scheme. When pumping takes place seaward of the toe, the interface there will rise and may contaminate wells if their screened portion is not a sufficient distance above it.

Instead of $h_f = 0$ at $x = L$, another more realistic boundary condition can be used for the confined aquifer at $x = L$, $h_f = \frac{\delta Q}{K}$ this will give the exact solution for the depth of the interface as derived by Glover (1959), assuming the interface has a parabolic shape. The same boundary condition with $Q = Q_{fL}$ = freshwater SGD can be used as approximation for the phreatic aquifer; Q_{fL} is the difference between total recharge and total pumping in the coastal aquifer strip. The length of the freshwater outflow face (x_o) in Fig. 12.3 can be estimated from the parabolic shape of a steady interface, which is given by the following expression (Glover 1959):

$$z^2 = \frac{2Q\delta x}{K} + \left(\frac{Q\delta}{K}\right)^2 \quad (12.10)$$

and the corresponding shape for the water table is given by:

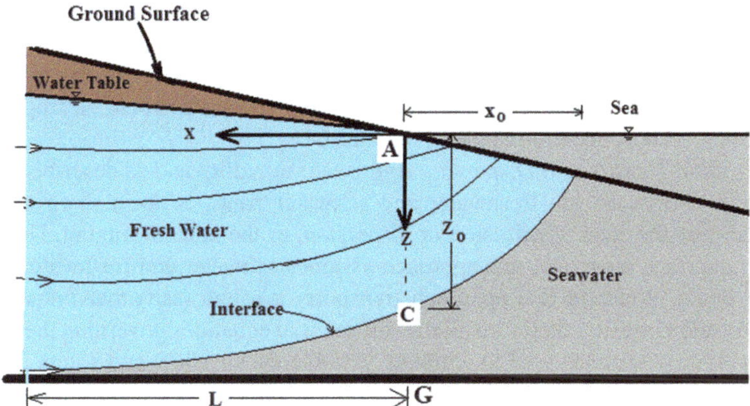

Fig. 12.3 Flow pattern along the saltwater freshwater interface in an unconfined coastal aquifer

$$h_f^2 = \frac{2Qx}{K(1+\delta)} \tag{12.11}$$

From Eqs. 12.10 and 12.11, it can be derived that when $x = 0$ the depth to the toe, $z_o = Q_o\delta/K$ and the length of the outflow face is $x_o = Q_o\delta/2K$ when $z = 0$ (Fig. 12.3). For a steady flow and a vertical outflow face, Henry (1959) obtained $z_o = 0.741\delta Q_o/K$. Equation 12.10 has the third term inside the bracket added to the Gyhben-Herzberg model by Glover (1959) to account for the missing seepage face that allows for the vertical components of flow and freshwater discharge into the sea floor (Fetter 2001). Since the fresh-saltwater interface intercepts the water table at the coastline in the Gyhben-Herzberg model it reduces to the following form of Hubert's equation:

$$z^2 = \frac{2Q\delta x}{K} \tag{12.12}$$

The Gyhben-Herzberg model, which does not allow for an outflow face, underestimates the depth to saltwater interface compared to the more exact solution by Glover (1959). The difference becomes more significant closer to the shoreline, especially at distances less than 20 m from the shoreline. On the other hand, Mays (2011) argues that the water table profile does not change significantly between two models.

In regional applications, the above approach has advantages because of its relative simplicity, ease of development and use, and less intense data requirements, however its main disadvantage for practical purposes is that it cannot evaluate chloride concentrations along the coastal aquifers. The sharp interface boundary between fresh and saline water as described in the Gyhben-Herzberg model and Glover's solution does not occur in field condition. Instead, a brackish transition zone of finite thickness separates the two flow zones. This zone develops from dispersion by flow of freshwater plus the unsteady movement of the interface by external influences such as tides, recharge variations, and pumping of wells. In general, greater thicknesses of transition zones are found in highly permeable coastal aquifers. Therefore, variable density flow approach should be employed to obtain chloride concentration along the interface.

Variable-Density and Dispersion Approach: in reality and as described earlier, the region between the freshwater and saltwater zones is not a sharp interface, instead it is the zone of diffusion or dispersion, or the zone of mixing. Unlike the sharp interface approach, this approach assumes saltwater and freshwater as miscible fluids. Water in this approach transports a solute (salt) that influences its density and viscosity; therefore, partial differential equations governing the groundwater flow system are used to simulate variable density flow and solute transport with a relationship defining the density as a function of solute concentration. Thus, this approach requires coupling density-dependent groundwater flow and solute transport equations. In this approach, the flow field is modeled using single fluid with variable density, which is a function of concentration and pressure.

The mass balance equation for variable density fluid can be derived by combining the continuity equation, in which the storage term is written as a function of change in pressure and concentration with pressure based density-dependent form of Darcy's law as follows (Voss 1984):

$$(\rho S_{op})\frac{\partial P}{\partial t}+\left(n\frac{\partial \rho}{\partial t}\frac{\partial C}{\partial t}\right)=+\nabla\left[\left(\frac{\rho k}{\mu}\right)\cdot(\nabla P+\rho g\nabla z)\right]+Q \qquad (12.13)$$

where ρ is fluid density [ML^{-3}], S_{op} is specific pressure storage [LT^2M^{-1}], n is porosity, C is mass based solute concentration [MM^{-1}], t is time [T], k is intrinsic permeability of the solid matrix [L^2], μ is the fluid viscosity [$ML^{-1}T^{-1}$], P is the pressure applied on the fluid [$ML^{-1}T^{-2}$], g is the gravitational acceleration [LT^{-2}], z is the upward coordinate direction [L], and Q is fluid mass source [$ML^{-3}T^{-1}$];

The mass balance for a solute stored in solution can be expressed as:

$$\frac{\partial(n\rho C)}{\partial t}=\nabla[n\rho D\cdot\nabla C]-\nabla(n\rho C)+QC^{*} \qquad (12.14)$$

where D is the dispersion tensor, which also includes molecular diffusivity of solutes [L^2T^{-1}], v is the fluid velocity (which is the specific discharge, q, divided by porosity, n) [LT^{-1}], and C* is the solute concentration of fluid sources [MM^{-1}].

Since there are three unknowns (P, C, and ρ) and two equations; i.e., flow (Eq. 12.13) and transport (Eq. 12.14) equations, another equation is needed to solve the problem. Voss and Provost (2002) provided an equation of state that relates density to concentration:

$$\rho = \rho(C) = \rho_o + \frac{\partial \rho}{\partial C}(C - C_0) \tag{12.15}$$

where ρ_o [ML^{-3}] is reference fluid density at reference solute concentration C_o [M_s M^{-1}]. Usually, $C_o = 0$, and the reference density is taken as that of freshwater. $\partial\rho/\partial C$ gives the change of the density with the change of concentration as a constant value.

The advantage of the variable density method is that the response of the system to stresses influencing the thicknesses of the transition zone can be analyzed and understood more realistically. This method represents the actual physical system more accurately than the sharp interface approach (Dogan and Fares 2008).

Langevin and Guo (2006) presented an alternative methodology to couple "constant-density groundwater flow code" with "solute transport code" to simulate variable-density groundwater flow and solute transport using equivalent freshwater head concept. In this concept, density effects can be converted into equivalent freshwater heads so that the density dependent flow model turns into a constant-density flow model. In this approach, all saltwater heads are converted to the equivalent freshwater heads that exert the same pressure applied by saltwater at that given point. Weiss (1982) was one of the first to reformulate the groundwater flow equation in equivalent freshwater head form. Following this approach, Langevin et al. (2003) coupled MODFLOW (McDonald and Harbaugh 1988; Harbaugh et al. 2000), and MT3DMS (Zheng and Wang 1999), on a platform called SEAWAT computer program (Guo and Langevin 2002). Langevin et al. (2003) reformulated the flow equation in terms of freshwater heads to be able to use MODFLOW's flow equation routines. The equivalent freshwater head formulation leads to a system of variable density flow equations that can be solved relatively easily using the existing constant density groundwater flow equations in MODFLOW. SGD can easily be identified from boundary flux values calculated by SEAWAT at ocean aquifer interface.

3.2 *Effects of Groundwater Use on SGD*

Main supply of freshwater in coastal areas is generally obtained from groundwater pumping from coastal aquifers. Excessive groundwater use in coastal areas may lead to saltwater intrusion. Groundwater development depletes the amount of groundwater in storage and causes reductions in groundwater discharge to streams, wetlands, and coastal estuaries and lowered water levels in ponds and lakes. Increased level of saltwater concentration in groundwater resources has resulted in degradation of some drinking and irrigation water supplies. Overuse and the proximity of production wells to ocean boundary create some problems with respect to groundwater sustainability in coastal regions. These problems are primarily those of saltwater intrusion into freshwater aquifers and changes in the amount and quality of fresh groundwater discharge to coastal saltwater ecosystems.

The natural balance between freshwater and saltwater in coastal aquifers is disturbed by groundwater withdrawals and other human activities, i.e., irrigation, deforestation, land use/land cover change etc., that lower groundwater levels, reduce fresh SGD to coastal waters, and ultimately cause intrusion of saltwater into coastal aquifers. Although groundwater pumping is the primary cause of saltwater intrusion, other hydraulic stresses that reduce freshwater flow in coastal aquifers, such as lowered rates of groundwater recharge in urbanized areas could also lead to saltwater intrusion, but the impact of such stresses on saltwater intrusion is estimated to be smaller than that of pumping and land drainage (Barlow 2003). If excessive groundwater pumping at coastal aquifers creates deep cone of drawdown that may eventually cause whether saltwater upcoming or reverse direction of groundwater flow from ocean towards the production wells so that the SGD flow rates and characteristics of flow (i.e., amount of fresh and saltwater components and directions) will change.

3.3 Precipitation and SGD Relationship

Precipitation is the primary source of recharge to the aquifer. Snow recharges the aquifers more efficiently than rainfall because melting rate of snow is generally much slower than rate of rainfall that results in more infiltration and less runoff. Rate of precipitation determines the amount of infiltration and surface runoff. Thickness of the unsaturated zone plays an important role in timing and the amount of recharge to aquifers. The value of the maximum possible infiltration rate asymptotically approaches to the value of the saturated hydraulic conductivity of the unsaturated zone. Therefore the relationship between the recharge and precipitation depends on hydrogeological characteristics of the unsaturated zone, the rate and temporal distribution of the precipitation, and the form of precipitation, i.e., snow or rainfall. The larger the amount of precipitation, the more freshwater SGD rate will be. Effective precipitation that contributes to the recharge causes the water table to rise quickly in shallow unconfined aquifers as more pores of the soil media get saturated. That creates steeper hydraulic gradient in phreatic surface towards the ocean, which in turn, intensifies the freshwater SGD rates towards the ocean until the water table gradient reaches at new equilibrium. Therefore, characteristics of precipitation (amount, frequency, intensity, duration, type) together with the soil characteristics determine groundwater recharge and consequently SGD rates at coastal zones.

3.4 Effects of Stream Flow at Coastal Zones on SGD

Important component of the hydrologic system at coastal zones are rivers or streams. The hydraulic connection between groundwater and streams allows water to move freely between surface water and groundwater systems. Streams, depending on its relative position to the aquifer hydraulic heads, may drain or recharge the aquifer such that if the piezometric level in the aquifer is higher than the water level of the stream, groundwater moves toward the stream; where as if the water level in the stream is higher than the piezometric level of the aquifer then the water in the stream recharges the aquifer. In coastal regions, generally, the topography is relatively flat and water levels in streams are higher than groundwater levels in wet seasons, and lower than groundwater levels in dry seasons. When water level in the stream is higher, consequently the freshwater SGD rates will also be higher. Nonetheless, the total amount of freshwater discharge to the ocean both from streams and aquifer would be more or less same regardless of the existence of stream under similar recharge conditions because the SGD rates are strongly correlated with the amount of recharge. Stream collects some of its water from surface runoff and some portion from subsurface flow and small amount from groundwater discharge. Eventually all the effective precipitation, after evapotranspiration demand is satisfied, will reach to ocean whether through the stream flow or through the aquifer discharge as part of the SGD. The length of coastline, where SGD occurs, is considerably long when compared to the length of the stream recharging the groundwater at coastal aquifers. Therefore, SGD will stay as a prominent discharge process whether rivers or streams are present or not.

Effects of drainage canals in Florida resemble river's mechanism on SGD, as an example the one in Figs. 12.4 and 12.5 (Barlow 2003). Saltwater intrusion in Florida has been caused by the construction of drainage canals in addition to groundwater withdrawals for water supply. Since 1946, control structures have been placed on the canals to prevent inland migration of seawater, as well as to provide flood protection and artificial recharge to the aquifer. During the wet season, the gates are opened to lower water levels and discharge excess surface and groundwater to the ocean to prevent flooding. During the dry season, the gates

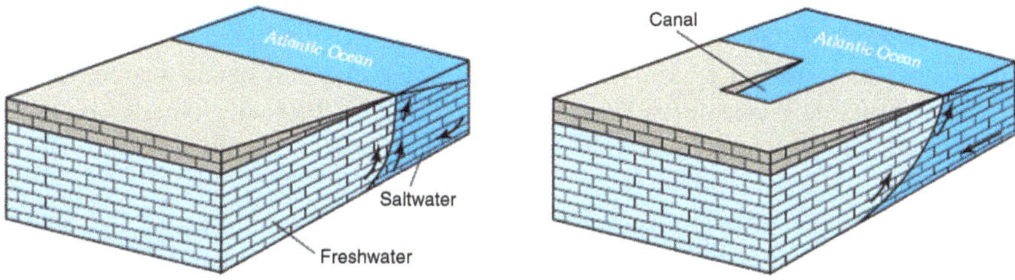

Fig. 12.4 The freshwater-saltwater interface was nearly stable before coastal canals were built (*left*). Uncontrolled tidal canals caused saltwater intrusion by lowering freshwater levels and providing open channels to the sea (*right*) (Barlow 2003)

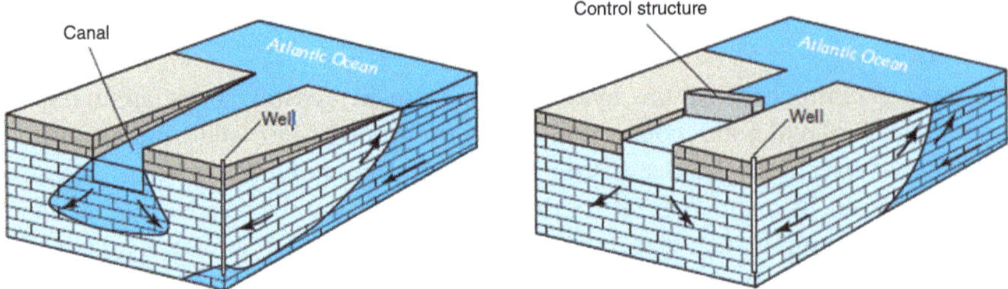

Fig. 12.5 An uncontrolled canal that extended into an area of heavy pumping could convey saltwater inland to contaminate freshwater supplies (*left*). In contrast, a controlled canal provides a perennial supply of freshwater from up-gradient areas to prevent saltwater intrusion and to recharge a well field (*right*) (Barlow 2003)

are closed to raise canal stages above groundwater levels near the coast, inducing water to seep from the canals into the aquifer and retarding saltwater intrusion. Figures 12.4 and 12.5 summarize the effects of controlled and uncontrolled coastal region drainage canals on saltwater intrusion.

4 Current Challenges in Identifying and Quantifying SGD

Exchange of groundwater between land and ocean is a major component of the global hydrologic cycle (Moore 2010), and groundwater circulation is perhaps the most difficult to identify and quantify among the components of the global hydrologic cycle (Zektser and Loaiciga 1993) because groundwater monitoring is not only difficult and challenging but also it is very expensive to implement (Loaiciga 1989). In addition, SGD has significant spatial and temporal variability (Li et al. 2009) in its rates of flux and storage in the subsurface environment, which extends from the ground surface to several kilometers deep in the earth's crust (Keller and Loaiciga 1991).

Discharge from the rivers and streams are visible, therefore, their contributions to the oceans are easily quantifiable. On the other hand, SGD occurs as springs and seeps on continental margins, usually at or below the water surface of oceans. SGD per unit length of coastline could be very significant as a discharge process, because the length of coastline where SGD occurs is large, and will occur whether or not rivers or streams are present (Taniguchi et al. 2002). Either way, since SGD flows below the ground surface and it is not visible, it is difficult to measure and identify it.

Moreover, the diffuse nature of SGD with significant temporal and spatial variability renders the identification, measurement, and assessment of SGD particularly more challenging and complicated. According to Burnett et al. (2003), SGD is characterized by low specific flow rates that make identification and quantification very difficult and challenging. However, since SGD flow occurs over very large areas, the total flux is significant to measure and identify. Identification and measurement were more challenging when the researchers were not certain whether or not recirculated seawater should be included into SGD. For example, Younger (1996) and recently Taniguchi et al. (2002) said that SGD with or without recirculated seawater is ambiguous in the literature and this ambiguity could lead to serious misunderstandings when comparing SGD with other freshwater discharges.

SGD has been neglected scientifically because of the difficulty in identifying and measuring these features and the reason that SGD has not been quantified in terms of the global water and material cycle on the earth (Taniguchi et al. 2002).

Since the identification and measurement of SGD is challenging and complicated, the number of SGD observations is very limited on a global scale. According to Taniguchi et al. (2002), many SGD measurements had been made in karst areas, where the hydraulic conductivity of the aquifers is large and thus significant amounts of groundwater discharge are expected under reasonable hydraulic gradients. They further concluded that the wide areas of the world (South America, Africa, and southern Asia) have little to no SGD assessments at all.

There are many direct and indirect indicators such as its color, temperature, salinity anomalies and growth of vegetation at coastal zones, which can help to detect possible SGD. However, accurate measurement of SGD can still be challenging as many uncertainties are associated with SGD flow.

5 Methods of Detecting and Quantifying SGD

Recently, several methods have been used to detect and quantify SGD. Zektzer et al. (1973) classified the methods into two large groups: study based on coastal drainage area and sea. Bokuniewicz et al. (2004) discussed three basic approaches; traditional measurements of hydraulic head and permeability, water budget estimation, and use of geochemical tracers to measure SGD. Burnett et al. (2001) listed three basic approaches to assess SGD; modeling, direct measurement, and tracer techniques. They further elaborated that there are several modeling approaches, however, the direct measurement is restricted to seepage meters whereas tracing techniques make use of either natural geochemical species or artificial tracers. Smith et al. (2003) and Smith and Nield (2003) quantified SGD in two ways; by inference from inshore water balance considerations and direct and indirect field measurements. Moore (2010) had presented the following methods to assess and measure SGD:

1. Thermal images to detect SGD
2. Electromagnetic techniques to discern subsurface porosity and/or pore water salinity
3. Seepage meters to directly measure discharge
4. Tracer techniques to integrate SGD signals on a regional scale
5. Groundwater flow modeling

On the other hand, Burnett et al. (2006) conducted a literature review and applied the following methods to detect and quantify SGD to five different coastal locations in the world:

1. Seepage meters
2. Piezometers
3. Natural and artificial tracers
4. Water balance approaches
5. Hydrograph separation techniques
6. Theoretical analysis and numerical simulations
7. Isotopic analyses
8. Radon and radium isotopes

The purpose of this chapter is not to cover all the methods that have used to detect and quantify SGD, instead to include brief descriptions of the most important methods that can be used to detect and quantify SGD as accurate as possible.

5.1 Local-Scale In Situ Approaches

5.1.1 Direct Measurements-Seepage Meter

In 1977, Lee (1977) designed and developed an inexpensive instrument to measure the movement of water between lakes and estuaries, which is named as manual seepage meter and has been used by many researchers including Bokuniewicz et al. (2004, 2008), Smith et al. (2003) and others to measure SGD. A simple seepage meter consists of 0.25 m^2 benthic chambers with a small plastic bag attached to a port on the top to collect SGD (Lee 1977; Moore 2010). In the beginning, the plastic bag was partially filled with seawater and volume change in the bag was measured as flux of water out of or into the sediment. A direct measurement, such as use of seepage meter, can be a good method to measure SGD, however, investigators rarely attempted to measure it using direct approach because they believe that this approach would require excessive investment of time and equipment (Lee 1977). Burnett et al. (2006) and Smith et al. (2003) supported this fact by stating "Perhaps the most serious disadvantage for coastal zone studies is that manual seepage meters are very labor intensive."

The major limitations of SGD studies using seepage meters are: many seepage meters are needed to capture natural spatial and temporal variability of SGD flow rates; the resistance of the used tube and bag should be minimized to the degree possible to prevent artifacts; collection bag cover might reduce the effects of surface water movements due to wave, current or stream flow activity; initially the collection bag should contain a measured volume of water, which can help to determine positive and negative seepage (Burnett et al. 2006 and others). In addition, caution is required in using seepage meters during rough conditions when they may be biased by currents and waves (Moore 2010; Cable et al. 2006; Shinn et al. 2003). However, seepage meters have the advantage of allowing direct measurement of groundwater discharge.

Later, with the rapid advancement of technology, many researchers (Sayles and Dickinson 1990; Taniguchi and Fukuo 1993; Krupa et al. 1998) developed automated seepage meters. The one developed by Taniguchi and Fukuo (1993), known as Taniguchi-type or heat-pulse type is based on the travel time of a heat pulse down a narrow tube (Fig. 12.6). This meter uses thermistors in a column positioned above an inverted funnel covering a known area of sediment. The system measures travel

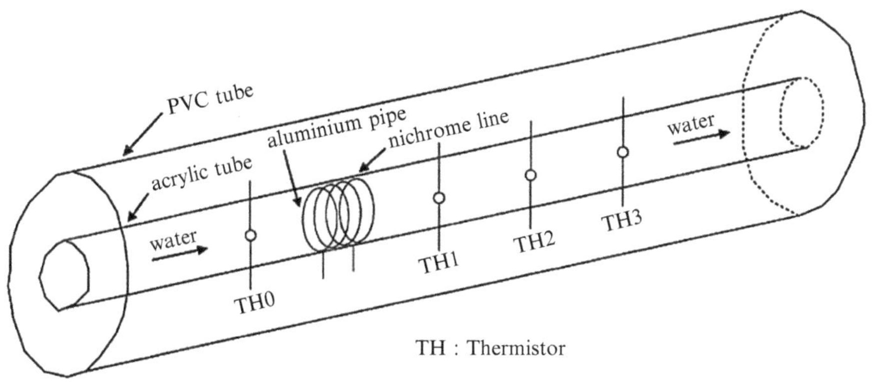

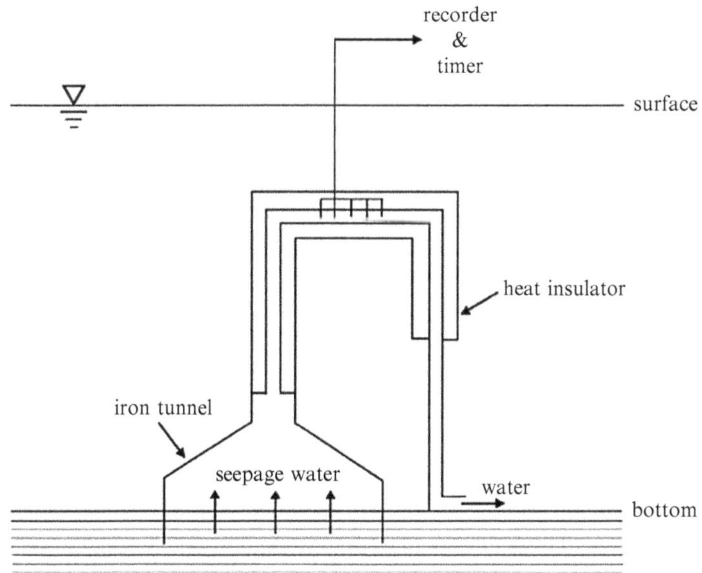

Fig. 12.6 Taniguchi-type or heat-pulse type seepage meter (Burnett et al. 2006)

time of a heat pulse generated within the column by a Nichrome wire induction heater, which is a function of the advective velocity of water flowing through the column. This meter can measure seepage at 5 min intervals (Burnett et al. 2006).

Taniguchi and Iwakawa (2001) developed a continuous heat type automated seepage meter, which is designed to measure temperature gradient of the water flowing between the downstream and upstream sensors in a flow tube of 1.3 cm diameter (Burnett et al. 2006). The temperature difference between two sensors is at its maximum in the absence of water flow and gradually decreases with increasing water flow velocity. Some other similar types of seepage meters were also developed, such as dye-dilution seepage meter at Woods Hole Oceanographic Institution that involves the injection of a colored dye into a mixing chamber attached to a seepage meter and the subsequent measurement of the dye absorbance in the mixing chamber over time. Typically, dye is injected every hour into a mixing chamber of known volume (usually 0.5 L), and the absorbance is recorded every 5 min. The rate at which the dye is diluted by the inflowing seepage water is used to calculate the flow-rate (Burnett et al. 2006). Paulsen et al. (2001) used to evaluate seepage flow based on ultrasonic measurements. The benthic chamber uses a commercially available acoustic flow meter to monitor SGD. Since the speed of sound depends on salinity, the same sensor output can be used to continuously calculate the salinity of SGD as well as the flow rates (Burnett et al. 2006). Rosenberry and Morin (2004) used an electromagnetic flow meter that calculates fluid velocity by measuring the voltage that is induced as it passes through an electromagnetic field. The flow meter uses Faraday's law of electromagnetic induction to measure the flow process. This is one of the simplest autonomous meters developed to date that simply requires off-the-shelf "upgrades" to the Lee-type meter (Zektser et al. 2007) and accurately measure SGD with no moving parts.

5.1.2 Hydrologic

There are two hydrologic approaches to estimate SGD: application of Darcy's law and the water mass balance method (Mulligan and Charette 2009). Darcy's law calculation uses two approaches; piezometer and flow nets methods. Piezometer method of estimating SGD uses multilevel piezometer nests that measure the groundwater potential in sediment at various depths. Using the known or measured hydraulic conductivity of the coastal aquifer, the SGD can be estimated using well known Darcy's equation ($\vec{q} = -K\nabla h$, where K is hydraulic conductivity, ∇h is the hydraulic gradient). This method is simple to install and monitor, however, the accuracy of the SGD measurements entirely depends on the accurate estimation of the hydraulic conductivity of the aquifer. Therefore, piezometer nests are often used in conjunction with seepage meters to estimate the hydraulic conductivity from observed seepage rates and the hydraulic gradient using Darcy's equation (Taniguchi et al. 2003).

The flow nets have been used to estimate direct SGD to oceans, however, this method is used to get preliminary estimates of SGD. In this method, a flow-net is plotted using stream lines forming imaginary stream tubes and equipotential lines perpendicular to the stream lines to form approximately squares. Then SDG can be calculated using the following equation:

$$q = K^*p^*dh^*\frac{b}{n} \tag{12.16}$$

where K is hydraulic conductivity, p is number of stream tubes in the flow net, dh is change in hydraulic head between two equipotential lines, b is thickness of the coastal aquifer perpendicular to the plane, and n is number of equipotential head drops in the flow net (Loaiciga and Zektser 2001).

The mass balance approach to estimate SGD requires all inputs and outputs of water flow, except SGD, through an imaginary control volume of the coastal aquifer. Assuming a steady-state condition over a specified time frame, the SGD rate is calculated as the difference between all inputs and outputs. Implementing this approach can be quite simple or can result in complex field works, but the quality of the data obviously affect the level of uncertainty. Even with extensive field sampling, accurate water budgets are seldom known with certainty and so should be used with precaution. Furthermore, the mass balance approach is not appropriate if the spatial and temporal variability of SGD is needed for a particular study (Mulligan and Charette 2009).

5.2 Medium-Scale Approaches

5.2.1 Chemical Tracers

A common approach for quantifying SGD to the coastal ocean is to use geochemical tracers that are naturally enriched in groundwater relative to seawater and that their chemistry is well-understood within the marine environment (Zektser et al. 2007). There are two types of tracers; natural geochemical tracers and artificial tracers that are frequently used to measure SGD. However, natural tracers have been much more widely used in SGD studies than artificial tracers (Burnett et al. 2006; Knee and Paytan 2011). Natural tracers, other than temperature, include salinity, silica, barium, methane, radon (e.g., ^{222}Rn) and the element radium, which has four naturally occurring isotopes, such as ^{223}Ra, ^{224}Ra, ^{226}Ra, and ^{228}Ra, are highly concentrated in the discharging groundwater relative to a coastal ocean to provide a detectable signal (Wilson and Rocha 2013).

Johannes (1980) used salinity as a groundwater tracer and found that salinity changes can be measured accurately to determine the fraction of observed changes in concentration of dissolved substances due to SGD. However, groundwater discharge to the coastal ocean may not be accompanied by a discernible freshening of receiving coastal waters. In fact, SGD may include a major portion of recirculated seawater and the effects of dilution, for example, may overshadow its nature as a tracer. Additionally, observed salinity differences could be the result of other freshwater inputs to the marine environment such as surface water discharge (Wilson and Rocha 2013).

In river water, Ra strongly adsorbs to particles whereas in seawater it primarily dissolves due to presence of salt. This difference in chemical behavior is due to a change in the adsorption coefficient of Ra between freshwater and saltwater.

Moore (2000) found more ^{226}Ra in the coastal waters than it could be supported by desorption from river-borne sediment. Rn and Ra quartet originate from the decay of uranium and thorium radioisotopes that are present in most rocks. According to Wilson and Rocha (2013), ^{222}Rn is an ideal tracer of SGD, as it behaves conservatively (as a noble gas), and is relatively easy to measure. In addition, its concentration in groundwater is several orders of magnitude higher than in seawater and its half-life of 3.82 days is comparable with the scale of coastal circulation (Cable et al. 1997; Dulaiova et al. 2008; Stieglitz et al. 2010).

Mostly, Radium concentrations show a distinct gradient being highest in the near-shore waters. By using an estimate of the residence time of these near-shore waters on the shelf and assuming steady-state conditions, one can calculate offshore flux of excess ^{226}Ra. If this flux is supported by SGD along the coast, then the SGD can be estimated by dividing radium flux by estimated ^{226}Ra activity of groundwater. A convenient enhancement to this approach is that one may use the short-lived radium isotopes, ^{223}Ra and ^{224}Ra, to assess the water residence time (Moore 2000; Burnett et al. 2006).

Groundwater discharge rates into the ocean can be determined with isotopes of radium and radon using a steady-state mass balance approach with the exception that atmospheric evasion must be taken into consideration (Zektser et al. 2007; Burnett et al. 2006). To close the mass balance, sources and sinks within the study area must be well known. The principal loss mechanisms are radioactive decay, advection, and dispersion by coastal currents, and for radon only, air-sea gas exchange. If inventory of ^{222}Rn is monitored over time by making allowances for losses due to atmospheric evasion and mixing with lower concentration water offshore, any changes observed can be converted to fluxes by a mass balance approach. Although changing radon concentrations in coastal waters could be in response to a number of processes, advective transport of groundwater (pore water) through sediment of Rn-rich solutions is often the dominant process. Thus, if one can measure or estimate the radon concentration in the advecting fluids, the ^{222}Rn fluxes may be easily converted to water fluxes, which requires source of groundwater. This is generally determined through direct measurements of groundwater samples collected from terrestrial wells (Zektser et al. 2007; Burnett et al. 2006).

Although radon and radium isotopes are very useful for assessment of SGD, they both clearly have some limitations. Radium isotopes, for example, may not be enriched in freshwater discharges such as from submarine springs, whereas Rn is subject to exchange with the atmosphere, which may be difficult to model under some circumstances (e.g., sudden large changes in wind speeds, waves breaking along a shoreline). The best solution may be to use a combination of tracers to avoid these pitfalls (Burnett et al. 2006).

Methane (CH_4) is another useful geochemical tracer used to detect SGD. Both ^{222}Rn and CH_4 were used to evaluate SGD in northeastern Gulf of Mexico (Cable et al. (1996)). The linear relationships between tracer inventories and measured seepage fluxes were found statistically significant. Cable et al. (1996) found that inventories of ^{222}Rn and CH_4 in the coastal waters varied directly with groundwater seepage rates and had a positive relationship (Burnett et al. 2006). The isotopes of helium (^3He and ^4He) were also used to detect SGD. A small flux of relatively old groundwater can be easily traced with helium (Zektser et al. 2007). For example, Top et al. (2001) used helium to quantify SGD in Florida Bay.

There are several other natural radioactive (^3H, ^{14}C,U, etc.) and stable (^2H, ^{13}C, ^{15}N, ^{18}O, $^{87/88}Sr$, etc.) isotopes and some artificial tracers (e.g., SF_6, ^{131}I, ^{32}P, fluorescein dye, and Chlorofluorocarbons (CFCs)) have been used for conducting SGD investigations, tracing water masses, and calculating the age of groundwater (Burnett et al. 2006; Knee and Paytan 2011). For example, Sulphur hexafluoride (SF_6) has been used successfully as a tracer in ocean-scale mixing experiments, demonstrating the volume of water that can be tagged with this tracer (Ledwell et al. 1993; Zektser et al. 2007). Reilly et al. (1994) used CFCs to calculate the recharge age of shallow groundwater and estimate the rate of groundwater movement. Although various tracers have been used successfully to identify SGD to estuarine and coastal zones, tracers may include negative environmental impacts due to the introduction of contaminants. Since SGD is spatially and temporally variable, tracers are particularly useful for integrating SGD on a larger, more synoptic time and space scales (Kim et al. 2011; Schubert et al. 2014).

5.3 Modeling Approaches

Modeling of groundwater flow that includes SGD estimate has become more common with the availability of computer based modeling packages. A simple water balance calculation can be useful in some situations to estimate SGD, whereas in some situations, more sophisticated hydrologic or hydrogeologic or numerical models can accurately estimate SGD. However, lack of data of hydraulic conductivity of aquifer materials over the range of scales can affect model results. Also, inconsistencies between modeling and direct measurement approaches may arise because different components of SGD are being evaluated or models do not include transient terrestrial (e.g., recharge cycles) or marine processes (e.g., tidal pumping, wave set up, storms, thermal gradients, seasonal sea level changes) that drive part or all of the SGD (Moore 2010). Nevertheless, a simple water balance approach, hydrologic/hydrogeologic modeling and numerical simulations, and hydrograph separation techniques are frequently used to identify and estimate SGD.

Water Balance Approach: Water balance approach depends on quantity of inputs (e.g., precipitation) and quantities of outputs (e.g., evapotranspiration, surface runoff, groundwater discharge, and change in storage) from the hydrological system (Knee and Paytan 2011). Generally, change in storage is assumed to be negligible when longer time periods (i.e., annual average rates) are considered for calculation purpose. Therefore, a water balance equation can be used to estimate fresh SGD. This method requires precise measurements of precipitation, evapotranspiration, surface runoff etc. to estimate SGD. However, this method is only suitable for fresh SGD component and does not address recirculated seawater, which is most likely an important source for nutrients and other chemicals. Although this method is relatively simple, it is often imprecise for SGD estimations because uncertainties associated with values used in the calculations are often of the same order of magnitude as the SGD being evaluated (Burnett et al. 2006; Dzhamalov 1996). In addition, water balance approach is generally applicable on a basin or larger scale, and it may not work well in situations where watershed boundaries do not correspond with groundwater recharge zones (Knee and Paytan 2011).

Hydrologic Modeling and Numerical Simulations: Hydrogeologic modeling using numerical simulations can be used to estimate the net SGD (Knee and Paytan 2011). The modeling approach is not only used to determine current SGD rate, but also used to predict future SGD rates for different hydrologic conditions. MODFLOW is commonly used numerical model to estimate freshwater component of SGD. Some studies have also used SEAWAT model, which unlike MODFLOW, allows modeling of variable-density groundwater flow to simulate seawater intrusion that would be expected within the subterranean estuary (Knee and Paytan 2011). SEAWAT can be used to calculate both salt and freshwater components of SGD successfully.

Although numerical models are widely used for analysis of basin-scale groundwater hydrology, they have some limitations. For example, aquifer systems are usually heterogeneous, and it is difficult to obtain sufficient representative values, such as hydraulic conductivity, hydraulic head, porosity and boundary conditions to adequately characterize this heterogeneity. All these parameters can vary considerably both in space and time. Hydraulic conductivity often varies over several orders of magnitude within short distances. Spatial and temporal variations for boundary conditions are also required for hydrological modeling, but this information is often hampered by our ability to acquire adequate field data within the time frame of a typical study (Burnett et al. 2006; Knee and Paytan 2011). In comparison, hydrologic modeling approach is very similar to water balance approach, which is best suited to regional or basin scale studies, but model could not capture variability of SGD at fine scale.

Hydrograph Separation Techniques: Integrated hydrological and hydrogeological method consists of using the procedure of separating the hydrograph of river runoff for a long-term period to determine SGD within coastal zone (Dzhamalov 1996; Zektser et al. 2007). Hydrograph separation technique is based on the assumption that the amount of fresh groundwater entering into streams can be estimated using hydrograph separation, which can be extrapolated by graphical or analytical means to coastal zone to estimate SGD (Burnett et al. 2006). River runoff hydrograph separation is widely used to calculate groundwater discharge from aquifers hydraulically connected with rivers. The technique is based on distinguishing baseflow component of river discharge during wet seasons. During dry seasons, groundwater discharge is assumed to be equal to the river discharge. The use of hydrograph separation of the total river runoff is mainly used for relatively small river basins under natural conditions (Zektser et al. 2007).

Generally, two approaches are used to separate the hydrograph for estimating the fresh groundwater flow component. The first method is simply to assign a base flow according to the shape of the hydrograph. This technique can be performed in several ways including the unit hydrograph method (Bouwer 1978; Zektzer et al. 1973; Burnett et al. 2006). This technique for large scale SGD estimates applies only to coastal areas with well-developed stream networks and to zones of relatively shallow, mainly freshwater aquifers (Taniguchi et al. 2003; Burnett et al. 2006). The second method uses geochemical tracers for hydrograph separation. Usually, water and geochemical mass balances in a river are calculated as follows:

$$D_T = D_S + D_G \quad (12.17)$$

$$C_T D_T = C_S D_S + C_G D_G \quad (12.18)$$

where D and C are discharge rate and geochemical concentrations, respectively, and subscripts T, S, and G represent the total, surface water and groundwater components. The two unknown values of D_S and D_G in Eqs. (12.17) and (12.18) can be solved using measured D_T, C_T, C_S, and C_G .

Similar to water balance approach and hydrological modeling and numerical simulation techniques, this technique also has same order of uncertainties. In addition, this technique excludes the groundwater discharge downstream of the gauging station because gauging stations for measuring river discharge are always located some finite distance upstream from the coast (Buddemeier 1996; Taniguchi et al. 2003; Burnett et al. 2006).

5.4 *Large-Scale Spatial and Remote Sensing approaches*

Remote sensing approach can be a good and an economical method for studying large-scale SGD study. Large-scale aerial photographs make it possible to study peculiarities of geologic and geomorphologic structure of the coastal sea and land in detail, and to single out faults and intensively fissured zones, where large submarine springs are usually confined. Change in seawater color or transparency caused by submarine springs can be identified from aerial photographs or satellite image (Zektser et al. 2007). While this technique is quite useful for identifying spatial discharge patterns, it has not yet been applied to estimating flow rates (Mulligan and Charette 2009).

There is substantial scope for further development and expansion of potential uses of additional available remote sensing datasets for SGD studies. Several studies have demonstrated that remote sensing can be used to detect SGD and map water quality parameters such as temperature, turbidity, particulate matter concentration (PMC), chlorophyll-a, and total suspended solids. These parameters can be derived from satellite images and can be used to further characterize the areas to identify SGD hot spots to expose the link between SGD and nutrient enrichment (Wilson and Rocha 2013).

5.4.1 Spatial Analysis

Geographical Information System (GIS) is a tool, which can store, display, analyze and generate spatial maps as vector and raster formats, and their attribute tables as data base or resultant reports. GIS tool has capabilities to integrate or overlay most of existing submarine and landscape features such as soil, geology, river, spring, lake, recharge area, well, seas, man-made features etc. which can help the researchers to detect and quantify SGD. GIS has many spatial analysis tools that can be used to extract information from spatial layers to identify SGD features (Zektser et al. 2007).

A GIS model can be used to estimate spatial distribution of coastal groundwater discharge and associated nitrogen loading to coastal waters. GIS model, however, requires numerous spatial data layers such as soil, land use, hydraulic conductivities, and hydraulic gradients. Hydraulic conductivities and hydraulic gradients can be used to calculate coastal groundwater discharge whereas land use and discharge patterns can be used to calculate contamination flux potential (Gallagher et al. 2001). GIS can be an alternative method of water balance approach and numerical simulation to estimate SGD. Gallagher et al. (2001) conducted a study to evaluate GIS-based approach to identify SGD and associated nitrogen loading at coastal waters of the southern Delmarva Peninsula, the Eastern Shore of Virginia. They found that the GIS-based approach was useful to predict SGD and associated nitrogen loading at large scale, however, this approach also had some significant limitations such as the need for reliable spatial data, and challenge to obtain high resolution data for hydraulic conductivities and hydraulic gradients, which have some degree of temporal and high spatial variability.

5.4.2 Thermal Remote Sensing

When you are walking along the beach barefoot and find the sand to be cold and wet on a warm day, it may represent a site of SGD, where cool groundwater is discharging from an aquifer having a temperature lower than the ocean. The same principle may be applied to aerial surveys using infrared images (Moore 2010).

A thermal infrared (TIR) imagery has proven to be a useful indicator of groundwater discharge locations. Infrared imaging has been used to identify the location and spatial variability of SGD by exploiting temperature difference between surface water and groundwater at certain times of the year (Portnoy et al. 1998; Miller and Ullman 2004; Mulligan and Charette 2006).

In the 1970s, the technology of airborne thermal sensors was mature enough to launch systematic exploration campaigns based on temperature contrasts between seawater and freshwater in regions where SGD were supposed to exist. Airplanes or helicopters flew with those sensors at few hundred meters of altitudes. The campaign was not all crowned with success due to difficulties in detecting small thermal anomalies because of sea surface instability. Nevertheless, some discoveries were recorded and ground resolution and thermal accuracy of the sensors allowed detection of small thermal anomalies. Nowadays, thermal sensors onboard of satellites such as LANDSAT 7 or ASTER can provide more accurate information on temperature contrasts (UNESCO 2004).

Kolokoussis et al. (2011) used TABI-320 (Thermal Airborne Broadband Imager) thermal image to study SGD at Agios Stefanos Bay, Evvoikos Gulf of Central Greece. According to Kolokoussis et al. (2011), the thermal images provided promising results as far as the coastal sub-aerial springs were concerned. They also revealed the different functions of coastal sub-aerial springs compared to submarine groundwater discharges. Freshwater flowing along the coast from small sub-aerial springs was distributed over the area as a thin surficial film of water having a different temperature, with a decreasing trend as the distance from the sources increased.

By contrast, submarine groundwater discharges caused generation of 3–5 m diameter circular shapes on the seawater surface, which were clearly visible by eyes. Rapid mixing of groundwater with seawater in combination with high heat capacity of seawater is the main cause of this situation, as thermal remote-sensing images capture only surface temperatures. As a result, small submarine groundwater discharges are not observed on thermal imagery, while even very small coastal sub-aerial springs are observed quite easily.

Obviously, it is important to acquire thermal images in periods when groundwater has a significant thermal difference with seawater. Based on TIR, groundwater appears 2–3 °C cooler than seawater in June (Fig. 12.7a) and 2–3 °C warmer than seawater in February (Fig. 12.7b). The images clearly show spatial variability in SGD along the beach face.

5.4.3 Hyperspectral Remote Sensing

Thermal remote-sensing is frequently used to detect fresh SGD (UNESCO 2004). Modern thermal sensors are capable of precisely mapping water surface temperature (Torgersen et al. 2001), but are not capable of sensing temperature underneath

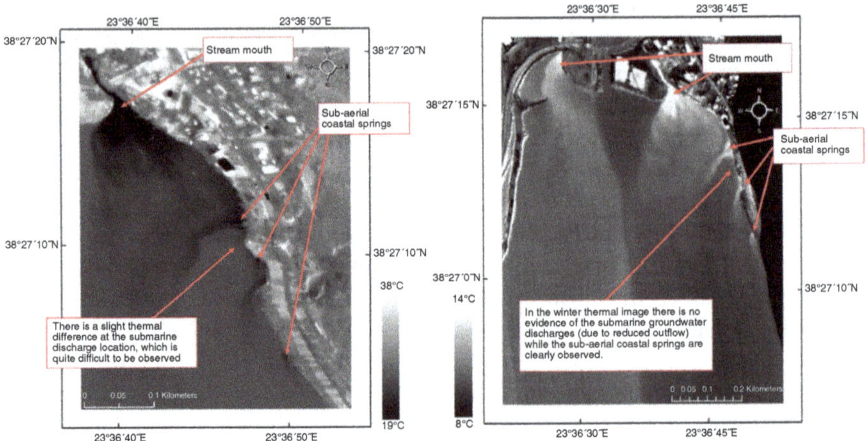

Fig. 12.7 TABI thermal image of Agios Stefanos Bay, (**a**) June 2005 (*left*), and (**b**) Feb 2006 (*right*) (Kolokoussis et al. 2011)

water surface. Therefore, relatively small submarine groundwater discharges may not be detected on thermal imagery due to high heat capacity of seawater (Kolokoussis et al. 2011). Hyperspectral data is most appropriate for detecting relatively small submarine groundwater discharges, which may not be detected on thermal imagery, due to the increase in turbidity that SGD causes. This is confirmed by strong correlations between hyperspectral data and in situ measured turbidity-related water inherent optical properties (Kolokoussis et al. 2011).

Kolokoussis et al. (2011) used CASI-550 hyperspectral imagery to study SGD at Agios Stefanos Bay, Evvoikos Gulf of Central Greece. They found higher digital values, mainly in the range of 530–580 and 440–490 nm, which were due to presence of high particulate matter concentration (PMC) and optical backscattering (OBS) at the SGD locations. As shown in Fig. 12.8, authors have shown the color composite image L_{490}, L_{560}, L_{690} nm (R, G, B) where L_{xxx} is the relevant hyperspectral image band, which is centered on the xxx nm wavelength. Fig. 12.8 shows a bright funnel shape (outlined in yellow), which represents the turbidity around the SGD. The funnel's head coincides with the SGD point and the location where turbidity is highest. Moving away, turbidity decreases and the funnel's wide mouth gradually vanishes.

Taking into consideration certain data acquisition related issues, the combined use of very high spatial resolution airborne thermal and hyperspectral sensors lead to a very efficient methodology for detection of relatively small sub-aerial coastal springs and submarine groundwater discharges (Kolokoussis et al. 2011).

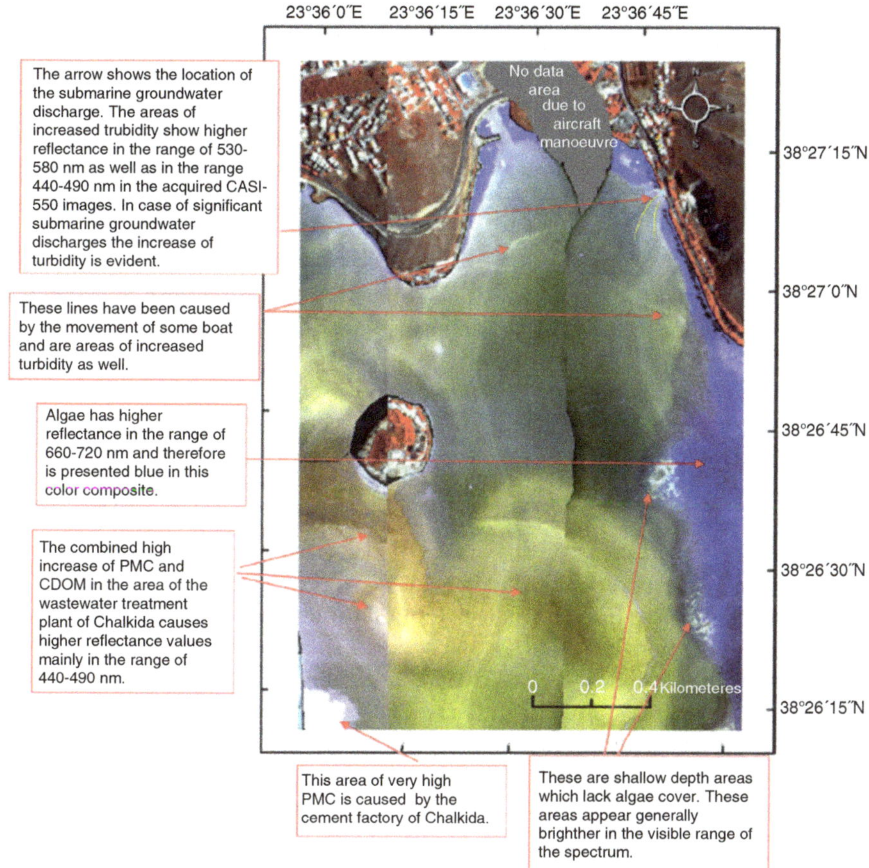

Fig. 12.8 CASI hyperspectral color composite L_{490}, L_{560}, L_{690} nm (*R*, *G*, and *B*) of Agios Stefanos Bay in June 2005. The funnel shaped area of increased turbidity around the main groundwater discharge is outlined in *yellow* (Kolokoussis et al. 2011)

6 Impacts of Temporal and Spatial Variability of SGD to the Coastal Environment

SGD is now recognized as an important water pathway between land and ocean. However, it is difficult to quantify SGD owing to its significant spatial and temporal variability (Li et al. 2009). For example, spatial variability of various parameters such as hydraulic head, salinity, and conductivity in one direction are much different than other directions. SGD is delivered into the oceans at various hydrogeological settings where coastal aquifers have spatial heterogeneous properties such as preferential flow paths, therefore, SGD is commonly heterogeneously distributed along the coast (Stieglitz et al. 2008).

The rate of SGD can also vary at any locations on different time scales from seconds to hours to weeks to months (Zektser et al. 2007). Wave actions can cause pore water to oscillate in its flow direction with periods of seconds to minutes, whereas tides act on time scales ranging from hours to weeks and can set-up equally persistent discharge patterns. Long-term patterns can also be established by other large-scale sea level variations and by changes in the onshore hydraulic gradients, for example, due to variations in recharge or the inverse effect of changes in barometric pressure on water tables. All these cause high spatial and temporal variability of SGD.

SGD is also recognized as important pathways for the transport of water and dissolved constituents such as contaminants, nutrients, carbon, and trace metals from land to ocean.

Therefore, it is important to evaluate the fluxes and biogeochemical characteristics of the groundwater interfacing with ocean, because the spatial and temporal variations of SGD may affect fauna and flora, which live in the coastal zones that ultimately impact the coastal environment as a whole (Taniguchi et al. 2003).

Spatial variability in SGD and recharge, control the preferential pathways and travel times for nutrients and contaminants if any to move from source (land) areas to the ocean. The presence of preferred flow paths mean that some areas of the aquifer are likely to contain less active-slower flowing-areas of SGD compared with other parts of the aquifer, which contain preferential flow paths (Smith et al. 2003). Spatiotemporal variability of SGD has considerable impact on mobilization of contaminants, dissolved solids, nutrients, and others from one location to other. Since spatial and temporal variations of SGD are directly linked with the variations in characteristics of biogeochemical, dissolved contaminants, nutrients of SGD; spatiotemporal variations of SGD has significant impact on coastal ecosystems and its environment.

7 Contribution of SGD to Nutrient Transport from Coastal Catchments

A substantial quantity of nutrients is delivered to ocean waters through SGD. Nutrient transport pathways and travel times from source to ocean are determined by the spatial and temporal patterns of groundwater recharge and flow. Nutrients are mobilized by local recharge and transported toward the coast by regional groundwater flow, which is dominated by channelized flows through high conductivity pathways (Smith et al. 2003). Quantitative studies on groundwater nutrient inputs to coastal zones have been carried out only in a few regions, and many of these are located in the United States (Slomp and Van Cappellen 2004). Nutrient input through SGD competes river inputs in certain regions and may play a significant role in nutrient cycling and primary productivity in coastal areas of ocean. Nitrogen (N) and phosphorus (P) concentrations of groundwater vary and depend on its inputs, soil and aquifer type, aquifer hydraulic conductivity, groundwater recharge rate, and climate (Tiessen 1995). Concentrations of N and P in coastal groundwater are often much higher than those in river water, compensating for the lower mass flux of groundwater relative to surface water (Slomp and Van Cappellen 2004). Continued residential and agricultural development of near-shore areas worldwide, for example, is leading cause to increased inputs of N and P from fertilizer and wastewater to groundwater, and part of these nutrients are released to coastal surface waters (Valiela et al. 1990). The primary factors controlling N and P flux through coastal aquifers and sediments to coastal waters are: the flow paths and rates of the groundwater as these determine the residence time and extent of contact with the aquifer solids, the supply rate and form of N and P from natural or anthropogenic sources, and the redox conditions in the subsurface, which strongly affect the transformation processes and mobility of N and P (Slomp and Van Cappellen 2004).

Since, groundwater usually has higher nutrient concentrations than the receiving seawater, and sometimes these concentrations are also significantly higher than those in rivers or streams, even small discharge of groundwater may make large contributions to coastal nutrient budgets (Burnett et al. 2003). Therefore, groundwater can play a significant role in nutrient budgets even when the volume contribution is small (Slomp and Van Cappellen 2004). For example, Hasaki Beach of Japan had about six times higher nitrate concentrations in shallow groundwater than in water from the nearby Tone River (Knee and Paytan 2011). Likewise, concentrations of nitrite, phosphate, and silica were also higher in groundwater than in the Tone River (Uchiyama et al. 2000). Similarly, in Chesapeake Bay, groundwater nutrient concentrations were up to two times higher than those found in estuarine surface waters (Charette and Buesseler 2004). In the Northeast region of the United States, highest fluxes of nitrogen from groundwater discharge were found in Massachusetts. For example, a study at Great Sippewisset Marsh showed groundwater nitrate and ammonium fluxes as high as 16,288 and 568 kg-N/ha/year, respectively (Zektser et al. 2007).

The contribution of SGD's nutrients to coastal budgets will likely increase as human activity on coastal watersheds increases (Bowen et al. 2007; Santos et al. 2013). The influences of SGD may be maximized in those regions where water exchange and fluvial discharge are limited and/or where groundwater is contaminated by agriculture or wastewater disposal. Variations of SGD-associated nutrient fluxes and N/P ratios in groundwater are large not only in a specific region, but also over different geological and environmental settings. Such large natural and artificial variations can hamper global estimate of N and P fluxes through SGD (Kim and Swarzenski 2010). All anthropogenic activity in the catchment region, including agriculture and tourism, has a potential impact on water quality of SGD. Terrestrial groundwater that is the source of fresh SGD only filters through a thin blanket of unconsolidated sediment before entering fractured bedrock aquifer system. The major input of dissolved nitrate and silicate is from terrestrially sourced fresh groundwater driven by inland aquifer head and from local run-off/discharge during periods of heavy rains. Phosphate and nitrate apparently may originate from domestic effluents, which have not been fully treated in local sewage systems. Dissolved nitrate, silicate and phosphate concentrations in groundwater wells, and in SGD indicate that water resources in the region are heavily influenced by agricultural activities and the release of domestic waste (Povinec et al. 2012).

Probably, SGD is quantitatively most important in shallow, permeable (sand and limestone) coastal aquifers with high rates of groundwater recharge. Estimates of N and P fluxes through SGD vary over orders of magnitude. Highest rates for N are observed in aquifers contaminated with sewage. Quantitative studies for areas in the US and Australia suggested that groundwater inputs of N and P may be larger than the river inputs on a regional scale (Slomp and Van Cappellen 2004).

8 Future Developments about SGD: Opportunities, Challenges and Management

SGDs are considered as strategic freshwater resources at the coastal regions of arid and semi-arid countries of the Mediterranean region (UNESCO 2004). In the 1980s, the European Communities financed a research project to evaluate and map the existence of submarine springs in the coastal areas of Spain, Sicily, and Greece. The Blue Plan program has estimated that the Mediterranean Sea receives a discharge of an average of 520×10^9 m^3 of freshwater each year from its basins; a quarter of this quantity is provided by submarine groundwater discharge.

Many professional organizations and scientific bodies involved in SGD research projects include the Intergovernmental Oceanographic Commission (IOC), the International Hydrological Program (IHP), the Scientific Committee on Oceanic Research (SCOR), which established its working group WG112 on SGD, Land-Ocean Interactions in the Coastal Zone (LOICZ), and the International Association of Physical Sciences of the Oceans (IAPSO). An initiative on SGD characterization was developed by the International Atomic Energy Agency (IAEA) and UNESCO in 2000 as a 5-year plan to assess methodologies and the importance of SGD for coastal zone management. SGD also has been taken up as one of the priority research issues of the GEOTRACES program (GEOTRACES Science Plan) (Zhang and Mandal 2012).

SGD can influence the ocean circulation structure and stratification of the ocean because it is an important route of freshwater and heat transport, especially in high latitudes. The temperature of groundwater is warm during winter and has small changes annually. In polar areas, higher temperature of SGD relative to the riverine water leads to increases in sea surface temperatures and freshening of surface salinity of seawater, when SGD flows into the ocean from late autumn to winter. This inhibits ice formation and interferes with the seawater subduction. Therefore, a quantitative understanding of SGD variation in the Polar Regions, including details of heat/freshwater transport is required, in particular as an important factor affecting the world ocean circulation in the future. Moreover, SGD is also an important carbon supply route to the ocean. Groundwater also plays an important role in providing temporary CO_2 storage from the atmosphere to the ocean because of its relatively long residence time (Zhang and Mandal 2012).

With new detection methods and developed field validation strategies, for example, in situ ^{222}Rn monitoring and delayed counting techniques, coastal scientists will be able to realistically identify and quantify SGD and associated fluxes into receiving coastal water bodies. Ideally, a thorough SGD study should begin with a reconnaissance survey that includes geophysical streaming resistivity and nearly continuous ^{222}Rn work to identify sites of enhanced fluid exchange across the sediment-water interface. Once such sites are established, direct measurements of this exchange via autonomous seepage meters and numerical modeling efforts to link coastal observations to a larger hydro-geologic framework should complement the use of geochemical tracers. This approach provides powerful diagnostic tools for regional scale SGD investigations (Swarzenski 2007).

Future research on SGD should include detailed experimental and modeling studies of the biogeochemical processes in the saltwater–freshwater mixing zone of coastal aquifers and near-coastal sediments. Particular insight in the redox conditions in coastal aquifers is essential since these strongly determine the transformation and mobility of nutrients in groundwater. Future study areas should include tectonically active coastal areas (e.g., Western coastline of the United States) and developing countries (e.g., Southeast Asia) (Slomp and Van Cappellen 2004).

Although numerous previous and current SGD studies have been carried out worldwide, knowledge of the significance and impact on oceans at global scale is still limited. This is because SGD fluxes are strongly controlled by many local and regional conditions, with the exception of the influences on the ocean circulation and CO_2 fixation via SGD (Zhang and Mandal 2012). Therefore, our knowledge of SGD is still insufficient, and so further research is needed. SGD is one of the factors that affect important coastal marine environments and material circulation. Better understanding of SGD will help to clarify global environmental changes due to human activity and climate change (Zhang and Mandal 2012).

References

Barlow PM (2003) Groundwater in freshwater-saltwater environments of the Atlantic coast. US Geological Survey Circulation. 1262:121 p

Bear J (1979) Hydraulics of groundwater. McGraw Hill, New York, 567 p

Bokuniewicz HJ, Kontar E, Rodrigues M, Klein DA (2004) Submarine groundwater discharge (SGD) patterns through a fractured rock aquifer: a case study in the Ubatuba coastal area, Brazil. Revista de la Asociacion Argentina de Sedimentologia 11(1):9–16

Bokuniewicz H, Taniguchi M, Ishitoibi T, Charette M, Kontar E (2008) Direct measurements of submarine groundwater discharge (SGD) over a fractured rock aquifer in Flamengo Bay Brazil. Estuar Coast Shelf Sci 76:466–472

Bouwer H (1978) Groundwater hydrology. McGraw-Hill, New York, 480 p

Bowen JL, Kroeger KD, Tomasky G, Pabich WJ, Cole ML, Carmichael RH, Valiela I (2007) A review of land-sea coupling by groundwater discharge of nitrogen to New England estuaries: mechanisms and effects. Appl Geochem 22(1):175–191
Buddemeier RW (1996) Groundwater discharge in the coastal zone. In: Proceedings of an international symposium, Texel, Russian Academy of Sciences, Moscow, p 179
Burnett W, Taniguchi M, Oberdorfer J (2001) Measurement and significance of the direct discharge of groundwater into the coastal zone. J Sea Res 46:109–116
Burnett W, Bokuniewicz H, Huettel M, Moore WS, Taniguchi M (2003) Groundwater and pore water inputs to the coastal zone. Biogeochemistry 66:3–33
Burnett WC, Aggarwal PK, Aureli A, Bokuniewicz H, Cable JE, Charette MA, Kontar E, Krupa S, Kulkarni KM, Loveless A, Moore WS, Oberdorfer JA, Oliveira J, Ozyurt N, Povinec P, Privitera AMG, Rajar R (2006) Quantifying submarine groundwater discharge in the coastal zone via multiple methods. Sci Total Environ 367:498–543
Cable J, Bugna G, Burnett W, Chanton J (1996) Application of ^{222}Rn and CH_4 for assessment of groundwater discharge to the coastal ocean. Limnol Oceanogr 41:1347–1353
Cable JE, Burnett WC, Chanton JP (1997) Magnitude and variations of groundwater seepage along a Florida marine shoreline. Biogeochemistry 38:189–205
Cable JE, Martin JB, Jaeger J (2006) Exonerating Bernoulli? On evaluating the physical and biological processes affecting marine seepage meter measurements. Limnol Oceanogr Methods 4:172–183
Charette MA, Buesseler KO (2004) Submarine groundwater discharge of nutrients and copper to an urban subestuary of Chesapeake Bay (Elizabeth River). Limnol Oceanogr 49:376–385
Dogan A, Fares A (2008) Effects of land use changes and groundwater pumping on saltwater intrusion in coastal watersheds. In: Fares A, ElKadi A (eds) Land management impacts on coastal watershed hydrology, Progress in water resources series. WIT Press, Southampton
Dulaiova H, Gonneea ME, Henderson PB, Charette MA (2008) Geochemical and physical sources of radon variation in a subterranean estuary—implications for groundwater radon activities in submarine groundwater discharge studies. Mar Chem 110:120–127
Dzhamalov RG (1996) Methodical approaches to regional assessment of groundwater discharge into the seas, Groundwater discharge in the coastal zone. In: Buddemeier RW (ed) LOICZ IGBP, 44-47, LOICZ Texel, Russian Academy of Sciences, Moscow, 179 p
Essaid HI (1990) The computer model SHARP, a quasi three dimensional finite-difference model to simulate freshwater and saltwater flow in layered coastal aquifer systems. USGS water resources investigations report 90-4130, Reston
Fetter CW (2001) Applied hydrogeology. Prentice Hall, Upper Saddle River, p 598
Gallagher DL, Wyne JW, Reay WG, Robinson M (2001) A Geographic information system analysis of submarine groundwater discharge on the eastern shore of Virginia. First international conference on saltwater intrusion and coastal aquifers monitoring, modeling, and management. Essaouira, 23–25 April 2001, 13 p
Ghyben WB (1888) Nota in verband met de voorgenomen putboring nabij Amsterdam, Tijdschrift van Let Koninklijk: Inst. Van Ing
Glover RE (1959) The pattern of fresh-water flow in a coastal aquifer. J Geogr Res 64:457–459
Guo W, Langevin CD (2002) User's guide to SEAWAT: a computer program for simulation of three-dimensional variable-density groundwater flow. Techniques of water-resources investigations Book 6 Chapter A7, 77 p
Harbaugh AW, Banta ER, Hill MC, McDonald MG (2000) MODFLOW-2000, the U.S. Geological Survey modular groundwater model: user guide to modularization concepts and the groundwater flow process. USGS open-file report 00-92. USGS, 121 p
Hays RL, Ullman WJ (2007) Direct determination of total and fresh groundwater discharge and nutrient loads from a sandy beachface at low tide (Cape Henlopen, Delaware). Limnol Oceanogr 52:240–247
Henry HR (1959) Salt intrusion into freshwater aquifers. J Geophys Res 64:1911–1919

Henry HR (1964) Effects of dispersion on salt encroachment in coastal aquifers, U.S. Geological Survey Water-Supply Paper, 1613-C, pp C71–C84
Herzberg A (1901) Die Wasserversorgung einiger nordseebader. J Gasbeleucht Wasserversorg 44:815–819
Hubbert MK (1940) The theory of groundwater motion. J Geol 48(8):785–944
Huyakorn PS, Anderson PF, Mercer JW, White JHO (1987) Saltwater intrusion in aquifers: development and testing of a three-dimensional finite element model. Water Resour Res 23:293–312
Hwang DW, Kim G, Lee Y-W, Yang H-S (2005) Estimating submarine inputs of groundwater and nutrients to a coastal bay using radium isotopes. Marine Chem 96:61–71
Johannes RE (1980) The ecological significance of the submarine discharge of groundwater. Mar Ecol Prog Ser 3:365–373
Johnson T (2007) Battling seawater intrusion in the Central & West Coast Basins. Technical Bulletin, Weather Replenishment District of Southern California. 13:1–2
Keller EA, Loaiciga HA (1991) Earthquakes, fluid pressure and mountain building: a model. Geol Soc Am 23(5):84–85, Abstracts with program
Kim G, Swarzenski PW (2010) Submarine groundwater discharge (SGD) and associated nutrient fluxes to the Coastal Ocean. In: Liu KK, Atkinson L, Quinones R, Talaue-McManus L (eds) Carbon and nutrient fluxes in continental margins, Global change-the IGBP Series, 757 p
Kim G, Kim J-S, Hwang D-W (2011) Submarine groundwater discharge from oceanic islands standing in oligotrophic oceans: Implications for global biological production and organic carbon fluxes. Limnol Oceanogr 56(2):673–682
King JN, Mehta AJ, Dean RG (2010) Analytical models for the groundwater tidal prism and associated benthic water flux. Hydrogeol J 18(1):203–215
Knee KL, Paytan A (2011) Submarine groundwater discharge: a source of nutrients, metals, and pollutants to the coastal ocean. In: Wolanski E, McLusky DS (eds) Treatise on estuarine and coastal science, vol 4. Academic, Waltham, pp 205–233
Kohout FA (1966) Submarine springs: a neglected phenomenon of coastal hydrology. Hydrology 26:391–413
Kolokoussis P, Karathanassi V, Rokos D, Argialas D, Karageorgis AP, Georgopoulos D (2011) Integrating thermal and hyperspectral remote sensing for the detection of coastal springs and submarine groundwater discharges. Int J Remote Sens 32(23):8231–8251
Krupa SL, Belanger TV, Heck HH, Brok JT, Jones BJ (1998) Krupaseep-the next generation seepage meter. J Coast Res 25:210–213
Kuan WK, Jin GQ, Xin P, Robinson C, Gibbes B, Li L (2012) Tidal influence on seawater intrusion in unconfined coastal aquifers. Water Resour Res 48:W02502. doi:10.1029/2011WR010678
Kumar M, Ramanathan AL, Neupane BR, Tu DV, Kim S (2010) Critical evaluation of the recent development and trends in submarine groundwater discharge research in Asia. In: Neupane BR, Ramanathan AL, Bhattacharya P, Dittmar T, Bala Krishna Prasad M (eds) Management and sustainable development of coastal zone environments. Springer, Dordrecht, pp 89–102
Langevin CD, Guo W (2006) MODFLOW/MT3DMS–based simulation of variable-density groundwater flow and transport. Groundwater 44(3):339–351
Langevin CD, Shoemaker WB, Guo W (2003) MODFLOW-2000, the U.S. Geological Survey modular groundwater model—documentation of the SEAWAT-2000 version with the variable-density flow process (VDF) and the integrated MT3DMS Transport Process (IMT). USGS Open-File Report 03-426. USGS, 43 p
LaRoche J, Nuzzi R, Waters R, Wyman K, Falkowski PG, Wallace DWR (1997) Brown tide blooms in Long Island's coastal waters linked to inter-annual variability in groundwater flow. Glob Change Biol 3:397–410
Ledwell JR, Watson AJ, Law CS (1993) Evidence for slow mixing across the pycnocline from an open-ocean tracer-release experiment. Nature 364:701–703

Lee DR (1977) A device for measuring seepage flux in lakes and estuaries. Limnol Oceanogr 22:140–147
Li L, Barry DA, Stagniti F, Parlange JY (1999) Submarine groundwater discharge and associated chemical input to a coastal sea. Water Resour Res 35(11):3253–3259
Li X, Hu BX, Burnett WC, Santos IR, Chanton JP (2009) Submarine groundwater discharge driven by tidal pumping in a heterogeneous aquifer. Ground Water 47(4):558–568
Loaiciga HA (1989) An optimization approach to groundwater quality monitoring network design. Water Resour Res 25:1771–1780
Loaiciga HA, Zektser IS (2001) Methods to estimate direct ground-water discharge to the ocean. J King Abdulaziz Univ Mar Sci 12:24–32
Mays L (2011) Ground and surface water hydrology. Wiley, New York, p 617
McDonald MG, Harbaugh AW (1988) A modular three dimensional finite-difference groundwater flow model. USGS techniques of water-resources investigations, Book 6, Chapter A1, USGS, 586 p
Miller DC, Ullman WJ (2004) Ecological consequences of groundwater discharge to Delaware Bay, United States. Ground Water 42:959–970
Moore WS (2000) Determining coastal mixing rates using radium isotopes. Cont Shelf Res 20:1995–2007
Moore WS (2010) The effect of submarine groundwater discharge on the ocean. Ann Rev Mar Sci 2:59–88
Mulligan AE, Charette MA (2006) Intercomparison of submarine groundwater discharge estimates from a sandy unconfined aquifer. J Hydrol 327:411–425
Mulligan AE, Charette MA (2009) Groundwater flow to the coastal ocean. In: John HS, Karl KT, Steve AT (eds) Encyclopedia of ocean sciences. Academic, Oxford, pp 88–97
Oude Essink GHP (2001) Improving fresh groundwater supply-problems and solutions. Ocean Coast Manag 44:429–449. doi:10.1016/S0964-5691(01)00057-6
Paulsen RJ, Smith CF, O'Rourke D, Wong T (2001) Development and evaluation of an ultrasonic groundwater seepage meter. Ground Water 39:904–911
Pinder GF, Cooper HH (1970) A numerical technique for calculating the transient position of the saltwater front. Water Resour Res 6(3):875–882
Portnoy JW, Nowicki BL, Roman CT, Urish DW (1998) The discharge of nitrate contaminated groundwater from developed shoreline to marsh-fringed estuary. Water Resour Res 34:3095–3104
Povinec PP, Burnett WC, Beck A et al (2012) Isotopic, geophysical and biogeochemical investigation of submarine groundwater discharge: IAEA-UNESCO intercomparison exercise at Mauritius Island. J Environ Radioact 104:24–45
Ravindran AA, Ramanujam N (2014) Detection of submarine groundwater discharge to coastal zone study using 2d electrical resistivity imaging study at Manapad, Tuticorin, India. Ind J GeoMarine Sci 43(2):224–228
Reilly T (1993) Analysis of groundwater systems in freshwater-saltwater environments (Chapter 18). In: Alley WM (ed) Regional groundwater quality. Van Nostrand Reinhold, New York, pp 443–469
Reilly TE, Goodman AS (1985) Quantitative analysis of saltwater-freshwater relationships in groundwater systems—a historical perspective. J Hydrol 80:125–160
Reilly TE, Plummer LN, Phillips PJ, Busenberg E (1994) The use of simulation and multiple environmental tracers to quantify groundwater flow in a shallow aquifer. Water Resour Res 30 (2):421–433
Rosenberry DO, Morin RH (2004) Use of an electromagnetic seepage meter to investigate temporal variability in lake seepage. Ground Water 42(1):68–77
Santos IR (2008) Submarine groundwater discharge driving mechanisms and biogeochemical aspects. Electronic Theses, Treatises and Dissertations, 145 p
Santos IR, Eyre BD, Huettel M (2012) The driving forces of porewater and groundwater flow in permeable coastal sediments: a review. Estuar Coast Shelf Sci 98:1–15

Santos IR, de Weys J, Tait DR, Eyre BD (2013) The contribution of groundwater discharge to nutrient exports from a coastal catchment: post-flood seepage increases estuarine N/P ratios. Estuar Coasts 36:56–73

Sayles FL, Dickinson WH (1990) The seep meter: a benthic chamber for the sampling and analysis of low velocity hydrothermal vents. Deep-Sea Res 88:1–13

Schluter M, Sauter EJ, Andersen CE, Dahlgaard H, Dando PR (2004) Spatial distribution and budget for submarine groundwater discharge in Eckernforde Bay (Western Baltic Sea). Limnol Oceanogr 49(1):157–167

Schubert M, Scholten J, Schmidt A, Comanducci JF, Pham MK, Mallast U, Knoeller K (2014) Submarine groundwater discharge at a single spot location: evaluation of different detection approaches. Water 6:584–601

Shinn EA, Reich CD, Hickey TD (2003) Reply to comments by Corbett and Cable on our paper, "Seepage meters and Bernoulli's revenge.". Estuaries 26:1388–1389

Simmons GM (1992) Importance of submarine groundwater discharge (SGWD) and seawater cycling to material flux across sediment/water interfaces in marine environments. Mar Ecol Prog Ser 84:173–184

Slomp CP, Van Cappellen P (2004) Nutrient inputs to the coastal ocean through submarine groundwater discharge: controls and potential impact. J Hydrol 295:64–86

Smith AJ, Nield SP (2003) Groundwater discharge from the superficial aquifer into Cockburn Sound Western Australia: estimation by inshore water balance. Biogeochemistry 66:125–144

Smith AJ, Turner JV, Herne DE, Hick WP (2003) Quantifying submarine groundwater discharge and nutrient discharge into Cockburn Sound Western Australia. CSIRO Land and Water. Perth, 185 p

Sonrel L (1868) Le fond de la mer. L. Hachette & Cie, Paris

Stieglitz T, Taniguchi M, Neylon S (2008) Spatial variability of submarine groundwater discharge, Ubatuba, Brazil. Estuar Coast Shelf Sci 76:493–500

Stieglitz TC, Cook PG, Burnett WC (2010) Inferring coastal processes from regional-scale mapping of 222Radon and salinity: examples from the Great Barrier Reef, Australia. J Environ Radioact 101:544–552

Swarzenski PW (2007) U/TH series radionuclides as coastal groundwater tracers. Chem Rev 107:663–674

Swarzenski PW, Bratton JF, Crusius J (2004) Submarine ground-water discharge and its role in coastal processes and ecosystems. USGS open file report, 2004–1226

Taniguchi M, Fukuo Y (1993) Continuous measurements of ground-water seepage using an automatic seepage meter. Ground Water 31:675–679

Taniguchi M, Iwakawa H (2001) Measurements of submarine groundwater discharge rates by a continuous heat-type automated seepage meter in Osaka Bay, Japan. J Groundw Hydrol 43:271–277

Taniguchi M, Burnett WC, Cable JE, Turner JV (2002) Investigations of submarine groundwater discharge. Hydrol Process 16:2115–2129

Taniguchi M, Burnett WC, Cable JE, Turner JV (2003) Assessment methodologies of submarine ground-water discharge. In: Taniguchi M, Wang K, Gamo T (eds) Land and marine hydrogeology. Elsevier, Amsterdam, pp 1–23

Tiessen H (1995) Phosphorus in the global environment, transfers, cycles and management. SCOPE, vol 54, Wiley, New York, 462 p

Todd DK (1980) Groundwater hydrology. Wiley, New York, 535

Top Z, Brand LE, Corbett RD, Burnett W, Chanton J (2001) Helium and Radon as tracers of groundwater input into Florida Bay. J Coast Res 17(4):859–868

Torgersen CE, Faux RN, Mcintosh BA, Poage NJ, Norton DJ (2001) Airborne thermal remote sensing for water temperature assessment in rivers and streams. Remote Sens Environ 76:386–398

Uchiyama Y, Nadaoka K, Rolke P, Adachi K, Yagi H (2000) Submarine groundwater discharge into the sea and associated nutrient transport in a sandy beach. Water Resour Res 36:1467–1479

UNESCO (2004). Submarine groundwater discharge: management implications, measurements and effects- prepared for international hydrological program (IHP), intergovernmental oceanographic commission (IOC) by scientific committee on oceanic Research (SCOR) and Land-Ocean Interactions in the Coastal Zone (LOICZ). Published in 2004 by the United Nations Educational, Scientific and Cultural Organization 7, place de Fontenoy, 75352 Paris 07 SP. 35 p

Valiela I, Costa J, Foreman K, Teal JM, Howes B, Aubrey D (1990) Transport of groundwater-borne nutrients from watersheds and their effects on coastal waters. Biogeochemistry 10 (3):177–197

Valiela I, Foreman K, LaMontagne M, Hersh D, Costa J, Peckol P (1992) Couplings of watersheds and coastal waters: sources and consequences of nutrient enrichment in Waquoit Bay, Massachusetts. Estuaries 15:443–457

Valiela I, Bowen JL, Kroeger KD (2002) Assessment of models for estimation of land-derived nitrogen loads to shallow estuaries. Appl Geochem 17:935–953

Volker RE, Rushton KR (1982) An assessment of the importance of some parameters for seawater intrusion and a comparison of dispersive and sharp-interface modeling approaches. J Hydrol 56:239–250

Voss CI (1984) SUTRA—a finite-element simulation model for saturated-unsaturated fluid density dependent groundwater flow with energy transport or chemically reactive single species solute transport. USGS water resources ınvestigations report 84-4369, USGS, 409 p

Voss CI, Provost AM (2002) SUTRA, a model for saturated-unsaturated variable-density groundwater flow with solute or energy transport. USGS water-resources ınvestigations report 02-4231, 250 p

Weiss E (1982) A model for the simulation of flow of variable-density groundwater in three-dimensions under steady state conditions. USGS, Reston, USGS open file report 82-352, 59 p

Wilson J, Rocha C (2013) Developing remote sensing as a tool for detection, quantification and evaluation of submarine groundwater discharge (SGD) to Irish Coastal Waters. EPA STRIVE Report Series No. 112. Environmental Protection Agency, Johnstown Castle, Wexford

Xia Y, Li H, Yang Y, Huang W (2012) The enhancing effect on todal signals of a submarine spring connected to a semi-infinite confined aquifer. Hydrol Sci J 57(6):1231–1248

Xin P, Robinson C, Li L, Barry DA, Bakhtyar R (2010) Effects of wave forcing on a subterranean estuary. Water Resour Res 46(12), W12505. doi:10.1029/2010wr009632

Xin P, Wang SSJ, Robinson C, Li L, Wang Y-G, Barry DA (2014) Memory of past random wave conditions in submarine groundwater discharge. Geophys Res Lett 41:2401–2410

Xin P, Wang SSJ, Robinson C, Li L, Wang Y-G, Barry DA (2015) Nonlinear interactions of waves and tides in a subterranean estuary. Geophys Res Lett 42:1–8. doi:10.1002/2015GL063643

Younger PL (1996) Submarine groundwater discharge. Nature 382:121–122

Zektser IS, Loaiciga HA (1993) Groundwater fluxes in the global hydrological cycle: past, present and future. J Hydrol 144:405–427

Zektser IS, Dzhamalov RG, Everett LG (2007) Submarine groundwater. CRC Press, Boca Raton, 466

Zektzer IS, Ivanov VA, Meskheteli AV (1973) The problem of direct groundwater discharge to the seas. J Hydrol 20:1–36

Zhang J, Mandal AK (2012) Linkages between submarine groundwater systems and the environment. Curr Opin Environ Sustain 4:219–226

Zheng C, Wang PP (1999) MT3DMS, a modular three-dimensional multispecies transport model for simulation of advection, dispersion and chemical reactions of contaminants ın groundwater systems. Waterways Experiment Station, U.S. Army Corps of Engineers, Vicksburg

Chapter 13
Management of Declining Groundwater Resources and the Role of Policy Planning in Semi-Arid Economies: The Case of Texas High Plains

Rachna Tewari

Abstract Recent decades have witnessed the expansion of irrigated agriculture in some of the most productive semi-arid economies in the world, such as the Texas High Plains. Interest in groundwater management continues to increase due to the excessive depletion of groundwater, which is the primary source of fresh water supply in these economies that critically depend on irrigated agriculture. In addition, the uncertainties posed through extreme climatic events such as drought exacerbate the challenge of managing this scarce yet vital resource. This chapter describes the problem of declining groundwater resources in a semi-arid economy exemplified by the Texas High Plains, discusses the management approach of water use restriction to handle the ever increasing demand for agricultural production with a limited supply in hand, and outlines the role of policy planning in groundwater management. For a semi-arid economy such as the Texas High Plains that is largely dependent on groundwater resources, an effective partnership between groundwater management bodies and local producers is an important step in planning towards the long term objective of ensuring adequate groundwater availability in the future.

1 Introduction

Recent decades have witnessed the expansion of irrigated agriculture in some of the most productive semi-arid economies in the world, such as the Texas High Plains. This in turn has led to an increasing interest in groundwater management, and has attracted much attention from producers, policy makers, and groundwater management bodies. Groundwater provides for the majority of fresh water supply in these economies that critically depend on irrigated agriculture for their sustenance in the

R. Tewari (✉)
Department of Agriculture, Geosciences, and Natural Resources, The University of Tennessee at Martin, 265 Brehm Hall, Martin, TN, USA
e-mail: rtewari@utm.edu

A. Fares (ed.), *Emerging Issues in Groundwater Resources*, Advances in Water Security, DOI 10.1007/978-3-319-32008-3_13

absence of other reliable sources of water. However, continued dependence has led to a decline of this limited resource due to heavy withdrawals and low recharge in several semi-arid agricultural and livestock production areas globally, as well as in the U.S. (Scanlon et al. 2006). In addition, the uncertainties posed through climatic events such as drought exacerbate the challenge of managing this scarce yet vital resource. This chapter describes the problem of declining groundwater resources in a semi-arid economy exemplified by the Texas High Plains, discusses the management approach of water use restriction to handle the ever increasing demand for agricultural production with a limited supply in hand, and outlines the role of policy planning in groundwater management.

2 Groundwater Resources and Agriculture

Among nature's valuable and partially renewable resources, groundwater is considered a vital resource globally because of its importance in food production by way of irrigation and for potable use by humans. The importance of groundwater in agriculture is manifested in several critical areas, such as irrigation that directly impact agricultural productivity. In agricultural production areas facing surface water deficit and low precipitation rates, groundwater provides for a reliable and continuous water supply thereby influencing the type of cropping systems and associated profitability in these high demand regions.

2.1 Importance of Groundwater in a Semi-Arid Agricultural Economy

Semi-arid regions comprise approximately 30 % of global terrestrial surface area, and have been expanding (Dregne 1991; Scanlon et al. 2006). With increase in population growth, water scarcity continues to remain a critical issue in these regions as compared to the more humid regions (Scanlon et al. 2006). Between 1960 and 2000, about 40 % of the population growth in the United States occurred in the semi-arid states in the south western region (US Census Bureau 2004). The most productive semi-arid regions in the world are agricultural economies that heavily rely on irrigated production for their sustenance. Irrigation consumed 90 % of global freshwater resources during the past century (Shiklomanov 2000; Jury and Vaux 2005), and represents 20 % of cropland and approximately 40 % of food production (Jury and Vaux 2005; Molden 2007).

Over the past few decades, groundwater has emerged as the primary irrigation source for 40 % of global irrigated acreage, and 60 % within the United States (Siebert et al. 2010). Besides irrigation, groundwater is the major source of about half of the United States' domestic and municipal water supply (Alley et al. 1999), and forms the backbone of several industrial economies in a majority of the states, besides contributing flow to rivers and wetland areas (Alley et al. 1999). Major reasons for the expansion of groundwater based irrigation are availability and ease of access to this resource, few infrastructure requirements for extraction, and consideration of groundwater as an alternate supply source to manage production in case of adverse climatic events such as drought (Giordano 2009).

The High Plains region in the United States is one of the most productive semi-arid regions in the country, and is often called the "grain basket" of the United States (Scanlon et al. 2012). This is enunciated by the fact that the market value of agricultural products was $35 billion in the High Plains relative to the United States total of $300 billion in 2007 (National Agricultural Statistics Services 2011). Groundwater has been the major source of irrigation particularly in the Texas High Plains region of the High Plains, and has largely supported the growth and expansion of irrigated agriculture as well as the livestock production in the area (Tewari et al. 2014). The Texas High Plains witnessed the expansion of irrigated agriculture as early as the 1950s (Colaizzi et al. 2009). Both irrigated area and volume pumped recorded their highest levels in mid 1970s and saw a steady decline in the next decade. In the early 2000s irrigated acreage was approximately the same as it was in the late 1950s, however volume pumped had shown a slight increase (Fig. 13.1).

In the present situation, with water demand far exceeding supply (Colaizzi et al. 2009), a semi-arid production economy such as the Texas High Plains faces increasing dependence on groundwater for irrigation resulting in high depletion rates for the region's groundwater resources.

2.2 *Withdrawals from a Partially Renewable Resource: The Ogallala Aquifer*

The Ogallala aquifer is the prime source of groundwater for irrigation purposes in the U.S. High Plains, and underlies parts of eight states: Texas, New Mexico, Oklahoma, Colorado, Kansas, Nebraska, South Dakota, and Wyoming (Alley

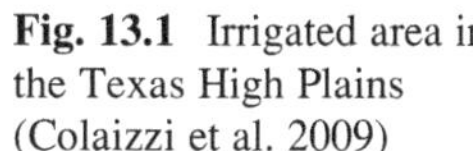

Fig. 13.1 Irrigated area in the Texas High Plains (Colaizzi et al. 2009)

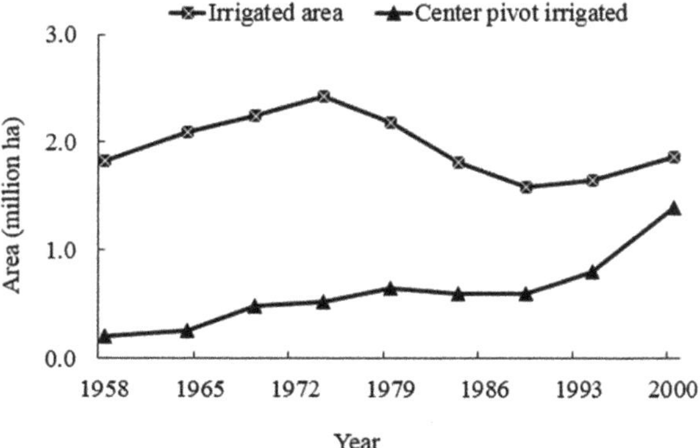

Fig. 13.2 Map depicting the location of Ogallala aquifer (USGS 2014)

et al. 1999). Figure 13.2 outlines the location of the aquifer underlying the above mentioned states.

Water table levels in the aquifer have been declining in certain locations over the years more specifically in the southern and central region of the aquifer (Tewari et al. 2014). This rate of decline for water levels in the aquifer is accelerated by the fact that recharge when compared to the rate of depletion is much lower in certain areas, and varies greatly from one region of the aquifer to the other (Birkenfeld 2003).

A study by Scanlon et al. (2012) for the High Plains region indicates that high recharge in the northern High Plains results in sustainable levels of groundwater being pumped, whereas in the central and southern High Plains higher depletion has occurred on account of lower recharge. In addition, depletion is highly localized with 4 % of the High Plains land area accounting for a third of total groundwater depletion in the area (Scanlon et al. 2012). Based on future predictions for recharge rates, this study indicates that more than one third of the southern High Plains will be unable to support irrigation within the next three decades (Scanlon et al. 2012).

In 1990, the Ogallala aquifer in the eight-state area of the Great Plains contained approximately three and half billion acre-feet of water, of which Texas had about 12 % in storage or approximately 417 million acre-feet of water (Tewari et al. 2014). A recent estimate of the volume of water in the eight-state Great Plains area was less than three billion acre-feet (Tuholske 2008). These changes in the groundwater resource supply will most likely have a significantly negative impact on the agricultural production areas that depend on the aquifer for their sustenance (Tewari et al. 2014).

In a study conducted by the Center for Geospatial Technology at Texas Tech University, changes in saturated thickness were observed over an 18 year interface for selected counties in the Texas High Plains, and estimates of the saturated thickness in the year 2030 were developed. The counties of study showed substantial change in the amount of water storage underlying the county over a study period of 18 years from 1990 to 2008 (Texas Tech Center for Geospatial Technology 2010). Figure 13.3 shows the saturated thickness of the aquifer underlying the counties of study as an estimate for the year 2030.

A saturated thickness level of 30 ft or less indicates a reduction in availability of water in the aquifer in the region for further use (Schloss and Buddemeier 2000). It is clearly observable that a majority of the counties in the study region will experience a steady decline in saturated thickness, and are anticipated to have a saturated thickness less than 30 ft in certain parts. As a result of this anticipated decline in the volume of water storage in the aquifer, continued availability of irrigation water will be interrupted. Irrigated production of economically important crops in the region such as cotton and corn may therefore experience reduction in terms of both productivity and profitability.

3 Management of Groundwater

Several studies such as Caswell and Zilberman (1986), Buller and Williams (1990), and Negri and Brooks (1990), have evaluated how irrigators make management decision choices with regard to the use of technology, and the use of available water. The outcome of these studies indicates that several factors such as commodity prices, energy prices, pump lift, and well capacity determine the use of technology, which in turn could affect the amount of water use. In addition, the amount of water use in agricultural production is also influenced by the water rights in the area, crop water requirement, and management policies. Above all, since irrigators operate with an objective of profit maximization, a long-run investment in irrigation technology influences the selection and planning of future crop-mix and the amount of water used for irrigation. Several past studies suggest that government intervention through policy measures is required to ensure that adequate groundwater stock is maintained in the aquifer for future use, and these will be discussed in the following section.

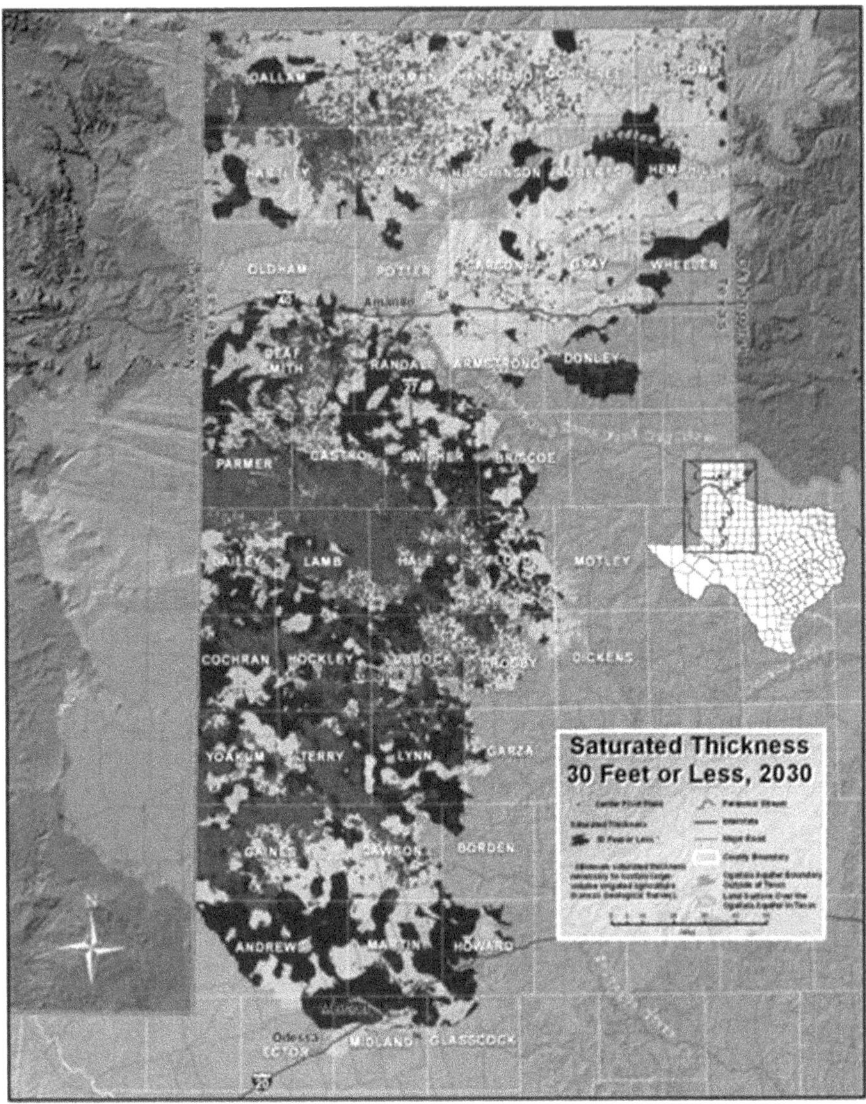

Fig. 13.3 Texas High Plains' counties with saturated thickness of 30 ft or less in 2030 (Texas Tech University Center for Geospatial Technology 2010)

3.1 The Role of Policy Planning and Groundwater Conservation Districts: The Case of Texas High Plains

It is imperative to first understand the water rights system in the state of Texas to understand policy implications for groundwater management. The Texas law of water rights has a complex structural framework which can be accounted for by inclusion of elements of the Hispanic water law, in combination with traditional English common law (Handbook of Texas Online 2009; Tewari et al. 2014). The existence of water-rights law paves the way for determining the entitlement of available water supply usage in respective quantities.

The Texas Judicial system divides water into different classes, which are governed by different set of rules regarding usage and ownership of each class. Broadly, there are three major legal classes of water (atmospheric moisture, surface water, and percolating groundwater) with several sub-classes for each class (Templer 1992). The basic assumption followed by the Texas courts is that all groundwater may be classified as percolating unless there is a distinct evidence of proof, about the source. With these defined laws in existence, the ownership of percolating groundwater in Texas is clearly articulated. For percolating water (percolating below the surface of the earth (Tex. Water Code §36.001(5) (Texas Constitution and Statues 2011))), the rule of capture also referred to as the "law of the biggest pump", is considered as the regulatory or guiding principle in the state of Texas (Tewari et al. 2014). This has been derived out of the English common law which was adopted in the year 1904 by the Texas Supreme Court in a historical ruling which is considered as a landmark in legal doctrines on groundwater. This specific ruling has been recorded as Houston and Central Texas Railway vs. East (Texas Water Development Board 2004). Following the rule of capture, the owner of the overlying land can pump and use the water with few restrictions, irrespective of the impact on adjacent landowners or more distant users of water (Tewari et al. 2014).

The rule of capture has been maintained as the case law for groundwater in the State of Texas, ever since the East ruling and has been modified with regard to groundwater management in different regions of the state (Tewari et al. 2014). A law passed in 1949 in the Texas Legislature provided for the voluntary establishment of local conservation districts for underground water. These are specifically called the underground water conservation districts (UWCD). Such local districts hold a strategic position in the regulation and management of groundwater in Texas. In the above context, a "district" is defined as an authority formulated to regulate the spacing of water wells, the production from water wells, or both, as defined in the Texas Water Code §36.001(1) (Texas Constitution and Statues 2011; Tewari et al. 2014).

By the late 1980s only 11 districts had been established under general law or by special legislation, and several areas with rapidly depleting groundwater levels still were devoid of any district and subsequent supervision. Currently there are 94 underground water conservation districts in Texas which have been confirmed by voters through local elections (Texas Water Development Board 2010). The Senate Bill 1 passed in 1997 recognized the importance of managing groundwater resources by suggesting a "grassroots" approach to be implemented through local UWCDs. Under the same bill, the Texas Water Development Board designated 16 regional water planning groups (Texas Water Development Board 2010). The primary objective followed by these planning groups was to formulate and submit regional water management plans addressing important ground and surface water management issues like regulation of water use under drought or severe water stress, maintenance of existing water rights and specific groundwater conservation district plans.

The second 5 year planning period began in 2001 with the approval of Senate Bill 2. The primary focus still was on UWCDs. Under the Senate Bill 2, UWCDs were awarded a provision for charging a fee on water use not to exceed an upper limit of $1 per acre foot and $10 per acre foot for agricultural and non-agricultural uses, respectively. The main purpose behind this was to apply charges on production and give the underground water conservation districts, authority to regulate water use. In the State of Texas, the rule of capture is still held as the foundational law governing water, however Senate Bills 1 and 2 provide authorization to regulate underground water pumping by the UWCDs (Johnson et al. 2004). The 80th Regular Session of the Texas Legislature in the year 2007, acknowledged the crucial role that water conservation plays in meeting the future demand via the passage of Senate Bill 3 and House Bill 4 in the year 2008. These proceedings were also the platform for creation of the Water Conservation Advisory Council, responsibilities of which include: monitoring trends in water conservation implementation and new technologies for possible inclusion as best management practices (Water Conservation Advisory Council 2008).

The on-going debate concerning groundwater conservation in Texas has resulted in studies that examine alternative policy options for groundwater management. Johnson et al. (2004) and Johnson et al. (2009) used a dynamic optimization model along with an input–output model and studied the impacts of different policy alternatives on the saturated thickness of the Ogallala Aquifer and economy of Texas High Plains. The study compared a baseline scenario with three policy alternatives for response to aquifer depletion using a planning horizon of 50 years. These policy alternatives were: introducing a fee on water extracted per acre-foot, an annual restriction of water use to 75 % of a 10 year average water use, and a restriction on the drawdown of the aquifer over a 50-year planning horizon to 50 % of the initial saturated thickness at the beginning of the period. The results showed that the baseline scenario resulted in the most rapid exhaustion of the water supply and caused the most dramatic decrease in net income for the economy over time. The production fee alternative showed little change from the baseline and the drawdown restriction resulted in slightly lower net income than the quota restriction. They further concluded that the aquifer drawdown restriction could be considered the most effective alternative because it projected the best equivalence between producer profit, water conservation, and subsequent effects on the regional economy.

Wheeler et al. (2008) evaluated the effectiveness and efficiency of temporary water rights buyout policies for 10 and 20 years. This study used county level optimization models to maximize net present value of net returns to land, management, groundwater, and irrigation systems over a 60 year planning horizon for a given county as a whole. They evaluated two voluntary incentive based policies which could feasibly be implemented under current Texas water law and concluded that the longer term 20-year water rights buyout is a more efficient and effective

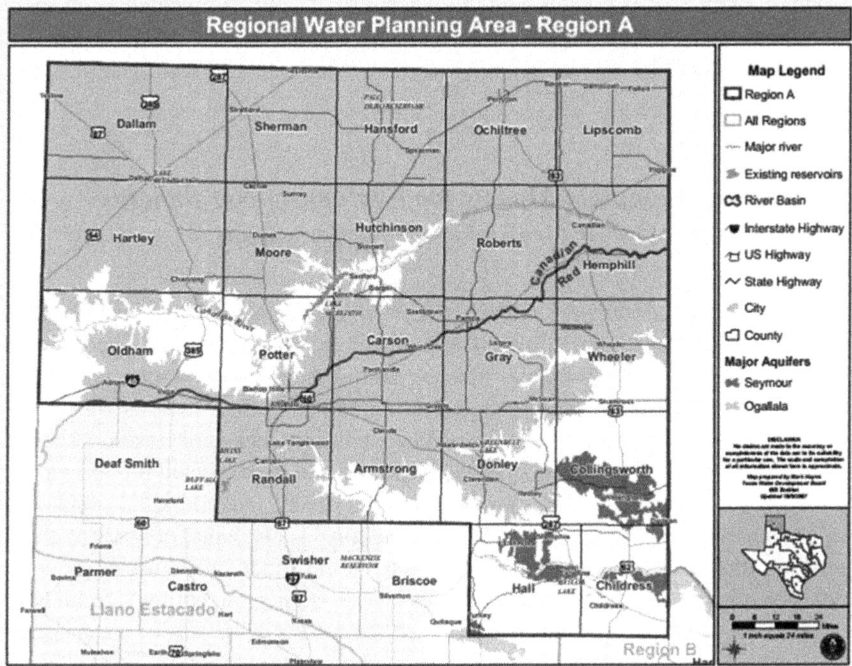

Fig. 13.4 Regional water planning area—Region A of Texas (Texas Water Development Board 2010)

water conservation tool than the 10 year water rights buyout for the Southern High Plains of Texas.

Recently, Tewari et al. (2014) evaluated the policy option of multi-year water allocation coupled with water-use restriction in the Regional Water Planning Area-Region A of Texas (Fig. 13.4), over a planning time frame of 60 years using an economic optimization model.

A water allocation system over multiple years will potentially reduce inefficient use of water during the allocated period by allowing for water stock (allocation) to accumulate for the judicious users, which could be rolled over into the next allocation period at an appropriate rate of the unused stock. An unconstrained baseline scenario with no restrictions was compared with water use restriction scenarios at successively increasing rates. Under the unconstrained baseline scenario with no policy implementation over 60 years, the counties of study showed a decline in saturated thickness that recommends the incorporation of water-use restriction alternatives at different rates. Increasing restrictions rates led to decline in water use per acre as well as total annual water use. The study suggested that such restrictions, if mandated by the water conservation districts, will result in individual irrigators bearing the cost of water savings in the form of reduction in net present value per acre. In addition, the decline in net present value will impact the regional economy, and therefore analyzing the socio-economic effects of implementing such a policy alternative is critical, and the feasibility of policy implementation should also be evaluated with regard to the existing legislative and political scenarios (Tewari et al. 2014).

3.2 Hydro-economic Models for Evaluating Net Benefits from the Use of Groundwater for Irrigation

Hydro-economic models are widely used to evaluate economic impacts from the use of groundwater in agriculture. These models can be adjusted for incorporation of water policies and subsequent impacts during the implementation time frames can be evaluated. Initial hydro-economic modeling studies such as Burt (1964) studied optimal allocation of nonrenewable or partially renewable resources such as water. The models estimate an optimal usage rate and the expected present value of groundwater operating under a socially optimal policy. The marginal social value of water at a point in time is equated to the marginal social value of water as a stock resource in the subsequent period. Extensions of this model made use of sequential decision theory, and formulated policies concentrating on optimal groundwater usage (Burt 1966). These models evaluated the net social benefit of water use and the issue of common property, and formulated the optimal groundwater extraction rate on the net present value of water in a specific area. Subsequently, these models were further expanded to incorporate institutional restrictions and their subsequent effect on storage of groundwater (Burt 1970).

In recent decades, a combination approach of utilizing diverse kinds of models to optimize the use of the remaining groundwater stock in the Ogallala aquifer has been used (Feng 1992; Johnson et al. 2004; Wheeler et al. 2008; Johnson et al. 2009; Tewari et al. 2014). This includes combining non-linear dynamic optimization models with bio-simulation models of crop growth, and including hydrological parameters of the groundwater source which directly impact costs of production from the perspective of irrigation.

The objective function of an economic model used in studies such as the above maximizes the Net Present Value of net returns over the study period. Because farm profitability is linked with parameters of crop production, the economic model provides a representation for quantifying the crop responses to water application in the form of net returns from irrigation. An objective function that maximizes net revenue over n years is expressed in Eq. 13.1.

$$MaxNPV = \sum_{t=1}^{n} NR_t(1+r)^{-t} \tag{13.1}$$

where NPV is the net present value of net returns; r is the discount rate; and NR_t is net revenue at time t. The bounds of summation for the net revenue are from one to n years.

The simplest form of evaluating these changes will be in the form of changes in net revenue, net present value of net returns over the study period, changes in irrigated acreages, and movement in crop-mix. NR_t is defined in Eq. 13.2 as:

$$NR_t = \sum_i \sum_k \Omega_{ikt} \{P_i Y_{ikt}[WA_{ikt}, (WP_{ikt})] - C_{ikt}(WP_{ikt}, X_t, ST_t)\} \quad (13.2)$$

where i represents crops grown; k represents irrigation systems used; Ω_{ikt} is the percentage of crop i produced using irrigation system k in time t, P_i is the output price of crop i, WA_{ikt} and WP_{ikt} are irrigation water application per acre and water pumped per acre respectively. Y_{ikt} is the per acre yield production function, C_{ikt} represents the costs per acre, X_t is pump lift at time t, ST_t represents the saturated thickness of the aquifer at time t. The model can be subjected to various constraints specific to the location, and the policy option evaluated, and further compared with an unrestricted status quo scenario to evaluate the impacts of policy implementation. The results of these models can be analyzed for the parameters of saturated thickness, annual net revenue per acre, pump lift, water applied per cropland acre, cost of pumping, net present value of net returns per acre (*NPV*), and for shifts in crop-mix over the planning time frames (Feng 1992; Johnson et al. 2004, 2009; Wheeler et al. 2008; Tewari et al. 2014).

Costs of production, crop prices, and energy prices are held constant thorough the time horizon. Costs of pumping irrigation water change as the pump lift increases with declining levels of saturated thickness in the aquifer. The location specific optimization model incorporates the initial values of crop acres, irrigated acres, average saturated thickness, and depth to water for a specific acreage unit such as a county, or a region. With these initial values, the model estimates the level of crop production and water use that optimizes farm net income for the location over the planning time frame.

4 Economic Value of Groundwater Management: Understanding Water Use Restriction Policies

The economic value of groundwater management in Texas can be studied using the case of optimal allocation of a common natural resource. In the state of Texas, the freedom extended by absolute ownership to landowners in making pumping decisions infers that groundwater is managed more as an open access resource than a common-property resource (Easter et al. 1998). Open access resembles a first-come, first-take, free for all that is barren of restrictions, whereas common property arrangements employ social expectations and rules of conduct governing resource use (Ciriacy-Wantrup and Bishop 1975). Given the above, groundwater remains a contentious issue for producers across Texas, particularly in high water use areas.

Among several suggested policy options for groundwater management, the water use restriction policy is a mandatory annual or multi-year limit that reduces the amount of water pumped from a common groundwater resource such as the Ogallala aquifer, for the purpose of agricultural irrigation. For the Texas High Plains region, the water use restriction policy along with various other water management alternatives was suggested in a study by Amosson et al. (2009) that compared the effectiveness of six different water management strategies in the Texas High Plains, including the policy of water use restriction. In order to implement this policy, a mandatory annual percentage reduction will be applied on the total water pumped for irrigation throughout the planning horizon (Amosson et al. 2009). Results suggest that while an annual water use restriction policy can lead to increasing saturated thickness levels in the aquifer, producer income will be negatively impacted. It is therefore implied that the underlying objective of a restriction policy on groundwater use in a region such as the Texas High Plains is to sustain the existing groundwater supply for use by future generations, and presents a trade-off that will have to be borne by existing producers in the form of reduction in net revenue.

The importance of this policy to the Texas High Plains is further emphasized by a few strategic modifications in the Texas Water Law in the recent decades. These modifications allow the regional groundwater conservation districts to set goals known as desired future conditions (DFC) for the district (Cook and Hope 2005), which are represented by an amount of groundwater remaining in the aquifer after a set period of time. Several conservation districts in the region have implemented pumping restrictions in order to ascertain a DFC of 50 % for the current water supply being available in 50 years, and these have been commonly addressed as the 50/50 management goal (Johnson et al. 2011). For example, the North Plains Groundwater Conservation District, in its Groundwater Management Plan for the years 2008–2018, set a maximum allowable production limit of 2 acre-feet per acre per-annum on water rights tracts not to exceed 1600 acres (North Plains Ground Water Conservation District 2008).

The economic impact of a water use restriction policy is depicted in Fig. 13.5.

The irrigation demand is depicted by AE, which is the derived demand for irrigation water from commodity prices. The policy places a certain percentage of annual restriction on the water pumped from the aquifer per acre. The quantity of water available per acre declines from Q1 to Q2, and because of decline in the average water application rate, production declines. This increases the marginal cost of water from C1 to C2, which also implies that use of irrigation water becomes more expensive for production.

Figure 13.6 shows the impact of a water use restriction policy on a single irrigated commodity. The point on the production curve corresponding to point W1 represents the level of output Y1 for current baseline water use. The water use restriction decreases the water use to point W2. This shift causes output on the panel above to decrease from Y1 to Y2, decreasing returns to producers. A water use restriction will encourage producers to adopt more efficient technology systems so they can continue to remain at a profitable level of output.

An evaluation of the costs and benefits associated with the implementation of the policy is critical to evaluate the feasibility of the policy in actual field conditions.

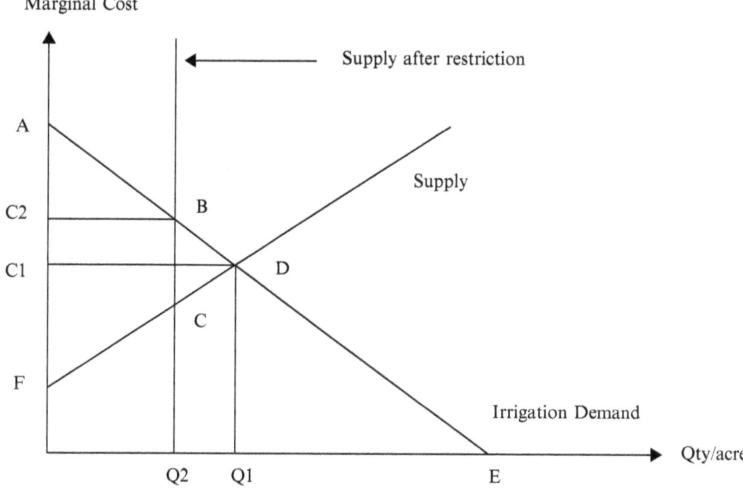

Fig. 13.5 Impact of a water use restriction policy on the marginal cost of irrigation water

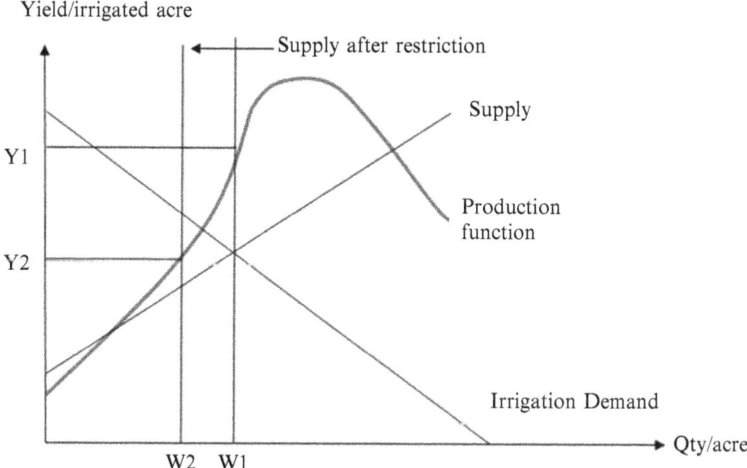

Fig. 13.6 Impact of a water use restriction policy on irrigated agricultural production

The costs of implementation are primarily calculated as the net social cost of each policy and comprise of private and social costs related to each policy measure. Private costs are calculated as the income loss borne by the producers on account of reduction in irrigation water per unit area as a result of policy implementation, as well as any other implementation costs incurred by the producer, for example technological advancements, and monitoring tools. Public costs refer to the administrative and operational costs of implementing the policy as incurred by the state, and management bodies such as the water districts.

5 Impacts of Climate Change

The future availability and sustainability of groundwater in several principal aquifers in the world is under threat on account of depletion by overuse and climatic stresses (Brekke et al. 2009; Alley et al. 2002). An in-depth understanding of climate change and variability is crucial for agricultural economies and ecosystems, especially in the context of complex changes in climatic variables that reduce chances of replenishment and pose questions regarding the sustainability of groundwater reserves in overused areas (Dragoni and Sukhija 2008). It is also crucial to incorporate how spatial changes in the available water supply and their temporal variability affect adaptation strategies to mitigate adverse impacts of climate change on irrigated production, as suggested in a study by Quiggin et al. (2010). This study evaluated the effects of climate change adaptation and mitigation in the Murray-Darling Basin in Australia, using a state-contingent simulation approach. This approach represents risk management by producers in the form of variations in land allocation between production activities. The results from the study suggested that in the absence of mitigation, climate change will adversely impact irrigated agriculture in the Murray-Darling Basin. The study also pointed out that these adverse impacts of climate change can be mitigated by land and water use changes in agricultural production.

Recently for the U.S. High Plains, Crosbie et al. (2013) used 16 global climate models (GCMs) and three global warming scenarios to investigate changes in groundwater recharge rates for a 2050 climate relative to a 1990 climate. Under a 2050 climate, median projections suggested increased recharge in the Northern High Plains, slight decrease in the Central High Plains, and a larger decrease in the Southern High Plains. There is however, considerable uncertainty in the magnitude and direction of these changes in recharge projections from the above study. In addition, the U.S. Global Change Research Program's report for the year 2009, predicts a temperature increase in the range of 2.5–13 °F from the 1960 to 1979 baseline for the Texas High Plains by the end of this century (Karl et al. 2009). The report also indicated that summer changes are projected to be larger than those in winter and precipitation will particularly change in winter and spring. Overall, the conditions are anticipated to become drier and hotter in the Texas High plains, as well as frequencies of extreme events such as heat waves and droughts are projected to increase. These changes in long-term climate could further increase the stress on the regions' already depleting water resources, as well as agricultural and ranching activities that form the backbone of the regional economy (Karl et al. 2009).

6 Conclusions

The demand for groundwater in semi-arid regions will continue to increase in the future, particularly under a changing climate that projects higher temperatures and lower precipitation rates. The uncertainty posed by climatic factors, as well as by continued expansion of irrigated agriculture in groundwater dependent semi-arid agricultural economies such as the Texas High Plains creates an impending need for incorporating water policy measures in the groundwater management plans. These measures will prove critical in managing the exhaustible supply of groundwater provided for by finite sources such as the Ogallala aquifer. However, the feasibility of implementing such policy measures is greatly influenced by the existing groundwater laws, as well as the agricultural production systems followed in the region.

References

Alley WM, Reilly TE, Franke OL (1999) Sustainability of ground-water resources. USGS Circular 1186. U.S. Geological Survey, Denver

Alley WM, Healy RW, LaBaugh JW, Reilly TE (2002) Flow and storage in groundwater systems. Science 296(5575):1985–1990

Amosson S, Almas L, Golden B, Guerrero B, Johnson J, Taylor R, Wheeler-Cook E (2009) Economic impacts of selected water conservation policies in the Ogallala Aquifer. Ogallala Aquifer Project

Birkenfeld D (2003) Reclaiming water as a commons. Southwest Chronicle 4(2)

Brekke LD, Kiang JE, Olsen JR, Pulwarty RS, Raff DA, Turnipseed DP, Webb RS, White KD (2009) Climate change and water resources management—a federal perspective. U.S. Geological Survey Circular, 1331, 65

Buller OH, Williams JR (1990) Effects of energy and commodity prices on irrigation in the high plains. Report of Progress No. 611, Agricultural Experiment Station, Kansas State University, December 1990

Burt OR (1964) Economic control of groundwater. J Farm Econ 48:632–647

Burt OR (1966) Economic control of groundwater reserves. J Farm Econ 48:632–647

Burt OR (1970) Groundwater control under institutional restrictions. Water Resour Res 3:1540–1548

Caswell M, Zilberman D (1986) The effects of well depth and land quality on the choice of irrigation technology. Am J Agric Econ 68:798–811

Ciriacy-Wantrup SV, Bishop RC (1975) "Common property" as a concept in natural resources policy. Nat Resour J 15(4):713–721

Colaizzi PD, Gowda PH, Marek TH, Porter DO (2009) Irrigation in the Texas High Plains: a brief history and potential reductions in demand. Irrig Drain 58(3):257–274

Cook R, Hope R (2005) Texas House Bill 1763—relating the notice, hearing, rulemaking, and permitting procedures for groundwater conservation districts. 79th Texas Legislature, Regular Session, 2005. http://www.legis.state.tx.us/BillLookup/Text.aspx?LegSess=79R&Bill=HB1763. Accessed 20 Mar 2014

Crosbie RS, Scanlon BR, Mpelasoka FS, Reedy RC, Gates JB, Zhang L (2013) Potential climate change effects on groundwater recharge in the High Plains Aquifer, USA. Water Resour Res 49:3936–3951. doi:10.1002/wrcr.20292

Dragoni W, Sukhija BS (2008) Climate change and groundwater—a short review. In Dragoni W, Sukhija BS (eds) Climate change and groundwater, 288. Geological Society, Special Publications, London, pp 1–12
Dregne HE (1991) Global status of desertification. Ann Arid Zone 30:179–185
Easter KW, Rosegrant MW, Dinar A (eds) (1998) Markets for water. Potential and performance. Springer, New York, p 352
Feng Y (1992) Optimal intertemporal allocation of groundwater for irrigation in the Texas High Plains. Ph.D. dissertation, Texas Tech University, Lubbock
Giordano M (2009) Global groundwater? Issues and solutions. Annu Rev Environ Resour 34:153–178
Handbook of Texas Online (2009) Water Law. Texas State Historical Association. https://www.tshaonline.org/handbook/online/articles/gyw01. Accessed 20 Aug 2013
Johnson JW, Johnson P, Segarra E, Willis D (2004). Water conservation policy alternatives for the Texas Southern High Plains. Paper presented at the 2004 Beltwide cotton conferences, San Antonio, 5–9 Jan 2004. http://www.aaec.ttu.edu/Publications/Beltwide%202004/D020.pdf. Accessed 20 Aug 2013
Johnson J, Johnson PN, Segarra E, Willis D (2009) Water conservation policy alternatives for the Ogallala Aquifer in Texas. Water Policy 11:537–552
Johnson JW, Johnson PN, Guerrero B, Weinheimer J, Amosson S, Almas L, Golden B, Wheeler-Cook E (2011) Groundwater policy research: collaboration with groundwater conservation districts in Texas. J Agric Appl Econ 43(3):345–356
Jury WA, Vaux H Jr (2005) The role of science in solving the world's emerging water problems. Proc Natl Acad Sci U S A 102:15715–15720
Karl RT, Melillo JM, Peterson TC (2009) Global climate change impacts in the United States. Cambridge University Press, New York
Molden D (2007) Water for food, water for life: a comprehensive assessment of water management in agriculture. Earthscan, London
National Agricultural Statistics Services (2011). National agricultural statistics services database. http://www.nass.usda.gov/index.asp. Accessed 21 July 2014
Negri D, Brooks D (1990) Determinants of irrigation technology choice. West J Agric Econ 15:213–223
North Plains Ground Water Conservation District, Ground water management plan (2008) http://www.npwd.org/Management%20Plan%20-%20Final2%20PDF.pdf. Accessed July 20 2014
Quiggin J, Adamson D, Chambers S, Schrobback P (2010) Climate change, uncertainty, and adaptation: the case of irrigated agriculture in the Murray-Darling Basin in Australia. Can J Agric Econ 58(1):531–554
Scanlon BR, Keese KE, Flint AL, Flint LE, Gaye CB, Edmunds M, Simmers I (2006) Global synthesis of groundwater recharge in semiarid and arid regions. Hydrol Process 20:3335–3370
Scanlon BR, Claudia CF, Laurent L, Robert CR, William MA, Virginia LM, Peter BM (2012) Groundwater depletion and sustainability of irrigation in the US high plains and central valley. Proc Natl Acad Sci U S A 109:9320–9325
Schloss JA, Buddemeier RW (2000) Estimated usable lifetime. In Schloss JA, Buddemeier RW, Wilson BB (eds) An atlas of the Kansas High Plains Aquifer. Kansas Geological Survey, Lawrence, Educational Series, 14(92)
Shiklomanov IA (2000) Appraisal and assessment of world water resources. Water Int 25:11–32
Siebert S, Burke J, Faures J, Frenken K, Hoogeveen J, Döll P, Portmann F (2010) Groundwater use for irrigation—a global inventory. Hydrol Earth Syst Sci 14:1863–1880
Texas Constitution and Statues (2011) Water code. http://www.statutes.legis.state.tx.us/Docs/WA/htm/WA.36.htm. Accessed 20 Aug 2013
Templer OW (1992) Hydrology and Texas water Law: a Logician's nightmare. Great Plains Res 2 (1):27–50

Tewari R, Almas LK, Johnson J, Golden B, Stephen AH, Guerrero BL (2014) Multi-year water allocation: an economic approach towards future planning and management of declining groundwater resources in the Texas Panhandle. Texas Water J 5(1):1–11

Texas Tech University, Center for Geospatial Technology (2010) Texas county water information. http://www.gis.ttu.edu/OgallalaAquiferMaps/TXCounties.aspx. Accessed 21 July 2014

Texas Water Development Board (2004) 100 years of rule of capture: from East to groundwater management. Austin, Texas Water Development Board, Report 361. http://www.twdb.texas.gov/publications/reports/numbered_reports/doc/R361/R361.pdf. Accessed 20 Aug 2013

Texas Water Development Board (2010). Regional water planning area—Region A http://www.twdb.state.tx.us/mapping/maps/pdf/rwpg/letter_size/Region%20A%208x11.pdf. Accessed 20 Aug 2013

Tuholske J (2008) Trusting the public trust: application of the Public Trust Doctrine to groundwater resources. Vermont J Environ Law 9(2):189–238

US Census Bureau (2004) Global Population Profile: 2002. International Population Reports WP/02. US. Government Printing Office, Washington, DC

United States Geological Survey (2014) High Plains aquifer. Cartography and Publishing Program. http://water.usgs.gov/ogw/aquiferbasics/ext_hpaq.html. Accessed 20 Feb 2015

Water Conservation Advisory Council (2008) http://www.savetexaswater.org/au/index.htm. Accessed 20 Aug 2013

Wheeler E, Golden B, Johnson J, Peterson J (2008) Economic efficiency of short-term versus long-term water rights buyouts. J Agric Appl Econ 40(2):493–501

Chapter 14
Groundwater Exploitation as Thermal Fluid in Very-Low Enthalpy Geothermal Plants in Coastal Aquifers

Rita Masciale, Lorenzo De Carlo, Maria Clementina Caputo, Giuseppe Passarella, and Emanuele Barca

Abstract Recently, in Italy, the interest for very low enthalpy geothermal resources ($T < 20$ °C) is growing. This is mainly because, these resources are widely available throughout the country and also unlike the other green energy sources (eg. solar and wind energy), and they do not need to be stored. Among the direct-use of geothermal resources, the open-loop groundwater heat pump (GWHP) system needs particular attention in terms of potential environmental impact. In coastal areas, that are generally densely populated, the installation of GWHP system is particularly appealing because the presence of shallow aquifers. This means significant savings of economic resources in terms of pumping energy and drilling costs. Nevertheless, vast areas of the Italian coastlines, as well as those of other Mediterranean countries, are often affected by seawater intrusion and hence are ruled by restrictive laws aimed to protect the groundwater quality and quantity.

In this chapter the environmental impacts, associated with the exploitation of low enthalpy geothermal resources, were assessed. For the purpose, a costal karst area in Southern Italy affected by seawater intrusion was investigated. A detailed characterization of the area was achieved in terms of geological, hydrogeological, geochemical and meteorological parameters. Moreover, the influence of an open-loop geothermal systems on the sea water intrusion was also studied by means of a long-term pumping test.

The investigated portion of aquifer was found to have a high hydraulic conductivity, as well as high and fast recharge rates, highlighting a good productivity of aquifer. The temperature of groundwater, reaching over 20 °C near the coast, was particularly useful for direct use especially for the space heating and cooling.

The long pumping test, lasted for 16 days, not affected the lowering of the water table that naturally occurs in the dry period. On the contrary, the reinjection of the extracted groundwater into the surface water drainage network partially restored the

R. Masciale • L. De Carlo • M.C. Caputo • G. Passarella (✉) • E. Barca
Department of Bari, IRSA-CNR—Water Research Institute of the National Research Council, Viale De Blasio, 5, Bari 70132, Italy
e-mail: giuseppe.passarella@ba.irsa.cnr.it

A. Fares (ed.), *Emerging Issues in Groundwater Resources*, Advances in Water Security, DOI 10.1007/978-3-319-32008-3_14

water table. The test also not detected any quality degradation of groundwater induced by pumping.

The quality of groundwater showed that the level of contamination in the investigated area was generally high both because of the presence of urban and industrial pollution and because of the presence of the seawater intrusion. The absence of a strong competition for use of groundwater makes them available for geothermal use. An extensive utilization of natural heat for the space cooling is also justified considering the local climate characteristics of the area that cause a peak of thermal energy demand in summer.

1 Introduction

Speaking about geothermal energy, Italy boasts an age-old and solid experience in the balneotherapy that has been practised in spas since ancient times and in the exploitation of geothermal energy for power generation. In fact, the first experiments of electricity generation from geothermal steam took place in Larderello (Tuscany) in 1904.

Conversely, the development of other direct uses of geothermal resources are still very limited compared to other countries in Europe and throughout the world. A large variety of possible direct uses of geothermal heat exists for example, in the residential, commercial, and industrial sectors, such as space heating and cooling, greenhouse heating, industrial processes, aquaculture pond heating and heat pumps. In Italy space heating is the second most important application of geothermal heat after thermal balneology. Nevertheless, from an economic and environmental point of view, the contribution of geothermal sources to space heating and cooling is almost negligible in comparison with the contribution of fossil sources, especially that of natural gas which remains the main source (UGI-CNG 2007; Lund et al. 2011).

Nowadays, interest in direct-use applications of geothermal energy is growing. Future scenarios foresee an exponential increase, especially in geothermal heat pumps used for heating and/or cooling (Bertani 2009). Compared to other renewable energies, like solar or wind ones, geothermal energy is more advantageous since it is continually available and therefore, overcomes the limitation of the other green energy sources, which are only available intermittently. Moreover, the use of natural heat for direct applications is very attractive, when considering the increasingly high cost of fossil fuels and the considerable reduction in CO_2 and sulphur emissions that should derive from it.

For these reasons, new political and legislative actions are needed in order to promote the consumption of renewable energy sources in space heating and cooling. Additionally, efforts should be made to reduce the thermal energy wastefulness by promoting new building criteria and energy-efficiency projects.

2 Open-Loop Groundwater Heat Pump: Development and Warning

The Italian geothermal potential is noteworthy and lying at economically accessible depths (e.g. 3–4 km) (UGI-CNG 2007). Particularly, very low temperature resources (T < 20 °C), suitable for both heating and cooling, are found almost everywhere. However, the heat stored in the ground requires the use of heat pumps and suitable heat exchangers to be extracted.

Among the basic ground source heat pump systems, the implementation of an open-loop groundwater heat pump (GWHP) is particularly suitable considering the great diffusion of shallow aquifer systems. An open-loop GWHP is a thermal machine which transfers into the environment the heat stored in the underground, by using groundwater as the geothermal fluid: during the winter, the GWHP extracts heat from the groundwater to provide space heating while during the summer, with reversible heat pumps, the groundwater absorbs heat from the living space and cools the air.

Besides the several advantages of this kind of applications, also the environmental impacts should be considered. In this case, impact assessment should not only focus on groundwater supply but also on its discharge into the receptor environments (i.e. aquifer, stream, lake) after the usage cycle, taking into account mainly the thermal effect. Nevertheless, the regulatory guidelines do not always take these issues into account, and rarely define how to carry out such assessments (Freedman et al. 2012; Donato et al. 2013). Particularly, in Italy, the installations of open-loop groundwater heat pumps are not yet regulated, while the use of closed-loop systems is regulated by Regional governments. However, only a few Regions have already adopted different guidelines so far while others are taking action in this direction (De Filippis et al. 2015).

Moreover, the use of renewable energy sources is strongly promoted by the 2009/28/EC Directive of the European Union, establishing a common framework about the energy production from renewable sources aiming at a 20 % increase by 2020.

Generally, GWHP systems employ a lot of pumping energy compared to closed-loop systems (Kevin and Rafferty 1998). However, in coastal areas, usually characterized by shallow aquifers, the installation of GWHP is economically convenient because it requires boreholes reaching a depth not more of 20–30 m, allowing a considerable saving in pumping energy. Against this advantage there is a drawback linked to seawater intrusion. Large areas of the Mediterranean coastline are affected by seawater intrusion, caused mainly by groundwater overexploitation for the water supply (Post 2005; Scheidleder et al. 2004). The Water Framework Directive (200/60/EC) considers seawater intrusion an issue of high relevance, and points out the importance of ensuring a quantitative balance between water extraction and aquifer recharge. As a consequence, each country has developed policies and regulations to protect groundwater against seawater intrusion, limiting water use by imposing the amount to be extracted (Giordana and Montginoul 2006).

This aspect could be a major impediment for the spread of geothermal resource use, particularly in regions characterized by 100 km of coastline, such as the Apulia region in Southern Italy. Commonly, coastal areas are the most densely populated areas in the world and host important tourist destinations and the largest industrial sites. These features make the exploitation of geothermal resources in these areas particularly appealing. This chapter reports a detailed hydrogeologic characterization of a coastal area in Southern Italy, suitable for the exploitation of a low enthalpy geothermal resource. In particular, the impact of a GWHP system on seawater intrusion was evaluated by means of a long-term pumping test, specifically designed for the test site and able to simulate the impact of the geothermal system itself on the aquifer.

3 Methods and Materials

3.1 Study Area

The study area (Fig. 14.1), extended for about 20 Km^2, includes the urban and periurban zone of the city of Bari (Southern Italy). This area is characterized by different kinds of settlements, such as residential, industrial and agricultural as well as hospitals, airports, shopping centres and sporting areas. These features make the application of geothermal resources for direct use particularly attractive, especially for winter heating and summer cooling. The study area includes the Water Research Institute (CNR-IRSA) experimental site, made up of five wells, that was used for monitoring.

From a geological point of view, the study area lies on the eastern edge of the Murge, the central part of the foreland of the Southern Apennines (Ciaranfi et al. 1988; Pieri et al. 2011), characterized by a thick Mesozoic sedimentary succession, overlain by relatively thin and discontinuous Quaternary deposits. Locally, the outcropped Mesozoic succession is represented by the 'Calcare di Bari' Formation (Fig. 14.2), made up of limestone with frequent intercalations of dolomitic limestones and grey dolostones. Different fields of tectonic stresses have acted on the limestone-dolostone bedrock, producing bland deformations and ruptures. Numerous karstic cavities of different shapes and sizes are also present, partially or completely filled by colluvial and eluvial deposits (terra rossa).

The Quaternary deposits are referable to different sedimentary units. The oldest one is represented by Calcarenite of Gravina (Pleistocene) made of litho-bioclasticcalcarenites. Inside of the little canyons, locally named 'lame', there are upper Pleistocene-Holocene deposits, made up of carbonate gravels with a reddish fine-grained matrix. The Quaternary sequence is completed by Holocene travertine deposits (crust) that appear intermittently in a narrow band, parallel to the coast.

A wide and thick aquifer resides in the carbonate bedrock which is characterized by high permeability resulting from the intense rock fracturing and karst dissolving

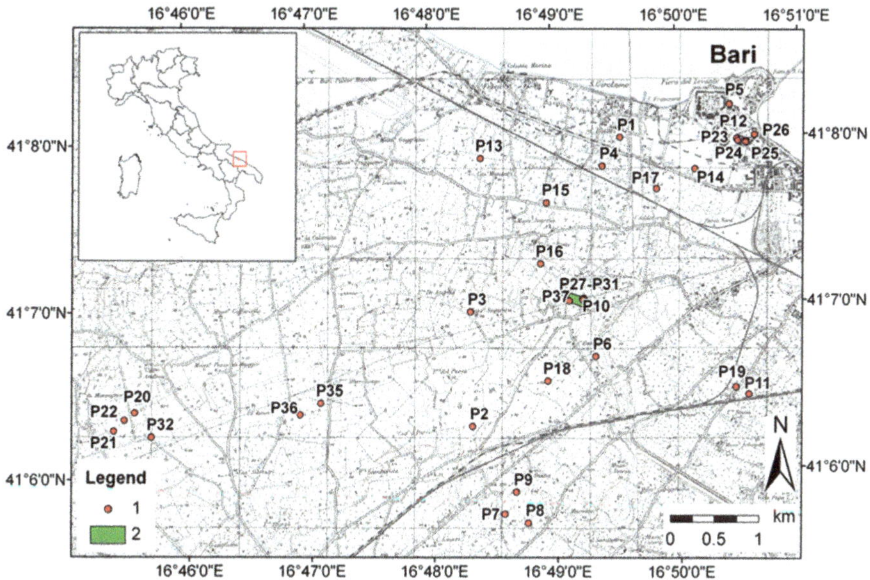

Fig. 14.1 Study area and local monitoring VIGOR project network defined in this study: (*1*) wells; (*2*) CNR-IRSA area

action. These specific features explain the absence of permanent surface water bodies in the Murge area, in contrast with the significant groundwater bodies. The irregular distribution of the fracture system and karstic channels makes the aquifer anisotropic, as its hydraulic conductivity values range from 10^{-3} to 10^{-7} m/s for vertical and horizontal component respectively (Maggiore 1993). The pumping tests, performed by Di Fazio et al. (1992) in the experimental site, have provided an average value for hydraulic conductivity of about 10^{-3} m/s, and transmissivity values between 9.8 and 12.1 m^2/s, highlighting the large water quantity that can be transmitted, and the high potential rate of aquifer recharge.

The groundwater flows toward the sea, mainly confined, fractionated into distinct levels separated by layers of dry rock, some of which are several hundred metres thick. Groundwater discharges into the sea through numerous subaerial and submarine coastal springs, which are very difficult to detect and to monitor. Karstic phenomena, which affect the carbonate rock, also occur below sea level, causing seawater intrusion into the coastal aquifer. The consequently intruding seawater underlies the fresh water because of its higher density. Aquifer is recharged only by rainfall that mainly infiltrates through the inner part of the Murge and along the 'lame'. A schematic hydrogeological cross-section of the carbonatic aquifer, oriented SW–NE, is shown in Fig. 14.3.

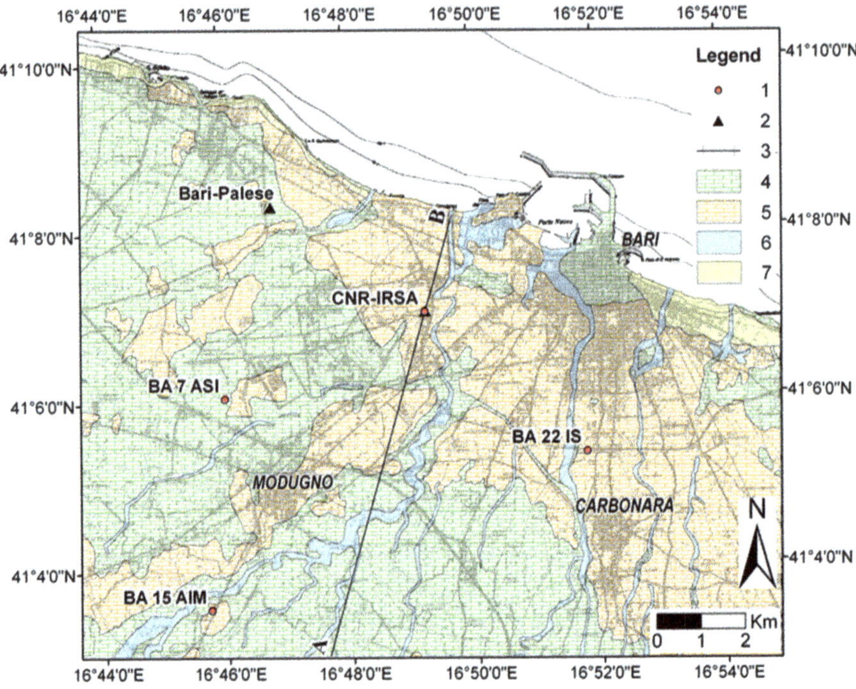

Fig. 14.2 Geological map and location of regional and national monitoring stations: (*1*) Regional Groundwater Monitoring Network; (*2*) Regional Meteorological Network; (*3*) Trace of hydrogeologicalsection; (*4*) Calcare di Bari (*Middle-Upper*Cretaceous); (*5*) Calcarenite di Gravina (Pleistocene); (*6*) Alluvial deposits (*Upper* Pleistocene-Holocene); (*7*) Coastal deposits (Holocene)

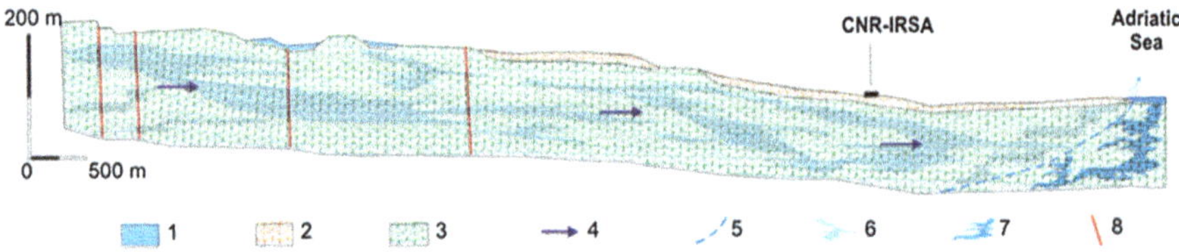

Fig. 14.3 Schematic hydrogeological cross-section. Trace of section is shown in Fig. 14.2: (*1*) Alluvial deposits; (*2*) Calcarenite; (*3*) Murge karstic aquifer (Carbonatic rock formation); (*4*) Groundwater flow direction; (*5*) Freshwater-seawater interface; (*6*) Fresh groundwater; (*7*) Salt water; (*8*) Fault

3.2 *Groundwater Monitoring Networks*

Hydrogeological, physical, and chemical data were collected from the existing Regional Groundwater Monitoring Network (Tiziano Network; Fig. 14.2) and from a new one, expressly defined for the purpose of this study in order to characterize groundwater in the study area (VIGOR Network; Fig. 14.1).

The VIGOR Network is composed by 35 wells, chosen from those existing in the area, and used for agricultural, industrial and domestic purposes.

The water level was measured in each monitoring well, while temperature and electrical conductivity (EC) profiles were acquired every 0.5 m, using an integrated down-hole probe. The measurements were collected from June to July 2011. Water levels and temperature data were interpolated by Kriging method to obtain the relative isoline maps.

3.3 Climate Data

A proper performance assessment of a geothermal system needs to take into account the local climate characteristic on which the evaluation of thermal energy demand depends. In fact, temperature and relative humidity (RH), which affects the perception of temperature, are among the major environmental factors that contribute to define thermal comfort. The ideal standard environmental conditions for the human body's thermal comfort vary, depending on the seasons. During summer the suggested temperature is between 24 and 26 °C, while the recommended temperature ranges between 18 and 22 °C during the winter. The recommended level of relative humidity is always in the range of 40–60 %. The climate of the study area was reconstructed acquiring air temperature and RH time series from two stations of the Regional Meteorological Network: Bari-Palese and CNR-IRSA, from 1994 to 2011 (Fig. 14.2).

To obtain the monthly distributions of each chosen parameter, the monthly average of all the daily mean values recorded during each month was first computed. Then was computed the average of all the monthly average values previously obtained for the same month, in the entire period.

The daily distribution was computed as average of all the daily mean values, recorded for the same day in the observation years. A detailed analysis of the variability of the temperature and the RH was performed on an hourly scale, using a time series acquired from the CNR-IRSA station. The averaged 24-h temperature and RH were obtained only for the hottest summer month and the coldest winter month.

3.4 Long-Term Pumping Test

For the purpose to study the influence of an open-loop geothermal systems on the sea water intrusion a long-term pumping test, lasted 16 days, was performed. The pumping rate, constantly equal to 50 m^3/h (about 13.9 L/s), was established on the basis of a hypothetical open-loop GWHP plant dimensioned to meet the space heating and cooling demand of the building hosting the CNR-IRSA and its equipment. The water was extracted from well P37 located in the CNR-IRSA area

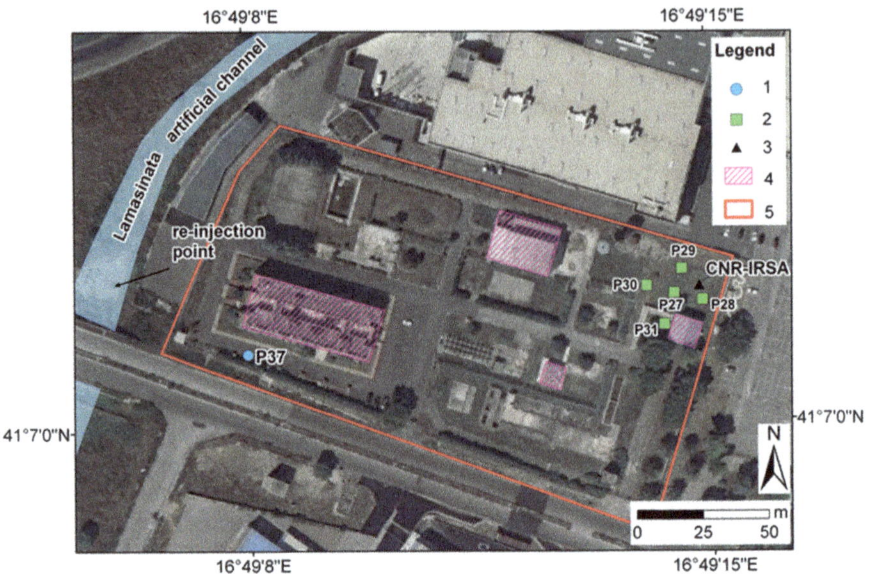

Fig. 14.4 Location of the long-term pumping test: (*1*) pumping well; (*2*) monitoring well; (*3*) regional meteorological station; (*4*) CNR-IRSA building and its appurtenance; (*5*) area owned by CNR-IRSA

(Fig. 14.4) and re-injected into the adjacent artificial channel, named Lamasinata, by means of a piping system. The Lamasinata channel is generally dry, and only conveys water during rain events, acting as a preferential way for groundwater recharge. For this reason, it can be asserted that the test really simulated an open-loop system. Water level, as well as water temperature and EC, were monitored during the test, as explained below.

Water level monitoring in the wells started few days before the beginning of the test and continued after the pumping stop, for a total of 37 days. Furthermore, water levels were manually recorded both in the extraction well (P37) and in five wells (P27–P31) of the experimental site, placed down-gradient at about 160 m, from the extraction well (Fig. 14.4). In addition, water levels were acquired continuously in the P27 well. Vertical thermal-conductivity logs were registered in some of the monitoring wells (P28, P29, P30), at different times: before, during and after the pumping test. The first profile was realized 10 days before the beginning of the test, while the others were recorded 1, 3, 9 and 16 days (a few hours before stopping the pump) after the beginning, respectively. Moreover, during the pumping test, temperature and EC were monitored continuously by means of thermal-conductivity probes, in both P37 and P27 wells. In the P37 well, temperature and EC were monitored by extracting water at a depth of about 5.5 m below the water table, and collecting it in a small-capacity container placed at the end of the piping system. In contrast, for the monitoring in P27 well a probe was placed into the well at about 24 m below the water table. This depth roughly corresponds to the beginning of the freshwater-seawater transition zone, as will be explained below (see Sect. 4.5). The limit of this zone is dynamic and hence the changes in time of EC are most noticeable here than anywhere else. During the monitoring period, rainfall data and air temperature were also acquired from the CNR-IRSA station.

4 Result and Discussion

4.1 Local Characteristics of Water Table

The piezometric head, mapped in Fig. 14.5, ranges from more than 30 m in the inner part, to 2–4 m above sea level (asl) in a large central portion of the investigated area, and tends to be zero near the coast. The groundwater flows through limestone-dolostone formation at depths ranging from more than 40 m from the ground to less than 1 m near the coast.

The density of piezometric contour lines reflects the slope of the water table. The steep slope of the water table—that is, high hydraulic gradients (0.9 %)—is visualized in the inner part of the investigated area, while the water table becomes almost flat towards the coast (hydraulic gradients of about 0.1 %). This anisotropy is strongly correlated to the different degree of fracturing and karstification of the bedrock reservoir. The average direction of groundwater flow is SW–NE, roughly perpendicular to the coastline. The groundwater moves under slight confining pressure owing to a local deeper and greater compactness of water-bearing formations. Daily groundwater levels, recorded in the wells CNR-IRSA and BA 22 IS, and the rainfall data from CNR-IRSA meteorological station (Fig. 14.2), are plotted in Fig. 14.6. The breaks in the plotted series are a result of maintenance work on sensors installed in the wells. The minimum groundwater levels were recorded in October, at the end of the summer season, when the aquifer gradually emptied. The maximum groundwater levels, instead, were recorded from November to April, when the groundwater resources tend to recover owing to winter recharge. This reflects the pluviometric regime of the area, characterized by rain mainly from October to March. Overall, the maximum excursion of the water table, which occurred in the two monitoring wells during the observation period, was about 2.0 m÷2.3 m.

4.2 Groundwater Quality

Quality data were collected from the Tiziano network in the period May–June 2010 were considered to evaluate the quality of groundwater. In particular, the data of the wells CNR-IRSA, BA 22 IS and BA 7 ASI were chosen because they are very close to the investigated area (Fig. 14.2). The chemical compositions of the three

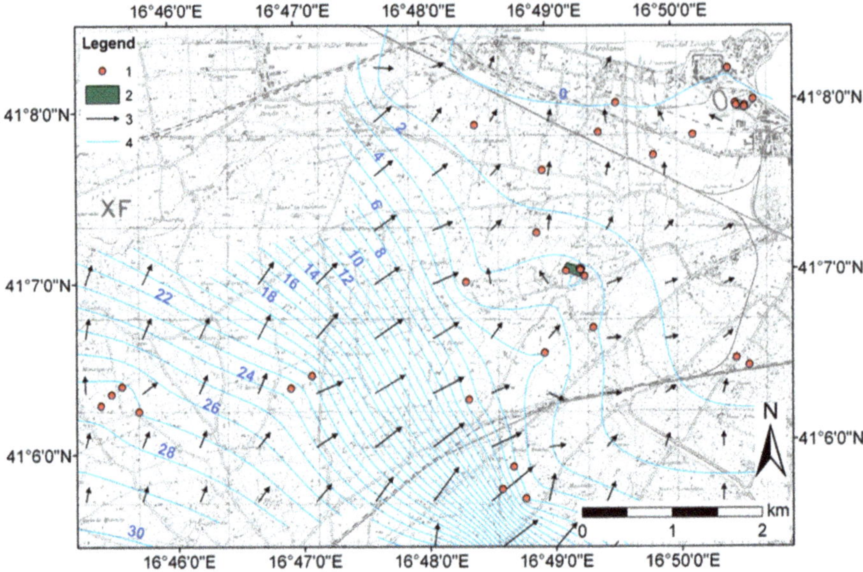

Fig. 14.5 Piezometric contour map based on measurements coming from Vigor Network: (*1*) wells; (*2*) CNR-IRSA area; (*3*) groundwater flow direction; (*4*) piezometric isolines (m asl), contour interval 1 m

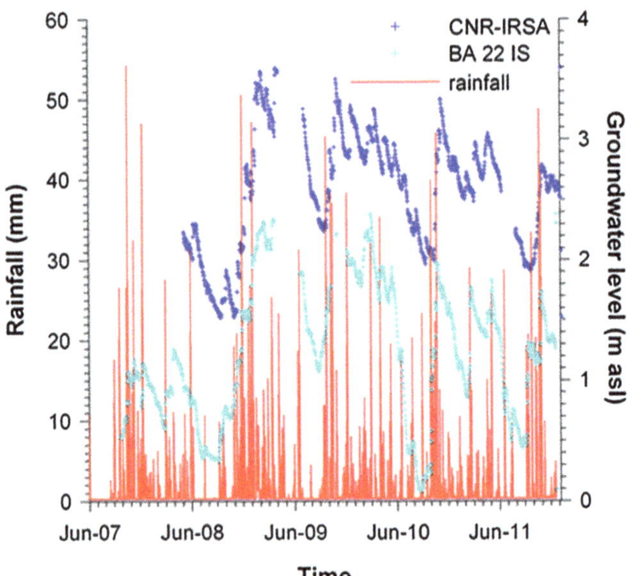

Fig. 14.6 Water table fluctuation and rainfall over time

Table 14.1 Chemical composition of groundwater sampled in some wells belonging to the Tiziano Network

Well	CNR-IRSA	BA 22 IS	BA 7 ASI
Date	10/05/2010	10/05/2010	09/06/2010
Sampling depth (m from ground)	18	53	–
Sampling mode	Static	Static	Dynamic
Temperature (°C)	19.92	17.74	22.7
pH	7.97	8.25	6.53
Electrical conductivity (mS/cm)	1.92	3.33	1.14
Redox (mv)	203	170	170
Water hardness (°F)	56.9	46.4	52.7
Bicarbonates (mg/L)	534	379	414
Calcium (mg/L)	115	72	109
Chloride (mg/L)	666	689	323
Magnesium (mg/L)	68.4	68.9	61.9
Potassium (mg/L)	25.3	31.7	6.12
Sodium (mg/L)	408	392	152
Sulfate (mg/L)	132	73.6	55.7
Ammonium (mg/L)	0.12	<0.030	<0.030
Nitrite (μg/L)	<50.00	<50.00	310
Nitrate (mg/L)	19.2	18.1	69.9
Ortho-Phosphate (μg/L)	<50.00	<50.00	<50.00
Total Organic Carbon (TOC) (mg/L)	2.49	1.48	0.35
SAR index	7.44	7.92	2.88
Lithium (μg/L)	16.7	13.5	<10.00
Bromide (μg/L)	2170	2330	1080
Fluoride (mg/L)	0.37	0.22	0.12
Iron (μg/L)	373.4	2200	22.8
Manganese (μg/L)	38.5	93.7	3.2

groundwater samples are shown in Table 14.1. On the basis of dominant cation and anion concentrations, the groundwater was classified as belonging to sodium-chloride hydrochemical facies, as shown in the Schoeller diagram of Fig. 14.7 (Schoeller 1962).

This means that the chemical composition of groundwater was influenced by two processes: the interaction with the carbonatic rock, through which the groundwater flow, and seawater intrusion, occurring almost in all the coastal aquifers of the Apulia region (Maggiore et al. 2001; Polemio 2005). This assertion was corroborated by the large variation in EC values (Table 14.1) among the samples that mainly (but not only) depends on the distance of the wells from the coast.

Particular attention must be paid to the water hardness that is a measure of the calcium and magnesium content of the water. Hard water is not a health risk, but it can lead to fouling in pipes, machines and in the heat pump's exchanger, reducing their performance. There is no univocal classification of the hardness of water. In

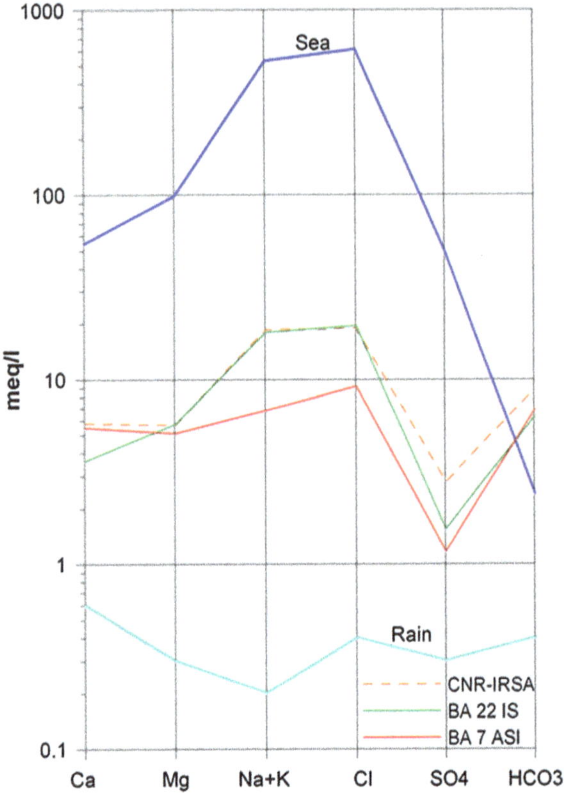

Fig. 14.7 Schoeller diagram of groundwater sampled in some wells belonging to the Tiziano Network

fact, it differs from country to country. In Italy, the most widely used is that defined by Celico (1986): soft water (<14 French Degreed—°F); moderately hard water (15–32 °F); hard water (33–54 °F); very hard water (>54 °F). The analysed samples had a hardness between 46.4 and 56.9 °F, corresponding to groundwater from hard to very hard, as would be expected for an aquifer with a prevailing carbonatic composition. These situations require some precautions, such as the use of water softeners, or carbon dioxide and other gases to keep calcium and magnesium in solution. Otherwise, different solutions, such as machines made of cupronickel and other materials that prevent fouling, or the introduction of chemicals into the groundwater, could be adopted.

The high values of some chemical parameters found in the water samples, such as total organic carbon, nitrate, iron, and bromide, reflect the presence of urban and industrial pollution. Basically, groundwater is not chemically suitable for drinking purposes and should also be used with caution for irrigation, owing to the salinity hazard. This means that there is no strong competition for the use of groundwater as a geothermal fluid for this area.

4.3 Temperature and Electrical Conductivity Variability of Groundwater

Temperature and EC data, acquired continuously over 3 years in the wells of the Tiziano Network (Fig. 14.2), are reported in Fig. 14.8.

No temperature variation was observed in each well throughout the monitoring period, while temperature values differed from one well to another, reaching a maximum of about 20 °C in well CNR-IRSA. On the contrary, EC presented a different behaviour, in terms of spatial and temporal variations according to the different distance of the wells from the coast. The logs, performed on two temporal horizons, are reported in Fig. 14.9. Due to the greater depth of well BA 15 AIM, a different vertical scale was used. Temperature in wells BA 22 IS and CNR-IRSA, except for the first metres in depth, was, on average, 17.7 °C and 20 °C, respectively. Logs repeated after 3 years in the well BA 22 IS, and after 1 year in the well CNR-IRSA, did not register any significant variation in temperature.

With regard to EC, an abrupt change was detected in well BA 22 IS at about 50 m depth, where the value was 3.5 mS/cm. In the well CNR-IRSA, EC values increased along a step profile, reaching values greater than 6.5 mS/cm at the bottom of the well itself. Marked EC variations were observed by comparing the logs performed at different times.

The trend of EC clearly reflected a zone of brackish groundwater, also called the 'transition zone', within which freshwater and saltwater are mixed. The extension of the transition zone can vary over time depending on the entity of groundwater withdrawals, combined with the changes in the rainfall regime, that both condition the landward movement (intrusion) and vertical movement (upconing) of saltwater. Temperature in well BA 15 AIM (Fig. 14.9), which is the furthest from the coast (about 10 km), was 16.5 °C, on average, throughout the log. Only slight increases in temperature were detected, between 80 and 90 m, and at around 260 m in depth, where it exceeded 17 °C. This temperature trend is typical of 'cold' geothermal zones where the geothermal gradient is less than 10–20 °C/km (Cataldi et al. 1995).

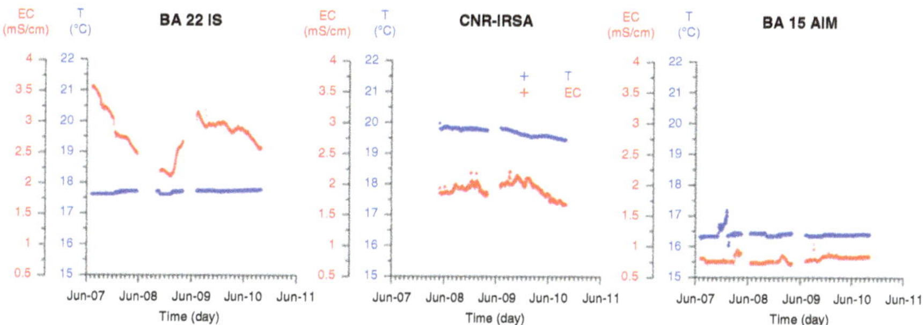

Fig. 14.8 T (*blue* cross) and EC (at 25 °C—*red* cross) values recorded in some wells belonging to Tiziano Network

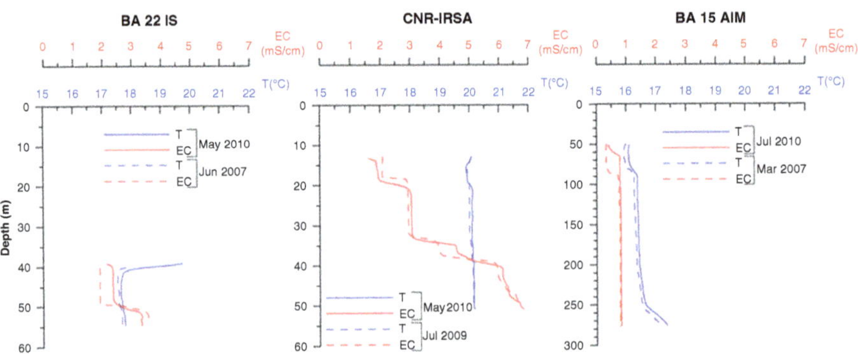

Fig. 14.9 T (*blue* curves) and EC (at 25 °C—*red* curves) logs performed in some wells belonging to Tiziano Network and on different temporal horizons. All depths are expressed as meters beneath ground

Always with reference to well BA 15 AIM, no significant change in temperature was observed between the two logs, and was the same for EC which showed a constant value, equal to 1 mS/cm. Only the upper part of the aquifer seemed to be affected by external variations while, in the deeper part, the water tended to reach equilibrium with the rocks through which it flowed. Temperature and EC values detected in the well BA 15 AIM, considering its greater distance from the coast, were considered as reference values for groundwater not affected by the process of seawater intrusion. Further details on the thermal and saline features of groundwater come from new thermal-conductivity logs performed in some wells of the Vigor Network, from June to July 2011. Figure 14.10 shows the location of the wells where logs were carried out. Logs in wells P21 and P22, differed both in terms of temperature and EC, although they have the same depth and are only 170 m apart. In particular, the temperature decreased slightly in the two wells, but its value was, on average, 0.2–0.3 °C higher in P22 than in P21. EC in P21 showed a jump at 52 m depth from the ground, reaching 2.5 mS/cm, and remaining constant up to the bottom of the hole. The lowest value of EC, never exceeding 1.6 mS/cm, was found in well P22. The variability in the logs of the two wells is consistent with the intrinsic characteristics of the aquifer that is fractionated in distinct levels because of the presence of compact dolomite benches. This means that the water level intercepted by two wells could be not the same. In the well P6, located in the central part of the investigated area, no temperature variation was registered. Instead, the EC showed a little increase, reaching a value of 3 mS/cm at 18 m depth. The log for well P30 reflected the same profile as the CNR-IRSA one (Fig. 14.9), which was only 20 m away. The temperature was constantly equal to 20 °C and an abrupt jump in EC was recorded close to the transition zone.

Maximum temperature variations were found in the wells P12 and P5, which are the closest to the sea. Water temperature decreased from 21 °C and 25 °C to 18 °C, respectively, in the first few metres. The shallow depth at which groundwater lies in this area (1–2.5 m below the ground) makes it strongly influenced by external

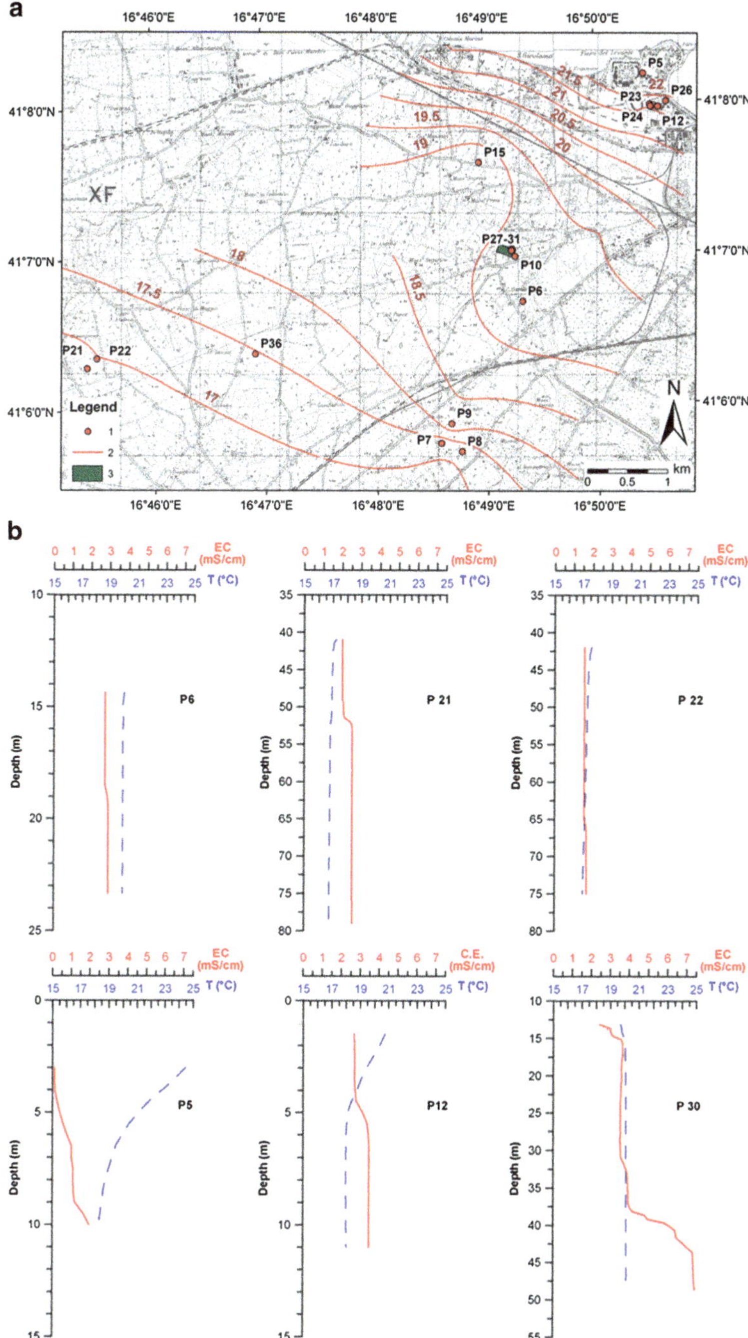

Fig. 14.10 (**a**) Map of groundwater temperature distribution at a depth of 1 m below the water table. (*1*) well where thermal conductivity logs were performed; (*2*) T isolines, contour interval 0.5 °C; (*3*) CNR-IRSA area. (**b**) More significant T (*dotted line*) and EC logs (*solid line*), referred at 25 °C, recorded in some well. The depths are expressed as meters beneath ground

variation. The discussed outcomes demonstrate that lithology, and fractures even more, are the main factors affecting groundwater flow and, in particular, its thermal regime and salinity.

A map of isotherm curves was drawn (Fig. 14.10) by interpolating the measured temperature in the monitoring wells at a depth of 1 m below water table. The temperature increased from 17 to 18 °C, in the inner part, up to values of more than 21 °C near the coast. The curve corresponding to 19 °C showed an anomalous shape with respect to the general trend of isotherms that is, generally parallel to the coast. Isotherm 19 °C delimits a sector of warmer waters from where the temperature increases, proceeding markedly towards the coast. This behaviour reflects the effect of the anthropogenic heat production (sewage network, underground structures, power lines, industry), known as the Urban Heat Island (UHI) effect (Allen et al. 2003; Huang et al. 2009; Zhu et al. 2010), that is strong in this area because the groundwater flows at shallow depths and the urban fabric is more dense. Nevertheless, the effect of mixing with seawater cannot be neglected since, as expected, the sea is warmer in the studied period (June–July).

4.4 Local Climate Conditions

According to the Köppen-Geiger Climate Classification System (Köppen 1900; Geiger 1954, 1961), which is the most widely used system to classify the world's climates, the study area belongs to the Warm Temperate Climate zone (named Csa) with dry and hot summers, known as the Mediterranean Climate. In Italy this climate is typical of the entire Southern coastline. In addition, to these general climate features, local ones were investigated and defined.

The analysis of monthly air temperature distribution (Fig. 14.11) showed that the mean temperatures range between 8 and 25 °C, with an annual average excursion of 17 °C. The highest values of temperature were found in July and August, ranging from 20 °C up to 30 °C, and the lowest in January and February, from 4 to 13 °C. Fluctuations in daily temperatures around the monthly averages were small, with a maximum value of 33 °C recorded in July, and the minimum equal to 2.5 °C recorded in February. Monthly and daily distributions of diurnal, nocturnal, and daily mean values of RH are reported in Fig. 14.12.

The RH values varied between the minimum of 64÷67 %, recorded in the months June-August, and maximum of 76÷77 %, in the period October-December. The largest difference of RH between day and night was in May and June (about 14 %), while the smallest was in December (about 7 %). The variability of daily RH values was larger than that of temperature, and only in few cases was lower than 60 %, which is the recommended value for the human body's thermal comfort.

The average 24-h distribution of temperature and RH are reported in Fig. 14.13, for the hottest month, July, and the coldest one, January. In July, the maximum temperature was reached at around 10:00 and lasted for 6 h. In January, the maximum temperature was reached from 12:00 to 13:00 and gradually declined

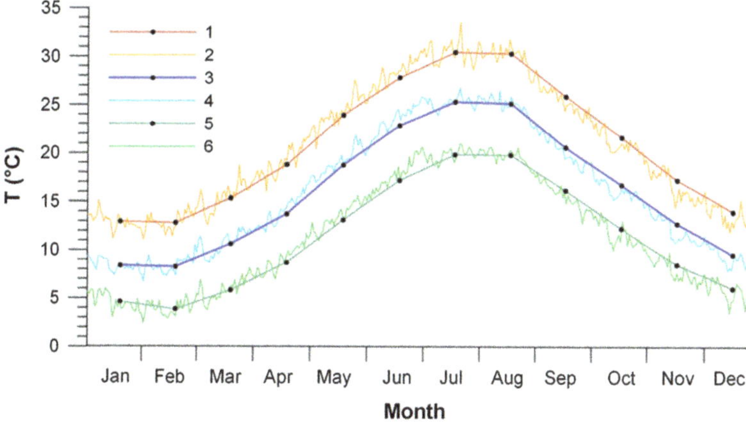

Fig. 14.11 Monthly and daily air temperature trends at Bari-Palese station. Climatic base period 1994–2011: (*1*) monthly average of daily maximum T; (*2*) average of daily maximum T; (*3*) monthly average of daily mean T; (*4*) average of daily mean T; (*5*) monthly average of daily minimum T; (*6*) average of daily minimum T

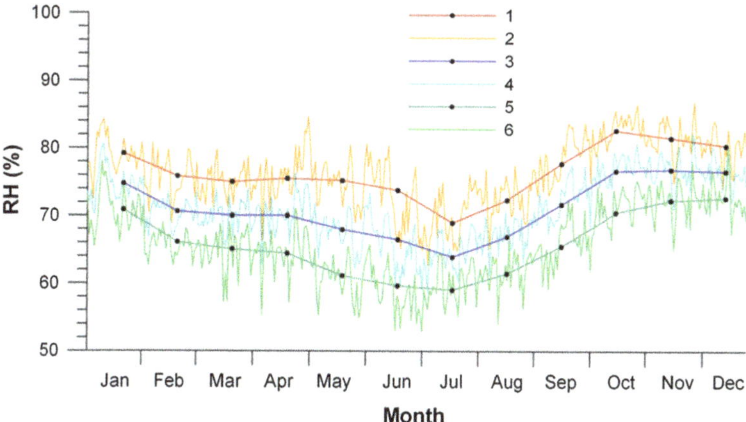

Fig. 14.12 Monthly and daily air relative humidity trends at Bari-Palese station. Climatic base period 1994–2011: (*1*) monthly average of night mean RH; (*2*) average of night mean RH; (*3*) monthly average of daily mean RH; (*4*) average of daily mean RH; (*5*) monthly average of diurnal mean RH; (*6*). average of diurnal mean RH

over the following hours, staying at a minimum value throughout the night. The daily temperature excursion was about 9 °C in July and 7 °C in January.

The 24-h RH distribution was exactly the opposite to that of temperature since, as is well known, the higher the temperature, the greater the maximum amount of water vapour that the air can hold. In July, the maximum value of RH (68 %) was reached just before sunrise, while the minimum values (44÷47 %) were recorded

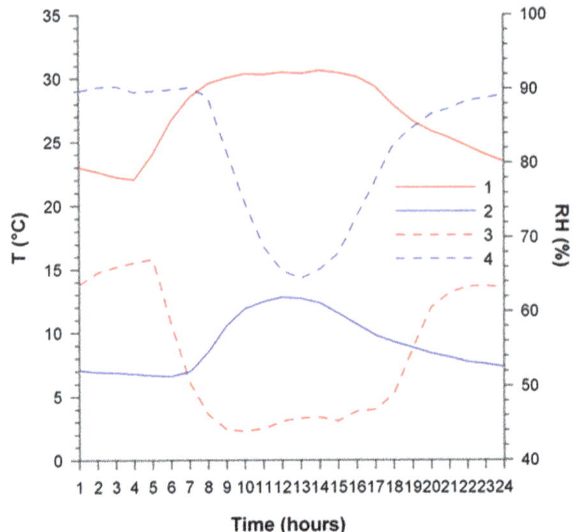

Fig. 14.13 Average 24-h air T (*solid lines*) and RH (*dotted lines*) distributions at CNR-IRSA station: (*1*) T of July; (*2*) T of January; (*3*) RH of July; (*4*) RH of January

from 8:00 to 17:00. In January, the minimum (64 %) was reached at around 13:00, while the maximum (90 %) values were recorded throughout the night.

Considering these local climate characteristics, a space-cooling system is required in July and August, when the average air temperature is around 25 °C÷26 °C, and a heating system is needed in the period from December to March, when the average temperature is less than 10 °C.

The direct comparison between air temperature and groundwater temperature (Fig. 14.14) reveals that against the air temperature, which ranges between 5 and 30 °C and exhibits a seasonal behaviour, the groundwater temperature is nearly constant at 20 °C throughout the period of observation. The maximum difference among temperatures is about 15 °C in winter and 10 °C in summer. The thermal characteristic of groundwater, compared to local climate conditions, represents an additional incentive to develop open-loop GWHP systems in this area, since the performance of the system is largely determined by the difference between the input and output temperature.

4.5 Analysis of the Long-Term Pumping Test Data

The graph in Fig. 14.15 shows the water levels measured in the wells used for the long-term pumping test (P37, P29, P30, P28, P31, and P27), and the rainfall during the monitoring period. The pumping was carried out in dry weather conditions, and significant rainfall events were recorded starting from the third day after pumping stopped and for the following days.

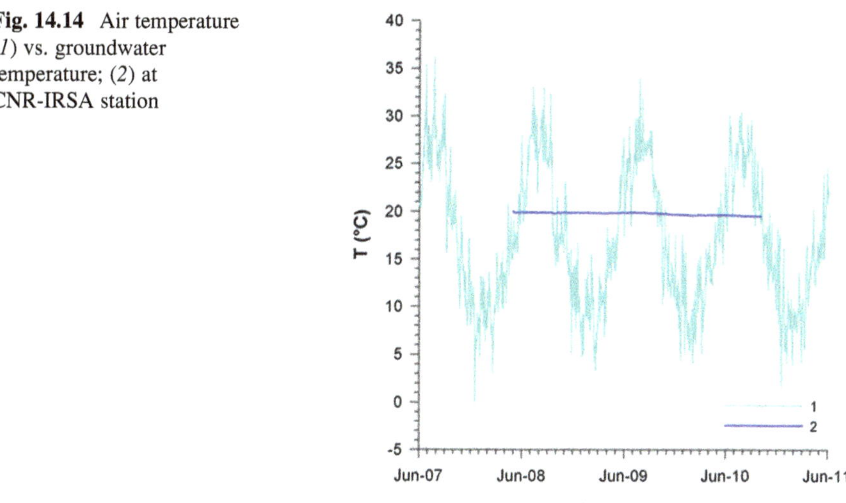

Fig. 14.14 Air temperature (*1*) vs. groundwater temperature; (*2*) at CNR-IRSA station

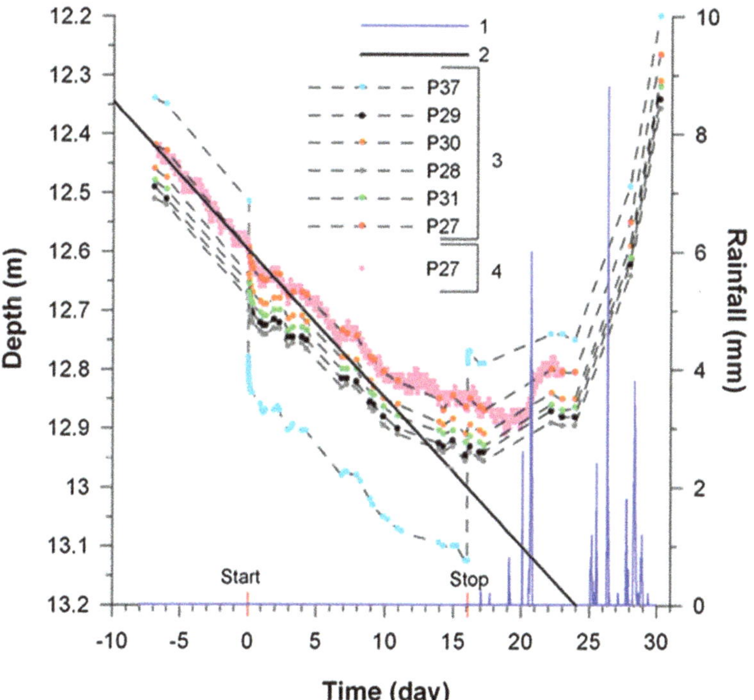

Fig. 14.15 Quantitative monitoring before, during and after the pumping test: (*1*) rainfall; (*2*) linear fit; (*3*) manual monitoring; (*4*) continuous monitoring. The time is expressed in days, assuming zero initial time as the start of the pumping test. Water levels are displayed as depth below ground

Soon after the start of pumping the water level dropped immediately in the extraction well, then continued to decrease gradually over the following days. Smaller and comparable water level decreases were recorded in all other monitoring wells, corroborating the fact that they belonged to the area of influence of the pumping well. By stopping the pumping, the level in P37 rose instantly and a slight rise was recorded simultaneously in the monitoring wells. However, the water levels never reached the values recorded immediately before the start of the test. This trend cannot be ascribed to pumping because the measured water levels showed a decreasing trend even prior to the beginning of the test. In fact, interpolating the data acquired prior to the start of pumping, a local negative trend owing to natural depletion of the aquifer was observed, with a drain rate of about 2.5 cm d^{-1}. Furthermore, the water levels recorded during the test were always above the local trend line, except for the first 2 days after the start of pumping. This means that the re-injection of the extracted water into the Lamasinata channel produced a partial restoration of the water level by mitigating the natural depletion. Moreover, the effect of slight rains on the water level showed that the aquifer is characterized by a quick response to rainfall events.

Figure 14.16 reports the results of the continuous monitoring of temperature and EC, carried out in the extraction well P37 and in the monitoring well P27. The EC values recorded in well P37 revealed very little variability in time, with an average value of 2 mS/cm.

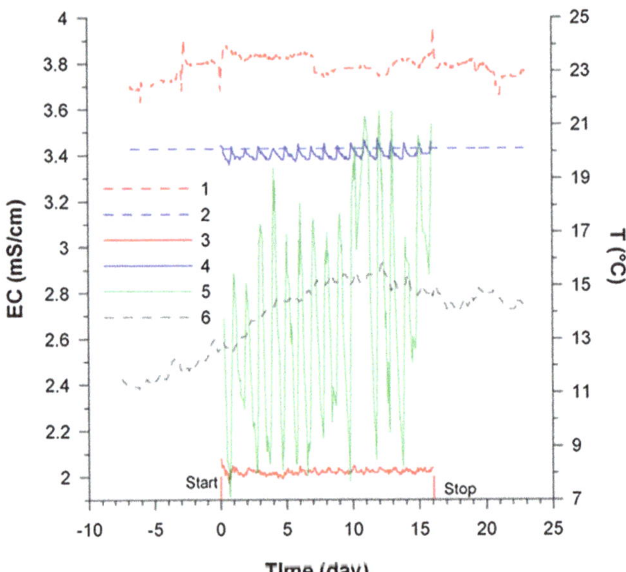

Fig. 14.16 Qualitative monitoring before, during and after the pumping test: (*1*). EC monitored in the well P27; (*2*) T monitored in the well P27; (*3*) EC monitored in the pumping well P37; (*4*) T monitored in the pumping well P37; (*5*) Air T; (*6*) Surface sea T

This becomes clear by considering the installation depth of the pump, at about 5.5 m below the water level, hence above the freshwater-seawater interface. In contrast, in well P27 the EC variability in time was much more pronounced, as a result of the position of the probe at 24 m below the water table, almost in the freshwater–seawater transition zone. Overall, the pumping did not increase groundwater salinity as corroborated by the constant trend of the EC values. In contrast, the temperature in well P27 was constant at 20 °C, while for well P37 there was a daily variation, around the average value of 19.8 °C. Comparing water temperature in P37 with the air temperature, a perfect overlapping of the cyclic nature of the temperature was observed. This is because the monitoring of temperature in P37 was done in an external container collecting extracted groundwater and placed at the end of the pipe. The capacity of the container is too small to ensure a certain thermal inertia so it is affected by the day/night thermal excursion.

No downward trend in groundwater temperature was recorded, owing to the mixing with seawater. In fact, as shown in Fig. 14.16, the sea surface temperature was lower than that of groundwater. For this reason, in case of mixing, an overall decrease in groundwater temperature should have occurred.

The results of vertical thermal-conductivity logs are shown in Fig. 14.17. Because of the low temporal variability of monitored parameters, only the first and last profile were reported.

Generally, all temperature profiles showed an increase of about 1 °C in the first 4 m below the water table. Starting from this depth, the temperature was almost constantly around 20 °C. In the same way, EC increased in the first few metres below the water table, moving from typical values of freshwater to more than 3.5 mS/cm. These variations of monitored parameters along the well water column can be ascribed to the recharge effect during rainfall events, both by diffuse infiltration and more commonly by concentrated infiltration along the close artificial Lamasinata channel. This hypothesis was corroborated by the rapid changes in

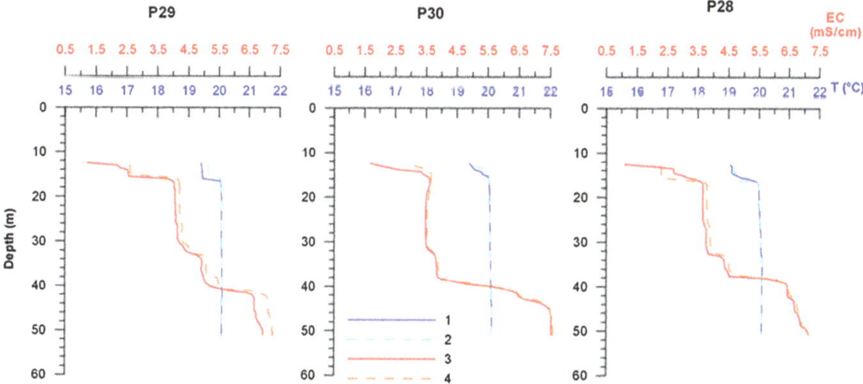

Fig. 14.17 EC and T profiles realized in the monitoring wells before and during the long-term pumping test: (*1*) T 3 days before pumping; (*2*) T 16 day after start of pumping; (*3*) EC 3 days before pumping; (*4*) EC 16 day after start of pumping (all depth are *below* the ground)

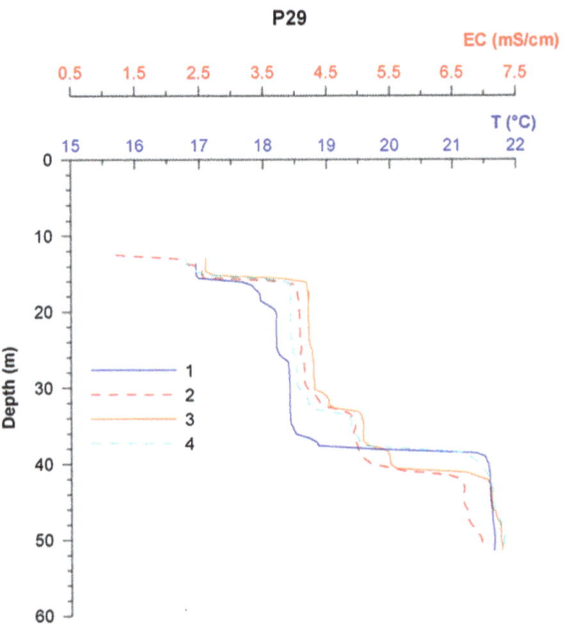

Fig. 14.18 EC profiles realized at different time interval in the monitoring well P29: (*1*) 160 days before pumping; (*2*) 10 days before pumping; (*3*) 16 days after start of pumping; (*4*) 50 days after stopping pumping (all depth are *below* the ground)

water level in the wells after rainfall events (Fig. 14.15). Unlike temperature, that remained constant over the whole well water column, except for the first metres, EC showed two more step increments, apart from the superficial one. The first was at about 33 m and the second at about 40 m depth, corresponding to the saltwater–freshwater transition zone.

The temporal comparison between the first and the last profile highlighted the substantial correspondence of temperature values in all wells. The EC values showed the same behaviour in well P30. In contrast, a slight increase of EC was observed at 13 m depth in well P28 and over 40 m depth in well P29. This trend was consistent with the SW–NE direction groundwater flow. Overall, these increases were not high enough to significantly affect the groundwater quality characteristics. By comparing the EC profiles of well P29, realized in different time intervals (Fig. 14.18), it can be argued that the EC variability is not attributable to the pumping. This evidence suggests that the long-term pumping test has not affected the groundwater qualitative characteristics. The rapid recharge of aquifer ensures the natural system recovery.

5 Conclusion

A detailed characterization of a coastal karst area was proposed, aimed at utilizing groundwater as a low enthalpy geothermal resource. A specifically defined monitoring network, consisting of 35 wells, was used to monitor groundwater quality and quantity. The impact of an open-loop GWHP system on seawater intrusion was studied by means of a long-term pumping test. The acquired information, together with the collection of existing data, allowed a detailed characterization of the area in terms of geological, hydrogeological, geochemical, and meteorological features.

The results showed that the investigated portion of the aquifer is characterized by high hydraulic conductivity and transmissivity values. Water enters this portion of aquifer very fast, highlighting good productivity. Groundwater temperature, reaching over 20 °C near the coast, is particularly useful for direct use, especially for space heating and cooling. In particular, extensive groundwater utilization for space cooling is advantageous and justified, considering the local climate characteristics that cause a peak of thermal energy demand in summer.

Groundwater contamination level is generally high, because of urban and industrial pollution and because of seawater intrusion. These characteristics determine the absence of a strong competition for groundwater use, making it available for geothermal purposes, even though particular attention must be paid to the hardness of water that can affect the performance of the GWHP system. The 16-day long-term pumping test did not caused a lowering of the water table greater than that naturally occurring in the dry period. On the contrary, the re-injection of the extracted water into the surface water drainage network through seepage, partially restored the water table. Moreover, the results showed that the long-term pumping test did not cause the worsening of groundwater quality that could have been expected, as a consequence of sea water intrusion caused by pumping.

In conclusion, the methodological approach adopted provides all the elements to correctly evaluate the environmental impact of an open-loop GWHP system in coastal areas. Therefore, it can be considered a useful tool for the feasibility study, also essential for the installation of a GWHP system.

Acknowledgments The present activity was performed in the framework of the VIGOR Project, aimed at assessing the geothermal potential and exploring geothermal resources of four regions in Southern Italy as part of the activities of the Interregional Programme 'Renewable Energies and Energy Savings: FESR 2007–2013 and Axes I Activity line 1.4-Experimental Actions in Geothermal Energy'.

Groundwater data used in the present work have been collected within the project "TIZIANO" for the qualitative and quantitative monitoring of the regional groundwater bodies. Thanks are due to the Water Protection Department of the "Regione Puglia" who kindly provided the permission for publishing the data.

References

Allen A, Milenic D, Sikora P (2003) Shallow gravel aquifers and the urban 'heat island' effect: a source of low enthalpy geothermal energy. Geothermics 32:569–578. doi:10.1016/S0375-6505(03)00063-4

Bertani R (2009) Geothermal energy: an overview on resources and potential. In: Proceedings of the international conference on national development of geothermal energy use, Slovakia

Cataldi R, Mongelli F, Squarci P, Taffi L, Zito G, Calore C (1995) Geothermal ranking of the Italian territory. Geothermics 24(1):115–129
Celico P (1986) Prospezioni idrogeologiche. vol I & II, Ed. Liguori, Napoli (in Italian)
Ciaranfi N, Pieri P, Ricchetti G (1988) Note alla carta geologica delle murge e del salento (puglia centromeridionale). Mem Soc Geol It 41:449–460 (in Italian)
De Filippis G, Margiotta S, Negri S, Giudici M (2015) The geothermal potential of the underground of the Salento peninsula (southern Italy). Environ Earth Sci 73:6733–6746
Di Fazio A, Maggiore M, Masciopinto C, Troisi S, Vurro M (1992) Caratterizzazione idrogeologica di una stazione di misura in ambiente carsico. It J Groundw. 34:17–26 (in Italian)
Donato A, Santilano A, Lombardo G, Bruno D (2013) Quadro normativo e iter autorizzativo per la ricerca e la coltivazione di risorse geotermiche. http://www.vigor-geotermia.it/index.php?option=com_content&view=article&id=16&Itemid=23&lang=it (in Italian)
Freedman VL, Waichler SR, Mackley RD, Horner JA (2012) Assessing the thermal environmental impacts of an groundwater heat pump in southeastern Washington State. Geothermics 42:65–77. doi:10.1016/j.geothermics.2011.10.004
Geiger R (1954) Klassifikation der Klimate nach W. Köppen, in Landolt-Börnstein—Zahlenwerte und Funktionen aus Physik, Chemie, Astronomie, Geophysik und Technik, Band III, alte Serie. Springer, Berlin, pp 603–607
Geiger R (1961) Überarbeitete Neuausgabe von Geiger, R.: Köppen- Geiger Klima der Erde. (Wandkarte 1:16 Mill.)—Klett-Perthes, Gotha
Giordana G, Montginoul M (2006) Policy instruments to fight against seawater intrusion in coastal aquifers: an overview. Life Environ 56:287–294
Huang S, Taniguchi M, Yamano M, Wang CH (2009) Detecting urbanization effects on surface and subsurface thermal environment-a case study of Osaka. Sci Total Environ 407:3142–3152. doi:10.1016/j.scitotenv.2008.04.019
Kevin D, Rafferty PE (1998) Well-pumping issues in commercial groundwater heat pump systems. ASHRAE Trans 104(Part 1B):927–931
Köppen W (1900) Versuch einer Klassifikation der Klimate, vorzugsweise nach ihren Beziehungen zur Pflanzenwelt. Geogr Zeitschr. 6:593–611, 657–679
Lund JW, Freeston DH, Boyd TL (2011) Direct utilization of geothermal energy 2010 worldwide review. Geothermics 40:159–180. doi:10.1016/j.geothermics.2011.07.004
Maggiore M (1993) Aspetti idrogeologici degli acquiferi pugliesi in relazione alla ricarica artificiale. Quad. IRSA, 94:1–32 (in Italian)
Maggiore M, Raspa G, Sabatelli L, Santoro D, Santoro O, Vurro M (2001) Monitoring of seawater intrusion in a karst aquifer (Apulia—Southern Italy). In: Proceedings of first international conference on saltwater intrusion and coastal aquifers—monitoring, modeling, and management. Essaouira, 23–25 April 2001
Pieri P, Sabato L, Spalluto L, Tropeano M (2011) Note illustrative della carta geologica dell'area urbana di Bari in scala 1:25.000. Rend Online Soc Geol It 14:26–36. doi:10.3301/ROL.2011.04 (in Italian)
Polemio M (2005). Seawater intrusion and groundwater quality in the Southern Italy region of Apulia: a multi-methodological approach to the protection. In: Proceedings oh the UNESCO IHP, Paris, 77, pp 171–178
Post VEA (2005) Fresh and saline groundwater interaction in coastal aquifers: is our technology ready for the problems ahead? Hydrogeol J 13:120–123
Scheidleder A, Grath J, Lindinger H (2004) The European Environment Agency indicator-based reporting about saltwater intrusion due to groundwater over-exploitation. In: Proceedings 18th SWIM, Cartagena 2004, Spain. Ed. Araguás, Custodio & Manzano. Instituto Geologico y minero de España, Madrid, pp 609–616
Schoeller H (1962) Les Eaux Souterraines. Ed, Masson, Paris
UGI-CNG (2007) Geothermal Energy: yesterday, today, tomorrow. Special Issue of UGI's Newsletter. ETS, Pisa. http://www.unionegeotermica.it/pdfiles/La-Geotermia-Studio-07.pdf
Zhu K, Blum P, Ferguson G, Balke KD, Bayer P (2010) The geothermal potential of urban heat islands. Environ Res Lett 5:044002. doi:10.1088/1748-9326/5/4/044002

Index

A. Fares (ed.), *Emerging Issues in Groundwater Resources*, Advances in Water Security, DOI 10.1007/978-3-319-32008-3

D

T

U

GPSR Compliance
The European Union's (EU) General Product Safety Regulation (GPSR) is a set of rules that requires consumer products to be safe and our obligations to ensure this.

If you have any concerns about our products, you can contact us on

ProductSafety@springernature.com

In case Publisher is established outside the EU, the EU authorized representative is:

Springer Nature Customer Service Center GmbH
Europaplatz 3
69115 Heidelberg, Germany

www.ingramcontent.com/pod-product-compliance
Ingram Content Group UK Ltd.
Pitfield, Milton Keynes, MK11 3LW, UK
UKHW021008290726
14059UKWH00001BA/26